Advanced Catalysts and Nanostructured Materials

Advanced Catalysts and Nanostructured Materials

Modern Synthetic Methods

EDITED BY

William R. Moser

Department of Chemical Engineering
Worcester Polytechnic Institute
Worcester, Massachusetts

ACADEMIC PRESS

San Diego London Boston New York
Sydney Tokyo Toronto

This book is printed on acid-free paper. ∞

Copyright © 1996 by ACADEMIC PRESS

All Rights Reserved.
No part of this publication may be reproduced or transmitted in any form or by any means, electronic or mechanical, including photocopy, recording, or any information storage and retrieval system, without permission in writing from the publisher.

Academic Press, Inc.
525 B Street, Suite 1900, San Diego, California 92101-4495, USA
http://www.apnet.com

Academic Press Limited
24-28 Oval Road, London NW1 7DX, UK
http://www.hbuk.co.uk/ap/

Library of Congress Cataloging-in-Publication Data

Advanced catalysts and nanostructured materials : modern synthetic
    methods / edited by William R. Moser.
        p.   cm.
    Includes bibliographical references and index.
    ISBN 0-12-508460-9 (alk. paper)
    1. Catalysts.   I. Moser, William R.
TP159.C3A34   1996
660'.2995--DC20   96-3066
                                                                   CIP

PRINTED IN THE UNITED STATES OF AMERICA
96   97   98   99   00   01   BC   9   8   7   6   5   4   3   2   1

Contents

8. *Sonochemical Preparation of Nanostructured Catalysts*   197

KENNETH S. SUSLICK, TAEGHWAN HYEON, MINGMING FANG,
AND ANDRZEJ A. CICHOWLAS

21. *The Preparation of Advanced Catalytic Materials by Aerosol Processes*

WILLIAM R. MOSER, JOHN D. LENNHOFF, JACK E. CNOSSEN,
KAREN FRASKA, JUSTIN W. SCHOONOVER, AND JEFFREY R. ROZAK

22. *Use of an Aerosol Technique to Prepare Iron Sulfide Based Catalysts for Direct Coal Liquefaction*   563

DADY B. DADYBURJOR, ALFRED H. STILLER,
CHARTER D. STINESPRING, AJAY CHADHA, DACHENG TIAN,
STEPHEN B. MARTIN, JR., AND SUSHANT AGARWAL

Contributors

Numbers in parentheses indicate the pages on which the authors' contributions begin.

Bruce D. Adkins (117), Center for Applied Energy Research, University of Kentucky, Lexington, Kentucky 40511.*

Sushant Agarwal (563), Department of Chemical Engineering, West Virginia University, Morgantown, West Virginia 26506.

M. Baerns (479), Lehrstuhl für Technische Chemie, Ruhr-Universität Bochum, D-44780 Bochum, Germany.†

J. E. Baumgartner (435), Exxon Research and Engineering Company, Annandale, New Jersey 08801.

J. S. Beck (1), Mobil Technology Company, Strategic Research Center, Paulsboro, New Jersey 08066.

A. Benedetti (143), Department of Physical Chemistry, University of Venice, 30123 Venice, Italy.

D. A. Berkel (91), Monsanto Enviro-Chem Systems, Inc., St. Louis, Missouri 63167.

J. L. Bonardet (395), Laboratoire de Chimie des Surfaces, Université Pierre et Marie Curie, Paris, France.

Helmut Bönnemann (165), Max-Planck-Institut für Kohlenforschung, 45466 Mülheim an der Ruhr, Germany.

Werner Brijoux (165), Max-Planck-Institut für Kohlenforschung, 45466 Mülheim an der Ruhr, Germany.

---

*Present address: Akzo Nobel, Pasadena, Texas 75707.

†Present address: Institut für Angewandte Chemie Berlin-Adlershof e.v., D-12484 Berlin, Germany.

M. F. Buehler (259), Pacific Northwest National Laboratory, Richland, Washington 99352.[‡]

G. W. Busser (213), Catalytic Processes and Materials, Faculty of Chemical Technology, University of Twente, 7500 AE Enschede, The Netherlands.

K. Carr (395), Department of Chemistry and the Guelph-Waterloo Centre for Graduate Work in Chemistry, University of Waterloo, Waterloo, Ontario, Canada N2L 3G1.

F. Cavani (43), Dipartimento di Chimica Industriale E dei Materiali, 40136 Bologna, Italy.

G. Centi (63), Dipartimento di Chimica Industriale E dei Materiali, 40136 Bologna, Italy.

G. Cerrato (143), Department of Inorganic, Physical, and Materials Chemistry, University of Turin, 10125 Turin, Italy.

Ajay Chadha (563), Department of Chemical Engineering, West Virginia University, Morgantown, West Virginia 26506.

C. Choi-Feng (453), Amoco Chemical Company, Amoco Research Center, Naperville, Illinois 60566.

Andrzej A. Cichowlas (197), Department of Chemistry, University of Illinois at Urbana-Champaign, Urbana, Illinois 61801.

Abraham Clearfield (345), Chemistry Department, Texas A&M University, College Station, Texas 77843.

Jack E. Cnossen (535), Department of Chemical Engineering, Worcester Polytechnic Institute, Worcester, Massachusetts 01609.

A. Colombo (43), Dipartimento di Chimica Industriale E dei Materiali, 40136 Bologna, Italy.

Dady B. Dadyburjor (563), Department of Chemical Engineering, West Virginia University, Morgantown, West Virginia 26506.

J. G. Darab (259), Pacific Northwest National Laboratory, Richland, Washington 99352.

Burtron H. Davis (117), Center for Applied Energy Research, University of Kentucky, Lexington, Kentucky 40511.

J. A. Donohue (453), Amoco Chemical Company, Amoco Research Center, Naperville, Illinois 60566.

Kirby Dwight (505), Brown University, Providence, Rhode Island 02912.

Mingming Fang (197), Department of Chemistry, University of Illinois at Urbana-Champaign, Urbana, Illinois 61801.

T. R. Felthouse (91), Monsanto Enviro-Chem Systems, Inc., St. Louis, Missouri 63167; and Huntsman Corporation, Austin, Texas 78752.

[‡]Current affiliation: Intel Corporation.

J. Fraissard (395), Laboratoire de Chimie des Surfaces, Université Pierre et Marie Curie, Paris, France.

Karen Fraska (535), Department of Chemical Engineering, Worcester Polytechnic Institute, Worcester, Massachusetts 01609.

Y-M. Gao (505), Brown University, Providence, Rhode Island 02912.

W. E. Gates (435), Exxon Research and Engineering Company, Annandale, New Jersey 08801.

F. Giuntoli (43), Dipartimento di Chimica Industriale E dei Materiali, 40136 Bologna, Italy.

A. Gutierrez (435), Exxon Chemical, Linden, New Jersey 07036.

Taeghwan Hyeon (197), Department of Chemistry, University of Illinois at Urbana-Champaign, Urbana, Illinois 61801.

S. R. Jost (91), Monsanto Enviro-Chem Systems, Inc., St. Louis, Missouri 63167; and Eli Lilly and Company, Indianapolis, Indiana 46285.

Joseph L. Katz (515), Department of Chemical Engineering, The Johns Hopkins University, Baltimore, Maryland 21218.

Robert Kershaw (505), Brown University, Providence, Rhode Island 02912.

Edmond I. Ko (21), Department of Chemical Engineering, Carnegie Mellon University, Pittsburgh, Pennsylvania 15213.

C. T. Kresge (1), Mobil Technology Company, Strategic Research Center, Paulsboro, New Jersey 08066.

T. La Torretta (63), Emirisorse, Centro Ricerche Venezia, Porto Marghera (VE), Italy.

John D. Lennhoff (535), Department of Chemical Engineering, Worcester Polytechnic Institute, Worcester, Massachusetts 01609.

M. E. Leonowicz (1), Mobil Technology Company, Strategic Research Center, Paulsboro, New Jersey 08066.

J. A. Lercher (213), Catalytic Processes and Materials, Faculty of Chemical Technology, University of Twente, 7500 AE Enschede, The Netherlands.

J. C. Linehan (259), Pacific Northwest National Laboratory, Richland, Washington 99352.

Mark C. Lovallo (307), Department of Chemical Engineering, University of Massachusetts Amherst, Amherst, Massachusetts 01003.

J. D. Lutner (1), Mobil Technology Company, Strategic Research Center, Paulsboro, New Jersey 08066.

M. Marella (63), Emirisorse, Centro Ricerche Venezia, Porto Marghera (VE), Italy.

Barbara Marshik-Guerts (285), Department of Chemical Engineering, Worcester Polytechnic Institute, Worcester, Massachusetts 01609.

Stephen B. Martin (563), Department of Chemical Engineering, West Virginia University, Morgantown, West Virginia 26506.

D. W. Matson (259), Pacific Northwest National Laboratory, Richland, Washington 99352.

S. B. McCullen (1), Mobil Technology Company, Strategic Research Center, Paulsboro, New Jersey 08066.

G. B. McGarvey (395), Department of Chemistry and the Guelph-Waterloo Centre for Graduate Work in Chemistry, University of Waterloo, Waterloo, Ontario, Canada N2L 3G1.

E. L. McGrew (91), Monsanto Enviro-Chem Systems, Inc., St. Louis, Missouri 63167.

J. B. McMonagle (395), Department of Chemistry and the Guelph-Waterloo Centre for Graduate Work in Chemistry, University of Waterloo, Waterloo, Ontario, Canada N2L 3G1.

G. B. McVicker (435), Exxon Research and Engineering Company, Annandale, New Jersey 08801.

L. Meregalli (63), Emirisorse, Centro Ricerche Venezia, Porto Marghera (VE), Italy.

Diane R. Milburn (117), Center for Applied Energy Research, University of Kentucky, Lexington, Kentucky 40511.

Daniel Miller (505), Brown University, Providence, Rhode Island 02912.

James B. Miller (21), Department of Chemical Engineering, Carnegie Mellon University, Pittsburgh, Pennsylvania 15213.

Philippe F. Miquel (515), Department of Chemical Engineering, The Johns Hopkins University, Baltimore, Maryland 21218.

S. Miseo (435), Exxon Research and Engineering Company, Annandale, New Jersey 08801.

J. B. Moffat (395), Department of Chemistry and the Guelph-Waterloo Centre for Graduate Work in Chemistry, University of Waterloo, Waterloo, Ontario, Canada N2L 3G1.

C. Morterra (143), Department of Inorganic, Physical, and Materials Chemistry, University of Turin, 10125 Turin, Italy.

William R. Moser (285, 535), Department of Chemical Engineering, Worcester Polytechnic Institute, Worcester, Massachusetts 01609.

G. G. Neuenschwander (259), Pacific Northwest National Laboratory, Richland, Washington 99352.

J. Paes (435), Exxon Chemical, Linden, New Jersey 07036.

S. Perathoner (63), Dipartimento di Chimica Industriale E dei Materiali, 40136 Bologna, Italy.

M. R. Phelps (259), Pacific Northwest National Laboratory, Richland, Washington 99352.

F. Pinna (143), Department of Chemistry, University of Venice, 30123 Venice, Italy.

M. Reiche (479), Lehrstuhl für Technische Chemie, Ruhr-Universität Bochum, D-44780 Bochum, Germany.

W. J. Roth (1), Mobil Technology Company, Strategic Research Center, Paulsboro, New Jersey 08066.

Jeffrey R. Rozak (535), Department of Chemical Engineering, Worcester Polytechnic Institute, Worcester, Massachusetts 01609.

K. D. Schmitt (1), Mobil Technology Company, Strategic Research Center, Paulsboro, New Jersey 08066.

Justin W. Schoonover (535), Department of Chemical Engineering, Worcester Polytechnic Institute, Worcester, Massachusetts 01609.

M. Seay (395), Department of Chemistry and the Guelph-Waterloo Centre for Graduate Work in Chemistry, University of Waterloo, Waterloo, Ontario, Canada N2L 3G1.

E. W. Sheppard (1), Mobil Technology Company, Strategic Research Center, Paulsboro, New Jersey 08066.

M. Signoretto (143), Department of Chemistry, University of Venice, 30123 Venice, Italy.

S. Soled (435), Exxon Research and Engineering Company, Annandale, New Jersey 08801.

Dennis E. Sparks (117), Center for Applied Energy Research, University of Kentucky, Lexington, Kentucky 40511.

Ram Srinivasan (117), Center for Applied Energy Research, University of Kentucky, Lexington, Kentucky 40511.

Alfred H. Stiller (563), Department of Chemical Engineering, West Virginia University, Morgantown, West Virginia 26506.

Charter D. Stinespring (563), Department of Chemical Engineering, West Virginia University, Morgantown, West Virginia 26506.

G. Strukul (143), Department of Chemistry, University of Venice, 30123 Venice, Italy.

Joseph E. Sunstrom IV (285), Department of Chemical Engineering, Worcester Polytechnic Institute, Worcester, Massachusetts 01609.

Kenneth S. Suslick (197), Department of Chemistry, University of Illinois at Urbana-Champaign, Urbana, Illinois 61801.

S. Termath (479), Lehrstuhl für Technische Chemie, Ruhr-Universität Bochum, D-44780 Bochum, Germany.

Dacheng Tian (563), Department of Chemical Engineering, West Virginia University, Morgantown, West Virginia 26506.

M. Tomaselli (63), Emirisorse, Centro Ricerche Venezia, Porto Marghera (VE), Italy.

F. Trifirò (43), Dipartimento di Chimica Industriale E dei Materiali, 40136 Bologna, Italy.

Michael Tsapatsis (307), Department of Chemical Engineering, University of Massachusetts Amherst, Amherst, Massachusetts 01003.

Andreas Tschöpe (231), Department of Chemical Engineering, Massachusetts Institute of Technology, Cambridge, Massachusetts 02139.

J. G. van Ommen (213), Catalytic Processes and Materials, Faculty of Chemical Technology, University of Twente, 7500 AE Enschede, The Netherlands.

J. C. Vartuli (1), Mobil Technology Company, Strategic Research Center, Paulsboro, New Jersey 08066.

A. Vavere (91), Monsanto Enviro-Chem Systems, Inc., St. Louis, Missouri 63178.

P. Vazquez (43), Dipartimento di Chimica Industriale E dei Materiali, 40136 Bologna, Italy.[§]

P. Venturoli (43), Dipartimento di Chimica Industriale E dei Materiali, 40136 Bologna, Italy.[¶]

Aaron Wold (505), Brown University, Providence, Rhode Island 02912.

Jackie Y. Ying (231), Department of Chemical Engineering, Massachusetts Institute of Technology, Cambridge, Massachusetts 02139.

Jin S. Yoo (453), Amoco Chemical Company, Amoco Research Center, Naperville, Illinois 60566.

[§]On leave from: Centro de Investigacion y Desarrollo en Procesos Cataliticos, La Plata, Argentina.
[¶]Present address: Lonza SpA, Scanzorosciate (BG), Italy.

Preface

With the increasing recognition of the importance of materials synthesis in heterogeneous catalysis, it is timely to assemble this book to provide an up-to-date statement on the status of novel methods for advanced catalyst synthesis. Several disciplines of modern materials sciences are essential to the discovery, development, and improvement of advanced catalysts for chemical, petrochemical, environmental, energy, commodity, and fine chemical processes. The modern catalytic scientist must have a detailed and current knowledge of sophisticated materials synthesis, materials properties, and characterization techniques. Optimally, a working knowledge of inorganic and organic reactivity theory, chemical process technologies, and reaction engineering is also needed. This book provides the catalytic scientist with the most important and current information on the key materials synthesis component of this required knowledge base. Although the synthetic techniques examined are about the preparation of advanced catalysts, these same techniques deal with those that the materials scientist might select for the fabrication of electronic and structural ceramics and superconductors. Indeed, each chapter emphasizes concepts, results, and equipment relevant to the preparation of advanced materials.

The objective of this book is to identify the newest and most crucial techniques for the controlled synthesis of advanced solid state materials used as catalysts. Authors were selected based on their recent discoveries of new synthesis technologies. All manuscripts were peer reviewed using archival journal standards to ensure high-quality descriptions of each synthesis technology. Each chapter contains two essential parts. The first part is a detailed description of the preparative technique, equipment used, range of materials synthesized, characterization and proof of structures, and in some cases catalytic data. The second part is a condensed review of the most recent literature and results relating to the synthetic method described.

Recent literature has shown the significance of nanostructured materials for both catalytic and ceramic applications. Many of the recent advances in materials synthesis reported here yield materials whose primary grain sizes are a few nanometers. Nanostructured materials not only are essential in electronic devices and for densification in ceramics, but also represent a starting point in the formulation of a finished catalyst. Thus, one of the several synthesis techniques described may be used to prepare nanometer grains of a desired catalyst composition, which are then formulated into the desired crystallite size and morphology through further calcination and processing.

Advanced Catalysts and Nanostructured Materials provides both industrial and academic researchers with a current description of the most valuable techniques available for the preparation of advanced catalysts. It is hoped that this will accelerate the development of new catalytic knowledge and chemical processes based on solid state materials.

William R. Moser

CHAPTER 1

Designed Synthesis of Mesoporous Molecular Sieve Systems Using Surfactant-Directing Agents

J. C. Vartuli,* C. T. Kresge,[†] W. J. Roth,[†] S. B. McCullen,*
J. S. Beck,* K. D. Schmitt,* M. E. Leonowicz,[†]
J. D. Lutner,* and E. W. Sheppard*

*Mobil Research and Development Corporation
Central Research Laboratory, Princeton, New Jersey 08543
[†]Paulsboro Research Laboratory
Paulsboro, New Jersey 08066

KEYWORDS: M41S mesoporous molecular sieves, MCM-41, MCM-48, MCM-50

1.1   Introduction

The use of cationic surfactants as structure-directing agents has resulted in the discovery of M41S, the first, ordered mesoporous molecular sieves [Kresge *et al.*, 1992; Beck *et al.*, 1992]. This new family of materials displays an array of structures that are thermally stable inorganic analogs of organic, lyotropic liquid crystalline phases [Beck *et al.*, 1992; Chen *et al.*, 1993; Monnier *et al.*, 1993; Vartuli *et al.*, 1994]. Furthermore, the chemistry of liquid crystal systems may be adapted in the synthesis of these mesoporous materials to further tailor structure and porosity. Herein we describe the ability of the surfactant molecules to interact with silicate counterions resulting in the formation of

*Advanced Catalysts
and Nanostructured Materials*

organosilicate–surfactant composite arrays that exhibit hexagonal, cubic, or lamellar structures. These composites may be further modified by choice of surfactant and/or auxiliary organics to produce uniform pore systems from 15 to greater than 100 Å. The final completely *inorganic* analogs of the liquid crystalline phases are isolated after removal of the organic liquid crystal template by air calcination. This resulting silicate structure, containing numerous silanol moieties, is amenable to a wide range of functionalization studies. The degree of control in tailoring these mesoporous molecular sieves has not been achieved in microporous systems [Davis, 1994].

1.2 Experimental

1.2.1 Materials

Sodium silicate (N brand, 27.8% silica, P.Q. Corp.), tetraethylorthosilicate (TEOS, Aldrich), and cetyltrimethylammonium chloride (CTMACl) solution (29 wt %, Armak Chemicals) were used as received. Batch exchange of a 29% by weight aqueous CTMACl solution with IRA-400(OH) exchange resin (Rohm and Haas) produced a cetyltrimethylammonium hydroxide/chloride (CTMAOH/Cl) solution with an effective exchange of hydroxide for chloride ion of ~ 30%.

1.2.2 Synthesis of MCM-41 (Hexagonal)

One hundred grams of the CTMAOH/Cl solution was combined with 12.7 g of sodium silicate and 40.1 g of 1N H_2SO_4 with stirring. This mixture was put in a polypropylene bottle and placed in a steambox at 100°C for 48 h. After cooling to room temperature, the solid product was filtered, washed with water, and air dried. The *as*-synthesized product was then calcined at 540°C for 1 h in flowing nitrogen, followed by 6 h in flowing air. X-ray diffraction revealed a high-intensity first peak having a *d*-spacing of 38 Å and several lower angle peaks having *d*-spacings consistent with hexagonal hk0 indexing (Fig. 1, top). Found in the *as*-synthesized product (wt %): C, 45.3; N, 2.83; Si, 11.0; Ash (1000°C), 24.1%.

1.2.3 Synthesis of MCM-48 (Cubic)

With stirring, 100 g of the CTMAOH/Cl solution were combined with 30 g of TEOS. This mixture was put in a polypropylene bottle and placed in a steambox at 100°C for 48 h. After cooling to room temperature, the sample was calcined as described previously. X-ray diffraction

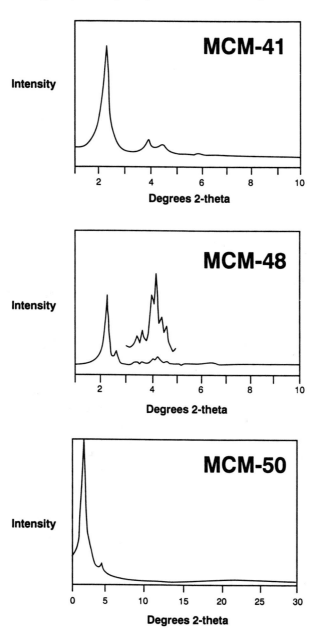

Figure 1. X-ray powder diffraction patterns of calcined samples of MCM-41, MCM-48, and MCM-50.

pattern of the calcined version (Fig. 1, middle) revealed a high-intensity first peak having a *d*-spacing of approximately 33 Å and several peaks having *d*-spacings consistent with a cubic indexing. Found in the *as*-synthesized product (wt %): C, 33.4; N, 1.88; Si, 11.7; Ash (1000°C), 25.9%.

1.2.4 Synthesis of MCM-50 (Stabilized Lamellar)

The lamellar structure is prepared by combining 100 g of the CTMAOH/Cl solution with 20 g of TEOS with stirring. This mixture was put in a polypropylene bottle and placed in a steambox at 100°C for 48 h. After cooling to room temperature, the sample was processed as described previously. The X-ray diffraction pattern of the *as*-synthesized lamellar material exhibited a high-intensity peak having *d*-spacings of approximately 36 Å and two higher angle peaks having *d*-spacings (18 and 12 Å) consistent with lamellar indexing of 001 reflections. The X-ray diffraction pattern of the calcined sample was essentially featureless. Found in the *as*-synthesized product (wt %): C, 45.1; N, 2.24; Si, 14.4; Ash (1000°C), 30.8%.

To form MCM-50, the stabilized lamellar structure, a sample of the *as*-synthesized lamellar material was treated with 1 g of TEOS per gram of *as*-synthesized material at room temperature for 16 h, washed in acetone, filtered, washed with water, filtered, and air dried. This treated sample was then calcined at 540°C as described, earlier. The X-ray diffraction pattern of this treated-calcined sample revealed two peaks at 39 and 20 Å consistent with retention of lamellar indexing (Fig. 1, bottom).

The silylation of MCM-41 using trimethylsilylchloride or hexamethyldisilazane has been previously described (Beck *et al.*, 1992).[1]

1.2.5 Instrumentation

X-ray powder diffraction was obtained on a Scintag XDS 2000 diffractometer using CuKα radiation. High-resolution transmission electron microscopy (TEM) images and electron diffraction patterns were obtained on a JEOL 200 CX transmission electron microscope operated at 200 kV. Samples were examined as microtomed sections.

1. In a typical trimethylsilylation experiment 0.50 g of calcined MCM-41, 10.0 g Me_3SiCl, and 15.0 g $(Me_3Si)_2O$ were refluxed overnight with magnetic stirring under N_2. The volatiles were stripped on a rotovap and the dry powder washed two or three times with 10 mL reagent grade acetone with centrifuging. Material recovery was typically >98%. For the hexamethyldisilazane reactions the same procedure was followed but 15 g hexamethyldisilazane were substituted for the trimethylsilylchloride.

Benzene sorption data were obtained on a computer-controlled 990/951 DuPont TGA system. The calcined sample was dehydrated by heating at 350°C or 500°C to constant weight in flowing He. Benzene sorption isotherms were measured at 25°C by blending a benzene saturated He gas stream with a pure He gas stream in the proper proportions to obtain the desired benzene partial pressure. Argon physisorption was used to determine pore diameters [Horváth *et al.*, 1983; Borghard *et al.*, 1991].

Si-nmr spectra were obtained in 9.5 mm zirconia rotors on the JEOL/Tecmag 200 MHz NMR spinning 4–4.2 kHz using 90° pulses at 1200 s intervals with high-power proton decoupling. Between 36 and 72 pulses (12–24 h) gave high-quality spectra. Air was used as the drive gas to attempt to obtain as much benefit as possible from O_2 paramagnetic relaxation.

1.3 Results and Discussion

1.3.1 Properties of M41S Structures

All three M41S structures, MCM-41 (hexagonal), MCM-48 (cubic), and MCM-50 (stabilized lamellar), are thermally stable in air to temperatures of $> 540°C$ and the mesopores are accessible to both gaseous and aqueous phases. Benzene and argon sorption isotherms of these three mesoporous molecular sieves are shown in Figs. 2 and 3. The benzene isotherms of MCM-41 and MCM-48 (Fig. 2, left, middle) exhibit three unique characteristics. The first is the exceptionally high hydrocarbon sorption capacity (55 wt % benzene at 50 torr at 25°C). The second characteristic is the sharp inflection of the isotherm indicative of capillary condensation within uniform pores [Gregg and Sing, 1982]. The third feature is the position of the inflection point at relatively high partial pressure (P/P_0) suggesting large-diameter pores. Argon physisorption (Fig 3, left, middle) confirms the uniformity (width at half-height = 4 Å for the MCM-41 sample, 5 Å for MCM-48) and pore size (~ 40 Å of this MCM-41 sample and 28 Å for MCM-48).

MCM-50 (the stabilized lamellar structure) is not similar to MCM-41 or MCM-48 as shown in Figs. 2 (right) and 3 (right). although the total benzene capacity is relatively high (37 wt % benzene at 50 torr at 25°C), there is no sharp inflection point in the isotherm, which suggests that the pore size of MCM-50 is not as uniform as that of the other two mesoporous materials. Argon physisorption data (Fig. 3, left) indicate a maxima at an approximate pore size of 27 Å, but the pore size distribution is broader than that of the other two samples (width at half-height = 9 Å). This lack of pore uniformity may be due to the pillared–layered structure of this material (Section 1.3.3).

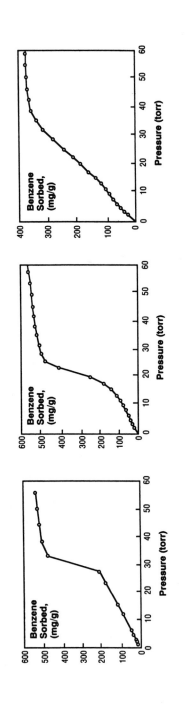

MCM-41   **MCM-48**   **MCM-50**

Figure 2. Benzene sorption isotherms of calcined samples of MCM-41, MCM-48, and MCM-50.

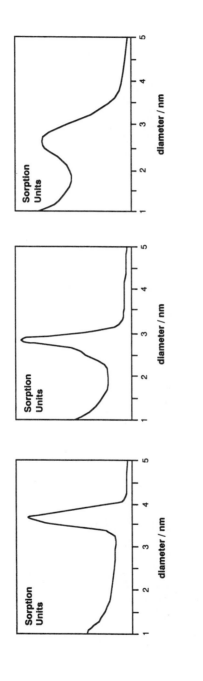

MCM-41   **MCM-48**   **MCM-50**

Figure 3. Argon Horváth-Kawazoe plots of calcined samples of MCM-41, MCM-48, and MCM-50.

MCM-41　　　　**MCM-48**　　　　**MCM-50**

Figure 4. Transmission electron micrographs of calcined samples of MCM-41, MCM-48, and MCM-50.

1.3.2 Diffraction Properties

The transmission electron micrograph of MCM-41 in Fig. 4 (left) shows the regular, hexagonal array of uniform channels with each pore surrounded by six neighbors. The repeat distance is approximately 40 Å consistent with the unit cell parameter calculation obtained from the X-ray diffraction pattern ($a_0 = 2d_{100}/\sqrt{3}$). The transmission electron micrograph of MCM-48 is somewhat more complex. We believe that the image shown in Fig. 4 (middle) represents the [111] projection of the Ia3d cubic structure consistent with the indexing of the X-ray diffraction pattern. Finally, the transmission electron micrograph of MCM-50 consists of layers with an approximate separation of 40 Å (Fig. 4, right).

1.3.3 Structure Considerations

It has been proposed that the individual pores of MCM-41 are obtained from silicate condensation about separate cylindrical micelles, whereas the ordered porous structure results from the hexagonal arrangement of the silica-encased micellar array [Beck *et al.*, 1992; Kresge *et al.*, 1992]. The resultant oxide structure mimics that of the hexagonal liquid crystal phase (Fig. 5, left).

The structure of MCM-48 presents a significantly more complex problem than the straightforward case of the hexagonal MCM-41. The struc-

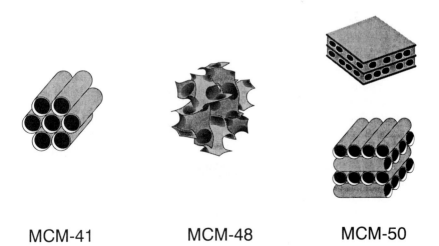

MCM-41   MCM-48   MCM-50

Figure 5. Proposed structures for MCM-41, MCM-48, and MCM-50.

ture for MCM-48 is also expected to be analogous to that of a liquid crystal with the cubic Ia3d symmetry. Proposed structures of the cubic liquid crystal phase vary from an independent, mutually intertwined arrangement of surfactant rods to a complex, infinite, periodic minimal energy surface structure [Fontell, 1990; Fontell, 1972; Luzzati et al., 1968]. A gyroid form of an infinite periodic minimal surface model (Q^{230}) for MCM-48 proposed in the recent work of Monnier et al. is consistent with both the X-ray diffraction data of the cubic liquid crystal and that of MCM-48 (Fig. 5, middle) [Monnier et al., 1993; Mariani et al., 1988].

The structure of MCM-50 (the stabilized lamellar structure) should also mimic the proposed structure of a lamellar liquid crystal phase. The lamellar liquid crystal phase can be represented by sheets or bilayers of surfactant molecules with the hydrophilic ends pointed toward the oil–water interface while the hydrophobic ends of the surfactant molecules face one another. Any silicate structure produced from this lamellar phase could be similar to that of two-dimensional layered silicates such as Magadiite or Kenyaite (Fig. 5, right). However, the lack of any observable peaks in the X-ray diffraction pattern of the lamellar material in the region of 20–25° 2θ (Fig. 1, right) suggests that the silicate layers of this lamellar phase are not as well ordered as those of layered silicates. This lack of order may be due to the higher concentration of silanol groups resulting in less condensation of the silicate species as suggested by the NMR data [Beck et al., 1992]. Removal of the surfactant from between the silicate sheets could result in a condensation of the layers, collapsing any structure and forming a dense phase with little structural order or porosity. This result is consistent with the lack of thermal stability of the lamellar material, which is unstable unless posttreated with TEOS.

The posttreatment with TEOS produced a thermally stable structure that retained a characteristic lamellar X-ray diffraction pattern. The TEOS may form stable inorganic pillars between the silicate layers that keep the layers separated after the removal of the surfactant [Landis et al., 1991]. This structure is like the classic, pillared, layered silicate (Fig. 5, right). Alternatively, MCM-50 may be composed of a variation on the stacking of surfactant rods similar to that of the structure of MCM-41 which would lead to a structure also shown in Fig. 5 (right). The TEOS posttreatment could stabilize this structure by reacting with nests of silanol groups created by the incomplete silicate condensation of the wall structure. The variation in the stacking of the surfactant rods to form MCM-50 is not inherently thermally unstable and should not, a priori, require the posttreatment of TEOS for thermal stability. Synthesis data cannot preclude either possible proposed structure.

1.3.4 Properties of Various Pore Size M41S

When quaternary ammonium surfactants $(C_nH_{2n+1}(CH_3)_3N^+)$ with $n = 8-16$ were used in the synthesis, MCM-41 materials exhibiting different X-ray diffraction pattern spacings were obtained [Beck et al., 1992; Beck et al., 1994; Stucky et al., 1994; Huo et al., 1994]. The location of the first X-ray diffraction lines (d_{100}) of the calcined products and approximate pore size as determined by argon physisorption increased with increased alkyl chain length. MCM-41 materials having pore size from 18 to 40 Å could be obtained using this synthesis method. Variation of the alkyl chain length of the surfactant have also produced various pore size MCM-48 and layered intermediates [Stucky et al., 1994; Huo et al., 1994]. Argon isotherms for three different MCM-41 preparations (where $n = 8$, 10, and 14) are shown in Fig. 6. The materials were synthesized under similar conditions while varying the alkyl chain length. As the surfactant chain length increases, the inflection point for capillary condensation increases to higher P/P_0 indicative of a larger pore size.

MCM-41 could also be prepared with pore sizes larger than 40 Å by the use of solubilization of auxiliary organic molecules within the surfactant micellar interiors. The addition of incremental amounts of a solubi-

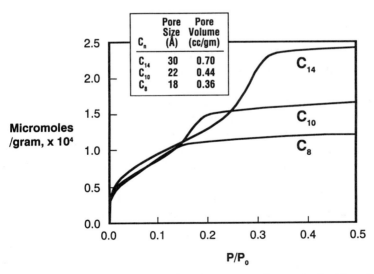

Figure 6. Argon physisorption isotherms for MCM-41 samples prepared with surfactant chain length ($n = 8$, 10, and 14).

lizate to the synthesis mixture produced a corresponding increase in the pore size of the resultant MCM-41 from 40 to greater than 100 Å [Beck *et al.*, 1992].

A variety of nonpolar organics, such as alkylated aromatics and straight or branched chain hydrocarbons, were effective as solubilizates in increasing the pore dimension of MCM-41 type materials. Organic solubilizates that produced the most well-defined X-ray powder diffraction patterns (3–4 peaks related by hexagonal constraints) were generally nonpolar aromatics possessing short aliphatic chains.

Straight and branched chain hydrocarbons in the C_5–C_{12} range were also effective in increasing pore size; however, the products often exhibited an apparent mixture of M41S phases and/or pore sizes. Polar organic species, including alcohols, aldehydes, ketones, and ethers, were generally found to be ineffective in increasing pore size in MCM-41, and in several cases were found to disrupt the synthesis resulting in the isolation of amorphous materials [Beck, 1991].

These results support a swelling mechanism in which the auxiliary organic is solubilized by surfactant micelles. Organics that are nonpolar and hydrophobic are susceptible to solubilization in the micellar interior and are found to be effective swelling agents. Those organics that have considerable polar character are insoluble in the micelle interior and are therefore ineffective for micellar swelling.

1.3.5 Properties of Functionalized MCM-41

Reaction of silylation reagents, such as trimethylsilylchloride or hexamethyldisilazane, with the silanols within the pore walls is another example of the synthesis variations available for the modification of these mesoporous materials [Beck *et al.*, 1992].[2] Silylation of the pore walls not only can be an effective method of reducing the pore diameter of the MCM-41 product, but can also affect the hydrophilicity of the treated mesoporous material.

Analysis of the CP-MAS Si-nmr of calcined and uncalcined MCM-41 suggested between 20 and 40% of the silicons were silanols, similar to the results found by others for amorphous silica heated to 500°C [Sindorf and Maciel, 1983]. Reaction with trimethylsilylchloride has been reported for a number of silanol-containing materials including amorphous silica [Sindorf

2. The reaction of a silanol with Me_3SiCl produces the TMS-OSi group and HCl. We were concerned that the evolution of HCl in the reaction of M41S silanols might cause additional attack of the framework to make the determination artificially high, so we also used hexamethyldisilazane, $(Me_3Si)_2NH$, as a silylating reagent. Hexamethyldisilazane produces NH_3 instead of HCl when it reacts with silanols. The TMS values were the same with the two reagents, which shows that there is no anomalous attack of the framework.

and Maciel, 1982], Kenyaite, and Magadiite [Yanagisawa *et al.*, 1979]. For amorphous silica, only about 9% of the silicons appear to be accessible; for Kenyaite and Magadiite, all of the 16–20% silanols can be derivatized.

Silicons in a trimethylsilyl (TMS) group show peaks at approximately 12 ppm in the Si NMR. The silicon that originally was a silanol (Q3) is converted to a silicon surrounded by four other silicons (Q4) and thus takes on the typical Q4 shift from −100 to −120 ppm and often is indistinguishable from "framework" silicons. In terms of stoichiometry, the TMS silicon concentration is equal to the original silanol concentration, so its peak area relative to the total of Q4 silicons after derivatization is the percent of silanol silicons, or at least the percent converted. Figure 7 shows typical before-and-after Si NMR spectra.

M41S mesopore materials show a strong linear sorption of benzene at ambient temperature up to 20–60 torr (depending on pore size) at which point there is a sudden increase in sorption due to capillary condensation. In comparing MCM-41 and its trimethylsilylated forms, Fig. 8, the linear

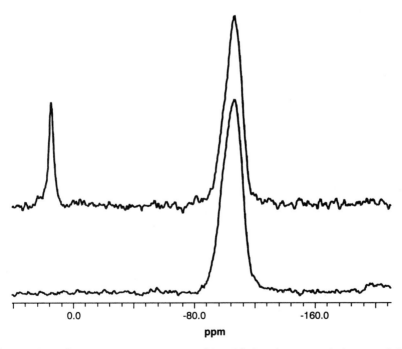

Figure 7. Typical 39.64 MHz Si-nmr spectra of M41S before (bottom) and after trimethylsilylation (top).

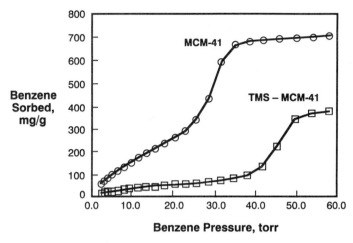

Figure 8. Room temperature benzene isotherms of calcined MCM-41, its TMS derivative, and the TMS–MCM-41 form as synthesized.

regime was strongly suppressed and a large shift in the position, but not amount, of capillary condensation was seen. It made little difference whether the TMS–MCM-41 was made by trimethylsilylation of the calcined form or direct reaction with the *as*-synthesized form. It is clear that the changes in the benzene isotherms are due to the presence of the TMS groups and these groups indeed reside within the pores of MCM-41. The sorption of benzene on these materials is strongly affected by polar interactions with the silanols, so the replacement of OH with TMS will greatly alter these interactions. In fact, the benzene isotherms can be completely interpreted in light of these polar interactions.

In a calcined sample the initial portion of the isotherm, that which occurs prior to the abrupt rise that indicates capillary condensation, shows a slight convexity with respect to the abscissa. This is typical of the Type II isotherm universally exhibited by sorption of benzene on hydroxylated silica surfaces. The same portion of the isotherm for the treated sample, however, is flatter. The sorbate–solid interaction is weaker and a Type III isotherm, commonly seen for sorption of benzene on dehydroxylated silica, results [Gregg and Sing, 1982].

The shift of the capillary condensation to higher pressure in the treated materials is also a result of altering the sorbate–solid interaction. From a consideration of the interaction energies, using a Kirkwood–Muller expression for the adsorption potential, Horváth and Kawazoe, 1983, developed a relationship between pore diameter and the reduced

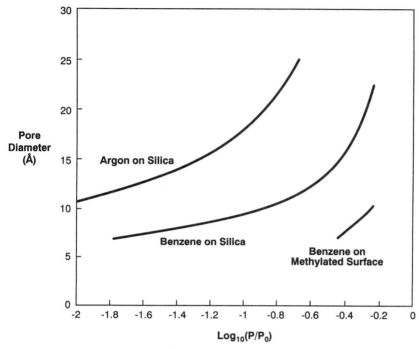

Figure 9. Horváth–Kawazoe calculation showing the effect of sorbate/sorbant interaction energy on the reduced pressure at which pore filling will occur.

pressure at which "capillary" condensation will occur in small-pore systems. A plot of this relationship is shown in Fig. 9 for three different cases: argon sorbing onto a silica surface (87 K), benzene sorbing onto the same surface (298 K), and benzene sorbing onto a surface dominated by methyl groups (also 298 K). It is clear from this graph that filling of pores of the same diameter occurs at higher reduced pressure for benzene at room temperature than for argon at its boiling point. The graph also shows that benzene filling of pores of equivalent diameters occurs at higher pressure when the surface effects are dominated by the presence of methyl groups, even though only a small portion of this latter curve can be calculated by the equations employed by Horváth and Kawazoe. The argon curve for Fig. 9 was calculated using energy parameters for argon and silica already reported [Borghard *et al.*, 1991]. For benzene the following values were used: polarizability $= 2.89^{-3}$ nm^3 [Haddon, 1979]; diamagnetic susceptibility $= 9.1^{-8}$ nm^3 [Weast, 1970]; molecular diameter $= 6.64$ Å [Gregg and Sing, 1982]. For methane, polarizability $= 2.62^{-3}$ nm^3 [Applequist *et al.*, 1972]; diamagnetic susceptibility $= 2.08^{-8}$ nm^3 [Weast, 1970]; molecular diameter $= 3.23$ Å;[3] surface coverage $= 6.99$ molecules/nm^2.[4]

The differences in uptake observed for the benzene isotherms can be explained by considering that the diameter of the pores has been reduced in the treated samples by the protrusion of the trimethylsilyl groups into the previously available pore space. The pore volume for untreated MCM-41 calculated from benzene total uptake (0.797 mL/g) agrees with the value obtained from argon physisorption (0.773 mL/g) and the same values for its treated analog also agree fairly well (0.433 and 0.489, respectively). The pore volumes calculated solely on the basis of benzene uptake in the region of capillary condensation are essentially identical for original and treated samples (0.455 and 0.405 for the two preceding materials). This leads us to conclude that the strong initial sorption of benzene in untreated materials occurs primarily as a monolayer on the cystallite and pore wall surfaces. This is corroborated by using the benzene uptake in this initial portion of the isotherm (~ 300 mg/g) to calculate a surface area. Using 43 Å^2/molecule [Gregg and Sing, 1982] as the surface coverage of benzene, we calculate 994 m^2/g. The total surface area calculated from argon physisorption is 1011 m^2/g, in good agreement with the benzene value.

After the initial monolayer sorption on the surfaces is complete, the capillary condensation then occurs in a pore already restricted in size by the 3.7 Å-thick surface layer of benzene [Gregg and Sing, 1982], which occludes a pore volume roughly the same as that occluded by the trimethylsilyl groups in treated analogs since the "thickness" of the two moieties is approximately the same. In the treated samples surface sorption is suppressed by this already present "hydrocarbon layer." This results in both the differences in uptake and the appearance of the initial portion of the isotherm, as discussed earlier. Argon is not expected to be affected by surface polarity effects. Figure 10 shows the Horváth–Kawazoe transformed Ar physisorption results on trimethylsilated M41S and its parent. The results show the TMS groups inside the pores. The pore diameter decreases from 39.4 to 30.4 Å upon conversion of the silanols to TMS groups. The 4.5-Å decrease in radius determined by the Horváth–Kawazoe method is what one would predict for replacement of a proton by a TMS group based on CPK models.

There is a small, low-pressure secondary peak also present in MCM-41 physisorption that is believed to represent monolayer sorption. Its mag-

3. Calculated from measurement of CPK models.
4. Calculated from the surface density of oxygens as given in footnote 1, the percentage of silanols calculated in this work, and the ratio of three methyl groups per silanol in the trimethyl-silylated samples.

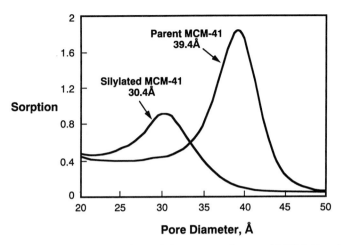

Figure 10. Horváth–Kawazoe transform of the argon isotherms of MCM-41 and its TMS derivative. Reprinted with permission from *J. Am. Chem. Soc.*, 114, 10834–10843 (1992), American Chemical Society.

nitude (area) can be used to calculate the interior surface area of the pores in these large-pore materials (using 16.2 Å^2/argon atom or 9.75^{-2} m^2/mmol in the monolayer sorption). For calcined and TMS-M41S the pore wall surface areas are 240 and 180 m^2/g. The ratio of these two numbers in 1:33. Cylindrical pores being assumed, the ratio of the pore wall surface areas should be the same as the ratio of their diameters, since the area will go as the circumference, which is just πd. The ratio of the pore diameters for these samples is 1.29. This is strong confirmation of the siting of the TMS groups (and hence the original silanols) and our assignment of the low-pressure physisorption peak in large-pore materials to monolayer sorption on the pore walls.

1.4   Conclusions

The M41S family of materials is both unique and diverse. These materials are thermally stable mesoporous molecular sieves that have demonstrated exceptionally high sorption capacity. These sieves have been synthesized in various structures: MCM-41, the hexagonal phase; MCM-48, the cubic phase; MCM-50, the stabilized lamellar phase. The pore size can be changed by varying the alkyl chain length of the surfactant or by adding various amounts of an auxiliary organic to the synthesis. Postsynthesis functionalization of the pore walls is another method for modifying the pore size and the sorption character of the material. The M41S family of materials represents an example of a molecular sieve system whose physical properties can be designed by controlled synthesis variations.

Acknowledgments

The authors are grateful to the staff at Mobil's Central and Paulsboro Research Laboratories for their invaluable discussion and effort. In particular, we acknowledge C. D. Chang, R. M. Dessau, J. B. Higgins, D. H. Olson, and H. M. Princen for helpful technical discussions. We thank N. H. Goeke, S. L. Laney, C. Martin, J. A. Pearson, and H. W. Solberg for their expert technical assistance. We also thank Mobil Research and Development Corporation for its support.

References

Applequist, J., Carl, J. R., and Fung, K.-K. (1972). *J. Am. Chem. Soc.*, **94**, 2952.

Beck, J. S. (1991). U.S. Patent #5 057 296, assigned to Mobil.

Beck, J. S., Vartuli, J. C., Roth, W. J., Leonowicz, M. E., Kresge, C. T., Schmitt, K. D., Chu, C. T-W., Olson, D. H., Sheppard, E. W., McCullen, S. B., Higgins, J. B., and Schlenker, J. L. (1992). *J. Am. Chem. Soc.*, **114**(27), 10834–10843.

Beck, J. S., Vartuli, J. C., Kennedy, G. J., Kresge, C. T., Roth, W. J., and Schramm, S. E. (1994). *Chem. Mater.*, **6**(10), 1816–1821.

Borghard, W. S., Sheppard, E. W., and Schoennagel, H. J. (1991). *Rev. Sci. Instrum.*, **62**, 2801–2809.

Chen, C. Y., Burkett, S. L., Li, H. X., and Davis, M. E. (1993). *Microporous Materials*, **2**, 27–34.

Davis, M. E. (1994). *Chem. Tech.*, 22–26.

Fontell, K. (1972). *J. Colloid Interface Sci.*, **43**(1), 156–164.

Fontell, K. (1990). *Colloid Polymer Sci.*, **268**, 264–285.

Gregg, S. J., and Sing, K. S. W. (1982). *Adsorption, Surface Area, and Porosity*, 2nd ed. Academic Press, London.

Haddon, R. C. (1979). *J. Am. Chem. Soc.*, **101**, 1722.

Horváth, G., and Kawazoe, K. (1983). *J. Chem. Eng. Jpn.*, **16**(6), 470–475.

Huo, Q., Margolese, D. I., Ciesia, U., Feng, P., Gier, T. E., Sieger, P., Leon, R., Petroff, P. M., Schüth, F., and Stucky, G. D. (1994). *Nature*, **368**, 317–321.

Kresge, C. T., Leonowicz, M. E., Roth, W. J., Vartuli, J. C., and Beck, J. S. (1992). *Nature*, **359**, 710–712.

Landis, M. E., Aufdembrink, B. A., Chu, P., Johnson, I. D., Kirker, G. W., and Rubin, M. K. (1991). *J. Am. Chem. Soc.*, **113**, 3189–3190.

Luzzati, V., Tardieu, A., Gulik-Krzywicki, T., Rivas, E., and Reiss-Husson, F. (1968). *Nature*, **220**, 485–488.

Mariani, P., Luzzati, V., and Delacroix, H. (1988). *J. Mol. Biol.*, **204**, 165–189.

Monnier, A., Schüth, F., Huo, Q., Kumar, D., Margolese, D., Maxwell, R. S., Stucky, G. D., Kishnamurthy, M., Petroff, P., Firouzi, A., Janicke, M., and Chmelka, B. F. (1993). *Science*, **261**, 1299–1303.

Sindorf, D. W., and Maciel, G. E. (1983). *J. Am. Chem. Soc.*, **105**, 1487.

Sindorf, D. W., and Maciel, G. E. (1982). *J. Phys. Chem.*, **86**, 5208.

Stucky, G. D., Monnier, A., Schüth, F., Huo, Q., Margolese, D., Kumar, D., Krishnamurthy, M., Petroff, P. M., Firouzi, A., Janicke, M., and Chmelka, B. F. (1994). *Mol. Cryst. Liq. Cryst.*, **240**, 187–200.

Vartuli, J. C., Schmitt, K. D., Kresge, C. T., Roth, W. J., Leonowicz, M. E., McCullen, S. B., Hellring, S. D., Beck, J. S., Schlenker, J. L., Olson, D. H., and Sheppard, E. W. (1994). *Zeolites and Related Microporous Materials: State of the Art 1994*, Proceedings of the 10th International Zeolite Conference, Garmisch-Partenkirchen, Germany, 7/17–22/1994 (J. Weitkamp, H. G. Karge, H. Pfeifer, and E., Hölderich, eds.), p. 53, Elsevier Science.

Weast, R D. (ed.) (1970). *Handbook of Chemistry and Physics*, pp. E-133ff, The Chemical Rubber Company, Cleveland, Ohio.

Yanagisawa, T., Kuroda, K., and Kato, C. (1988). *React. Sol.*, **5**, 167.

CHAPTER 2

The Role of Prehydrolysis in the Preparation of Zirconia–Silica Aerogels

James B. Miller and Edmond I. Ko
Department of Chemical Engineering
Carnegie Mellon University
Pittsburgh, PA 15213-3890

KEYWORDS: Prehydrolysis, mixed oxide homogeneity, zirconia–silica

2.1 Introduction

Sol–gel chemistry has long been successfully used in the preparation of glasses and ceramics for structural, optical, and electronic applications [1,2]. There have also been a lot of research activities on the sol–gel preparation of catalytic materials [3,4]. The key advantage of sol–gel processing is its considerable control over material properties. For a single-component oxide, process variables including precursor concentration, pH, and hydrolysis ratio have been used to control important textural and chemical properties of the calcined oxide product, including surface area, pore size distribution, crystallite size, and surface functionality [1,4–10]. Aging before drying to remove the solvent [11,12] as well as the type (evaporative versus "supercritical")[13] and conditions of drying [14] provide yet another level of property control.

For multicomponent catalytic materials, sol–gel preparation allows the introduction of several components in a single synthetic step. Such an approach has been demonstrated for specific systems such as zirconia–sulfate [15] and Li/MgO [16] and general classes of materials such as metal–oxide and oxide–oxide [4]. In the case of a multicomponent oxide preparation, sol–gel chemistry provides an important additional benefit: control of the quality of molecular scale component mixing, or homogeneity, to a degree that is not available with traditional catalyst preparation techniques [1,17]. Homogeneity affects both the textural and catalytic properties of a mixed oxide, most notably surface acidity. According to models proposed by Tanabe *et al.* [18] and others [19,20], acid sites form upon combination of dissimilar oxides as a result of charge imbalances that develop along heterolinkages (M—O—M', where M and M' represent different cations). High acid site densities in a mixed oxide are, therefore, expected of samples in which the components are intimately mixed. In fact, high activity in a reaction that requires an acid catalyst has often been cited as evidence for homogeneous component oxide mixing [21–23].

Homogeneity in a sol–gel mixed oxide preparation is related to the relative reactivity of the alkoxide precursors—good component mixing is expected when precursor reactivities are evenly matched and poor mixing when they are not. In cases where a significant reactivity difference exists, sol–gel chemistry offers several approaches to minimize the effects of the reactivity mismatch and promote homogeneous mixing. We have previously shown that (1) prehydrolysis of the silicon precursor, (2) chemical modification of the zirconium precursor with acetylacetone, and (3) replacement of tetra*ethyl* orthosilicate silicon precursor with the more reactive tetra*methyl* orthosilicate can each be employed to improve component mixing in the preparation of 95 mol % zirconia–5 mol % silica aerogels [22]. Toba *et al.* have demonstrated the use of a hexylene glycol "complexing agent" to promote homogeneous component mixing in zirconia–silica [24], titania–silica [25], and alumina–titania [26] mixed oxides.

Prehydrolysis, in which the less reactive precursor (usually a silicon alkoxide) is given a "head start" in the reaction sequence that forms the oxide network by allowing it to have first contact with water, is by far the most common reactivity matching strategy. The technique was first reported in the 1970s as a means for promoting homogeneity in sol–gel–derived multicomponent silicate glasses [27,28]. Prehydrolysis has also been widely used in sol–gel preparations of catalytic mixed oxides. For example, we have reported that zirconia–silica [29] and titania–silica [23] aerogels prepared by silicon precursor prehydrolysis possess higher acid site densities and display higher catalytic activities for isomerization of 1-butene than comparable samples prepared without prehydrolysis. Baiker and co-workers [30,31,32] have applied prehydrolysis in their preparation of titania–silicas used to support vanadia as catalysts for the selective catalytic reduction (SCR) of nitrogen oxides. These researchers employed spectroscopic techniques to illustrate the improvements in molecular-scale homogeneity that are attributable to prehydrolysis. They linked the mixing differences to changes in the textural properties and catalytic performance of their materials in the SCR application.

Regardless of the end-use of the product, reports of silica-containing mixed-oxide preparation usually specify a fixed set of prehydrolysis conditions. Prehydrolysis ratios (mol H_2O/mol Si)—arguably the most important prehydrolysis variable—as low as 0.8 [33] and in excess of 10.0 [34] have been reported, with ratios between 1.0 to 2.0 in most common use. Despite the facts that water content has been identified as a key variable in single-oxide syntheses [1,5,6], and that it has the potential to profoundly affect mixing in a multicomponent preparation, little attention has been given to the impact of changing prehydrolysis ratio in a mixed-oxide preparation.

In this work we use a set of 50 mol % zirconia–50 mol % silica aerogels to demonstrate that varying the prehydrolysis ratio over the range 0.00 to 3.22 mol/mol provides a fine degree of control over the mixed oxides' textural and acidic properties. There are two reasons for our selection of zirconia–silica as a prototypical mixed-oxide pair. First, it builds on our work with pure zirconia aerogels [5]. Zirconia itself displays a wide range of catalytic properties [35] and has received much attention as a support material [15,36–38]. Second, incorporation of silica to form a mixed oxide results in property changes that depend not only upon composition but also upon the intimacy of component mixing [22,29,39]. Even small amounts of silica can retard zirconia's sintering, thereby stabilizing surface area and low-temperature structural forms [22,39]. Zirconia–silica possesses acid sites that are among the strongest of common oxide pairs [40–43]; it has been studied as a catalyst for alcohol dehydration [40,41], cumene dealkylation [40], and alkene isomerization [22,29].

2.2 Methods

2.2.1 Preparation of the Zirconia–Silica Aerogels

We prepared zircona–silica alcogels from zirconium *n*-propoxide (Johnson Matthey, 70 wt % in *n*-propanol) and tetraethylorthosilicate (TEOS, Aldrich) precursors in an *n*-propanol solvent (Fisher certified) in the presence of a small amount of nitric acid (Fisher, 70 wt %) by a method similar to that reported by Nogami and Nagasaka [44]. The sol–gel parameters used to make the gels appear in Table 1; Fig. 1 illus-

TABLE 1
Sol–Gel Parameters Used in Preparing the Zirconia–Silica Aerogels[a]

Sample ID (AZS50-)[b]	PH water ratio (mol H$_2$O/mol Si)[c]	PH acid ratio (mL HNO$_3$/mol Si)	PH TEOS conc'n. (mmol Si/ml n-PrOH)	PH time (min)
Base series				
N03	0.00	N/A	N/A	N/A
P04	0.65	6.0	0.8	10
P03	1.13	6.0	0.8	10
P17	2.68	6.0	0.8	10
P28	3.22	6.0	0.8	10
Long prehydrolysis time				
P32	3.22	6.0	0.8	30
High prehydrolysis concentration				
P33	3.22	6.0	1.1	10

[a]Parameters that are held constant: 10 ml HNO$_3$/mol, 2.24 mol H$_2$O/mol(Zr + Si), 1.0 mmol (Zr+Si)/ml n-PrOH, 32 ml total n-PrOH solvent.
[b]Sample nomenclature: A = aerogel; ZS50 = 50 mol % zirconia, 50 mol % silica; Nxx = non-prehydrolyzed sample xx; Pyy = prehydrolyzed sample yy.
[c]PH = prehydrolysis.

trates the key features of the procedure. In a nonprehydrolyzed (NPH) preparation, the silicon and zirconium precursors (represented as Zr and Si for brevity in Fig. 1) are first added to a solution of acid and n-propanol in a 100-ml beaker. Next, a second solution of water and n-propanol is added. We measure gel time as the elapsed time between addition of the second solution and the point at which a magnetic stir-bar can no longer maintain a vortex in the reaction mixture.

For a prehydrolyzed (PH) preparation, a portion of the total hydrolysis water and a portion of the total nitric acid are combined with a solution of silicon precursor in n-propanol. *Prehydrolysis ratio* (PH ratio) is defined as the molar ratio of water to silicon precursor (mol H$_2$O/mol Si) in this step. After being stirred for a 10-min prehydrolysis period, the remainder of the acid, followed by the zirconium precursor, is added. Finally, the contents of a second breaker containing the remaining water and additional solvent are poured into the mixed precursor solution. Gel time is measured as in the NPH case. An important feature of our synthetic protocol is that the *overall* acid ratios, water ratios, and concentrations are kept constant (see Fig. 1). In other words, we used the same amount of every chemical in each preparation, changing only the order of addition. By varying the fraction of the total water used in the prehydrolysis step, we

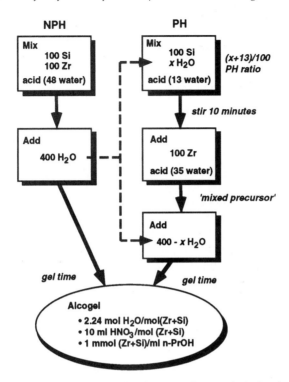

Figure 1. Differences between prehydrolyzed (PH) and nonprehydrolyzed (NPH) preparations of 50 mol % zirconia–50 mol % zirconia alcogels. Unless noted otherwise, numerical values represent relative number of mols of a reagent.

prepared a series of samples having prehydrolysis ratios from 0.00 to 3.22 (see Fig. 1 and Table 1).

At 3.22 mol/mol PH ratio we also prepared gels with (1) higher prehydrolysis water, precursor, and nitric acid concentrations (by moving some of the prehydrolysis solvent to the final step of the preparation) and (2) a longer prehydrolysis time (30 versus 10 min).

All samples gelled within 30 min. After an aging period of 2 h, each alcogel was dried by contact with supercritical (343 K, $\sim 2.2 \times 10^4$ kPa) carbon dioxide in a semicontinuous supercritical screening system (Autoclave Engineers Model 08U-06-60FS). At a CO_2 flow rate of 1400 ccm (ambient conditions) the alcohol solvent was removed in approximately 2 h. Product aerogels were ground to pass 100 mesh, vacuum dried at 383 K and 523 K to remove physisorbed water and residual organics and then calcined to 773 K in flowing oxygen (~400 ccm) in a tube furnace. Selected samples were calcined to 973, 1173, and 1373 K; high-temperature calcinations were performed on samples that had been previously heated to 773 K.

2.2.2 Characterization of the Mixed-Oxide Aerogels

Surface areas and pore size distributions of the calcined aerogels were measured by nitrogen adsorption–desorption experiments performed on an Autosorb-1 instrument (Quantachrome, Inc.). Samples were first outgassed in vacuum at 473 K for approximately 3 h. X-ray diffraction (XRD) patterns were obtained with a Rigaku D/Max diffractometer (Cu K_α radiation). Differential thermal analysis (DTA) was performed at 10 K/min in flowing helium on samples vacuum dried at 523 K using a Perkin-Elmer 1700 high-temperature thermal analyzer.

Samples calcined at 773 K were evaluated for their activity as catalysts for 1-butene isomerization, a reaction requiring a weak Brønsted catalyst [45]. Approximately 0.2 g of sample was used as catalyst in a downflow, fixed-bed reactor. Each sample was first pretreated in 50 sccm He (Matheson HP) at 473 K for 1 h. The bed temperature was then lowered to 423 K and the feed stream switched to a mixture of 5 sccm 1-butene (Matheson research grade) and 95 sccm He. Reaction products (*cis*-2-butene and *trans*-2-butene) were quantified by gas chromatography (Gow Mac 550P with thermal conductivity detector; Supelco 23% SP 1700 on 80/20 Chromasorb column). All samples deactivated at low on-stream times but had stabilized by 95 min on stream. We report isomerization activity as the reaction rate *per unit area* at 95 min.

The total acid site densities of selected samples were measured by ammonia temperature-programmed desorption (TPD) experiments performed on an Altamira AMI-1 instrument. Samples (~0.04 g) were first pretreated in flowing helium (30 sccm) for 1 h at 473 K to simulate pretreatment conditions used for 1-butene activity testing. The temperature of the sample was then lowered to 423 K, where it was exposed to ammonia flow (30 sccm) for 90 min. While still at 423 K, the sample was again purged with helium (30 sccm for 1 h) to remove weakly bound ammonia. The temperature of the sample was ramped from 423 to 773 K at 10 K/min, then held at 773 K for an additional 30 min. Desorbed ammonia was quantified with a thermal conductivity detector. The total amount of desorbed ammonia was determined from the area under the TPD curve (thermal conductivity detector output versus time) as compared with areas computed from known amounts of ammonia. We report the average result of two TPD experiments.

Calcined samples (773 K) were also characterized by diffuse reflectance infrared Fourier transform (DRIFT) spectroscopy using a Mattson Galaxy 5020 FTIR with a Harrick diffuse reflectance attachment. A DTGS detector was used to collect spectra between 400 and 4000 cm^{-1} with 2 cm^{-1} resolution. Samples diluted in KBr (~ 5 wt % mixed oxide) were placed inside a Harrick reaction chamber, where they were pretreated in helium (~ 50 sccm) at 473 K for 1 h and then cooled 423 K to simulate pretreatment conditions used for isomerization activity tests. While under helium flow at 423 K, a spectrum was collected for characterization of the sample's hydroxyl inventory (above ~ 3000 cm^{-1}) and "skeletal vibration" region ($\sim 700-1200$ cm^{-1}). The sample was then exposed to pyridine by diverting the helium flow through a pyridine saturator for 15 min. Thirty minutes after termination of pyridine exposure, with the sample still under helium flow at 423 K, a second spectrum was collected. From the integrated intensities of peaks centered at approximately 1445 and 1490 cm^{-1}, we estimated the relative population of Brønsted and Lewis acid sites using the method of Basila and Kantner [46]. We report the results as "fractional Brønsted acidity," defined as (Brønsted sites)/(Brønsted sites + Lewis sites).

^{29}Si NMR experiments were performed on solutions of *n*-propanol, TEOS, nitric acid, and water having the same relative compositions as those of the prehydrolysis step of our gel preparations. Solutions were prepared immediately before placement in a Bruker AM-300 instrument. Inverse gated decoupling experiments (each having 32 scans with 12 delay and 7 μs pulse width) were performed, providing a series 6.5 min time-averaged spectra at mean sampling times between approximately 5 and 30 min after sample preparation. Assignments of spectral features were based on published sources [1,47]. We report the results as relative product distribution at about 10 min after solution preparation, a period comparable with our standard prehydrolysis time.

2.3 Results and Discussion

2.3.1 Catalytic–Chemical Properties

Figure 2 illustrates that activity for 1-butene isomerization, a reaction requiring a weak Brønsted acid catalyst [45], increases continuously with increasing PH ratio. We have recently reported that high relative l-butene isomerization activities are characteristic of well-mixed titania–silica [23] and 95 mol % zirconia–5 mol % silica [22] aerogels. Therefore, the data in Fig. 2 suggest that, by increasing PH ratio over the 0.00 to 3.22

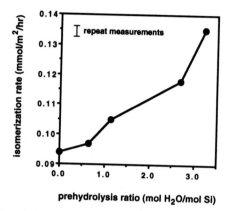

Figure 2. Prehydrolysis ratio's effect on the catalytic activities of 50 mol % zirconia–50 mol % silica aerogels for 1-butene isomerization. Activities reported as reaction rates at standard conditions (see text) and 95 min time on stream. Error bar shows range of activities measured in repeat trials. All samples calcined at 773 K in oxygen for 2 h before activity testing.

mol/mol range, we gain access to a "continuum of mixing states." Of particular significance is our observation that component mixing continues to improve beyond the 1.0–2.0 mol/mol PH ratio region that is cited most often in literature reports of sol–gel preparations of silica-containing mixed oxides.

As shown in Fig. 3, isomerization activity improvements result from increases in both (1) total (Lewis + Brønsted) acid site density and (2) the fraction of Brønsted sites that occur as PH ratio rises. These trends are consistent with results we recently reported for titania–silica aerogels [23].

The hydroxyl regions of the DRIFT spectra of our samples appear in Fig. 4. At low prehydrolysis ratios (0.00 and 0.65 mol/mol), the high wavenumber "spike," a feature attributable to the stretch of "isolated" SiO—H [1,48], dominates the broad feature associated with "hydrogen-bonded" hydroxyl groups [1,48]. Within a series of well-mixed zirconia–silica aerogels of varying composition, we have observed dominance of the high wavenumber spike at *high silica content* [29]. Together these observations suggest that the surfaces of the low PH ratio 50 mol % zirconia–50 mol % silica samples are "silica-like." We therefore associated segregation (poor mixing) in the low-PH-ratio samples with surface segregation of silica, as we previously did for 95 mol % zirconia–5 mol % silica [22] and titania–silica [23] aerogels.

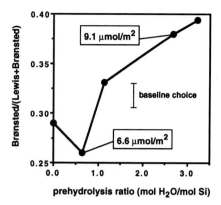

Figure 3. Fractional Brønsted population [defined as Brønsted sites/total (Lewis + Brønsted) acid sites] as a function of prehydrolysis ratio. Effect of baseline choice for integration of IR intensities shown by error bar. Figures in boxes are total acid site densities as measured by ammonia TPD experiments. All samples calcined at 773 K in oxygen for 2 h before evaluation.

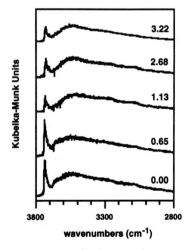

Figure 4. DRIFT spectra hydroxyl regions for the 50 mol % zirconia–50 mol % silica aerogels. Numerical values are prehydrolysis ratios (mol H_2O/mol Si). Samples originally calcined at 773 K in oxygen for 2 h.

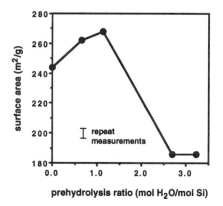

Figure 5. BET surface areas of 50 mol % zirconia–50 mol % silica aerogels as a function of prehydrolysis ratio. Error bar shows range of repeat measurements. All samples calcined at 773 K in oxygen for 2 h.

2.3.2 Textural–Structural Properties

The dependence of our samples' textural properties upon PH ratio also provides evidence for a relationship between high PH ratio and homogeneous component mixing. As shown in Fig. 5, samples having the highest surface area are those prepared at the lowest PH ratios. This result parallels our observation of high relative surface areas in 95 mol % zirconia–5 mol % silica aerogels that, by surface segregation of their silica component, are *poorly mixed* [22]. Figure 6 illustrates how, at higher PH ratios, the pore size distributions of the calcined aerogels broaden and shift to higher mean pore size. This behavior is also characteristic of *well-mixed* 95 mol % zirconia–5 mol % silica samples [22].

The ability of silica to delay zirconia's crystallization to higher heat treatment temperatures in mixed-oxide catalysts and glasses has been well documented [22,33,39,40,49,50]. To assess the impact of component mixing on the stabilization of X-ray amorphous zirconia, we obtained XRD patterns for the 0.65 and 2.68 mol/mol samples, representative of low and high PH ratios, after calcination to 773, 973, 1173, and 1373 K. Figure 7 shows that zirconia's metastable tetragonal phase [5] begins to crystallize between 973 and 1173 K in both samples; for reference, the tetragonal phase is observed in pure zirconia aerogel calcined at only 773 K [5]. We note that the XRD peak at 1173 K is slightly narrower at 0.65 mol/mol PH ratio. Indeed, zirconia's crystallite size, estimated by Scherrer's line-broadening method [51], is larger in the low PH ratio sample ($\sim$ 78 versus 43Å). These observations suggest that crystallization may have begun at a

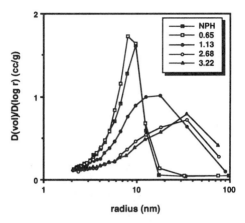

Figure 6. Prehydrolysis ratio's effects on the pore size distributions of 50 mol % zirconia–50 mol % silica aerogels. Prehydrolysis ratios (mol H_2O/mol Si) appear in the legend. All samples calcined at 773 K in oxygen for 2 h.

lower temperature in the 0.65 mol/mol sample; in other words, low-ratio prehydrolysis may be *less* effective for stabilization of amorphous zirconia. Differential thermal analysis (DTA) data of our samples provide a clearer picture of PH ratios' effect on crystallization. For each PH ratio the

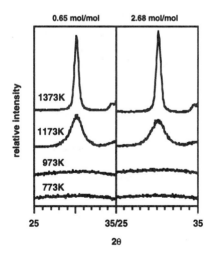

Figure 7. X-ray diffraction (XRD) patterns of the 0.65 and 2.68 mol H_2O/mol Si aerogels as functions of heat treatment temperature. Samples were calcined in oxygen for 2 h at the temperatures indicated.

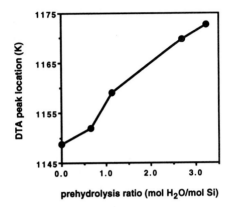

Figure 8. Differential thermal analysis (DTA) peak location as a function of prehydrolysis ratio for 50 mol % zirconia–50 mol % silica aerogels. DTA experiments were performed on samples vacuum dried at 523 K.

DTA scan had only a single exothermic peak in the temperature region over which the crystallization of zirconia occurred (as verified by the XRD experiments). Figure 8 illustrates how the location of the DTA peak moves to higher temperatures with increasing PH ratio, clearly demonstrating the ability of high-ratio prehydrolysis to enhance the stability of amorphous zirconia. We note that this behavior contrasts with what we have previously reported for 95 mol % zirconia–5 mol % silica aerogels [22]. In this case, *poorly mixed* samples (nonprehydrolyzed) were more effective for stabilization of amorphous zirconia. We shall return to address the synergistic effects of composition and mixing on zirconia's crystallization later in this paper.

In the skeletal vibration portion of the DRIFT spectrum (700–1300 cm^{-1}), our samples display a single main feature, a peak assigned to a Si—O—Si asymmetric stretch at ~ 1020 cm^{-1} [1,40,41,50,52]. As we show in Fig. 9, this peak's location moves to higher wavenumbers at high PH ratio, indicative of a larger, more densely interconnected siloxane network in those samples [1,40,41,50].

2.3.3 The Link between Prehydrolysis Chemistry and Calcined Mixed-Oxide Properties

The results of ^{29}Si NMR experiments performed on simulated prehydrolysis solutions allow us to relate the properties we observe in the calcined aerogels as a function of PH ratio to the chemistry that occurs in the

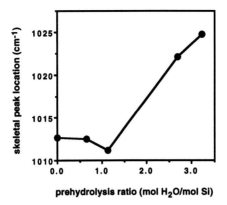

Figure 9. Prehydrolysis ratio's effect on the location of the asymmetric Si—O—Si stretch feature in the DRIFT spectra of 50 mol % zirconi–50 mol % silica aerogels. All samples calcined at 773 K in oxygen for 2 h.

prehydrolysis step of the preparation. Figure 10 illustrates how the prehydrolysis product distribution changes as a function of PH ratio. At low ratios prehydrolysis products are primarily hydrolyzed monomers $[Si(OH)_x(OR)_{4-x}]$ and unreacted silicon precursor. As the PH ratio increases: (1) unreacted monomer disappears, (2) hydrolyzed dimers

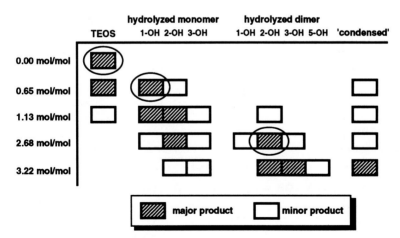

Figure 10. Relative prehydrolysis product distributions, measured by ^{29}Si NMR, at different prehydrolysis ratios. "Condensed" products are TEOS trimers and higher oligomers. 'Mixed precursors' that form from the circled species are illustrated in Fig. 11.

[(Si—O—Si)(OH)$_x$(OR)$_{6-x}$] and more highly condensed species (trimers and higher) appear, and (3) the extent of both monomer and dimer hydrolysis increases (x becomes larger).

We have used the NMR results as a basis for postulating molecular architectures of typical "mixed precursors"—the species that exists after addition of the zirconium precursor but before addition of the "final water" (see Fig. 1). We "form" the mixed precursor assuming that (1) zirconium precursor condenses only with silanol (Si—OH) groups (in other words, nonhydrolyzed Si—OR groups are *unreactive*) and (2) zirconium precursor adds to the prehydrolysis product until a 1:1 Zr:Si atomic ratio (matching the overall mixed-oxide composition) is achieved. Figure 11 illustrates prototypical mixed precursors that form from hydrolyzed monomers and hydrolyzed dimers.

In our nonprehydrolyzed preparation (PH ratio = 0.0), the absence of any hydrolyzed silicon species (Fig. 10) implies that no mixed precursor forms (Fig. 11). Because TEOS is relatively unreactive, we expect it to par-

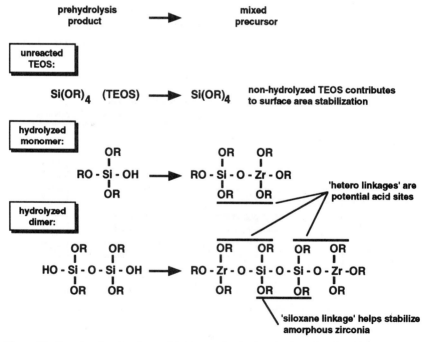

Figure 11. Postulated molecular architectures of mixed precursors that form upon addition of zirconium alkoxide to representative prehydrolysis products (see Fig. 10).

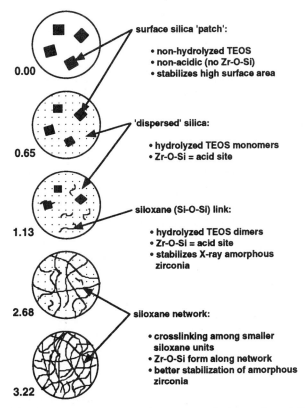

surface silica 'patch':
- non-hydrolyzed TEOS
- non-acidic (no Zr-O-Si)
- stabilizes high surface area

'dispersed' silica:
- hydrolyzed TEOS monomers
- Zr-O-Si = acid site

siloxane (Si-O-Si) link:
- hydrolyzed TEOS dimers
- Zr-O-Si = acid site
- stabilizes X-ray amorphous zirconia

siloxane network:
- crosslinking among smaller siloxane units
- Zr-O-Si form along network
- better stabilization of amorphous zirconia

Figure 12. Models of component mixing in the zirconia–silica aerogels at different prehydrolysis ratios. Silica–siloxanes are shown in black, zirconia in white. Numerical values are prehydrolysis ratios (mol H_2O/mol Si).

ticipate in the network-building reactions only after substantial condensation of zirconia-rich "primary particles" has taken place. Consequently, silicon atoms are located preferentially at the surface of the mixed oxide, most likely in the form of nonacidic silica-rich patches [22,23], as shown in our model of the calcined aerogel surfaces, Fig. 12. The low frequency of the Si—O—Si IR feature (Fig. 9) displayed by the nonprehydrolyzed sample indicates that the surface patches are not densely interconnected; they may be only a few molecular layers thick—essentially two-dimensional. We believe that surface silica patches play an important role in stabilizing high surface areas of all low PH ratio samples (Figs. 5 and 12), perhaps by interfering with the sintering processes that would otherwise lead to surface area loss. Finally, we note that the modest isomerization activity of the nonprehydrolyzed sample is likely related to acidic $Zr-O-Si$ linkages that develop at the *perimeter* of the patches upon calcination.

As PH ratio is increased to 0.65 mol/mol, prehydrolysis produces hydrolyzed TEOS monomers (Fig. 10) that can form mixed precursors having embedded $Zr-O-Si$ linkages (Fig. 11). These heterolinkages, which are formed at low temperature, eventually confer acidity, and therefore isomerization activity, upon the calcined mixed oxide as "dispersed silica" (Fig. 12). However, much of the total silicon content at 0.65 mol/mol continues to exist as nonhydrolyzed TEOS. Therefore, while smaller in size, or perhaps fewer in number, silica-rich surface patches are still created at this PH ratio (Fig. 12), which accounts for the high relative surface area of this sample (Fig. 5).

The most significant change as PH ratio is increased from 0.65 to 1.13 mol/mol is the appearance of hydrolyzed dimers among the prehydrolysis products (Fig. 10). As illustrated in Fig. 11, hydrolyzed dimers from mixed precursors that contain both (1) $Zr-O-Si$ linkages (potential acid sites) and (2) embedded siloxane linkages. We believe that the siloxane linkage plays an important role in stabilizing X-ray amorphous zirconia at 1.13 mol/mol and higher PH ratios (Fig. 8), perhaps by limiting the mobility of zirconium atoms within the mixed-oxide matrix. Some nonhydrolyzed TEOS remains after prehydrolysis at 1.13 mol/mol (Fig. 10), which results in surface silica patches that continue to stabilize high surface area (Fig. 12).

At 2.68 mol/mol, no unreacted TEOS exists at the mixed precursor stage (Fig. 10) and the surface area of the calcined mixed oxide is markedly lower (Fig. 5). These results suggest that the silica patches are no longer present on the surface of this sample (Fig. 12). In addition, with no unreacted TEOS remaining after prehydrolysis, more silicon atoms are available to participate in heterolinkages, accounting in large part for the higher acidity and activity of mixed oxides prepared at this and higher prehydrolysis ratios (Figs. 2 and 3). High populations of hydrolyzed dimers among the prehydrolysis products (Fig. 10) suggest that many of the mixed precursors now contain internal siloxane $(Si-O-Si)$ bridges between zirconium atoms (Fig. 11). At 2.68 mol/mol, homocondensation of $Si-OH$ groups in neighboring mixed precursors may begin to occur to form a longer, more complex siloxane network embedded within the mixed oxide's matrix (Fig. 12). This view is supported by an increase in the frequency of the asymmetric $Si-O-Si$ vibration in the IR spectrum (Fig. 9), a measure of an increase in the interconnectivity, or length, of the embedded silica network [1,41,50]. The continued increase in the temperature of the DTA peak (Fig. 8) indicates that the network is also more effective than less interconnected siloxane groups for stabilizing X-ray amorphous zirconia.

As PH ratio is increased further to 3.22 mol/mol, prehydrolysis produces dimers (and, presumably, trimers and higher) that contain more Si—OH and fewer Si—OR functionalities (Fig. 10). The bias toward Si—OH groups provides opportunities for (1) additional crosslinking to further strengthen the siloxane network (Fig. 12) and (2) formation of a higher density of Zr—O—Si (acid) linkages along the network. Indeed, relative to the 2.68 mol/mol sample, the 3.22 mol/mol aerogel displays (1) greater stability of amorphous zirconia (Fig. 8), (2) a higher frequency silica network vibration (Fig. 9), and (3) improved isomerization activity (Fig. 2).

We now return to the issue of differences in silica's ability to inhibit crystallization of tetragonal zirconia at different compositions. We view both "surface segregated silica" from nonhydrolyzed silicon precursor (low PH ratio) and the "embedded siloxane network" (high PH ratio) as contributors to stabilization of amorphous zirconia. At 95 mol % zirconia–5 mol % silica, there is too little silica present to "anchor" a significant number of zirconium atoms with siloxane linkages in PH samples. Therefore, at this composition, we observed that NPH samples containing "surface segregated silica" were more effective than their PH counterparts for stabilizing amorphous zirconia [22]. On the other hand, at the 50 mol % composition, stabilization through the siloxane network becomes the more important mechanism, and prehydrolysis is observed to promote stability of amorphous zirconia.

2.3.4 Other Prehydrolysis Variables

Finally, we note that while water ratio is not the only prehydrolysis step sol–gel parameter that can be varied for control of mixed-oxide properties, it is by far the most important. We have changed both the prehydrolysis time (from 10 to 30 min) and concentration (from 0.8 to 1.1 mmol Si/ml *n*-PrOH) in our 3.22 mol/mol preparation. As shown in Table 2, the effects of these changes on the properties of the calcined aerogels are small, especially when compared with the effects of varying PH ratio. However, it is instructive to consider the direction of the effects within the framework of our surface homogeneity model (Fig. 12).

At long prehydrolysis times our NMR experiments show a significant increase in the population of condensed (trimers and higher) prehydrolysis products. When silicon atoms have primarily —O—Si nearest neighbors, we anticipate the type of component segregation shown at the top of Fig. 13, in which small, silica-rich "kernels" are embedded in a zirconia ma-

TABLE 2
Effects of Time and Concentration on Properties of 3.22 mol H_2O/mol
Si Prehydrolyzed 50 mol % Zirconia–50 mol % Silica Aerogels

Preparation type	Base	Long time	High concentration
Sample ID (AZS50-)	P28	P32	P33
prehydrolysis concentration (mmol Si/ml n-PrOH)	0.8	0.8	1.1
prehydrolysis time (minutes)	10.0	30.0	10.0
BET surface area (m^2/g)	186	183	178
1-butene isomerization activity (mmol/m^2/hr)	0.135	0.127	0.145
location of DTA peak (K)	1173	1170	1168
location of asymmetric stretch in IR (cm^{-1})	1024.7	1023.7	1022.3

trix. With a fewer silicon atoms participating in heterolinkages, a long pre-hydrolysis time sample would be expected to be both less acidic (i.e., lower isomerization activity) and less capable of stabilizing noncrystalline zirconia than its short PH time counterpart. As shown in Table 2, this is indeed how these properties change with increasing prehydrolysis time. We also note that the lower Si—O—Si IR wavenumber displayed by the long pre-hydrolysis time sample probably reflects a less well-developed siloxane network (Fig. 13).

As shown in Table 2, increasing prehydrolysis concentrations by removal of some of the alcohol solvent improves isomerization activity but *lowers* the temperature at which zirconia crystallizes. This contrasts with the results of our low-concentration preparations in which high-isomerization activities are associated with high-crystallization temperatures (Figs. 2 and 8). One explanation for this difference is that higher water and precursor concentrations may favor formation of smaller (monomers and dimers versus dimers and higher) and more highly hydrolyzed siloxane products. We expect a lower degree of siloxane connectivity within prehydrolysis products to result in a less mature network—a network that is less effective for stabilizing amorphous zirconia—in the calcined mixed oxide (Fig. 13). The lower wavenumber for the Si—O—Si IR feature in the "high concentration" aerogel supports this view. An increased density of Si—OH (rather than Si—OR or even Si—O—Si) functional groups within the prehydrolysis products, on the other hand, provides additional

long prehydrolysis time:

- high density silica 'kernels'

- fewer Zr-O-Si (acid sites)

- poorer stabilization of
  amorphous zirconia

high concentration:

- less siloxane crosslinking =
  poorer stabilization of
  amorphous zirconia

- higher density of Zr-O-Si
  along the network = more
  acid sites

Figure 13. Models of component mixing in long prehydrolysis time (top) and high prehydrolysis concentration (bottom) samples.

opportunities for creation of heterolinkages during gelation, perhaps accounting for the increased catalytic activity of the high-concentration sample.

2.4   Conclusions

Prehydrolysis (PH) ratio is an important variable for controlling the textural and acidic properties of zirconia–silica aerogels. The effects of PH ratio on the properties of the calcined materials can be linked to differences among the types of hydrolyzed silicon intermediates that form during the initial step of the preparation. This observation suggests that the central role of PH ratio is not specific to zirconia–silica but is general in the preparation of silica-containing mixed oxides. Of particular note to catalytic researchers is our observation that homogeneity and activity enhancements attributed to prehydrolysis extend beyond the 1.0–2.0 mol/mol range currently in most common use. This work illustrates the importance of a fundamental understanding of solution chemistry for effective use of sol–gel technology in preparing catalytic materials of desirable properties.

Acknowledgments

We thank Drs. Timothy Flood and Bulent Yoldas of the PPG Chemicals Group Technical Center for their assistance with the ^{29}Si NMR experiments. This work is supported by the Division of Chemical Services, Office of Basic Energy Sciences, Office of Energy Research, U.S. Department of Energy (Grant No. DE-FG02-93ER14345).

References

1. Brinker, C. and Scherer, G. (1990). *Sol-Gel Science: The Chemistry and Physics of Sol-Gel Processing*, Academic Press, Boston.
2. Yoldas, B. E. (1993). *J. Sol-Gel Sci. and Tech.*, 1, 65.
3. Cauqui, M. A., and Rodriguez-Izquierdo, J. M. (1992). *J. Non-Cryst. Solids*, **147/148**, 724.
4. Ward, D. A., and Ko, E. I. (1995). *I&EC Research*, 34, 421.
5. Ward, D. A., and Ko, E. I. (1993). *Chem. Mater.*, 5, 956.
6. Campbell, L. K., Na, B. K., and Ko, E. I. (1992). *Chem. Mater.*, 4, 1329.
7. Iler, R. K. (1979). *The Chemistry of Silica*, Wiley, New York.
8. Ying, J. Y., and Benziger, J. B. (1980). *J. Non-Cryst. Solids*, **147/148**, 222.
9. Ying, J. Y., Benziger, J. B., and Navrotsky, A. (1993). *J. Am. Ceram. Soc.*, 76, 2571.
10. Duran, A., Fernandez-Navarro, J. M., Casanego, P, and Joglar, A. (1989). *J. Non-Cryst. Solids*, **82**, 69.
11. Smith, D. M., Davis, P. J., and Brinker, C. J. (1990). *Proc. Mater. Res. Soc.*, 180, 235.
12. Glaves, C. L., Brinker, C. J., Smith, D. M., and Davis, P. J. (1989). *Chem. Mater.*, 1, 34.
13. Maurer, S. M., and Ko., E. I. (1992). *J. Catal.*, 135, 125.
14. Brodsky, C. J., and Ko, E. I. (1994). *J. Mater. Chem.*, 4, 651.
15. Ward, D. A., and Ko, E. I. (1994). *J. Catal.*, 150, 18.
16. Lopez, T., Gomez, R., Ramirez-Solis, A., Poulain, E., and Novaro, O. (1994). *J. Mol. Catal.*, 88, 71.
17. Courty, P., and Marcilly, C. (1976). *Preparation of Catalysts* (B. Delmon, P. A. Jacobs, and G. Poncelet, eds.), p. 119, Elsevier, Amsterdam.
18. Tanabe, K., Sumiyoshi, T., Shibata, K., Kiyoura, T., and Kitagawa, J. (1974). *Bull. Chem. Soc. Jpn.* , 47(5), 1064.
19. Thomas, C. L. (1949). *Ind. Eng. Chem. Res.*, 41(11), 2564.
20. Kung, H. H. (1984). *J. Solid State Chem.*, 52, 191.
21. Tanabe, K., Itoh, M., Morshige, K., and Hattori, H. (1976). *Preparation of Catalysts* (B. Delmon, P. A. Jacobs, and G. Poncelet, eds.), p. 65, Elsevier, Amsterdam.
22. Miller, J. B., Rankin, S. E., and Ko, E. I. (1994). *J. Catal.*, 148, 673.
23. Miller, J. B., Johnston, S. T., and Ko, E. I. (1994). *J. Catal*, 150, 311.
24. Toba, M., Mizukami, F., Niwa, S., Sano, T., Maeda, K., Annila, A., and Komppa, V. (1994). *J. Mol. Catal*, 94, 85.
25. Toba, M., Mizukami, F., Niwa, S., Sano, T., Maeda, K., Annila, A., and Komppa, V. (1994). *J. Mol. Catal*, 91, 277.
26. Toba, M., Mizukami, F., Niwa, S., Kiyozumi, Y., Maeda, K., Annila, A., and Komppa, V. (1994). *J. Mater. Chem.*, 4(4), 585.
27. Thomas, I.M. (1974). U.S. Patent 3 791 808.
28. Yoldas, B. E. (1977). *J. Mater. Sci.*, 12, 1203.
29. Miller, J. B., and Ko, E. I. (1996). *J. Catal.*, 159, 58.

30. Handy, B. E., Maciejewski, M., Baiker, A. and Wokaun, A. (1992). *J. Mater. Chem.*, 2(8), 833.
31. Schraml-Marth, M., Walther, K. L., Wokaun, A., Handy, B. E. and Baiker, A. (1992). *J Non-Cryst. Solids*, 143, 93.
32. Handy, B. E., Baiker, A., Schraml-Marth, M., and Wokaun, A. (1992). *J. Catal.*, 133, 1.
33. Monros, G., Marti, M. C., Carda, J., Tena, M. A., Escribano, P., and Agnlada, M. (1993). *J. Mater. Sci.*, 28, 5852.
34. Brautigam, U., Meyer, K., and Burger, H (1992). *Eurogel 91*, (S. Vilminot, R. Nass, and H. Schmidt, eds.), p. 335, Elsevier, Amsterdam.
35. Tanabe, K. (1985). *Mater. Chem. Phys.*, 13, 347.
36. Amenomiya, Y. (1987). *Appl. Catal.*, 30, 57.
37. Iizuka, T., Tanaka, Y., and Tanabe, K. (1982). *J. Catal*, 76, 1.
38. Hino, M., and Arata, K. (1980). *J. Chem. Soc. Chem. Commun.*, 851.
39. Soled, S., and McVicker, G. B. (1992). *Catal. Today*, 14, 189.
40. Sohn, J. R., and Jang, H. J. (1991). *J. Mol. Catal.*, 64, 349.
41. Bosman, H. J. M., Kruissink, E. C., van der Spoel, J., and van den Brink, F. (1994). *J. Catal.*, 148, 660.
42. Dzisko, V. (1964). *Proc. Third Int. Congr. Catal.*, 1(19), 422.
43. Shibata, K., Kiyoura, T., Kitagawa, J., Sumiyoshi, T., and Tanabe, K. (1973). *Bull. Chem. Soc. Jpn*, 46, 2985.
44. Nogami, M., and Nagaska, K. (1989). *J. Non-Cryst. Solids*, 109, 79.
45. Goldwasser, J., Engelhardt, J., and Hall, W. K. (1981). *J. Catal.*, 71, 381.
46. Basila, M. R., and Kantner, T. R. (1966). *J. Phys. Chem.*, 70, 1681.
47. Pouxviel, J. C., Boilot, J. P., Beloeil, J. C., and Lallemand, J. Y. (1987). *J. Non-Cryst. Solids*, 89, 345.
48. Little, L. H. (1966). *Infrared Spectra of Adsorbed Species*, p. 230, Academic, London.
49. Campaniello, J., Rabinovich, E., Revcolevschi, A., and Kopylov, N. (1990). *Mater. Res. Soc. Symp. Proc.*, 180, 541.
50. Miranda Salvado, I., Serna, C., Fernandez Navarro, J. (1988). *J. Non-Cryst. Solids*, 100, 330.
51. Klug, H. P., and Alexander, L. E. (1974). *X-ray Diffraction Procedures*, Wiley, New York.
52. Nogami, M. (1985). *J. Non-Cryst. Solids*, 69, 415.

CHAPTER 3

The Chemistry of Preparation of V–P Mixed Oxides

Effect of the Preparation Parameters on the Catalytic Performance in n-Butane and n-Pentane Selective Oxidation

F. Cavani, A. Colombo, F. Giuntoli, F. Trifirò,
P. Vazquez,* and P. Venturoli†
Dipartimento di Chimica Industriale e dei Materiali,
40136 Bologna, Italy
*On leave from the Centro de Investigacion y Desarrollo en Procesos Cataliticos,
La Plata, Argentina.
†Present address: Lonza SpA, 24020 Scanzorosciate (BG), Italy.

KEYWORDS: Preparation of V/P/O-based systems, selective oxidation, n-butane oxidation, maleic anhydride, (VO) $HPO_4 \cdot 0.5H_2O$ precursor, n-pentane oxidation, phthalic anhydride, vanadyl pyrophosphate, vanadium overoxidation

3.1   Introduction

Different methods of preparation for the V/P/O catalysts have been reported in the scientific and patent literature [1–4]. All of them achieve the ultimate active phase via the following main stages:

1. Initial preparation of $(VO)HPO_4 \cdot 0.5H_2O$—the active phase precursor

2. Thermal decomposition of the hemihydrate vanadyl orthophosphate, with partial or total loss of the hydration water, formation of new phases, and elimination of precursor impurities (chlorine ions, organic

43

compounds) as well as of additives employed for powder tableting when the tablets are prepared before the dehydration stage

3. Activation or aging inside the reactor; phase and morphological transformations, recrystallization, creation or elimination of structural defects, selective poisoning by high-boiling compounds; migrations of vanadium and phosphorous species occur at this stage, which can last from a few days to 1 month (This stage is necessary to achieve catalysts with optimum catalytic performances.)

4. Formation of the catalysts in such a way as to achieve the best mechanical resistance for use in fixed-, fluidized-, or transport-bed reactors

In this chapter the main features of the first three stages will be examined.

3.2   Literature Survey

3.2.1   The Preparation of the Precursor

Two main methods of preparation of the precursor can be singled out:

1. Reduction of V^{5+} compounds (V_2O_5) to V^{4+} in water by either HCl or hydrazine, followed by addition of phosphoric acid and separation of the solid by either evaporation of water or by crystallization.
2. Reduction of V^{5+} compounds in a substantially anhydrous medium with either an inorganic or an organic reducing agent, addition of dry phosphoric acid, and separation of the solid obtained either by filtration, by solvent evaporation, or by centrifugation.

The addition of phosphorous compounds before V^{5+} reduction has also been claimed for both methods of preparation, but it does not seem to be the preferred procedure.

A substantially anhydrous medium means the use of a dry organic solvent, of dry metal salts and components, as well as the use of phosphoric acid containing more than 98% of H_3PO_4; moreover, the water formed by vanadium reduction and by digestion is removed by azeotropic distillation during the preparation. The organic solvent must possess the properties to dissolve, but not to react with, the phosphoric acid, eventually to reduce the vanadium species and not dissolve the precursor. In the preparation in organic solvent, intercalated or occluded organic materials may represent 25% by weight of the precursor.

The aqueous solvent must be capable of dissolving the components of the precursor and the reducing agent, but unfortunately, at the same time, it also dissolves the precursor. In the preparation in aqueous medium, the anions of the metal, i.e., sulfates or chlorines, can be incorporated into the structure of the precursor.

3.2.1.1 Preparation in Aqueous Medium

In the preparation in an aqueous medium the following steps for the formation of the precursor can be proposed:

1. Reduction of V_2O_5 to soluble V^{4+}
2. After addition of H_3PO_4 no precipitation occurs, due to the strong acid conditions
3. Development of $(VO)HPO_4 \cdot 0.5H_2O$ with another spurious amorphous phase only after complete evaporation of the solvent
4. Alternatively, crystallization of pure $(VO)HPO_4 \cdot 0.5H_2O$ by addition of water when the solution is highly concentrated (when it is very viscous), or by seeding under hydrothermal conditions (high temperature and steam pressure

Two examples of preparation of catalysts in aqueous medium are reported here, to illustrate the main features of this type of preparation.

In the preparation developed by Mitsubishi [5], the following main stages are proposed:

1. Reduction of V_2O_5 with hydrazine in an aqueous solution, under reflux, and formation of the V^{4+} species
2. Addition of chelating agents possessing two ligand groups (ethylene glycol or oxalic acid), and heating under reflux
3. Addition of the phosphorous compound, and introduction of the solution together with seed crystals of the precursor in a closed vessel at $120-200°C$ under steam pressure
4. Development of the precursor, filtration of the slurry, drying, and activation in N_2 at $500°C$ and then in air

In the preparation developed by Alusuisse [6], the following main stages can be singled out:

1. Suspension of V_2O_5 in a concentrated solution of HCl and heating under reflux at $100°C$
2. Addition of oxalic acid and phosphoric acid
3. Concentration of the solution until a viscous solution is obtained
4. Addition of excess water to the viscous solution, obtaining a bright blue crystalline compound (the precursor)

46 / F. Cavani et al.

3.2.1.2 Preparation in Anhydrous Medium

The following steps for the formation of the precursor in organic media can be proposed:

1. Formation of colloidal V_2O_5 at the water–alcohol interface, proposed by some authors [7] but according to others an unimportant step [8]
2. Solubilization of V^{5+} through the formation of vanadium alcoholates or of $VOCl_3$ when HCl is used as a reductant
3. Reduction of the alcoholate in the liquid phase to solid V_2O_4 by the organic compound (the solvent itself or another more reactive alcohol such as benzyl alcohol) or by an inorganic reducing agent, such as HCl
4. Reaction at the surface of V_2O_4 with H_3PO_4 to form $(VO)HPO_4 \cdot 0.5H_2O$ at the solid–liquid interface
5. Separation of the precursor by filtration, centrifugation, decantation, and evaporation or by extraction of the solvent with a more volatile solvent followed by distillation under vacuum. Alternatively, the precursor is washed with water to allow an organic layer to separate from an aqueous layer, followed by recovery of the precursor by drying.

A less likely alternative or parallel route is the solubilization of V^{4+} in an aqueous emulsion (water formed by vanadium reduction is not easily removed) and formation of $(VO)HPO_4 \cdot 0.5H_2O$ in water droplets.

The type of aliphatic alcohol used modifies the temperature at which vanadium is reduced. The reduction is kinetically controlled and complete only when benzyl alcohol is present (forming benzaldehyde and benzoic acid), when a long reduction time is used, and after the addition of phosphoric acid. The type of alcohol may affect the morphology of $(VO)HPO_4 \cdot 0.5H_2O$ [9].

In the preparation in the presence of benzyl alcohol many authors report the formation of platelets with stacking faults [deduced from the preferential line broadening of the (001) reflection] attributed to the trapping of the alcohol between the layers of the precursor and its release during activation [1,8,9].

Some examples, taken from the patent literature, are given here to illustrate this kind of preparation. In a patent from Chevron [10] the following main steps can be distinguished:

1. Suspension of V_2O_5 in anhydrous alcohol (isobutanol is the most preferred, which acts also as a mild reducing agent)
2. Dissolution and reduction of V^{5+} by bubbling gaseous HCl through the solution at temperatures lower than 60°C

3. Addition of phosphoric acid and digestion under reflux
4. Stripping of the alcohol under vacuum at temperatures lower than 170°C

In Amoco patents [11,12] the following steps can be singled out:

1. V_2O_5 and salt promoters are introduced in an organic solvent based on ethers (tetrahydrofuran is the most preferred) in the presence of hydrogen donor compounds (ethanol or water are the most preferred).
2. $POCl_3$ is hydrolyzed by the H donor (temperature is raised), with formation of anhydrous phosphoric acid and HCl, which dissolves all the metal compounds and reduces the V^{5+}.
3. Organic modifiers are added, along with aromatic acids or anhydrides, or aromatic hydrocarbons such as benzoic acid, phthalic anhydride, or xylene, added during reflux of the solvent.
4. The thick syrup obtained by solvent evaporation is dried under vacuum at 130–200°C to develop the precursor.

In BP preparation [13] the following stages are singled out:

1. Addition of the V^{5+} compound to an organic solvent selected from alcohols and glycols (isobutanol and ethylene glycols)
2. Addition of phosphoric acid
3. Reduction of the vanadium by heating the solution under distillation and by removing 1.5 moles of organic liquid (including organic by-products) per mole of vanadium reduced

The preparation in organic medium leads to microcrystalline $(VO)HPO_4 \cdot 0.5H_2O$ with a preferential exposure of the *(001)* crystallographic plane, and with higher specific surface area. There is large consensus [1,8,9,14–18] that the preparation in organic medium leads to an optimal precursor for the development of a catalyst active and selective in the oxidation of *n*-butane to maleic anhydride.

3.2.2 The Transformation of the Precursor to the Active Phase

Activation of the precursor is usually realized with a multistage procedure. The first stage is heating at temperatures lower than 300°C to eliminate the organic impurities or chlorine ions from the precursor without causing dehydration to occur.

After this treatment different types of thermal dehydration and activation have been proposed:

1. Dehydration inside the reactor starting from a low temperature

(280°C) in a flow of a lean reactant mixture and at low flow rate until standard operating conditions are reached in approximately 1 day

2. Dehydration in an oxygen-free atmosphere at temperatures higher than 400°C, followed by introduction of the reactant mixture (*n*-butane in air) (With this procedure, after the first step, crystalline $(VO)_2P_2O_7$ is obtained which, after the introduction of the reactant mixture, can remain substantially unmodified or be partially or totally reoxidized to a V^{5+}-containing phase [19,20].)

3. Single or multistep calcination in air until a temperature lower than 400°C is reached, and then introduction of the reactant mixture [21,22]

4. Activation during time-on-stream, i.e., in the presence of the reactants

After calcination at 280°C, the precursor is still present during release of the trapped benzyl alcohol, and this release leads to disruption of the structure, causing an increase in the surface area.

Figure 1 shows the evolution of the X-ray diffraction patterns of the precursor prepared in an organic medium when it is treated in air at high temperature. When the precursor is maintained at 380°C in air, the reflections typical of vanadyl orthophosphate progressively decrease in intensity, while evident amorphization occurs. After some hours at 380°C in air, the sample is mostly amorphous, and broad reflections relative to the vanadyl pyrophosphate are observed. Transformation to the well-crystallized $(VO)_2P_2O_7$ occurs in the reactor, after several hundreds of hours of time-on-stream.

Scheme 1 summarizes the possible evolutions of the precursor with temperature. The degree of crystallization of the precursor and its morphology remarkably affect the thermal structural evolution and the final properties of the vanadyl pyrophosphate.

Figure 2 shows different precursors, ranging from a highly crystalline compound (sample 1, prepared in aqueous medium) to compounds prepared in organic medium (samples 2–4), that are characterized by very different XRD patterns. When the extent of crystallinity is decreased, we move from patterns where the *(001)* reflection is the most intense to patterns where the only feature left is that one corresponding to the *(220)* crystallographic plane.

Factors that affect the crystallinity and morphology of the precursor during the preparation are the method of phosphoric acid addition [23], the amount of water present in the system (water is formed by vanadium

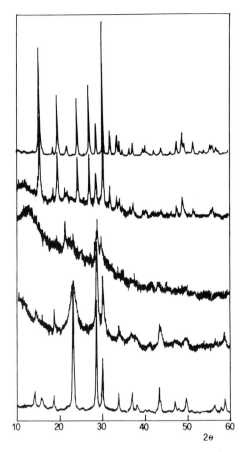

Figure 1. Structural evolution of the $(VO)HPO_4 \cdot 0.5H_2O$ (top) to the active phase in equilibrated catalyst (bottom), through intermediate compounds obtained at increasing times of calcination at 380°C in air.

reduction and can be removed from the system by distillation), and the nature of the alcohol selected to carry out the preparation [24]. The water is needed to crystallize the vanadyl orthophosphate and when it is removed from the organic medium a less crystalline compound is obtained.

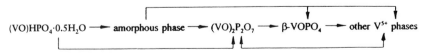

Scheme 1. Structural evolution of $(VO)HPO_4 \cdot 0.5H_2O$ as a function of temperature.

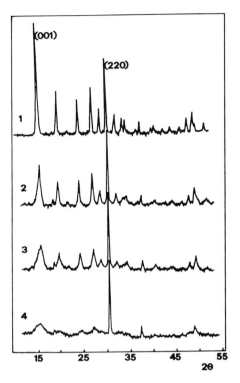

Figure 2. XRD patterns of precursors prepared in aqueous (1) and organic (2–4) media (from Ref. 23).

The thermal evolution of the compounds is represented in Fig. 3, which shows the corresponding Differential Scanning Calorimeter (DSC) patterns. Two different endothermal transformations can be singled out, occurring at around 300–400°C and 450–550°C, respectively. These can be accompanied by an exothermal peak associated to the combustion of organic compounds trapped in the layered structure of the precursor.

The corresponding amount of heat relative to each endothermal transformation depends on the nature of the precursor. In the case of the very crystalline sample 1, the only peak is that one occurring in the high-temperature range, while in samples 3 and 4 the prevailing peak is that one at low temperature. Both transformations are associated with the loss of water, and the amount of water released is a function of the precursor features. Characterization of the samples by means of FT-IR spectroscopy after the first endothermal peak indicates that the functional groups typical

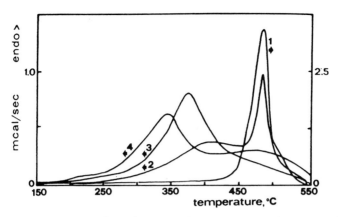

Figure 3. DSC patterns in air of samples prepared in aqueous (1) and organic media (2–4).

of the vanadyl acid orthophosphate are left; the second peak is associated with the formation of the vanadyl pyrophosphate.

Two kinds of water molecules are formally lost when the orthophosphate is transformed into the pyrophosphate:

1. $2\,(VO)HPO_4\cdot0.5H_2O \rightarrow 2\,(VO)HPO_4 + H_2O$.
2. $2\,(VO)HPO_4 \rightarrow (VO)_2P_2O_7 + H_2O$.

Therefore, the two endothermal peaks can be associated to the two different losses of water. The first loss leads to the formation of an amorphous or a microcrystalline intermediate compound that yet retains the functional groups typical of the vanadyl orthophosphate, while the second loss corresponds to the condensation of the orthophosphate groups and to the formation of the vanadyl pyrophosphate. The two transformations can occur either simultaneously at high temperature (as in the case of the highly crystalline compound 1, prepared in aqueous medium) or in two more distinguished steps, with samples prepared in organic medium. Also in this case, the degree of crystallinity of the precursor affects the temperature at which each transformation occurs and the amount of heat to be furnished.

In the former case (one endothermal peak at high temperature) it is possible to talk of a topotactic evolution, with the epitaxial transformation of the *(001)* face in the precursor to the *(100)* basal plane in the vanadyl pyrophosphate [25], while in the latter case the formation of the compound is rather a multistep transformation [4], occurring through the formation of an amorphous or a microcrystalline intermediate compound but with retention of the main morphological features of the precursor.

3.2.3 Catalyst Aging: Modifications during Time-on-Stream

After the stage of dehydration the catalyst has to be activated. During prolonged exposure to the reactant atmosphere changes occur with time-on-stream both in catalytic behavior and in the physicochemical properties of the catalyst.

In catalysts calcined in air the transformation from a partially amorphous, possibly oxidized compound to an almost completely crystalline vanadyl pyrophosphate inside the reactor and in the presence of the reactant mixture requires more than 100 h, depending on the features of the fresh catalyst, i.e., the calcination conditions employed [26]. If the fresh catalyst is highly oxidized (because of oxidizing treatments of calcination at high temperatures), the permanence in the reaction environment has to lead to vanadium reduction and to $(VO)_2P_2O_7$ crystallization; in this case more than 500 h are necessary to complete the transformations. When the fresh catalyst is reduced (contains only V^{4+}, having been treated either in nitrogen, calcined under milder conditions, or activated under reactants atmosphere), the final crystalline compound is obtained in a shorter period of time (200–300 h). These structural transformations occur with modifications in the catalytic performance. The surface P/V ratio, as determined by XPS, was found to remain approximately constant during the aging procedure [21,26].

A fresh catalyst has been designated as *nonequilibrated* and a catalyst after prolonged time-on-stream (i.e., after activation) as *equilibrated*. A more precise definition of an *equilibrated* catalyst has been given by Ebner and Thompson [27]. According to these authors, an *equilibrated* catalyst is one that has been kept in a flow of *n*-butane with a concentration of 1.4–2% in air and at least GHSV 1000 h^{-1} for approximately 200–1000 h. It has an average oxidation degree for vanadium in the range 4.00–4.04, a bulk phosphorus-to-vanadium ratio of 1.000–1.025, and an XPS surface P/V atomic ratio of 1.5–3.0. The catalyst has a BET surface area ranging from 16–25 m^2g^{-1}, and its X-ray diffraction pattern only shows the reflections typical of the vanadyl pyrophosphate. TEM analysis shows rectangular platelets and rodlike structures. One of the main properties of an *equilibrated* catalyst is the formation of stable V^{4+} (average valence state 4.00–4.03) [28,29]. Equilibrated catalysts can no longer be reoxidized in air at 400°C, whereas freshly prepared $(VO)_2P_2O_7$ or *nonequilibrated* catalysts can be oxidized at this temperature.

Monsanto recently claimed a procedure to accelerate the catalyst equilibration, thus shortening the initial period of activation during which the catalyst does not possess the optimal catalytic performance [30]. A procedure of activation was proposed that consists of the thermal treatment of the precursor in an air/steam mixture under controlled heating conditions. It is likely that these hydrothermal conditions may favor the development of the well-crystallized vanadyl pyrophosphate.

3.3 Results and Discussion

3.3.1 A Study of the Effect of the Procedure Employed for Thermal Activation and Effect of Aging

The importance of the dehydration treatment in affecting the structural composition of the catalyst, its evolution in the reaction environment, and finally its catalytic properties are illustrated in the following examples, where the reactivity of three samples prepared under different calcination or aging treatments, starting from the same precursor, are compared.

The reactions of *n*-butane oxidation to maleic anhydride and of *n*-pentane oxidation to phthalic and maleic anhydrides have been studied. The *n*-butane oxidation constitutes the industrial application for this catalyst, and the *n*-pentane oxidation is more suitable as a probe to characterize the nature of the surface sites [31,32].

The precursors have been prepared following the procedure in alcoholic medium. The three samples have been obtained according to three different methods of calcination:

1. Sample a has been calcined in air, with gradual heating from room temperature up to 400°C for several hours. Then, the sample has been subjected to a treatment in the reaction environment (1.7% *n*-butane in air, flow rate 3 g·s/ml, temperature 340°C) for 100 h, in order to achieve a fairly stable catalytic performance. In fact, immediately after the calcination treatment the sample shows a rather unstable catalytic activity.

2. Sample b has been dehydrated in a nitrogen flow at 450°C. In this case the obtained sample also has been left in the reaction medium for approximately 100 h in order to achieve reproducible results.

3. Sample c has been obtained by "equilibration" of sample b by permanence in the reaction medium for 1000 h. It also underwent a reactivation procedure.

XRD patterns and FT-IR spectra of the three samples after reaction are given in Fig. 4 and 5, respectively.

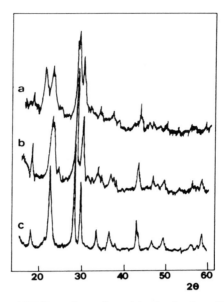

Figure 4. XRD patterns of V/P/O catalysts after calcination in air at 400°C of the precursor prepared in organic medium and unloaded after 100 h reaction (*n*-butane oxidation) (sample a), after treatment in N_2 at 450°C and 100 h reaction (sample b), and after 1000 h reaction (sample c).

Sample a is characterized by a low surface area (11 m²/g) and an average oxidation degree of 4.37 (as determined by chemical analysis), thus indicative of the presence of oxidized vanadium. The XRD pattern and FT-IR spectra confirm the presence of $(VO)_2P_2O_7$ and of a $VOPO_4$ phase of uncertain attribution, most likely a γ-$VOPO_4$ phase [33]. Sample b is completely reduced (average oxidation degree 4.00) and has a surface area of 24 m²/g. The XRD pattern and FT-IR spectrum show a more amorphous compound, with reflections relative to poorly crystallized vanadyl pyrophosphate. In agreement with the chemical analysis, no reflections or absorptions relative to crystalline oxidized phases are observed. Sample c is completely reduced (average oxidation degree ≤ 4.00; the chemical analysis also indicates the likely presence of small amounts of V^{3+}) with a surface area of 26 m²/g. The equilibrated sample is highly crystalline, with the reflections and IR absorptions relative only to the $(VO)_2P_2O_7$.

Figure 6 shows the *n*-butane conversion as a function of the temperature and the corresponding values of selectivity to maleic anhydride for the three catalysts. Sample a is far less active than the other two catalysts; sample c, the equilibrated catalyst, is less active than the corresponding nonequilibrated one (sample b). The low activity of sample a can be ex-

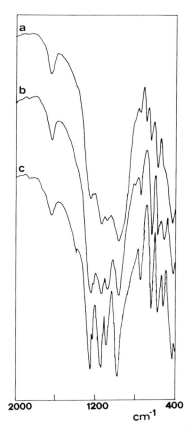

Figure 5. FT-IR spectra of samples a, b, and c (same activation procedure as in Fig. 4).

plained by taking into account its lower surface area; samples b and c possess instead comparable surface areas, and therefore the observed differences must be related either to structural or to surface characteristics of the two materials.

Sample b is the most selective when we compare the values of selectivity either at the same temperature or at comparable conversion levels. In addition, sample a shows a lower, but rather constant selectivity in the range of temperature examined, while samples b and c show a decline of selectivity. This indicates differences among samples in the values of the apparent activation energy relative to *n*-butane and maleic anhydride combustion to CO_x with respect to the selective conversion of *n*-butane to maleic anhydride.

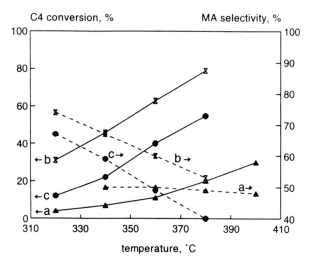

C4 conversion, %   MA selectivity, %

temperature, °C

Figure 6. *n*-Butane conversion (solid lines) and selectivity to maleic anhydride (dotted lines) as a function of temperature for samples a, b, and c. Conditions: W/F (catalyst weight/total flow rate) 3 g∗s/ml, 1.7% *n*-butane in air.

Figure 7 shows the *n*-pentane conversion as a function of residence time for the three catalysts, and Fig. 8 the corresponding values of selectivity to maleic anhydride and to phthalic anhydride as functions of the conversion. The trend of activity observed with the *n*-butane is here confirmed: sample a is the least active; sample b is the most active.

The products distribution indicates the following:

1. In all the samples the trend of selectivities clearly indicates that the formation of the two anhydrides occurs through independent parallel pathways. It is possible that the surface centers responsible for the phthalic anhydride formation may differ from those that give rise to the maleic anhydride formation.

2. The samples exhibit different behaviors. Sample a shows a rapid decline of the selectivity to both anhydrides when the conversion is increased, which indicates the presence of consecutive reactions of combustion. On the contrary, sample b shows constant selectivities to the anhydrides. Sample c exhibits an intermediate behavior, with a decline of the selectivity to phthalic anhydride when the conversion is increased.

3. Sample c is the only one that exhibits a higher selectivity to phthalic anhydride than to maleic anhydride. This means that the crystallization of the vanadyl pyrophosphate leads to the preferential formation of surface sites that are specific for the formation of the phthalic anhy-

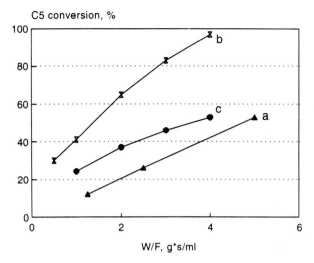

Figure 7. *n*-Pentane conversion as a function of the residence time for samples a, b, and c. Conditions: temperature 340°C, 1% *n*-pentane in air.

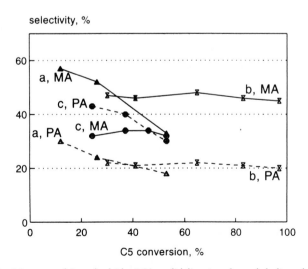

Figure 8. Selectivity to maleic anhydride (MA, solid lines) and to phthalic anhydride (PA, dotted lines) as functions of *n*-pentane conversion for samples a, b, and c. Conditions: temperature 340°C, 1% *n*-pentane in air (as in Fig. 7).

dride. It is possible that the ordered structure of the well-crystallized vanadyl pyrophosphate guarantees the surface geometric constraints necessary to address the reaction pathway toward the formation of the phthalic anhydride.

The effect if the *n*-pentane content in the feed is summarized in Figs. 9 and 10 for the three catalysts examined. The differences in activity are here confirmed. Moreover, it is shown that when the hydrocarbon concentration is increased, the selectivity to phthalic anhydride is remarkably increased, while the selectivity to maleic anhydride either decreases or remains approximately constant. This indicates that the rate of phthalic anhydride formation has a higher order of reaction with respect to the hydrocarbon partial pressure than the rate of maleic anhydride formation. This is expected, because the mechanism of phthalic anhydride formation must necessarily involve a bimolecular step where two hydrocarbon species adsorbed on the catalyst surface interact, via dimerization or condensation, which yields the precursor of the phthalic anhydride.

It is possible to make a correlation between the catalytic behavior and the structural features of the compounds examined. In particular, it is possible to make the following remarks:

1. A higher extent of vanadium reduction is necessary in order to achieve an active and a selective catalyst. Calcination in an oxidizing at-

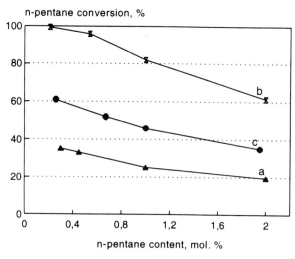

Figure 9. *n*-Pentane conversion as a function of the *n*-pentane content in feed for samples a, b, and c. Conditions: temperature 340°C, W/F 3 g$_*$s/ml.

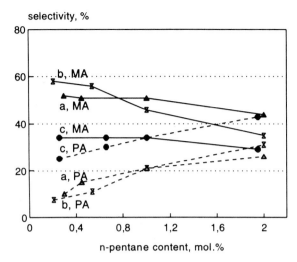

Figure 10. Selectivity to maleic anhydride (MA) and to phthalic anhydride (PA) for samples a, b, and c. Conditions: temperature 340°C, W/F 3 g∗s/ml (as in Fig. 9).

mosphere, if not properly carried out, can give rise to a large extent of vanadium oxidation and formation of $VOPO_4$ phases, and to a catalyst characterized by a low surface area and a worse catalytic performance. This catalyst may require a very long aging treatment (1000 h) to be reduced and crystallized to the $(VO)_2P_2O_7$, the stable active phase. On the contrary, a precursor transformation in a nonoxidizing atmosphere (nitrogen or even the reaction medium itself) allows a better control of the precursor dehydration.

2. In reduced samples the activity may be conditioned by other features, such as the degree of crystallinity. When the catalyst is characterized by a low degree of crystallinity, because of the presence of both a crystalline $(VO)_2P_2O_7$ and an amorphous or microcrystalline $V^{4+}/P/O$ phase, the activity in both n-butane and n-pentane oxidation is higher. In addition, this sample possesses centers that are slightly more selective to the maleic anhydride from n-butane. Nevertheless, it has to be considered that these peculiar features are of low utility, as the catalyst fate is to be slowly converted in the reaction medium to the *equilibrated* catalyst, i.e., to the well-crystallized vanadyl pyroposphate. The latter is instead characterized by an high specificity to the phthalic anhydride formation from n-pentane. It can be hypothesized that the formation of a C_8 compound starting from C_5 hydrocarbons (the n-pentane or, more likely, some olefinic-like intermediate species formed by n-pentane activation on the catalyst surface) re-

quires surface geometrical constraints or arrays of neighboring active sites that are obtained only when the compound is well crystallized.

3.4 Conclusions

These data evidence that the catalytic behavior of a fresh catalyst can be remarkably affected by the conditions employed for the precursor dehydration. The behavior of the fresh catalyst is important only in the first 200–300 h of time-on-stream, as nonequilibrated catalysts are slowly converted to the equilibrated form. Nevertheless, it is very important to perform activation procedures that can guarantee the obtainment of a structural composition as close as possible to the equilibrated form, in terms of vanadium valence state and degree of crystallinity, in order to shorten the unstable catalytic performance at the beginning of lifetime. The most important aspect is to avoid vanadium overoxidation, performing the dehydration in either nonoxidizing atmosphere or in air but with a careful control of the calcination conditions.

References

1. Centi, G., Trifirò, F., Ebner, J. R., and Franchetti, V. M. (1988). *Chem. Rev.*, **88**, 55.
2. Cavani, F., and Trifirò, F. (1994). *Chem.-Tech.*, **24**, 18.
3. Cavani, F., and Trifirò, F. (1994). *Catalysis Volume 11*, p. 246, Royal Society of Chemistry, Cambridge.
4. Cavani, F., and Trifirò, F. (1995). *Preparation of Catalysts VI* (G. Poncelet, J. Martens, B. Delmon, P. A. Jacobs, and P. Grange, eds.), p. 1, Elsevier Science, Amsterdam.
5. Hatano, M., Masayoshi, M., Shima, K., and Ito, M. (1992). U. S. Patent 5 128 299, assigned to Mitsubishi Kasei.
6. Stefani, G., and Fontana, P. (1979). U. S. Patent 4 178 298, assigned to Lonza.
7. O'Connor, M., Dason, F., and Hodnett, B. K. (1990). *Appl. Catal.*, **64**, 16.
8. Cornaglia, L. M., Sanchez, C. A., and Lombardo, E. A. (1993). *Appl. Catal. A:General*, **95**, 117.
9. Horowitz, H. S., Blackstone, C. M., Sleight, A. W., and Teufer, G. (1988). *Appl. Catal.*, **38**, 193.
10. Katsumoto, K., and Marquis, D. M. (1979). U. S. Patent 4 132 670, assigned to Chevron Research.
11. Edwards, R. C. (1990). U. S. Patent 4 918 201, assigned to Amoco.
12. Taheri, H. (1991). U. S. Patent 5 001 945, assigned to Amoco.
13. Bremer, N. J., Dria, D. E., and Weber, A. M. (1984). U. S. Patent 4 448 893, assigned to The Standard Oil.
14. Ebner, J. R., and Thompson, M. R. (1991). *Structure-Activity and Selectivity Relationships in Heterogeneous Catalysis* (R. K. Grasselli and A. W. Sleight, eds.), p. 31, Elsevier Science, Amsterdam.
15. Matsuura, I. (1993). *Catal. Today*, **16**, 123.

16. Okuhara, T., and Misono, M. (1993). *Catal. Today*, **16**, 61.
17. Bordes, E. (1993). *Catal Today*, **16**, 27.
18. Ye, D., Satsuma, A., Hattori, A., Hattori, T., and Murakami, Y. (1993). *Catal. Today*, **16**, 113.
19. Meisel, M., Wolf, G. U., and Bruckner, A. (1992). *Proceedings DGMK Conference on Selective Oxidations in Petrochemistry* (M. Baerns and J. Weitkamp, eds.), p. 27, Tagungsbericht.
20. Bergeret, G., David, M., Broyer, J. P., Volta, J. C., and Hecquet, G. (1987). *Catal. Today*, **1**, 37.
21. Cornaglia, L. M., Caspani, C., and Lombardo, E. A. (1991). *Appl. Catal.*, **74**, 15.
22. Contractor, R. M., Ebner, J. R., and Mummey, M. J. (1990). *New Developments in Selective Oxidations* (G. Centi and F. Trifirò, eds.), p. 553, Elsevier Science, Amsterdam.
23. Busca, G., Cavani, F., Centi, G., and Trifirò, F. (1986). *J. Catal.*, **99**, 400.
24. Ellison, I. J., Hutchings, G. J., Sananes, M. T., and Volta, J. C. (1994). *J. Chem. Soc., Chem. Commun.*, 1093.
25. Bordes, E., Johnson, J. W., and Courtine, P. (1984). *J. Solid State Chem.*, **53**, 270.
26. Sola, G. A., Pierini, B. T., and Petunchi, J. O. (1992). *Catal. Today*, **15**, 537.
27. Ebner, J. R., and Thompson, M. R. (1993). *Catal. Today*, **16**, 51.
28. Kubias, B., Rodemerck, U., Wolf, G. U., Meisel, M., and Schaller, W. (1992). *Proceedings DGMK Conference on Selective Oxidations in Petrochemistry* (M. Baerns and J. Weitkamp, eds.), p. 303, Tagungsbericht.
29. Trifirò, F. (1993). *Catal. Today*, **16**, 91.
30. Ebner, J. R., and Andrews, W. J. (1991). U. S. Patent 5 137 860, assigned to Monsanto.
31. Centi, G., Lopez Nieto, J., Pinelli, D., and Trifirò, F. (1989). *Ind. Eng. Chem. Res.*, **28**, 400.
32. Fumagalli, C., Golinelli, G., Mazzoni, G., Messori, M., Stefani, G., and Trifirò, F. (1994). *New Developments in Selective Oxidation* (V. Cortes Corberan, and S. Vic Bellon, eds.), p. 221, Elsevier Science, Amsterdam.
33. Ben Abdelouahab, F., Olier, R., Guilhaume, N., Lefebvre, F., and Volta, J. C. (1992). *J. Catal.*, **134**, 151.

CHAPTER 4

Gel-Supported Precipitation
An Advanced Method for the Synthesis of Pure and Mixed-Oxide Spheres for Catalytic Applications

G. Centi,* M. Marella,† L. Meregalli,† S. Perathoner,*
M. Tomaselli,† and T. La Torretta†

*Dip. Chimica Industriale e dei Materiali, 40136 Bologna, Italy
†EniRisorse, Centro Ricerche Venezia, Porto Marghera (VE), Italy

KEYWORDS: Mixed-oxide microspheres, mixed-oxide spheres, TiO_2–Al_2O_3 mixed oxides, ZrO_2–Al_2O_3 mixed oxides

4.1 Introduction

Control of the physical and chemical characteristics of oxide supports is receiving increasing attention because of the role these properties play in determining the reactivity and stability of supported active components. Furthermore, industrial application of advanced catalytic materials requires a fine control of characteristics such as texture, mechanical strength, and shape and density of the pellets that can critically determine the industrial use of these catalysts.

A large number of studies and commercial samples exist for oxides such as SiO_2 and Al_2O_3, but not for oxides such as TiO_2 and ZrO_2, despite their interesting catalytic properties [Matsuda and Kato, 1983; Yamaguchi, 1994]. Their wider use is limited by the difficulty in obtaining thermally stable samples with textural and mechanical strength properties comparable with those of alumina and silica. The possibility of tuning the porosity and mechanical strength characteristics of TiO_2 by creating a thin layer film of TiO_2 over preformed silica has been shown [Beeckman and Hegedus, 1991]. Besides difficulties in obtaining a truly homogeneous overlayer [Wauthor *et al.*, 1991], this coating method does not allow control of the local structure of TiO_2 and it is not possible to exploit the advantages in terms of tuned surface properties (e.g., acidity) offered by the preparation of a homogeneous mixed oxide.

It has been shown (Brambilla *et al.*, 1992) that by using a preparation method called gel-supported precipitation (GSP), homogeneous, spherical TiO_2–Al_2O_3 pellets with controlled textural and dimension characteristics and good mechanical resistance properties suitable for fluid or mobile bed reactor applications can be synthesized. Alumina acts as a skeletal matrix for TiO_2, promoting its properties without negatively affecting the surface characteristics as in coated samples but, instead, promoting resistance to deactivation of the supported active vanadium component [Brambilla *et al.*, 1992; Centi *et al.*, 1991a and 1991b]. Improved properties of zirconia samples prepared by this method have also been recently reported [Marella *et al.*, 1994a].

GSP method is based on precipitation of the hydroxide(s) of the element(s) in a basic solution but in the presence of an organic additive that immediately gelifies in the basic solution, thus acting as a support for the precipitation of the hydroxide. The main steps of the preparation are: (1) mixing of inorganic salts of the elements (Ti, Al, Zr, etc.) with a thickening agent, (2) formation of a pseudocolloidal solution, (3) dispersing the colloidal solution, through nozzles or a rotary atomizer, into controlled size droplets, (4) solidification of the liquid droplets by sol–gel conversion, (5) aging and washing, and (6) drying and calcination.

Although the GSP method shows several analogies with sol–gel methods, it differs in two significant aspects: (1) GSP allows oxide spheres to be obtained directly with characteristics suitable for their use in reactors with solid transport and (2) the GSP method uses inorganic salts rather than alkoxides as precursors. Oxides and mixed oxides prepared by controlled hydrolysis of alkoxide solutions, in fact, have been widely studied [Haas, 1989; Hubert-Pfalzgraf, 1987; Johnson, 1985; Sanchez and Livage, 1987], but (1) the high cost of the alkoxides, (2) the difficulty in obtaining truly homogeneous mixed oxides when alkoxides with different reactivities are present, (3) the difficulty in the scale-up of the process, and (4) the necessity of secondary forming processes to obtain spherical pellets have hampered the wider commercial application of alkoxide-based methods for the synthesis of catalyst supports. Methods aimed at obtaining microspheres directly by alkoxide sol–gel processing have been proposed [Ding and Day, 1991; Haas, 1989], but control of the form, dimensions, and other characteristics of the spheres is rather difficult.

In this work, after a brief survey on the preparation of oxide carriers for catalytic application and an outline of the GSP method, some examples of the preparation of pure and mixed ZrO_2, TiO_2, and Al_2O_3 oxide supports and of samples already containing an active component such as copper oxide are shown, to illustrate their specific innovative characteristics.

4.2 Brief Survey of the Industrial Preparation Methods of Oxide Carriers for Catalytic Applications

Two main methods are used for the preparation of industrial oxide carriers used in catalytic applications: (1) precipitation (usually of a hydroxide or mixed hydroxide) from an aqueous solution by changing the pH and (2) gelation or flocculation of a hydrophilic colloidal solution to form a hydro- or an alko-gel (depending on the reaction medium) or agglomerated micelles, respectively. The second procedure and especially sol–gel methods based on the gelation of alkoxides [Doelbear, 1993], offers better control over surface area, pore volume, and pore size distribution but, however, with higher costs of production. The characteristics of the oxide obtained depend considerably on the procedure of solvent elimination from the gel which can contain up to 90% solvent, particularly for hydrogels. During the process of solvent elimination, in fact, the gel mass shrinks, collapses, and breaks. Vacuum drying at low temperature is widely used to reduce the negative effect of such behavior on the textural properties of the final oxide. Alternatively, large capillary stresses can be avoided by operating at temperatures higher than the critical temperature of the solvent. The solid structure undergoes only a small modification and retains the textural properties of the wet gel. The latter drying process leads to highly porous dry solids (aerogels) [Pajonk, 1991] with interesting properties, but the procedure is complex and expensive for the production of large amounts of solid.

The solid precursor after filtration or solvent removal is then calcined to give a fine powder which must be further processed to obtain the final catalyst. The forming and shaping of supports is an important step (usually underestimated) in their industrial production, which can have considerable influence on (1) the catalyst activity, (2) the particle resistance to crushing and abrasion, (3) the reactor pressure drop, (4) the formation of dust during operations, and (5) the uniformity of the catalyst performances in the reactor bed, etc. [Fulton, 1986].

For applications of the catalysts in fluid-bed reactors, spray drying is the most commonly used technique to obtain particles in the 20–100 μm range. A sol or hydrogel (obtained from the calcined solid by adding suitable additives or by direct feeding of the aqueous solution before precipitation or gelification, when possible) is sprayed through nozzles into a heated zone. The control of all parameters is rather difficult, however, and solid microspheres with a relatively wide range of dimensions and not uniform spherical shapes are obtained in several cases. It also is quite difficult to obtain homogeneous mixed-oxide supports using this technique.

For larger particles (above 1 mm) various methods can be used such as (1) extrusion and wet pressing, (2) dry tableting, (3) granulation, and (4) drop coagulation. The first two methods are used to form cylindrical (or other shapes such as rings, trilobates, hollow cylinders) pellets for applications in fixed bed reactor operations. Suitable additives, however, are usually necessary to act as binders.

Granulation is a common method used to obtain particles with more or less spherical shapes. A round dish rotating on an inclined axis is used. Small particles are fed into the dish and at the same time a cohesive slurry is sprayed onto the particles. The small particles develop layer by layer into larger particles. However, also in this case a relatively large distribution of particle diameters is obtained, the shape of the particles is not uniform, and evident lack of homogeneity in the pore structure exists along the particle profile.

Drop coagulation results from sols suspended in an immiscible liquid by simultaneous gelation, ripening, and forming. The aqueous sol is forced, through a sparger, in a water-immiscible solvent (oil, usually), the temperature of which is raised to 100°C or above. The surface tension created on the droplets permits the formation of gel spheres that are then ripened and dried. This method allows rather uniform and spherical-shaped pellets to be obtained. The method shows strong analogies with the GSP method, but the simultaneous gelation, ripening, and drying during the drop do not allow fine control of important parameters such as texture.

The preparation of oxide carriers such as alumina and silica has been widely discussed in the literature [Doesburg and van Hoof, 1993; Oberlander, 1994; Richardson, 1989; Twigg, 1989]. However, because of the lack of data on the synthesis of oxide carriers such as ZrO_2 and TiO_2, more specific aspects of the preparation of these oxides are discussed here.

4.2.1 Zirconia

Zirconia is an efficient catalyst support for chromium oxide, copper oxide, Pd, Pt, and Rh for processes such as the oxidation of CO [Yamaguchi *et al.*, 1991], hydrogenation of CO [Guglielminotti *et al.*, 1993], selective hydrogenation of organic molecules [Lancia *et al.*, 1994], removal of SO_2 and NO from flue gas [Centi et al., 1993], propane oxidation [Hubbard et al., 1993], combustion of low-temperature fuel gas [Tsurumi et al., 1993], and low-temperature steam reforming [Igarashi et al., 1993]. Zirconia itself acts as a catalyst in the isomerization of l-butene [Pajonk and Tanany, 1992], dehydration of alcohols [Araki and Hibi, 1987], hydrogenation of dienes by H_2 and hydrogen donor molecules such as cyclohexadiene [Nakano et al., 1983], hydrogenation of carboxylic acids [Yokoyama et al., 1992], and hydrogenation of carbon monoxide [Maruya et al., 1993].

Zirconia supports can be prepared by traditional ceramic technologies, e.g., shaping of a mixture of organic binders and ZrO_2 powder [Hashimoto et al., 1991], or granulation of ceramic powder and combustible particles [Sasai and Hiraishi, 1989], or formation of compacts (outer diameter about 2–15 mm) by a pourable powder [Deller et al., 1989]. The drawbacks of these methods are low surface areas and dishomogeneity of the composition particularly when doping agents are present in small amounts, e.g., Ce, La, Nd.

Other much more utilized techniques for the preparation of supports involve chemical methods and particularly precipitation techniques of a zirconyl nitrate solution in NH_3, NaOH, or KOH solutions which finally result in the formation of a very fine powder (Davis, 1984] or a voluminous gel that after vacuum drying is crushed to a grain size of 50–150 μm [Koeppel et al., 1991], or thin cakes of gel fragmented into grains of around 1 mm [Dachet et al., 1991]. The precipitation of zirconyl chloride in ammonia has been used by several authors [Mercera et al., 1992; Ozawa et al., 1987 and 1991]. Again, fine powders with a mean diameter less than 100 μm are obtained by grinding the gel. The major drawback of these procedures is the difficulty of obtaining the final particles with diameters of 1–2 mm and proper shapes by a secondary shaping procedure.

Several alkoxide sol–gel techniques for the preparation of zirconia samples are also reported in the literature [Aynal et al., 1990; Lerot et al., 1991; Wolf and Russel, 1992]. They consist in the hydrolysis and polycondensation of (1) zirconium propoxide under acidic conditions in an alcoholic medium, (2) zirconium tetra-n-propoxide in the presence of long-chain carboxylic acids, and (3) zirconium n-propoxide in acetic acid, acetylacetone, and isopropanol. With these methods Lerot et al. (1991) obtained monodisperse spherical zirconia particles tuned from 0.1 to 2.5 μm which in some cases were processed by spray drying processing. The drawbacks of these methods are the cost of the reagents, their stability, and again the fine grain size of the powders. Indeed, these were developed for structural ceramics rather than for catalystic carriers.

4.2.2 Titania

Titania, in the anatase form, is a commercial oxide carrier for vanadia, the active component in applications such as the selective oxidation of o-xylene [Centi *et al.*, 1991c] and the selective reduction of NO with ammonia in flue gas [Matsuda and Kato, 1993], because of its superior per-formances over other oxides in maintaining the vanadium oxide in a pseudo-monolayer active form. TiO_2, furthermore, is an excellent oxide carrier which promotes the catalytic performances of both noble metals and transition metal oxides such as Cu, Ni, and Mo [Breysse *et al.*, 1991; Matsuda and Kato, 1993]. TiO_2 also has interesting photocatalytic properties and is widely used as such or as a catalytic support in various photocatalytic reactions of complete oxidation of organics in aqueous solution [Palmisano *et al.*, 1994]. TiO_2 is also an excellent catalyst for the hydrolysis of CS_2, COS, and HCN in Claus tail gas [Huisman *et al.*, 1994].

Commercial samples of TiO_2 are usually prepared by calcination of hydrous titania obtained by precipitation or flocculation from various Ti-salts, $TiCl_4$, $Ti(SO_4)_2$, and $TiOSO_4$ [Matsuda and Kato, 1983], but contain significant amounts of impurities and normally a mixture of phases (anatase and rutile). Advanced methods to synthesize titania with higher purity and better control of the composition include (1) alkoxide-based sol–gel techniques [Montoya *et al.*, 1992; Rodriguez *et al.*, 1992], (2) hydrolysis of $Ti(OEt)_4$ in hydroxypropyl cellulose to obtain ultrafine sterically stabilized titania particles [Nagoal *et al.*, 1992], (3) hydrothermal treatment of amorphous hydrous titania powder synthesized by controlled hydrolysis and polymerization from titanium tetraethoxide solutions [Ogura *et al.*, 1988], (4) hydrolysis of titanium tetraisopropoxide [Sasamoto *et al.*, 1993], (5) sol–gel techniques starting from $TiCl_4$ and the autoclave treatment with water vapor [Chertov *et al.*, 1992], and (6) pH-controlled hydrolysis of $TiCl_4$ [Cavani *et al.*, 1988]. The presence of doping elements such as phosphate, sulphate, or borate ions stabilize the crystal matrix and prevent the formation of rutile from anatase [Hums, 1986] but have negative effects on the catalytic activity.

4.2.3 Mixed Oxide

Mixed-oxide carriers or catalysts have received much less attention than pure oxides, apart from Al_2O_3–SiO_2 mixed oxides for their analogy with zeolite materials and their application in acid-catalyzed reactions such as catalytic cracking [Courty, 1974]. The preparation method of mixed-oxide carriers is usually made by alkoxide-based sol–gel techniques [Klimova and Solis, 1994], because co-precipitation methods usually do not allow homogeneous dispersions to be obtained. However, a pretreatment is necessary to either increase the reactivity of the less reactive alkoxide through prehydrolysis or decrease the reactivity of the more reactive one, by complexation [Doelbear, 1993; Hubert-Pfalzgraf, 1987].

4.3   Gel-Supported Precipitation (GSP) Method

4.3.1 General Features

The GSP process was originally developed for the preparation of ceramic nuclear fuel microspheres [Brambilla, 1970; Gerontopoulos, 1970a and 1970b] and then adapted for the preparation of pure and mixed-oxide spheres, microspheres, and washcoats for catalytic applications [Marella *et al.*, 1994b]. Basically the process consists of the following steps:

1. The sol is prepared from relatively inexpensive raw materials. The sol is usually obtained from a nitrate-based solution of the refractory metals denitrated by solvent extraction up to an extent just below that which would compromise the integrity of the final spheres or microspheres. Long-chain organic activities are used as thickening agents to tailor the viscosity to the dripping apparatus.

2. Depending on the range of diameters of the spheres, either an assembly of dripping capillaries or a rotary cup atomizer is used [Gerontopoulos, 1970a, 1970b]. With the former, spheres in the range of 0.5–2 mm or even larger are obtained, while microspheres in the range of 20–200 μm are obtained with the latter.

3. The sol is dripped into a gelation bath usually made of a diluted ammonia solution in H_2O and aged for some minutes. In this stage a gel with a spherical shape is obtained instead of a precipitate in the form of a fine powder. In the case of ZrO_2, for example, the zirconium polycation structures, α, β, or γ-type, usually involving four zirconium ions [Murase, 1978], are probably linked by hydrogen bonds of their terminal hydroxyl groups to $-OH$, $-OCH_3$, or $-CH_2OCH_3$ groups of the long-chain organic additives. The term "gel supported precipitation" is related to these phenomena.

4. The gel is rinsed with water, dried preferentially by azeotropic dehydration and finally calcined. A schematic flow sheet of the preparation steps for the synthesis of spheres and microspheres by the GSP method is shown in Fig. 1.

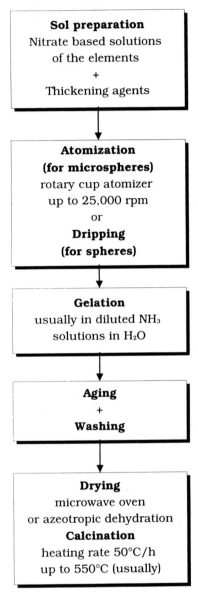

Figure 1. Schematic flow sheet of the preparation steps in the synthesis of spheres and microspheres by the GSP method.

The advantages of this process can be summarized as follows:

1. Relatively inexpensive raw materials, in comparison, e.g., with alkoxide-based sol–gel techniques
2. Flexibility of the process (the possibility of use for a wide variety of refractory oxides with minor modifications in the procedure)
3. Easy scale-up for large productions
4. Perfect control of sphere diameters with suitable mechanical strength and fluidizability properties for fluid- or mobile-bed reactor technologies (Furthermore, mixed oxides obtained by the GSP method are highly homogeneous in composition, a rather important factor for both the preparation of mixed-oxide carriers and final catalysts, when the active component is added directly during GSP preparation.)

4.3.2 Details of the Method

4.3.2.1. Zirconia-Based Samples

Pure zirconia, zirconia–alumina (1–10% wt) and zirconia–alumina (10% wt)–copper oxide (8%) samples were prepared. An example of the preparation procedure for pure zirconia is given below. This procedure is readily transferred to the other compositions, starting from suitable nitrate solutions of the elements.

A zirconium oxynitrate solution 2.4 M was denitrated by solvent extraction (volume ratio 1:3) with an alkyl tertiary amine diluted in an organic solvent (Solvesso 100 by Exxon) with a volume ratio 1:10. The final Zr^{4+} content was 2.28 M and the viscosity was about 40 mPa·s. The extractant was regenerated by a continuous process. The solution was diluted to 120 g·dm^{-3} (as ZrO_2), and the thickening agent (approximately 1%) and the surface-active agent (e.g., iso-octyl-phenoxy-polyethoxy-ethanol) were added.

For the preparation of mm-size spheres, the solution was dropped by vibrating or nonvibrating dripping capillaries into a concentrated ammonia solution containing the surface-active agent. The inner diameter of the needles can be varied from 0.5 to 4.0 mm; the type, molecular weight, and concentration of the thickening agent must be optimized with respect to the diameter. After aging, the ammonia was filtered and the spheres were washed with water. Azeotropic dehydration by toluene was preferentially performed. The calcination was carried out in a muffle furnace with air flowing at a rate of 1 dm^3·min^{-1}. The heating schedules were optimized on the basis of the thermal analysis data with maximum temperatures of 573, 723, 823, and 1073 K held for 3 h.

Alternatively, for the preparation of microspheres a solution of similar composition was fed at a rate of 2.5 $dm^3 \cdot h^{-1}$ into a rotary cup atomizer rotating at 12,000 rpm. The droplets were dripped into a gelation bath as previously reported. After the microspheres were aged in concentrated ammonia and washed with water, they were azeotropically dried and calcined at the various temperatures for 3 h. For the zirconia–alumina (10% w)–copper oxide (8%) catalyst a modification of the gelation bath was necessary because ammonia is a complexing agent for copper. In this case, a strong organic base was used.

4.3.2.2 Titania Samples

To date, two procedures have been studied:

1. A titania solution was prepared by hydrolysis of $TiCl_4$ and further evolution of HCl by distillation under vacuum at 333 K. During the hydrolysis about 60% of the initial chlorine was liberated and during the subsequent treatment 15% more was eliminated. The Cl/Ti ratio was thus reduced from 4 to 1. After sol preparation by addition of the thickening agent and surface-active agent, e.g., iso-octyl-phenoxy-polyethoxyethanol, the sol was dripped by the same procedure experimented for zirconia. After rinsing, drying in a microwave oven and calcining at 823 K, spheres with diameters 1.2 mm were obtained.

2. The sol was prepared by addition of the thickening agent and surface-active agent to a nitrate-based solution of titania. After dripping into diluted ammonia with 2.5 mm (i.d.) capillaries and rinsing, the gel spheres were calcined directly at 823 K at 2 $K \cdot min^{-1}$ in air without intermediate azeotropic dehydration. The spheres had a diameter of 1 mm.

The second procedure is preferred basically for these two reasons: (1) better control over the process and better reproducibility; (2) absence of chlorine in the final product.

4.3.2.3 Alumina-Based Samples

Pure alumina, alumina–zirconia (1–20% wt), and alumina–copper oxide (2–25%) samples were prepared. Pure alumina was prepared from a solution of $Al(NO_3)_3$. The solution was denitrated by solvent extraction (volume ratio 1:3) with the same procedure used for zirconia. The final $[NO_3]^-/[Al^{3+}]$ molar ratio was 0.5 and the pH was 3.81. The extractant was regenerated by a continuous process. The solution was diluted to 50 $g \cdot dm^{-3}$ (as Al_2O_3), and the thickening agent (approx. 0.5%) and surface-active agent (iso-octyl-phenoxy-polyethoxyethanol) were added. After dripping into diluted ammonia 1:1 with i.d. 2.5 mm capillaries, the spheres were aged for 30 min and rinsed with a diluted solution of ammonia 1:3 and then with H_2O/ethanol 1:1 and were finally calcined in air at 773K giving a final diameter of 1 mm.

An Al_2O_3–CuO (25%) catalyst was prepared from a sol of commercial water-dispersible alumina hydrate with addition of a copper acetate solution and an organic thickening agent. After dripping into a strong organic base with i.d. 3 mm capillaries, the spheres were aged for 30 min and rinsed with water to pH 7, dried by azeotropic dehydration in toluene, and calcined in air at 773 K giving a final diameter of 1.4 mm.

4.4   Properties of Oxides Prepared by the GSP Method

4.4.1 Thermal Stability

Pure oxides prepared by the GSP method show a higher thermal stability in comparison with conventional samples prepared by calcination of corresponding hydroxide compounds obtained by base precipitation from an aqueous solution or by calcination of gels obtained by sol–gel methods. Reported in Fig. 2 are typical results for ZrO_2 1.4 mm spheres prepared by the GSP method. The surface area decreases with increasing calcination temperature, but at 773 K and 1073 K values of about 100 and 40 $m^2 \cdot g^{-1}$, respectively, are still present in contrast to values of about 50 and 10

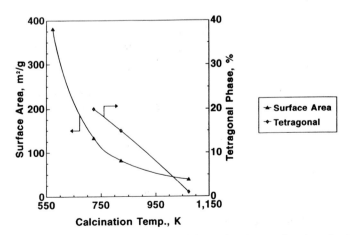

Figure 2. Change in the surface area and crystal form of ZrO_2 as a function of calcination temperature for pure zirconia GSP samples.

$m^2 \cdot g^{-1}$ or lower, respectively, found when conventional preparation methods were used [Norma, 1994; Yamaguchi, 1994]. Similar good thermal stability in the surface area was observed for TiO_2 GSP samples [Brambilla et al., 1992].

As the calcination temperature increases, the metastable tetragonal phase of zirconia progressively transforms to the monoclinic form as also shown in Fig. 2. The GSP method does not significantly affect this transformation in comparison with conventional methods [Norma, 1992], but in the case of TiO_2 a shift to temperatures about 100 K higher in the anatase to rutile transformation was observed [Brambilla et al., 1992].

The surface area, thermal stability, and crystallinity of zirconia can be further modified by preparing a zirconia–alumina mixed oxide with low alumina content ($< 10\%$ wt). Reported in Fig. 3 is the effect of increasing amounts of alumina up to 10% wt on the surface area at different calcination temperatures. The addition of alumina leads to an increase in the surface area and the stabilization of high surface areas even for the higher calcination temperatures.

At 823 K the zirconia–alumina sample with 10% alumina is only poorly crystallized differently from the pure ZrO_2, but X-ray diffraction patterns, even though characterized by broad diffraction lines, indicate that zirconia is preferentially present with its higher symmetry phases (tetragonal and cubic) (Fig. 4).

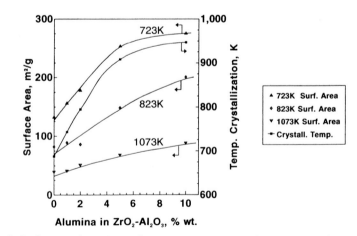

Figure 3. Surface area for various calcination temperatures and temperature of crystallization (DTA/DTG tests) as a function of alumina content in ZrO_2–Al_2O_3 GSP samples.

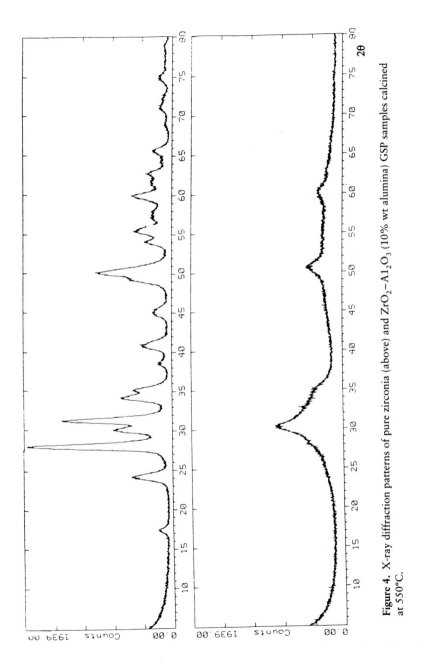

Figure 4. X-ray diffraction patterns of pure zirconia (above) and $ZrO_2-Al_2O_3$ (10% wt alumina) GSP samples calcined at 550°C.

The increase in alumina content also affects the crystallization temperature of zirconia. DTA/DTG analysis for pure ZrO_2 shows, in fact, an exothermic peak at 688 K caused by a phase change from the amorphous phase to tetragonal crystalline phase. This transition progressively shifts to higher temperatures with increasing alumina content in ZrO_2–Al_2O_3 mixed oxides (Fig. 3).

Scanning electron micrograph mapping of the Al distribution within the pellets shows its homogeneous distribution within the zirconia matrix, without any noticeable bulk or surface segregation of alumina intergrowth microparticles, although SEM analysis do not allow establishment of the homogeneity on a molecular scale. The homogeneity, however, was verified using infrared spectroscopy and CD_3CN as probe molecule (see Section 4.4.4), this method being useful to evidence the possible presence of small surface intergrowth structures of alumina, e.g., in the titania or zirconia matrix. Addition of alumina to zirconia thus allows textural and structural effects to be obtained analogous to those shown by sulphate modification of zirconia [Norma, 1992] but without causing significant modification of surface properties as in the case of sulphate additions. It is thus possible to tailor the properties of zirconia in a broad range without having large surface concentrations of dopants.

4.4.2 Textural Properties

Oxide samples prepared by the GSP method are characterized by a unimodal mesopore size distribution differently from the much broader distribution found in samples prepared with conventional methods. Reported in Fig. 5a is an example of the porosity distribution obtained for the sample ZrO_2–Al_2O_3 (10% wt alumina) calcined at 550°C which shows a unimodal pore distribution with r^{max} at 2.8 nm and a peak broadening at half-width of about 0.5 nm. Reported in Fig. 5b is the dependence of pore radius on maximum differential pore volume (r^{max}) for pure ZrO_2–Al_2O_3 GSP samples calcined at increasing temperatures and with variable alumina content. In all cases a narrow unimodal pore size distribution is found. The r^{max} shifts progressively to higher values with increasing calcination temperature, but the increase is much more limited when even low amounts of alumina are present in the ZrO_2 matrix.

4.4.3 Form and Mechanical Resistance Properties

A suitable control of parameters in the preparation steps allows nearly perfect spherical pellets of the pure or mixed oxides to be obtained in a wide range of controlled dimensions (from about 20 μm to up to 3–4 mm depending on the specific dripping capillaries or rotary cup atomizer used).

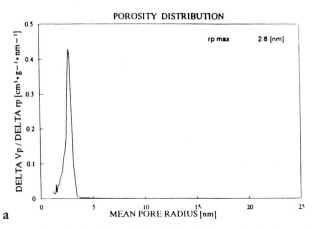

a

Figure 5a. Porosity distribution for a ZrO_2–Al_2O_3 (10% wt alumina) sample calcined at 550°C.

However, for selected dimensions of pellets, samples characterized by a narrow distribution in pellet dimension can be obtained. Reported in Fig. 6 is an example of the uniform diameter of the spheres of pure zirconia obtained after the drying stage. The mean diameter of the spheres shown in Fig. 6 is 1.6 mm with a dispersion in pellet diameter of about 0.1 mm.

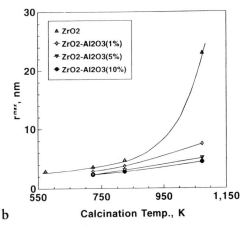

b

Figure 5b. Effect of calcination temperature on the pore radius for max. differential pore volume (r^{max}) in ZrO_2–Al_2O_3 GSP samples. r^{max} determined from the desorption branch of the isotherms obtained using the BET N_2 adsorption method.

Figure 6. Photograph of zirconia spheres prepared by the GSP method (dripping apparatus) after the drying stage; mean diameter of the spheres 1.6 mm.

After calcination the mean diameter of the spheres slightly decreases, but a heating rate lower than 20°C/h during calcination avoids the formation of cracks or inhomogeneous contractions in the spheres. The absence of surface roughness in the spheres makes the oxide samples prepared by the GSP method very suitable for reactor applications with solid transport.

Single pellet crushing strength and attrition resistance results for 1.4 mm pure zirconia and zirconia–alumina (10% wt) calcined at 823 K are shown in Table 1. The data show the good mechanical properties of both the supports. However, the ZrO_2–Al_2O_3 (10%) composition shows remarkably higher crushing strength properties.

Fluidizability properties of the oxide pellets obtained by the GSP method (rotary cup atomizer) are also excellent because of the absence of surface roughness, the good attrition resistance properties of the samples, and their low density. Reported in Fig. 7 is an example of the fluidizability properties of TiO_2 particles obtained by the GSP method and rotary atomizer apparatus. Characteristics of TiO_2 microspheres used in these tests are summarized in Table 2. The data reported in Fig. 7 show the good characteristics of fluidizability of TiO_2 microspheres prepared by this method due

TABLE 1
Single Pellet Crushing Strength[a] and Attrition Resistance (AIF)[b] Data for
GSP Samples

Composition	ZrO_2	ZrO_2–Al_2O_3 (10%)
Average crushing strength [N]	25.1	56.3
95% reliability [N]	23.6–26.6	45.8–66.8
%AIF	99.9	99.9

[a] Crushing strength tests: on single spheres using a Wolpert equipment following ASTM D 4179-88.

[b] Attrition resistance measurements (AIF) on 10 g of spheres using equipment assuring a displacement of 40 mm to an ampulla of length = 72 mm and internal diameter = 34 mm. The speed, variable in the range 0–1400 rpm, was fixed at 700 rpm and the test duration was 5 min.

to both their uniform spherical shape (Table 2) without significant roughness and the low density of the spheres due to the high surface area (Table 2). No significant formation of fines was noted after these tests as well as after Forsythe tests to determine attrition resistance properties that were found to be comparable with those of commercial alumina samples for fluid bed reactor applications [Brambilla, 1992].

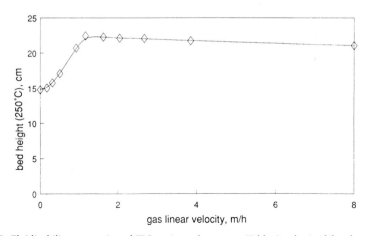

Figure 7. Fluidizability properties of TiO_2 microspheres (see Table 2) obtained by the GSP method. Tests at 250°C, bed apparent density at the minimum fluidization gas velocity, 0.55 g·cm^{-3}.

TABLE 2
Summary of the Characteristics of TiO$_2$
Microspheres Prepared by the GSP Method Used for
the Fluidizability Tests Shown in Fig. 7

Calcination temperature, °C	500
Surface area m^2·g^{-1}	133
Anatase crystalline form, %	100
C residual, %	0.054
Cl residual, %	0.070
Mean particle	58
Dimensions, μm	—
Dispersion of particle	22
Dimensions, μm	—
Apparent density, g·cm^{-3}	0.5
Fine particles (<20 μm) after 310 min	<5
In accelerate Forsythe tests, % wt	—

All these data thus show that carriers or catalysts prepared by the GSP method are well suited for fluid or mobile bed reactor applications or technologies where pneumatic transport of the solid is required.

4.4.4 Surface Acidity Properties

The surface acidity properties in mixed oxides can be modified and tuned by the addition of the second element. This is shown, for example, for alumina pellets prepared by the GSP method and containing variable amounts of zirconia.

Pure GSP Al$_2$O$_3$ is characterized by four detectable stretching frequencies for isolated OH groups at 3768, 3727, 3677, and 3585 cm^{-1} because of the different surface coordination modes, in good agreement with that expected for γ-Al$_2$O$_3$. Pure GSP ZrO$_2$ is instead characterized by three νOH modes at 3773, 3739 (shoulder), and 3674 cm^{-1}. With increasing zirconia content up to 20% wt in ZrO$_2$–Al$_2$O$_3$ mixed oxides, however, the IR spectrum in the νOH stretching frequency region remains nearly unchanged as compared with pure alumina. Tests by pyridine adsorption also showed that the strength of the Brønsted acid sites is not significantly affected by the zirconia content (up to 20% wt ZrO$_2$).

Lewis acidity, on the contrary, was found to depend significantly on the zirconia content. This can be seen in Fig. 8a, which shows the change in the νCN stretching frequency of CD_3CN adsorbed at r.t. (5 torr) and evacuated at 298 K and 373 K for ZrO_2–Al_2O_3 samples with increasing zirconia content. The shifts to higher frequencies with respect to νCN in liquid (2259 cm^{-1}) are proportional to the strength of the Lewis acid sites. Data reported in Fig. 8a thus show a progressive decrease in the strength of the Lewis acid sites from the very strong value for pure alumina to the medium value for pure zirconia.

No evidence, on the contrary, was found of the coexistence of two Lewis acid site characteristics for separate alumina and zirconia surface microdomains, as demonstrated by the spectrum reported in Fig. 8b showing the presence of a single, defined νCN for the ZrO_2–Al_2O_3 (10% wt alumina) sample. This clearly shows the homogeneous surface composition of the mixed oxide and the absence of surface domains of the two single oxides.

The data, therefore, indicate that the GSP method allows mixed oxides to be obtained where the surface Lewis acid properties of alumina can be tuned by the zirconia component, but without significantly affecting the alumina Brønsted acid characteristics and without having surface domains of zirconium oxide.

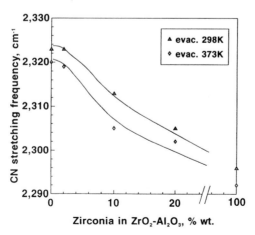

Figure 8a. νCN stretching frequency for CD_3CN (5 torr) adsorption at room temperature on ZrO_2–Al_2O_3 GSP samples after evacuation at 298 K and 373 K.

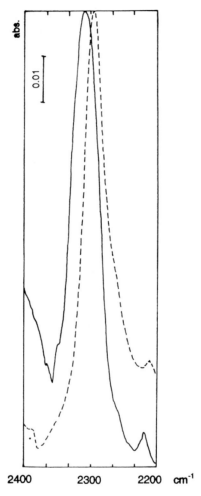

Figure 8b. Infrared spectrum of CD_3CN (5 torr) adsorbed at room temperature on ZrO_2–Al_2O_3 (10% wt alumina) and on Al_2O_3–ZrO_2 (10% wt zirconia) GSP samples and evacuated at 373 K. The background spectrum of the sample evacuated at 500°C has been subtracted.

4.4.5 Dispersion of an Active Component

Samples containing an active component and prepared by the GSP method show a higher resistance to sintering of the active component dispersed in the oxide matrix. This is exemplified by the preparation of copper oxide–alumina samples directly by the GSP method. In samples prepared by the conventional impregnation method on preformed alumina, copper can be deposited in a highly dispersed form only up to loadings of about 5–6% wt (as CuO) for a γ-Al_2O_3 sample with a surface area of about 100–150 $m^2 \cdot g^{-1}$ [Centi *et al.*, 1995]. For higher loadings, copper oxide macrocrystallites appear with a mean diameter of 1000 Å. On the contrary, the preparation of CuO–Al_2O_3 GSP mixed oxide allows samples without CuO crystallites to be obtained up to loadings of about 20% wt (Fig. 9, top), even after calcination at high temperature (800°C). In subsequent redox cycles, CuO crystallites tend to sinter (Fig. 9, bottom) as shown by the increase in the diffraction lines at about $2\theta = 35°$ and 38° (more intense for CuO), but their dimensions remain smaller (about 200–400 Å) than in comparable samples prepared by impregnation on preformed γ-Al_2O_3.

4.5 Experimental and Apparatus Section

4.5.1 Preparation of the Samples

Details on the sample preparation were reported in Section 4.3.2. Further details have been reported previously by Brambilla [1992]. A schematic drawing of the laboratory apparatus used for the synthesis of oxides by the GSP method using the rotary atomizer or the dripping system is shown in Fig. 10.

4.5.2 Characterization of the Samples

Differential thermal analysis was carried out in a Netzsch STA 409 apparatus on azeotropically dried samples. The measurements were made either in air or in nitrogen with a flow rate of 0.3 $L \cdot min^{-1}$, from room temperature to 800°C at a heating rate of 2°C$\cdot min^{-1}$.

The morphology was studied by scanning electron microscopy (SEM; Philips 505) and optical microscopy (Reichert Polyvar 2). EDAX element mappings were made on the cross sections of the microspheres.

X-ray diffraction (XRD) patterns were obtained using a powder diffractometer Siemens D500 equipped with a graphite crystal monochromator using a copper K_α radiation source. Samples were run with the diffractometer in step scan mode with a step interval of 0.02° 2θ and a count rate of 1 s per step over the range 5–90° 2θ. The relative fractions of the tetragonal and monoclinic forms were determined by the Rietveld method, which takes into account the measured intensities of the whole diffraction pattern instead of a few selected reflections. In the estimation of the relative ratio between the tetragonal and monoclinic crystalline phases was assumed to be absent the cubic form of zirconia which XRD reflections overlaps with those of the tetragonal phase.

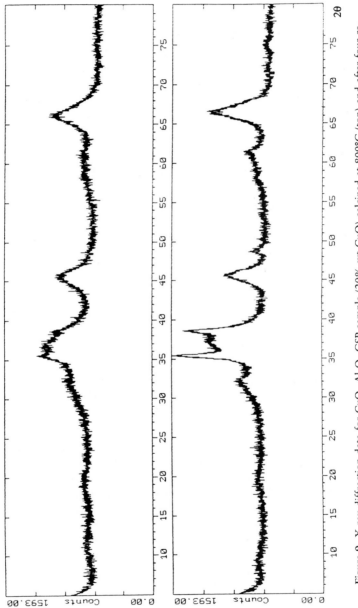

Figure 9. X-ray diffraction data for CuO–Al$_2$O$_3$ GSP sample (20% wt CuO) calcined at 800°C (top) and after four redox cycles of reduction with propane at 500°C followed by reoxidation with air at the same temperature (bottom).

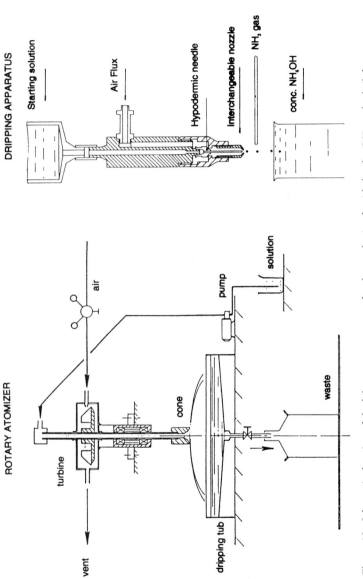

Figure 10. Schematic drawing of the laboratory apparatus for the synthesis of oxide by the GSP method: (left) rotary atomizer used for the synthesis of microspheres in the 20–200 μm range and (right) dripping apparatus for the preparation of spheres in the 0.5–3 mm range.

Nitrogen adsorption and desorption were measured at 77 K using a Carlo Erba Sorptomatic 1800 instrument and following ASTM D3663 and ASTM D4541 indications. The samples (0.4 g) were pretreated at 300°C under vacuum (1.33 Pa).

The crushing strength was determined on single spheres using Wolpert equipment following the ASTM D4179-88 method. Attrition resistance measurements (AIF) were made on 10 g of spheres with equipment assuring a displacement of 40 mm to an ampulla of length = 72 mm and internal diameter = 34 mm. The speed, variable in the range 0–1400 rpm, was fixed at 700 rpm and the test duration was 5 min. Attrition resistance characteristics of microspheres for fluid bed reactor applications were evaluated using the accelerated Forsythe method, based on the increase in the fraction of fine particles (particularly those under 20 μm) after some time. In the Forsythe method a gas jet passes through the specimen at supersonic speed, which causes impacts among the specimen particles, with the container walls, and with the shock wave produced by the gas. The performances of TiO_2 samples were evaluated by comparing the increase in the fraction of particles with a diameter below 20 μm after 310 min of accelerated tests with respect to that of a commercial sample of alumnia for fluid bed reactor applications.

Surface acid properties of the mixed oxides were evaluated on the basis of IR spectroscopy data for the adsorption of pyridine and deuterated acetonitrile at room temperature followed by evacuation up to temperatures of 300°C. The samples were pretreated in vacuum at 500°C before the adsorption. The IR experiments were performed using a Fourier transform Perkin Elmer 1750 instrument, the self-supporting disk technique, and conventional gas and sample manipulation apparatus.

4.6 Conclusions

The gel-supported precipitation (GSP) method is a versatile technique to prepare pure and mixed oxides to be used as supports or catalysts in various applications ranging from fluid bed to mobile bed to fixed bed reactors.

The characteristics of the samples are analogous to those obtained by the xerogel method using alkoxide precursors in terms of purity, homogeneous dispersion of the components, and the textural and thermal resistance properties. The GSP method, however, has the advantage of allowing (1) the direct synthesis of spheres or microspheres from about 20 μm up to about 3 mm and (2) the synthesis of spherical pellets with shape and mechanical strength characteristics suitable for applications in reactors with solid transport. Furthermore, the method is tailored to avoid the use of expensive raw materials such as alkoxides.

The possibility of tuning the textural, thermal resistance, and surface acidity properties by synthesis of $ZrO_2-Al_2O_3$ and $TiO_2-Al_2O_3$ mixed oxides has also shown. These mixed oxides are of considerable potential interest for new catalytic materials with modulated performances.

Finally, the possibility of improving the dispersion of an active component such as copper oxide over alumina has been shown. The potential interest of this preparation method for the synthesis of more-active samples has thus been indicated.

Acknowledgments

Thanks are due to F. Gerolin for sample preparations, to F. Danieli for BET, G. Pannocchia for DTA/DTG, M. Battagliarin for XRD, and B. Burtet Fabris for SEM/EDS measurements.

References

Aynal, A., Assih, T., Abenoza, M., Phalippou, J., Lecomte, A., and Dauger, A. (1990). *J. Mater. Sci.*, **25**, 1268.
Araki, M., and Hibi, T. (1987). Eur. Patent 22 235 6A1.
Bachet, J. C., Tilliette, M. J., and Cornet, D. (1991). *Catal. Today*, **10**, 507.
Beeckman, J., and Hegedus, L. L. (1991). *IEC Research*, **30**, 969.
Brambilla, G., Gerontopoulos, P., and Neri, D. (1970). *En. Nucl.*, **17**, 217.
Brambilla, G., Centi, G., Perathoner, S., and Riva, A. (1992). *Proceedings 2nd European Conference on Advanced Materials and Processes* (T. W. Clyne, and P. J. Withers, eds.), Vol. 3, p. 287, The Institute of Materials, London.
Breysse, M., Portefaix, J. L., and Vrinat, M. (1991). *Catal. Today*, **10**, 489.
Cavani, F., Foresti, E., Parrinello, F., and Trifirò, F. (1988). *Appl. Catal.*, **38**, 311.
Centi, G., Militerno, S., and Perathoner, S. (1991a). *J. Chem. Soc. Chem. Commun.*, 88.
Centi, G., Grange, P., Matralis, H., Ruwet, M., and Trifirò, F. (1991b). *Proceedings JECAT'91* (Hirashi, eds.), p. 135.
Centi, G., Pinelli, D., Trifirò, F., Ghoussoub, D., Guelton, M., and Gengembre, L. (1991c). *J. Catal.*, **130**, 238.
Centi, G., Nigro, C., Perathoner, S., and Stella, G. (1993). *Catal. Today*, **17**, 159.
Centi, G., Giamello, E., Murphy, E., and Perathoner, S. (1995). *J. Catal.*, **152**, 75.
Chertov, V. M., Makovskaya, T. F., Tsyrina, V. V., and Kaganovskii, V. A. (1992). *Neorg. Mater.*, **28**, 1134.
Courty, Ph., and Duhaut, P. (1974). *Rev. Inst. Franc. du Petrole*, **24-6**, 861.
Davis, B. H. (1984). *J. Am. Ceram. Soc.*, **67**, 8.

Deller, K., Ettlinger, M., Klingel, R., and Krause, H. (1989). DE Patent 3 803 898.

Ding, J. Y., and Day, D. E. (1991). J. Mater. Res., 6, 168.

Doelbear, G. E. (1993). "Sol-gel Preparation of Catalytic Materials," Novel Approaches to Catalysts Preparation, Catalytica Studies Division.

Doesburg, E. B. M. and van Hoof, J. H. C. (1993). Stud. Surf. Sci. and Technol., 79.

Fulton, J. W. (1986). Chem. Eng., 12, 97.

Gerontopoulos, P., Rotoloni, P., and Fava, R. (1970a). Patent IT 8 120 167.

Gerontopoulos, P. (1970b). Patent IT 8 120 168.

Guglielminotti, E., Giamello, E., Pinna, F., Strukul, G., Martirengo, S., and Zanderighi, L. (1993). Stud. Surf. Sci. Catal., 75, 2761.

Hashimoto, T., Segawa, T., Sakurada, S., and Go, A. (1991). JP Patent 03 249 942.

Hass, P. A., (1989). Chem. Eng. Prog., 4, 44.

Hubbardd, C. P., Otto, K., Gandhi, H. S., and Ng, K. Y. S. (1993). J. Catal, 139, 268.

Hubert-Pfalzgraf, L. G. (1987). New J. Chem., 11, 663.

Huisman, H. M., van der Berg, P., Mos, R., van Dillen, A. J., and Geus, J. W. (1994). Environmental Catlaysis (Armor, J., ed.), p. 393, American Chemical Society: Washington, DC.

Hums, E. (1986). Patent DE 86 633 229.

Igarashi, A., Ohtaka, T., Honnma, T., and Fukuhara, C. (1993). Stud. Surf. Sci. Catal., 75, 2083.

Klimova, T. E., and Solis, J. R. (1994) Material Science Forum, Vol. 152–153, p. 309, Trans Techn., Switzerland.

Koeppel, R. A., Baiker, A., Schild, Ch., and Wokaum, A. (1991). Stud. Surf. Sci. Catal., 63, 59.

Yamaguchi, T., Tan-No, M., and Tanabe, K. (1991). Stud. Surf. Sci. Catal., 63, 567.

Yokoyama, T., Setoyama, T., Fujita, N., Nakajima, M., Maki, T., and Fuji, K. (1992). Appl. Catal., 88, 149.

Johnson, D. W. (1985). Am. Ceram. Soc. Bull., 64, 1597.

Lancia, R., Fumagalli, C., Arnbruster, E., and Vaccari, A. (1994). Patent IT M I94 A00 317.

Lerot, L., Legrand, F., and de Bruycker, P. (1991). J. Mater. Sci. Lett., 26, 2353.

Marella, M., Tomaselli, M., Meregalli, L., Battagliarin, M., Gerontopoulos, P., Pinna, F., Signoretto, M., and Strukul, G. (1994a). Preprints 6th International Symp. Scientific Bases for the Preparation of Heterogeneous Catalysts, Vol. 1, p. 335, Louvain-la-Neuve, Belgium.

Marella, M., Meregalli, L., and Tomaselli, M. (1994b). Patent IT MI 94A 002 588.

Maruya, K., Takasawa, A., Haraoka, T., Aikowa, M., Arai, T., Domen, K., and Onishi, T. (1993). Stud. Surf. Sci. Catal., 75, 2733.

Matsuda, S., and Kato, A. (1983). Appl. Catal., 8, 149.

Mercera, P. D., van Ommen, J. G., Doesburg, E. B., Burggraaf, A. J., and Ross, J. R. (1992). J. Mater. Sci., 27, 4890.

Montoya, I. A., Viveros, T., Dominguez, J. M., Canales, L. A., and Schifter, I. (1992). Catal. Lett., 15, 207.

Murase, Y., and Kato, E. (1978). Nippon Kagaku Kaishi, 367.

Nagoal, V., Davis, R., and Riffle, J. (1992). Polym. Mater. Sci. Eng., 67, 235.

Nakano, Y., Yamaguchi, T., and Tanabe, K. (1983). J. Catal., 80, 307.

Norma, C. J. (1994). Catal. Today, 20, 313.

Oberlander, R. K. (1984). Applied Industrial Catalysis, p. 63, Academic Press, New York.

Ogura, Y., Riman, R. E., and Bowen, H. K. (1988). J. Mater. Sci., 23, 2897.

Ozawa, M., Kimura, M., and Hasegawa, H. (1987). JP Patent 62 168 544.

Ozawa, M., and Kimura, M. (1991). J. Less-Common Metals, 171, 195.

Palmisano, M., Schiavello, A., Sclafani, G., Martra, E., Borello, E., and Coluccia, S. (1994). Appl. Catal. B, 3, 117.

Pajonk, G. M. (1991). *Appl. Catal.*, **72**, 217.

Pajonk, G. M., and El Tanany, A. (1992) *React. Kinet. Catal. Lett.*, **47**, 167.

Richardson, J. T. (1989). *Principles of Catalyst Development*, Plenum Press, New York.

Rodriguez, O., Gonzalez, F., Bosh, P. Portilla, M., and Viveros, T. (1992). *Catal. Today*, **14**, 243.

Sanchez, C., and Livage, J. (1990). *New J. Chem.*, **14**, 513.

Sasai, A., and Hiraishi, H. (1989). JP Patent 01 072 979.

Sasamoto, T., Enomoto, S., Shimoda, Z., and Saeki, Y. (1993). *J. Ceram. Soc. Jpn.*, **101**, 230.

Tsurumi, K., Sasaki, M., and Yamamoto, T. (1993). JP Patent 05 103 983.

Twigg, M. U. (1989). *Catalyst Handbook*, 2nd ed., Wolfe Publishing.

Yamaguchi, T. (1994). *Catal. Today*, **20**, 199.

Wauthor, P., Ruwet, M., Machej, T., and Grange, P. (1991). *Appl. Catal.*, **69**, 149.

Wolf, C., and Russel, C. (1992). *J. Mater. Sci.*, **27**, 3749.

CHAPTER 5

Platinum-Catalyzed Sulfur Dioxide Oxidation Revisited

Assembly of Acid- and Sintering-Resistant Honeycomb
Washcoat and Catalytically Active Phase Using Sols of
Silica, Zirconia, and Platinum

T. R. Felthouse,[*,†] D. A. Berkel,[*] S. R. Jost,[*,‡]
E. L. McGrew,[*] and A. Vavere[§]

[*]Monsanto Enviro-Chem Systems, Inc., St. Louis, Missouri 63167
[†]Huntsman Corporation, Austin, Texas 78752
[‡]Eli Lilly and Company, Indianapolis, Indiana 46285
[§]Monsanto Enviro-Chem Systems, Inc., St. Louis, Missouri 63178

KEYWORDS: Metal oxide sols, honeycomb composition, washcoat composition, $Pt(ZrO_2-SiO_2)$/mullite

5.1 Introduction

Monolithic or honeycomb catalysts were commercialized in large numbers about 20 years ago for application in auto exhaust emissions control. Both early [DeLuca and Campbell, 1977] and recent [Irandoust and Andersson, 1988; Cybulski and Moulijn, 1994; Armor, 1994] applications of honeycomb catalysts have been reviewed. One intrinsic advantage of honeycomb catalysts over particulates (extruded or tableted shapes) is high geometric surface area per unit volume. Figure 1 compares geometric surface areas calculated per liter of catalyst [particulates were commercial Monsanto Enviro-Chem Systems catalysts with an average number of par-

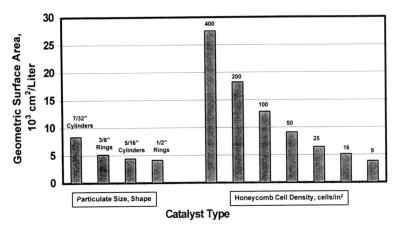

Figure 1. Comparison of geometric surface areas for some common particulate catalyst shapes with those for honeycomb substates presented on a per volume (liter) basis.

ticulates per 50 cm^3 volume scaled to 1 liter; honeycomb sample volumes include both void and solid volumes with the geometric surface area calculated for a square cell geometry using the equations given (see DeLuca and Campbell, 1977, Table I)] and clearly show the honeycomb advantage at cell densities of 100 or higher cells/in^2. Further advantages of honeycomb catalysts over conventional particulate catalysts include greatly reduced (about 10^{-3} times) pressure drop, high heat and mass transfer, efficient use of the active phase, and improved industrial hygiene and safety. These advantages come at a higher materials cost (namely, the honeycomb substrate) than particulate catalysts, so it is essential that active and promoter phases are loaded and activated on honeycomb substrates to produce high-activity catalysts that maintain their activity for years of use.

One of the oldest large-volume industrial chemicals is sulfuric acid. In 1831 an Englishman by the name of Peregrine Philips discovered that supported platinum catalyzed the air oxidation of sulfur dioxide, SO$_2$, to sulfur trioxide, SO$_3$ [Fairlie, 1936]. Since that time supported platinum catalysts were used in some commercial operations through about the end of World War II. In the early 1900s silica-supported alkali–vanadium sulfates were discovered as new catalysts for SO$_2$ oxidation [Donovan *et al.*, 1983] and have become the particulate catalyst of choice since that time. Supported alkali–vanadium sulfate particulates offer long life and, compared with platinum catalysts, less sensitivity to gas phase impurities due to their molten salt active phase mechanism of operation and lower cost per unit volume.

This work developed out of a program to identify new ways by which SO_2 is catalytically oxidized to SO_3. Catalysts with both platinum– and alkali–vanadium active phases on honeycomb substrates were developed and evaluated in laboratory differential and integral reactors. Development of high-performance honeycomb catalysts for SO_2 oxidation having platinum as the active phase is described in this chapter. Platinum was reconsidered for use on a honeycomb substrate because of higher activity at reaction temperatures compared with alkali–vanadium sulfate catalysts. Lower pressure drops by honeycomb catalysts allow operation at higher gas flows, which affords reactor downsizing. In this work novel techniques were developed for washcoat and active phase loadings using sols of silica, zirconia, and platinum that serve as models for how other metal oxide particles related at silica can be bonded to a ceramic honeycomb substrate. Those techniques are reported along with comparative reactor data. Thermal and chemical aging studies demonstrate superiority of these newly developed Pt honeycomb catalysts for activity maintenance compared with commercial $Pt(Al_2O_3)$/honeycomb catalysts. Use of electron microscopy and auxiliary techniques to define the Pt-promoter-support structure provides a consistent picture that accounts for the high activity, thermal stability, and acid resistance.

5.2   Literature Survey

Development of platinum-containing particulate catalysts for the oxidation of SO_2 ceased about 50 years ago. Examination of typical patents from that time [Streicher, 1940; Rosenblatt, 1947] shows that important concepts in supported catalysts such as metal dispersion, effective use of promoter elements to improve activity and life, and nanotechnology in synthesis and characterization were unavailable for catalyst preparations. Deposition of a high-surface-area washcoat onto a low-surface-area ceramic honeycomb substrate was first described about 40 years ago in the patent literature [Houdry, 1956; Benbow and Lord, 1974], but application of this technology to various chemical catalytic processes including sulfuric acid manufacture has occurred slowly. In the 1970s Johnson Matthey introduced [Pratt and Cairns, 1977] a $Pt(Al_2O_3)$/metal honeycomb catalyst that used a fabricated metal honeycomb substrate developed at the United Kingdom Atomic Energy Authority in Harwell [Cairns and Noakes, 1978]. The honeycomb substrate was composed of "Fecralloy" ferritic steel. Alumina washcoats were used to bond to the alumina-rich surface of the thermally activated

Fecralloy substrate [Nelson et al., 1981]. The Pt(Al$_2$O$_3$)/metal honeycomb catalyst was used as prototype catalyst in a process patent issued to Davy McKee [Parish et al., 1982] that claims use of at least 50 cells/in^2 honeycombs with gas velocities of 1500–2500 actual ft/min. No honeycomb catalyst preparations or evaluations were disclosed in this Davy McKee patent.

In the 1980s Degussa reported development of a ceramic Pt(Al$_2$O$_3$)/honeycomb catalyst [Brand et al., 1988] for use in an integrated process for the conversion of both SO$_2$ and NO$_x$ ("DESONOX" process) [Ohlms, 1990]. Weak (77–80%) sulfuric acid is made in this process, which was demonstrated in a pilot plant located in Muenster, Germany. The SO$_2$ oxidation catalyst consists of 0.119 ft^3 blocks of honeycombs having 100 cells/in^2 mullite coated with an α-AL$_2$O$_3$ washcoat and Pt deposited with a loading of 70.8 g-Pt/ft^3. Gas is passed over the honeycomb catalyst at 7500 h^{-1} space velocity at 420–460°C containing no more than about 50 mg/m^3 dust particles. Three layers of these 0.119 ft^3 blocks of honeycomb catalysts were used to make up the catalyst bed. After 2000 h of operation, the % SO$_2$ conversion remained above 91%.

The technology described in the preceding literature formed the basis for initiation of this work on development of a high-activity, long-lived Pt honeycomb catalyst for the oxidation of SO$_2$ to SO$_3$. In particular, one goal was to develop a silica (and not alumina) washcoat that forms a more chemically resistant composition toward dissolution by sulfuric acid. As a second goal, the amount of Pt used in the Degussa Pt(α-Al$_2$O$_3$)/honeycomb catalyst should be reduced through the use of a high-surface-area (>100 m^2/g) silica washcoat combined with an effective Pt-loading technique. With these two goals in mind, the following assumptions were made at the outset of this work: (1) due to poisoning of a Pt active phase by volatile metal oxides such as AsO$_x$, a guard reactor column is needed; (2) a filter for the incoming gas stream is needed to protect the fine honeycomb cells from plugging with fly ash deposits; and (3) the Pt honeycomb catalyst developed was studied only for use in dry SO$_2$-containing gas streams—no "wet acid process" is considered. After the completion of this work, a patent appeared assigned to Engelhard [Deeba et al., 1992] that closely describes (see Example 7 in Deeba et al., 1992) the washcoat composition developed here. However, the bonding of the washcoat to the honeycomb surface made use of a ball-milled slurry of 2% Pt(ZrO$_2$–SiO$_2$) and washcoat adhesion upon sulfuric acid treatment was not reported. Use of a silica precursor sol described in this work provides adhesion of the washcoat to the ceramic honeycomb substrate. The entire catalyst preparation process reported here is covered by a patent [Felthouse, 1992].

5.3   Experimental Section

5.3.1 Materials

All chemicals were of reagent grade and used as received. Honeycomb supports of mullite (Celcor 9494, 200 cells/in², cpsi), cordierite (Celcor 9475, 400 cpsi), and silica (HOT-1000, 200 cpsi) were obtained from Corning. For some evaluations these samples were cut (about 0.9 in. diameter and 1–3 in. long or 0.6 in. diameter and 0.5 in. long) in the Monsanto glass shop using diamond-tipped hole saws with glass sleeves and water-coated circular saws. Silica spheres of 1.7 mm (S 980 G 1.7, 1.0 cm³/g, 80 m²/g) were obtained from Shell, Houston, TX. Metal oxide promoters included nitrate salts (Fisher), colloidal alumina (Nalco 1SJ-614), and zirconia (PQ, Nyacol Zr10/20). Hydrogen peroxide (Mallinckrodt) was used as a 30% aqueous solution. Tetraethylorthosilicate (TEOS) and ethanol were from Fisher. Silica powders for washcoats were all below 20 mm particle size and obtained from W. R. Grace (Syloid 74, 350 m²/g, 1.1 cm³/g; Grade 56—milled, 300 m²/g, 1.2 cm³/g) or Aldrich (Davisil Grade 710, 480 m²/g, 0.75 cm³/g). For work reported here, all platinum active phases were prepared using Strem Chemicals $H_3Pt(SO_3)_2OH$ solution (15.3% Pt). A commercial $Pt(Al_2O_3)$/cordierite sample having 400 cpsi and 18.0 g-Pt/ft³ was obtained from Johnson Matthey, Catalytic Systems Division, Wayne, PA, and cut to desired dimensions.

5.3.2 Catalyst Preparations

5.3.2.1 0.1% Pt(metal oxide)/SiO₂ Spheres

Initial identification of preferred metal oxide promoters for use with Pt used 1.7 mm SiO_2 spheres (Shell S980G) to which were added water soluble metal oxide precursors by impregnation using a rotary evaporator at 90°C. Metal oxide loadings of 2 wt % (based on the metal) were used. After air calcination at 600°C, platinum was deposited [Petrow and Allen, 1978] onto the metal oxide-washcoat layer by means of immersion of the washcoated honeycomb in a stirred aqueous solution of $H_3Pt(SO_3)_2OH$ resulting in a 0.1 wt % Pt loading in the final catalyst. The Pt phase was affixed to the metal oxide/SiO_2 spheres by addition of a few drops of 30% H_2O_2. After 10 min the spheres were recovered, air dried at 115°C and air calcined to 350°C.

5.3.2.2 Pt(ZrO$_2$–S74–SiO$_2$)/Mullite

All laboratory-prepared Pt honeycomb catalysts used the following general procedure illustrated with ZrO$_2$ as the metal oxide promoter and Syloid 74 (S74) SiO$_2$ as the particulate SiO$_2$ source. A clear, coatable TEOS sol was prepared by sequential addition of 27 g water, 33 g ethanol, 40 g TEOS, and 0.08 mL HNO$_3$ to a beaker containing a magnetic stirrer and watch glass cover. After being stirred for 1 h, the cooled sol was transferred to a plastic bottle, a few drops of Antifoam B were added, and the mixture was shaken vigorously after addition of 15 wt % S74 SiO$_2$. The slurry was poured into a stirred graduated cylinder as part of the apparatus displayed in Fig. 2. Mullite substrates suspended by stainless steel wires were dip coated in the slurry, blown free of excess washcoat (whenever channel blockage was observed), and air dried for 30 min in a Class C hood. Samples were then dried in a forced air oven at 115°C and air calcined in a box furnace to 600°C. S74 SiO$_2$/mullite samples were soaked in Nyacol Zr 10/20 ZrO$_2$ sol, dried again at 115°C, and calcined at 600°C. Washcoat loadings of about 17 wt % were produced with these procedures. Pt loadings again made use of H$_3$Pt(SO$_3$)$_2$OH solutions containing Pt to produce loadings based on g-Pt/ft^3

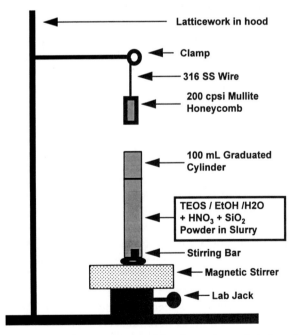

Figure 2. Diagram of laboratory apparatus used for dip coating and platinum loading of ceramic honeycombs.

assuming solid cylindrical honeycomb volumes. Excess (about 1 mL) 30% H_2O_2 was added to the Pt salt solution with ZrO_2-S74-SiO_2/mullite sample immersed. After about 1 h (O_2 bubbles observed through H_2O_2 treatment) the sample was recovered, blown free of solution, and dried at 115°C, and air calcined to 350°C (final activation *in situ* in SO_2/air at 300–475°C).

5.3.3 Catalytic Reactor Systems and Analytical Methods

Several reactor systems provided evaluations of the Pt particulate and Pt honeycomb catalysts. Lower temperature activity and thermal aging effects were judged by use of a differential conversion reactor system known as the thermal catalyst aging tester (TCAT) reactor. The reactor system consisted of eight dip-tube-style reactors immersed in a common furnace with quartz tubing (25 mm o.d. outside tube and 19 mm inside reactor tube). Samples were either loaded between quartz wool plugs or for honeycombs wrapped with fiberglass cloth tape (Fisher) around the exterior to provide a seal between honeycomb and reactor wall. Mass-flow-controlled gases at 100 cm^3 (NTP)/min (9% SO_2 12% O_2, 79% N_2) were fed to each sample through the entire run. Typically temperatures in 25°C intervals were selected from 350 to 450°C and the SO_2 conversions measured, samples aged for 24 h at 750°C, and then the SO_2 conversions remeasured at the selected temperatures. Effluent gases were scrubbed of SO_3 (sulfuric acid solutions or solid state scrubber columns) and the % SO_2 and % O_2 analyses made with a modified gas analyzer now commercially available from Monsanto Enviro-Chem Systems [Monsanto Enviro-Chem Bulletin, 1991]. A Spectra Physics ChromJet integrator with appropriate programming reported gas analyses and conversion levels of SO_2 and O_2.

5.3.4 Activity Tester Reactor System

Rapid differential SO_2 conversions over a wide gas flow range were obtained by an "activity tester" reactor. Honeycomb samples were fit into a 50-mm i.d. quartz sample holder with a quartz grid plate on the bottom. Honeycomb catalysts were sealed into the holder through multiple wrappings of fiberglass cloth tape between the sample and sample holder wall. The sample holder was lowered into an all-quartz reactor (55–60 mm i.d.) immersed in a molten lead bath at 475°C. The reactor was capped by a 71/60 ground quartz fitting. The gas stream was adjusted to a flow of 9.26 L(NTP)/min. (10% SO_2, 11% O_2 79% N_2) through the catalyst sample. Downstream analytical provided feed and reactor gas analyses using another modified gas analyzer and integrator as described for the TCAT system [Monsanto Enviro-Chem Bulletin, 1991].

5.3.5 Physical Measurements

All analytical measurements were performed in the Monsanto Analytical Sciences Center. Pore volumes were measured on a Micromeritics Autopore 9220-II by mercury intrusion porosimetry. Surface areas were obtained on a Micromeritics Digisorb 2500. Electron microscopy measurements used a Philips EM430ST instrument with auxiliary nanoprobe analytical techniques. Fourier analysis and reconstruction were done using a Zeiss-Kontron image analysis system.

5.4　Results and Discussion

5.4.1 Honeycomb Composition

Three compositions of honeycomb ceramic are applicable to this work. Cordierite, available in either the $Mg_2Al_4Si_5O_{18}$ composition or a silica-surface-rich, acid-leached version, is the most common commercial honeycomb material due to its use in automobile emissions control. However, because cordierite is readily leached by acids, it was dropped early in this work as a possible support.

Silica represents a second composition of honeycomb and is a natural choice because commercial sulfuric acid particulate catalysts use silica-based (i.e., diatomaceous earths) supports for the alkali–vanadium sulfate active phase [Donovan et al., 1983]. Developmental samples of silica-based honeycombs were received and tested based on proprietary formulations developed at Corning. These silica honeycomb supports feature a "washcoat-in-the-wall" design that combines a porous silica with a low-surface-area silica to produce a composite material that possesses both the porosity needed for effective use as a catalyst support and mechanical strength conferred through the use of a low-surface-area silica. Typical materials for the preparation of these silica composite honeycombs include a low-density, high-porosity silica powder with below 20 μm average particle size; a low-surface-area silica with particles smaller than 74 μm such as SUPERSIL silica from Pennsylvania Glass Sands Co.; and a silicone resin such as Dow Corning Resin Q6-2230. A "dough" mixture suitable for extrusion is made through the addition of water and isopropyl alcohol. Further details on these composite honeycomb supports of this type are given in Corning patents [Lachman and Nordlie, 1986; Lachman et al., 1986; DeAngelis and Lachman, 1987]. The preferred material for honeycomb catalysts developed here is silica extruded in nominally 200 cpsi with square cells. These composite silica supports have pore volumes from 0.25 to 0.50 cm^3/g with surface areas from 15 to 50 m^2/g. Higher pore volumes

(0.50–0.75 cm³/g) can be obtained, but the resulting silica honeycombs lack sufficient mechanical strength. Adequate mechanical strength results from a modulus of rupture greater than 500 psi [DeAngelis and Lackman, 1987]. The porous silica component to the silica composite honeycomb support can be selected from several silica powders with high surface areas (100–500 m²/g) or silicas with low surface areas (below 10 m²/g) but high pore volumes such as diatomaceous earths. These silica honeycombs can be used with either platinum or alkali–vanadium active phases.

In spite of silica honeycomb advantages, the present work focused on a third honeycomb composition that is readily available in commercial quantities, mullite. Mullite is an aluminosilicate ceramic that has the nominal composition Al_6SiO_{11}. Mullite has been successfully identified for use in heat exchanger applications where cordierite with its susceptibility to acid leaching is not suitable [Day, 1978]. Various chemical tests conducted as part of this work verified that mullite was not leached by sulfuric acid. As noted in Section 5.3, mullite honeycombs used in this work were Corning Celcor 9494 having 200 cpsi. Figure 3 displays the pore volume distribution for Celcor 9494 mullite honeycomb. All macroporosity in the mullite is centered around a pore diameter of about 0.66 μm. This pore diameter affords good uptake rates of washcoat from coating slurries and locates washcoat particles in pores having highly efficient contact with the gas stream for gas phase reactions. Mullite is the honeycomb composition of choice for this work.

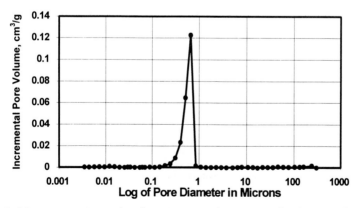

Figure 3. Mercury porosimetry data for a Corning Celcor 9494 mullite honeycomb substrate plotted as incremental pore volume against the logarithm of the mean pore diameter in microns. Note that all macroporosity is centered around a pore diameter of about 0.66 μm.

5.4.2 Washcoat Composition

Development of a Pt(washcoat)–mullite honeycomb catalyst depended upon use of sols of silica, zirconia, and platinum to assemble the catalyst. Sols of silica are novel in their ability to form adhesive coatings to ceramic surfaces at room temperature. A key development was made with use of partially hydrolyzed TEOS as a film-forming silica adhesive. Use of partially hydrolyzed TEOS (i.e., primarily linear TEOS oligomers) precludes the need for the slow and chemically undefined step of ball milling some or all of the washcoat particles so as to generate mechanically reactive silica-containing oligomers that act as a washcoat adhesive. The TEOS oligomer composition needed was developed through reference of the ternary phase diagram for TEOS-ethanol-water shown in Fig. 4 [Sakka *et al.*, 1984; Klein, 1985]. Both acid- and base-catalyzed hydrolysis of TEOS occurs but the acid-catalyzed method is best suited for this application. Note that three types of materials are obtained from TEOS depending on the phase composition: (1) high water promotes bulk castable forms, (2) intermediate TEOS-ethanol-water composition produces films, and (3) high-TEOS compositions afford spinnable silica fibers. Intermediate TEOS-ethanol-water compositions that produce films are needed for washcoat bonding to the mullite honeycomb. An "X" in Fig. 4 shows the TEOS-ethanol-water composition used in this work. Through partial acid hydrolysis a TEOS-based silica sol forms oligomeric species of the

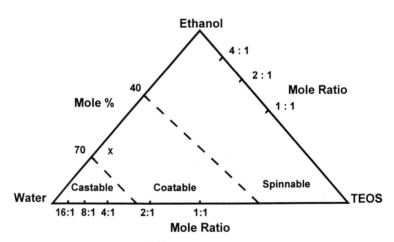

Figure 4. Ternary phase diagram for the water-ethanol-tetraethylorthosilicate (TEOS) system. Three regions are shown that afford castable, coatable, or spinnable silica precursor solutions. An "X" marks the composition used in this work for coatable silica precursor solutions having molar ratios of water/ethanol/TEOS = 8 : 4 : 1.

type $[Si(O)_x(OH)_y(OC_2H_5)_z]_n$ as shown in the formula that follows. With a starting $H_2O/C_2H_5OH/TEOS$ mole ratio of $8:4:1$, the silica sol produced is in the "coatable" regime [Sakka *et al.*, 1984; Klein, 1985] with a t_{gel} time of approximately 100 h, and primarily linear chains are produced based on ^{29}Si NMR and molecular weight measurements [Sakka *et al.*, 1984].

$$Si(OC_2H_5)_4 + n\,H_2O \xrightarrow[C_2H_5OH]{[HNO_3]=0.01} [Si(O)_x(OH)_y(OC_2H_5)_z]_n + (2n)\,C_2H_5OH$$

TEOS   **Silica Precursor Sol**

Further development of the Pt(washcoat)/mullite honeycomb catalyst required that several catalyst development steps be performed to generate a prototype catalyst that was then evaluated through a combination of physical property tests (i.e., weight loss through acid leaching, thermal sintering by means of an accelerated aging test—TCAT), and differential reactor data for % SO_2 conversion. Before the issue of washcoat powder choice could be resolved, the use of promoters needed to be demonstrated and a selection made for the preferred promoter. Initially the TCAT reactor system was used to survey a set of binary metal oxide promoters to stabilize a Pt active phase on SiO_2 particulates in the form of 1.7-mm spheres. The results for five oxide promoters are summarized in differential % SO_2 conversion data in Fig. 5. From these data the preferred thermal stability of Pt/SiO_2 is shown by the following sequence of metal oxides: $ZrO_2 > Fe_2O_3 > Al_2O_3 > CeO_2 > Cr_2O_3 > None$ (unpromoted). Note that the unpromoted silica sample in Fig. 5 showed low initial activity compared with all other samples and upon thermal aging the activity declined to a very low level. Iron and zirconium oxides were the top two promoters of choice. Both ZrO_2 and Al_2O_3 promoters showed enhanced activity after thermal aging. Further development of the Pt honeycomb catalyst made use of ZrO_2 derived from a commercial ZrO_2 sol (i.e., Nyacol Zr 10/20 having zirconia phases of $50-100$ Å).

Variations in silica particle type were made to determine the highest per unit volume activity toward SO_2 conversion. Among Syloid 74, Grade 56, and D710 silica particles used in $Pt(ZrO_2-SiO_2)$/mullite catalysts and a $Pt(Al_2O_3)$/mullite catalyst, all containing 20 g Pt/ft^3, comparative % SO_2 conversions are displayed in Fig. 6. The activity ranking in this series is Syloid $74 > Al_2O_3 > $ Grade $56 > D710$. From the surface area and pore volume data given in Fig. 6, the preferred Syloid 74 (S74) powder has intermediate surface area and pore volume to both of the other silica powders, Grade 56 and D710.

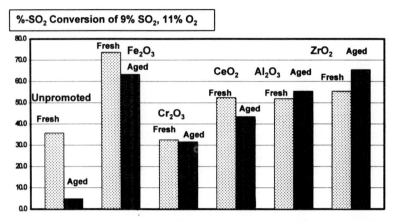

Figure 5. Effect on % SO$_2$ conversion of binary oxide promoters added at a 2 wt % level to 1.7-mm silica spheres having a 0.1% Pt loading. For each sample the fresh activity was measured at 450°C (and five other temperatures down to 300°C) at 2300 gas hourly space velocity at the constant feed gas composition shown. All samples were then subjected to aging at 750°C under feed gas flow for 24 h. The aged conversions measured afford selection of the Fe$_2$O$_3$- and ZrO$_2$-containing catalysts as preferred promoters.

5.4.3 Platinum Deposition

Although the use of H$_3$Pt(SO$_3$)$_2$OH/H$_2$O$_2$ to form a platinum sol has been described in the patent literature [Petrow and Allen, 1978] and an article [Allen and Larson, 1984], specific application of this platinum(II) salt to form a platinum(IV) precursor phase that deposits onto washcoated honeycombs is reported here for the first time. During initial development it was observed that surface acidity (such as that imparted by ZrO$_2$) is critically important for efficient deposition of the Pt precursor phase. Use of only silica on the honeycomb substrate leads to solution deposition of a Pt colloid in about 30 min after H$_2$O$_2$ addition to the H$_3$Pt(SO$_3$)$_2$OH solution. Successful Pt deposition through facilitated absorption using H$_3$Pt(SO$_3$)$_2$OH/H$_2$O$_2$ and a (ZrO$_2$–S74–SiO$_2$)/mullite sample [*hereafter denoted as (ZrO$_2$–SiO$_2$)–mullite, where SiO$_2$ implicitly denotes both S74 and TEOS-based silica*] results in no solution coloration, the onset of O$_2$ bubbles in a few minutes (excess H$_2$O$_2$ decomposition), and formation of a straw-colored coating on the washcoat. The Pt active phase is then generated through drying and activation in an SO$_2$/air gas stream at about 400–475°C. The entire sequence facilitated adsorption of the Pt phase, shown diagrammatically in Fig. 7.

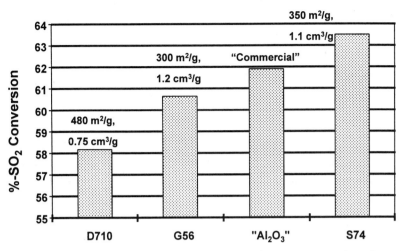

Figure 6. Comparative differential reactor data showing the % SO$_2$ conversion of a 10% SO$_2$, 11% O$_2$ gas stream at 9.26 standard liters per minute and 480°C when passed through samples of 200 cell/in^2 honeycomb substrates coated with different washcoat compositions of silica and alumina. Each sample contains a ZrO$_2$ promoter and Pt active phase derived from H$_3$Pt(SO$_3$)$_2$OH prepared with Pt loadings of 20 g Pt/ft^3. The powder type codes are: D710 = Davisil 710 silica, G56 = Davison Grade 56 silica, Al$_2$O$_3$ = Catapal G alumina, S74 = Davison Syloid 74 silica.

5.4.4 Final Assembly

Figure 8 gives the final assembly sequence for the honeycomb catalyst developed in this work. The first step consists of a silica washcoat application to a mullite honeycomb that next undergoes drying and calcination. In a second dip coating cycle zirconia sol is added to the washcoat and the resulting promoted washcoat is again dried and calcined. The (ZrO$_2$–SiO$_2$)/mullite sample is then treated with H$_3$Pt(SO$_3$)$_2$OH solution to give a final washcoat composition having the mole ratio of SiO$_2$:ZrO$_2$:Pt of 250:115:1 for a Pt loading of 20 g Pt/ft^3.

5.4.5 Pt(ZrO$_2$–SiO$_2$)/Mullite Honeycomb Catalyst Features

The final Pt(ZrO$_2$–SiO$_2$)/mullite sample shows gray-brown coloration along the washcoated channels. Honeycomb catalysts were evaluated in laboratory reactors using calcined silica tape as a "gasket" to prevent bypassing of the gas stream and allow the sample to move smoothly along the reactor tubes. Three important features were demon-

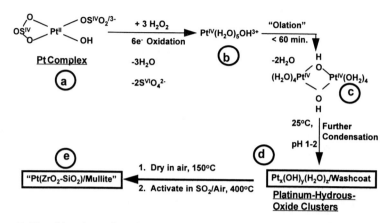

Figure 7. Plausible scheme for adsorption onto the ZrO_2-SiO_2 washcoat of hydrous oxide platinum clusters from hydrogen peroxide addition to a solution of $H_3Pt(SO_3)_2OH$. The elementary chemical steps are still not elucidated with respect to the Pt complex and its intermediates.

strated through a series of tests on the $Pt(ZrO_2-SiO_2)$/mullite samples: sintering resistance, Pt active phase loading optimization, and acid resistance.

Table 1 presents comparative reactor data at seven temperatures for three fresh and aged catalysts samples: $Pt(SiO_2-ZrO_2)$/mullite, $Pt(ZrO_2-SiO_2)$/mullite, and $Pt(Al_2O_3)$/cordierite. The first two differ by washcoat adhesive. $Pt(SiO_2-ZrO_2)$/mullite denotes use of ZrO_2 sol as the ceramic adhesive used to bond particles of S74 SiO_2 to mullite. $Pt(ZrO_2-SiO_2)$/mullite mentioned above denotes use of hydrolyzed TEOS solution for bonding S74 SiO_2 particles onto the mullite substrate with ZrO_2 added as a promoter. Commercial $Pt(Al_2O_3)$/cordierite performed very well in this test (Table 1) and all three catalysts showed good activity retention after aging.

The loading of platinum on the washcoat was investigated through a series of samples where the ZrO_2 sol was used to bond S74 SiO_2 to the mullite. In this $Pt(SiO_2-ZrO_2)$/mullite series, the activity per unit volume was found to reach a local maximum around 20 g Pt/ft^3. Comparative reactor data are presented in Table 2. Space–time yield values remove variations in honeycomb catalyst volume from the comparison. Further catalyst development always used a Pt loading around 20 g Pt/ft^3.

As seen in Table 1, thermal stability alone of the Pt phase on the washcoat is insufficient to distinguish the catalysts developed here from com-

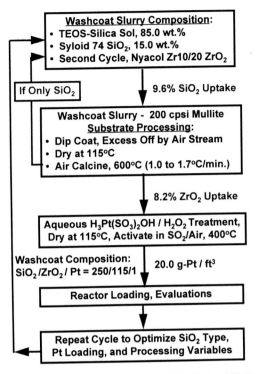

Figure 8. Final assembly sequence for the preparation of $Pt(ZrO_2-SiO_2)$/mullite honeycomb catalysts. Note the gray-lined recycle loop needed to apply two dip coatings of first silica powder and then zirconia as a sol to the washcoat. Another of these gray recycle loops connects the final catalyst with continuing attempts to optimize the silica powder, Pt loading, and processing variables.

mercial samples of $Pt(Al_2O_3)$/cordierite. In light of the solubility of Al_2O_3 in acid solutions, chemical stability tests were devised to make distinctions between catalysts developed here and $Pt(Al_2O_3)$/cordierite [Felthouse, 1992]. An acid-leaching test was developed to assess the ability of washcoated catalysts to remain unaffected by contact with wet acid. Two pairs of honeycomb catalysts, $Pt(ZrO_2-SiO_2)$/mullite, and $Pt(Al_2O_3)$/cordierite, were tested in the TCAT reactor system before and after acid leaching. The results are compiled in Table 3. Excellent activity retention is displayed by the $Pt(ZrO_2-SiO_2)$/mullite sample, both before and after leaching in heated dilute sulfuric acid solution. Quite different results were found for the commercial $Pt(Al_2O_3)$/cordierite sample. As seen for aging data presented in Table 1, $Pt(Al_2O_3)$/cordierite shows good thermal stability (see

TABLE 1
Sintering Resistance Demonstrated through Accelerated Aging
Reactor Evaluations for Three Pt(Washcoat)/Honeycomb Catalysts.

Sample[a] (aged state)	% SO_2 Conversion[b] at temperature (°C)						
	300	325	350	375	400	425	450
Pt(SiO_2–ZrO_2)/mullite[c]							
(Fresh)	4.7	11.0	20.9	35.8	52.0	67.5	77.7
(Aged)	3.6	8.3	16.8	29.1	43.2	57.6	68.8
Pt(ZrO_2–SiO_2)/mullite[d]							
(Fresh)	6.3	14.4	26.4	43.6	61.6	75.2	83.2
(Aged)	12.3	26.6	46.6	66.3	79.8	87.3	90.2
Pt(Al_2O_3)/cordierite[e]							
(Fresh)	6.3	15.9	29.3	45.8	60.2	71.1	76.9
(Aged)	6.7	17.3	31.8	49.3	65.8	77.8	83.9

[a]Samples occupy volumes of about 2.5 cm^3.
[b]Conversion of 9% SO_2, 12% O_2 gas stream flowing at 100 standard cm^3 per min. Aged conversions were measured after a hold at 750°C in an SO_2, O_2 gas stream for 24 hr (TCAT reactor).
[c]Mullite has 200 cpsi and a washcoat of S74-SiO_2 and ZrO_2 with ZrO_2 serving as both washcoat adhesive and promoter. The Pt loading is 18.5 g Pt/ft^3 and the sample weighs 1.44 g.
[d]Mullite has 200 cpsi and a washcoat of S74-SiO_2, ZrO_2, and hydrolyzed TEOS as washcoat adhesive. The Pt loading is 20.5 g Pt/ft^3 and the sample weighs 1.64 g.
[e]Commercial sample having 400 cpsi from Johnson Matthey. The Pt loading is 18.0 g Pt/ft^3 and the sample weighs 1.34 g.

TABLE 2
Differential Reactor Data for Pt(SiO_2–ZrO_2)/Mullite Catalysts
Showing the Pt-Loading Effect on the % SO_2 Conversion[a]

Pt loading, g Pt/ft^3	% SO_2 conversion[b]	Space time yield,[c] $\times 10^6$
4.8	26.8	1.27
10.1	42.2	1.96
19.4	58.5	2.87
31.7	59.9	2.67
45.8	62.7	2.91

[a]Honeycombs have 200 cpsi, an average volume of 10.5 cm^3, and 9.6 wt % Syloid 74-ZrO_2 washcoat.
[b]Conversions at 475°C of a 10% SO_2, 11% O_2 gas stream fed at 9.26 standard cm^3 per min (activity tester reactor).
[c]Space time yield = (% SO_2 conversion/100)(volumetric flow rate, SO_2)/(space velocity), where the volumetric flow rate of SO_2 is 2.48 mol/h, and the space–time yield is given in units of moles of SO_2 converted or moles of SO_3 produced.

TABLE 3
Effect of Acid Leaching on Pt Honeycomb Catalyst
Properties of % SO_2 Conversion and % Weight Loss

Sample[a] (aged state)	% SO_2 Conversion[b] at temperature (°C)					% Weight loss[c]
	350	375	400	425	450	
Pt(ZrO_2–SiO_2)/mullite[d]						
(Fresh)	11.6	21.0	34.6	51.9	70.3	—
(Aged)	15.4	27.3	41.5	56.4	68.7	—
Pt(ZrO_2–SiO_2)/mullite Acid leached[d]						
(Fresh)	33.4	52.7	69.6	80.4	85.9	1.0
(Aged)	14.6	26.1	41.4	57.6	70.5	—
Pt(Al_2O_3)/cordierite[e]						
(Fresh)	37.0	54.0	68.1	77.1	80.5	—
(Aged)	29.2	46.8	64.0	76.9	83.8	—
Pt(Al_2O_3)/cordierite Acid leached[e]						
(Fresh)	8.4	14.6	22.9	34.4	47.1	59.5
(Aged)	1.8	3.6	6.2	9.9	13.3	—

[a]Samples occupy volumes of 2.5–2.6 cm³ and fresh weights of 1.41–1.52 g. Pt loadings range from 18 to 20 g Pt/ft³.
[b]Conversion of 9% SO_2, 12% O_2 gas stream flowing at 100 standard cm³ per min. Aged conversions were measured after a hold at 750°C in an SO_2, O_2 gas stream for 24 h (TCAT reactor).
[c]Weight losses calculated starting with a dry, fresh catalyst and leached in 30% H_2SO_4 for 24 h at 95°C, rinsed with water, and dried at 150°C to a constant weight.
[d]Sample of 200 cpsi having a washcoat of ZrO_2 and S74-SiO_2 that was bonded to mullite with hydrolyzed TEOS solution. Mullite honeycomb was cored and cut into small cylinders.
[e]Commercial sample having 400 cpsi from Johnson Matthey and cored and cut into a cylindrical shape.

Table 3, third sample entry). However, once Pt(Al_2O_3)/cordierite is leached in acid, the activity toward SO_2 conversion rapidly declines (Table 3, entry 4). As evidenced by the recorded weight loss of nearly 60% (visually observable), a significant amount of active phase, washcoat, and cordierite honeycomb is dissolved by the acid treatment. On the other hand, the Pt(ZrO_2–SiO_2)/mullite catalyst developed here showed less than 1% weight loss when subjected to acid leaching.

5.4.6 Demonstration of Reactor Downsized Sulfuric Acid Plant

In a process patent issued to Davy McKee the possibility of a reactor downsized sulfuric acid plant was claimed without any supporting experimental data. Based on the $Pt(ZrO_2-SiO_2)$/mullite catalyst developed here, samples of about 1 in. in diameter by 3 in. long with 200 cpsi were used in an integral reactor operated at close to adiabatic conditions. Details appear in a process patent based on this technology [Felthouse and Vavere, 1993]. Integral reactor data were collected for 10–11% SO_2-containing gas streams to give estimated catalyst loadings per pass for a full-sized plant operated under adiabatic reaction conditions. Figure 9 displays a 2:2 interpass absorption (IPA) plant design based on 200 cpsi $Pt(ZrO_2-SiO_2)$/ mullite catalysts. Figure 9 illustrates a process flow diagram for a sulfuric acid plant that includes two reactor vessels of different sizes, one of which contains honeycomb catalysts for passes 1 through 3 and the other contains particulate catalyst for pass 4. The sulfuric acid process is well known; thus, details of this sulfuric acid plant design and operation is found in the process patent [Felthouse and Vavere, 1993].

5.4.7 Transmission Electron Microscopy Study

A reactor evaluated $Pt/(ZrO_2-SiO_2)$ powder was removed from the mullite honeycomb and investigated using transmission electron microscopy (TEM). TEM observations represented by Figs. 10 and 11 record an average Pt particle size of 90 Å, average cubic ZrO_2 crystallite sizes from 70 to 500 Å, and amorphous SiO_2 phases. Figure 11 provides a high-resolution view of a Pt crystallite surrounded by the ZrO_2 lattice fringes that suggests a possible barrier to Pt crystallite thermal sintering. The ZrO_2 phase is revealed from the data presented in Fig. 12. A cubic ZrO_2 phase is suggested by analysis of the lattice spacings and angles in Fig. 12. As seen in Fig. 13, no evidence exists from electron microscopy for compound formation between the Pt and ZrO_2 phases.

Measured electron diffraction d-spacings (TEM) are assignable to cubic ZrO_2 and a mixture of platinum phases existing as either Pt metal, PtO, or PtO_2. The nine d-spacings include 1.18 Å (spot; 1.17 Å, cubic ZrO_2; 1.18 Å Pt), 1.38 Å (spot; 1.36 Å PtO, 1.39 Å Pt, 1.397 Å PtO_2), 1.51 Å (ring; 1.53 Å, spot for cubic ZrO_2; 1.54 Å, rings for PtO and PtO_2), 1.69 Å (spot; 1.65 Å, PtO_2; 1.67 Å, PtO), 1.79 Å (spot; 1.80 Å, spot for cubic ZrO_2), 2.23 Å (spot; 2.25 Å, PtO_2, 2.27 Å, Pt), 2.46 Å (ring; 2.55 Å, ring for cubic ZrO_2), 2.71 Å (spot; 2.67 Å for both PtO and PtO_2), and 2.93 Å (ring; 2.93 Å, ring for cubic ZrO_2). As judged by Pt crystallite sizes ranging from

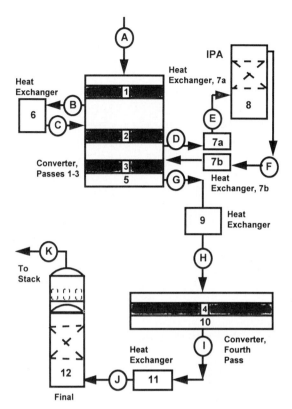

Figure 9. Diagram of a 2:2 interpass absorption plant design based on integral reactor data from the 200 cpsi Pt(ZrO$_2$–SiO$_2$)/mullite honeycomb catalyst developed in this work. The first three passes contain honeycomb catalyst and a fourth pass and separate reactor or converter contains Cs-110 (a Monsanto Enviro-Chem catalyst of 3/8 in. rings containing cesium) catalyst. This design achieved 99.7% conversion of a 10% SO$_2$, 11% O$_2$ gas stream entering the first pass (A).

60–120 Å, the Pt phase is likely a mixture between Pt metal and an oxide of Pt [Nandi et al., 1982]. According to wide-angle and small-angle X-ray scattering (i.e., WAXS and EXAFS) results by Nandi et al. [Nandi et al., 1982], when supported platinum crystallites approach 100 Å in size, the phases present exist as a Pt metal core with an oxide overlayer. The stability of the oxide phases in supported Pt crystallites is strongly influenced by the temperature and the gaseous atmosphere in contact with the supported Pt crystallites. *In situ* measurements may be added to infer accurately the phases of platinum that catalyze SO$_2$ oxidation in air.

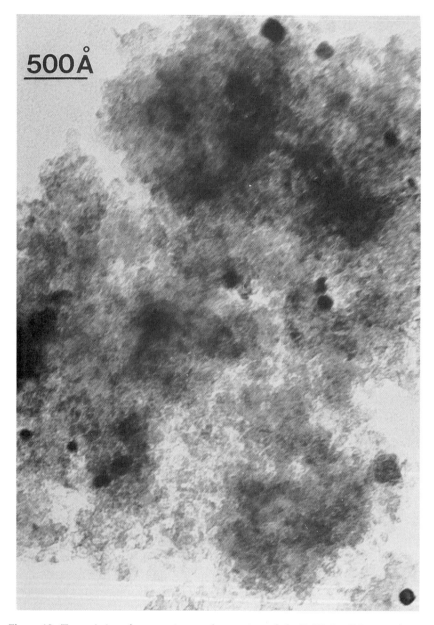

Figure 10. Transmission electron micrograph overview of the Pt/(ZrO$_2$–SiO$_2$) powder removed from a mullite honeycomb after testing for catalytic oxidation of SO$_2$. The image shows the Pt active phase as dark regions averaging about 90 Å in diameter.

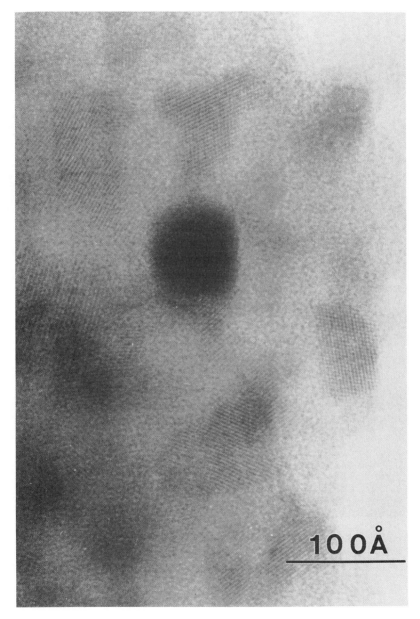

Figure 11. High-resolution electron micrograph of the Pt nanophase supported on the ZrO_2-SiO_2 washcoat. The ZrO_2 crystallites exhibit lattice fringes varying in size from 70 to 500 Å in diameter. The silica washcoat material remains amorphous.

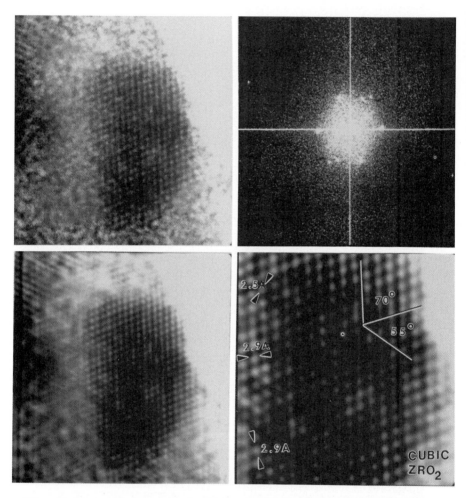

Figure 12. Image analysis sequence for a ZrO_2 crystallite found on the ZrO_2–SiO_2 washcoat. The upper left panel shows a digitized image of the ZrO_2 crystallite. To the right of this upper image is a two-dimensional power spectrum of the digitized image. At the lower left is a reconstructed digitized image from the power spectrum. The lower right panel gives a closeup of the reconstructed image to the left with lattice spacings and angles given that are appropriate for a cubic ZrO_2 lattice.

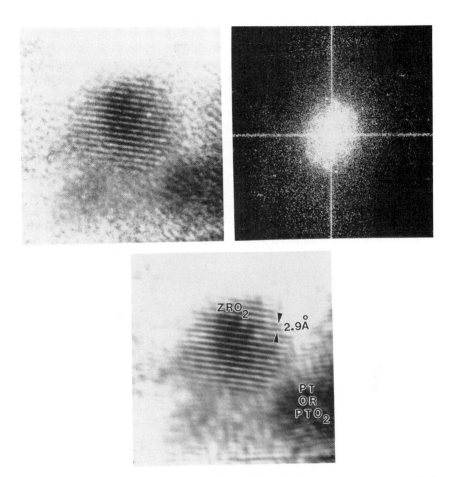

Figure 13. Image analysis sequence used to define the Pt phase interaction with the ZrO_2–SiO_2 washcoat. The top left image shows lattice fringes for both ZrO_2 and Pt crystallites. To the right of this image is a two-dimensional power spectrum generated from this image. The bottom image is a reconstructed from the two-dimensional power spectrum. No evidence exists for formation of a phase between ZrO_2 (or SiO_2) and Pt. The Pt phase is present as either a Pt metal, a Pt oxide, or a mixture of Pt metal and oxide.

5.5 Conclusions

Metal oxide sols play key roles in development of novel Pt honeycomb catalyst preparations. In this work a silica sol derived from partially hydrolyzed TEOS forms an effective washcoat adhesive for silica powders of about 10–20 µm particle size. A commercially available sol of zirconia can also serve as a washcoat adhesive but finds effective use as a promoter for thermal stabilization of a Pt active phase. A sol of platinum derived by solution oxidation of $H_3Pt(SO_3)_2OH$ adsorbs onto the ZrO_2–SiO_2 washcoat to provide a uniformly loaded Pt honeycomb catalyst. All of the catalyst development here requires comparative reactor data to make progress toward catalyst improvement. Useful features imparted to this new class of $Pt(ZrO_2$–$SiO_2)$/mullite catalyst include thermal stability to at least 750°C and acid resistance for both chemical stability and regeneration. Washcoats having the molar composition of $Pt/ZrO_2/SiO_2 = 1/115/250$ have a nanophase of Pt crystallites ranging from about 60 to 120 Å, a cubic ZrO_2 phase that ranges from about 70 to 500 Å, and silica phases from both washcoat powder and adhesive that remain entirely amorphous.

Acknowledgments

The authors gratefully acknowledge the contributions of the following individuals to this project: Irwin M. Lachman (Corning, for honeycomb substrate samples), Gail K. Fraundorf (formerly with Monsanto Company and now with MEMC Electronics Materials for TEM data), Russell H. Kloess and Allan J. Wootten (Monsanto Company, Glass Shop, for cutting honeycomb substrates into laboratory size pieces), and Stephanie K. Camden (Monsanto Company, mercury porosimetry data).

References

Allen, R. J., and Larson, H. R. (1984). *The Strem Chemiker*, X(1), 12–14.
Armor, J. N. (1994). *Chem. Mater.*, 6, 730–738.
Benbow, J. J., and Lord, L. W. (1974). U. S. Patent 3 824 196.
Brand, R., Engler, B., Kleine-Moellhoff, P., Koberstein, E., Voelker, H. (1988). U. S. Patent 4 744 967.
Cairns, J. A., and Noakes, M. L. (1978). U. S. Patent 4 098 722.
Cybulski, A., and Moulijn, J. A. (1994). *Catal. Rev.—Sci. Eng.*, 36, 179–270.
Day, J. P. (1978). *Trans. ASME (J. Eng. Power)*, Paper No. 78-GT-118, pp. 1–5.
DeAngelis, T. P., and Lachman, I. M. (1987). U. S. Patent 4 637 995.
Deeba, M., Chen, J. M., Lui, Y. K., and Speronello, B. K. (1992). U. S. [Patent] 5 145 825.
DeLuca, J. P., and Campbell, L. E. (1977). *Advanced Materials in Catalysis* (J. J. Burton and R. L. Garten, eds.), pp. 293–324, Academic Press, New York.

Donovan, J. R., Stolk, R. D., and Unland, M. L. (1983). *Applied Industrial Chemistry* (B. E. Leach, ed.), pp. 245–286, Academic Press, New York.
Fairlie, A. M. (1936) *Sulfuric Acid Manufacture*, Reinhold, New York.
Felthouse, T. R. (1992). U. S. Patent 5 175 136.
Felthouse, T. R., and Vavere, A. (1993). U. S. Patent 5 264 200.
Houdry, E. J. (1956). U. S. Patent 2 742 437.
Irandoust, S., and Andersson, B. (1988). *Catal. Rev.—Sci. Eng.*, 30, 341–392.
Klein, L. C., (1985). *Ann. Rev. Mater. Sci.*, 15, 227–248.
Lachman, I. M., Bardhan, P., and Nordlie, L. A. (1986). U. S. Patent 4 631 268.
Lachman, I. M., and Nordlie, L. A. (1986). U. S. Patent 4 631 267.
Monsanto Enviro-Chem Bulletin (1991). "Portable Gas Analyzer (PeGASyS) Update."
Nandi, R. K., Molinaro, F., Tang, C., Cohen, J. B., Butt, J. B., and Burwell, R. L., Jr. (1982). *J. Catal.* 78, 289–305.
Nelson, R. L., Ramsey, J. D. F., Woodhead, J. L., Cairns, J. A., and Crossley, J. A. A. (1981). *Thin Solid Films*, 81, 329–337.
Ohlms, N. (1990). *Umwelt*, 20(3), L23–L28; *Chem. Abstr.* (1990). 113, 120–129.
Parish, W. R., Nicholson, N. E., Scarlett, J. (1982). U. K. Patent Application GB 2 081 239A.
Petrow, H. G., and Allen, R. J. (1978). U. S. Patent 4 082 699.
Pratt, A. S., and Cairns, J. A. (1977). *Plat. Met. Rev.*, 21(3), 2–11.
Rosenblatt, E. F. (1947). U. S. Patent 2 418 851.
Sakka, S., Kimiya, K., Makita, K., and Yamamoto, Y. (1984). *J. Non-Cryst. Solids*, 63, 223–235.
Streicher, J. S. (1940). U. S. Patent 2 200 522.

CHAPTER 6A

Applications of Supercritical Drying in Catalyst Preparation

Diane R. Milburn, Bruce D. Adkins, Dennis E. Sparks,*
Ram Srinivasan, and Burtron H. Davis
Center for Applied Energy Research
University of Kentucky
Lexington, Kentucky 40511
*Presently at Akzo Nobel, Pasadena, Texas 77507

KEYWORDS: Surface forces, solvation forces, capillary forces, critical pressure solvent evaporation, methanol

6A.1 Introduction

The use of supported catalysts is widespread. The ability to support an expensive metal, platinum, at a low loading on a high-surface-area support, alumina, made it possible to develop a bifunctional catalyst needed for naphtha reforming [1]. In catalysis it is desirable in many instances for the solid catalyst to remain in a stationary bed and for the reactant to move through the bed [1]. In other instances, such as fluid catalytic cracking [2], fast-fluid-bed Fischer–Tropsch synthesis [3], and the H-Oil process [4], both the reactant and the catalyst move. However, even in the latter case it is desired that the catalyst particle be sufficiently large so that separation of the catalyst and the product can be easily accomplished.

In some instances, however, it may be desirable to use highly dispersed catalysts as is done in hydrogenation of fats, in the upgrading of heavy resid, or in direct coal liquefaction. One approach is to form the catalyst in a highly dispersed state in the reactant medium. This has been done, e.g., in the Dow Process for coal liquefaction [5]. In this case an aqueous solution of ammonium heptamolybdate is added together with the coal slurry, and the dispersed molybdenum oxide–sulfide particles are formed in the reactor. However, not all catalysts are amenable to *in situ* synthesis at or near the reaction conditions. Furthermore, supported catalysts cannot be formed easily using this approach.

In the preparation of a supported catalyst an aqueous solution is commonly employed. Evaporation of the water used as solvent may cause problems. An early example of poor distributions of the supported material and the reasons for this were described by Maatman and Prater [6]. Three of the major problems encountered in using an aqueous solvent are: (1) the concentration of the impregnating salt in the solution that is left in the smaller pores ultimately leads to the precipitation of salt crystals; (2) the concentration of the salt at the evaporating solvent surface because of a chromatographic effect; and (3) the fusing of the support particles by forces generated at the contact points during the solvent evaporation. This last problem will cause agglomeration of the dispersed particles, and the ability to subsequently form a highly dispersed suspension of the catalyst is lost. Supercritical drying provides a means to decrease the impact of these three factors or even to eliminate the problem. Thus the work described in the paper is directed toward illustrating the usefulness of this technique to prepare highly dispersed, supported catalysts.

6A.1.1 Supercritical Techniques in Catalyst Preparation

An early, if not the first, application of supercritical solvent evaporation was the preparation by Kistler of silica gel with a low bulk density [7–9]. The silica that results from the sol-gel-produced matrix leads to the lightest inorganic solid material known [10]. In addition, this silica is nearly transparent and has excellent thermal barrier properties. For this reason many studies have been directed toward using this material as highly insulating glazings to be used for windows, skylights, and solar collector covers.

Horn [11] points out that the term "surface forces" is gaining currency in the scientific literature although it is not readily found in textbooks. In the broad sense this term describes the forces that act between two surfaces when they are in close proximity. Surface forces are especially important in colloid science, because the properties of the dispersion depend on the forces acting between particles suspended in a medium, typically a liquid. Several of these forces operate in circumstances encountered in the course of catalyst preparation using commonly practiced techniques.

1. Solvation forces involve the interaction of the liquid with the surface and usually increase as the polarity of the solvent increases. Water, with its high polarity, is used frequently in catalyst preparation. Knowledge of solvation forces is based almost entirely upon experimental data. In aqueous dispersions the force may extend to 10 water molecule diameters or more from the solid. Whereas this force is recognized to be important in determining interactions between dispersed particles, there is currently no theory capable of predicting what the solvation forces will be in a given system [10].

2. Capillary forces result from the condensation of a vapor in a narrow gap or capillary (Fig. 1). A liquid bridge may form in the contact region of two particles in the presence of a condensable vapor whose liquid wets the surface, a situation that is almost always encountered with water and metal oxide catalytic materials. The critical distance between the two surfaces is related through the Kelvin radius, r_K:

$$r_K = \gamma \, V/R \, T \log (P/P_s),$$

where γ is the surface tension, V is the molar volume, R is the gas constant, T is the absolute temperature, and P/P_s is the relative pressure of the vapor. When a liquid bridge forms, a negative curvature of the meniscus is associated with a negative Laplace pressure in the liquid [12]:

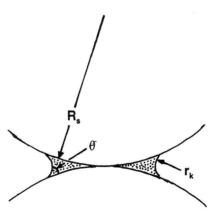

Figure 1. Schematic of capillary condensation that will occur at contact points of spheres (radius of R_s) for a condensed liquid that wets the surface (contact angle $\theta < 90°$) to produce a meniscus (Kelvin radius r_K).

$$\Delta P = \gamma/r_K.$$

which acts to pull the surfaces of the solid particles together. The result is that there is a net attractive force that, for two contacting spheres of radius R_S, with a small condensed meniscus is given approximately by [13]:

$$F = 2\pi R_S \gamma \cos \theta.$$

3. Concentration of the solute in the liquid bridge shown in Fig. 1 may lead to the formation of large particles of the supported material or of a neck of the precipitated material. This occurs because as the solution becomes more concentrated, crystals precipitate from the saturated solution. Thus there is a mechanism for the concentration of the supported material at the contact points to produce a poor distribution as well as a "cement" to prevent the particles from being able to be dispersed during subsequent use as a catalyst.

The primary ultrafine particles may "stick together" and form secondary particles. Colloidal particles may form the secondary particles in two ways, and the one that prevails may depend upon the chemical composition, shape, and size of the primary particles. One of these involves aggregation in which the ultrafine particles are loosely bound to each other; this assemblage is relatively easily dispersed. The aggregate can be converted to an agglomeration in which the primary ultrafine particles are held together with very strong linkages and are therefore not readily redispersed. With spherical particles the linkage usually involves the formation of necks connecting the primary particles. A number of pathways may effect the formation of these necks. High-temperature sintering and compression at high pressures are two techniques whereby these necks may be formed. In catalyst preparation the aggregate may be held together by the surface tension of the adsorbed water film; in this case the aggregate will have a reasonably flexible structure. During drying, however, the combined effects of surface tension, solution–dissolution, and the temperature used to effect the drying may convert the flexible porosity to a more rigid structure as the neck structure changes from the adsorbed water film to chemical bonding of the material that makes up to ultrafine primary particles. The strength of the linkages that form during drying will depend upon the amount of material that is transported during the solution–dissolution and/or surface–bulk diffusion processes that deposit material in the neck region. The objective of the supercritical pressure solvent removal is to eliminate, or at least greatly decrease, the effect of the neck contacts such as shown in Figs. 1 and 2 and therefore to prevent the formation of the strong linkages that lead to strongly bonded agglomerates that cannot be easily redispersed.

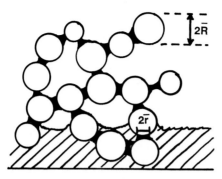

Figure 2. Schematic of spherical particles (diameter=$2R$) assembled into aggregates by condensed liquid (solid regions) between the contact points.

In the supercritical state the physical properties of water differ from those normally encountered at ambient conditions (374.2°C versus 20°C; 22.1 MPa versus 0.1 MPa) [14]. Dell'Orco *et al.* enumerate several differences as follows [14]. In the temperature range of 370–700°C the density and viscosity assume gaslike values [15] and the dielectric constant assumes values similar to nonpolar solvents at ambient conditions [16]. At the critical conditions the ion product of water decreases below 10^{-20} mol/kg [17], and hydrogen bonding assumes a lesser role than is encountered at ambient temperature [18]. Because of the changes in the physical properties of water, solubilities are also altered dramatically. Gases, such as nitrogen and oxygen, are relatively insoluble at ambient conditions but approach or become completely miscible with supercritical water [19, 20]. Under these conditions hydrocarbons also become miscible [21]. On the other hand, those salts that are very soluble under ambient conditions assume limited solubility at supercritical conditions [22]. Dell'Orco *et al.* [14] found that sodium chloride and sodium sulfate could be obtained as a solid near the solubility limit near 500°C. They were able to obtain greater than 96% recovery of sodium nitrate from a brine.

Thus the use of supercritical water for catalyst preparation may not provide a promising technique to obtain low-density-supported catalyst, because limited salt solubility will make it difficult to obtain a uniform distribution of the supported material. Furthermore, the high temperatures needed to obtain supercritical water most likely will have deleterious effects, such as hydration, upon the metal oxides that are normally used in catalyst preparations.

Another approach to utilize supercritical solvent removal is to dilute the water used for the impregnation step with a material that has a lower supercritical temperature than water. In utilizing this approach, there are essentially two variables: the choice of solvent and the speed of solvent removal. The primary effect of these two variables are:

6A.1.1.1 Structuring of the Support

Supports that consist of ultrafine particle materials are structured by the solvent (Fig. 2). For typical supports (e.g., alumina and silica) the surfaces of the particles have a high concentration of hydroxyl groups, which depends upon the temperature of heating and the extent of subsequent hydration. For this reason a strong bridging solvent such as water should have a greater effect than a lesser bridging solvent such as ethanol. Also, the rate of solvent removal should have an effect on the structuring; a rapid removal presents less time for interparticle motion and subsequent compaction.

Structuring of the ultrafine particle support material is directly responsible for catalyst porosity (Fig. 2). Pore volume and pore size distributions in turn can be controlling factors in mass transport of reactants and products to catalytic sites, as well as resistance to deactivation through pore plugging. Structuring is also strongly responsible for the "secondary particle size," defined here as the size and shape of the agglomerates of the primary ultrafine particles. In a colloidal catalyst system, where the catalyst is required to remain in a highly dispersed state in a liquid medium, the secondary particle size and the size distribution of the secondary particles are extremely important.

6A.1.1.2 Dispersion of the Active Phase

Catalyst impregnation is an ion exchange process that essentially becomes a precipitation process following the addition of an amount of the solute sufficient to approach saturation of the ion exchange sites. Thus both thermodynamic and kinetic factors play a role in determining the deposition of the active component and its distribution. Ideally, deposition will occur uniformly on the surface of the support; this optimizes the activity while minimizing the cost of the active materials. In addition, support–metal interactions may have a strong effect on desirable catalytic properties such as selectivity.

Precipitation of the salt on the surface (heterogeneous precipitation) must compete with the formation of homogeneous precipitates. Thermodynamic factors, such as the surface tensions of the various interfacial phases and the equilibrium phases of the solute–solvent system, obviously have an effect. However, the precipitation event is kinetic in nature; time-dependent occurrences such as formation of a heterogeneous nucleus of critical size and diffusion to nearby surfaces must also play a part. Thus the chemical properties of the solvent itself, as well as the rate of removal of the solvent, must play a part in determining the deposition of the active phase material.

To overcome some of the problems with conventional catalyst drying and the formation of nondispersible agglomerates, we have investigated an approach using drying at supercritical conditions that employs a nonaqueous solvent.

6A.1.2 Catalyst Preparation

In the present work we have utilized a modification of the supercritical technique in an effort to effect the deposition of an active phase upon a solid metal oxide support. Two types of catalysts were prepared with this technique. In one, molybdenum oxide was added to a Degussa alumina, an ultrafine low-density aluminum oxide, by using an aqueous or ethanol solution of ammonium heptamolybdate. For the other catalyst, a zirconia support prepared in-house, was impregnated with an ethanol solution of ammonium heptamolybdate or a zirconia–molybdena material was formed by co-precipitation and the solvent was removed by using the supercritical approach.

The alumina used was Degussa Aluminoxide C. It contains discrete spherical particles of 15–20 nm diameter and has a surface area of 110 m^2/g. The bulk density of the *as*-received alumina is approximately 0.05 g/cm^3.

Four impregnation techniques have been used in this investigation to prepare the alumina catalysts. They are as follows:

6A.1.2.1 Incipient Wetness (IW)

A salt of the active material is dissolved in distilled water and added dropwise to the catalyst support. This continues until the support is at an "incipient wetness" state, i.e., the point where unbound moisture is about to become available and the material will no longer pour. Prior to the preparation of the catalyst, the amount of water needed to reach the incipient wetness state is determined and this amount of solvent is used to add the active phase to the support. The catalyst is dried at about 120°C for 12–24 h and then ground using a mortar and pestle. The material is then calcined at 500°C for 4–6 h.

This approach introduces at least two problems. The distribution of the various species of the impregnating salt may depend upon concentration. For these situations where irreversible chemisorption may occur (e.g., low levels of $PtCl_6^{2-}$ on alumina), the salt concentration may not be a factor, but where adsorption follows, e.g., Henry's law adsorption, the solution concentration is important. In those instances of Henry's law adsorption, nucleation and precipitation of large crystals of the impregnating salt may occur as the solvent is evaporated and the concentration of the solution increases.

6A.1.2.2 Ethanol Slurry, Vacuum Dried (EtOH V)

The salt is dissolved in ethanol and a slurry is made by adding the support material while stirring, which is continued for several hours. The material is dried in a vacuum oven at approximately 100°C for 36–48 h. The dry solid is then ground and calcined at 500°C for 4–6 h.

6A.1.2.3 Ethanol Slurry, Rotary Evaporated (EtOH RV)

The preparation of this material parallels that of EtOH V except that a rotary evaporator is used to remove the solvent. This permits mixing of the suspension during solvent evaporation, which is completed more quickly (about 4 h).

6A.1.2.4 Critical Pressure Solvent Evaporation (CPSE)

The impregnated alumina catalyst is prepared as above for EtOH V. Following preparation the slurry is placed in a pressure vessel under a pressure of an inert gas that is higher than the vapor pressure of the solvent at a temperature slightly lower than the critical temperature. The slurry is then heated to a temperature that is above the critical point of the pure solvent; the presence of solids and dissolved salts may impact the actual critical temperature. Depending upon the heating rate, very rapid evaporation (a few minutes or less) of the solvent occurs as the critical point is reached. Subsequently the pressure is quickly released by opening a relief valve so that the inert gas and solvent are quickly vented. Following this the vessel is cooled and the catalyst is removed and then ground and calcined at 500°C for 4–6 h.

A 0.3 M solution (0.55 L) was prepared by dissolving anhydrous zirconium tetrachloride in water. Concentrated ammonia (2:1 = ammonia/ zirconium volume) solution was rapidly added into the vigorously stirred zirconium chloride solution to produce a suspension with a final pH of 11.3. A second sample was prepared in the same manner with the zirconium chloride solution's containing sufficient ammonium molybdate tetrahydrate to produce a material that would contain 4 wt % MoO_3. Each sample was washed seven times with 1 L distilled water; the last filtrate of each gave a negative test for chloride ion when a silver nitrate solution was added to it. The solid was then slurried in 1 L of methanol, stirred for 15 min, and then collected by filtration; this step was repeated. The resulting solid was then dispersed in 0.15 L methanol and added to 0.3 L autoclave. The sealed autoclave was heated to produce an autogenic pressure of 1620 psig (critical pressure 1170 psia). The stirrer was stopped, the heater disconnected, and the pressure released as quickly as possible without losing solid (0 psig was attained in 2 min or less). The solid was lightly ground in a mortar and pestle to produce a fluffy white powder; this is in contrast to the alumina samples, which provided visual evidence for carbon deposition following the removal of ethanol. A similar sample was prepared using ammonium metatungstate instead of ammonium molybdate to produce a material with a nominal 1.5 wt % WO_3.

Tap densities are measured by observing the volume of a known weight of the catalyst that was settled by tapping the side of the container. Nitrogen sorption is measured with a Quantachrome Autosorb 6 instrument. Prior to analysis the samples are outgassed at 100°C and <5 mtorr for at least 12 h.

Nitrogen pore size distributions are calculated by the Cohan method [23], using the Frankel–Halsey–Hill (FHH) multilayer expression with $a = 3242$ J/mol and $r = 3$. The mercury penetration plots are a composite of three pressure ranges: 0–24 psi, ambient–1,200 psi, and 1,200–60,000 psi. dV/dR distributions are obtained by a numerical differentiation of the composite penetration curves; this allows for a visual pore size distribution from hundreds of microns down to 10 nm or less. R_p, the mean pore size, was calculated from the maximum of the peak of the dV/dR distribution; the suffix e indicates values calculated from nitrogen desorption curves, f from nitrogen adsorption curves, and h from mercury penetration curves. ΣS_p values represent the surface areas obtained by adding the surface area of all pores in each distribution range. Plots of ΣS_p for all pores greater than or equal to R_p in size are useful in visualizing the discrepancies such as "negative surface areas" for certain pore size groups. ΣV_p is the corresponding additive volume of all pore sizes [23–25].

Mercury penetration is measured with a Quantachrome Autoscan instrument. Outgassing of the sample at <50 mtorr is performed at ambient temperature. Penetration is measured from ambient pressure to 60,000 psig.

A test reaction, the dehydration of 2-octanol, is utilized as a means of estimating the surface coverage of molybdena. The catalyst sample, held between quartz wool plugs in a plug flow reactor, is pretreated in air at 450°C for 21 h or reduced in flowing hydrogen for 4 h, also at 450°C. Following the calcination the catalyst is flushed with nitrogen to remove the air while cooling to reaction temperature (250°C); the reduced sample is cooled in flowing hydrogen. To provide a conversion in the 10–40%

range, 2-octanol is pumped over the catalyst. Products are analyzed for dehydration (octenes) and dehydrogenation (2-octanone) product content using gas chromatography (GC) with a DB-5 column. The selectivity for the dehydration products (1-, *cis*-2-, and *trans*-2-octenes) is obtained from the GC analysis. Previously a curve had been obtained for samples where the molybdena loading was varied to provide surface coverages of molybdena that range between 0 and 1; the dehydration selectivity is compared with this calibration curve.

6A.2   Results

Six different catalysts based upon the Degussa alumina were prepared. The results for the catalysts are compiled in Table 1. SEM examination and the data in Table 1 indicate that:

1. The IW preparation produces a material with a tap density similar to the wetted and dried alumina. The secondary particles are irregular in shape and have a wide distribution of sizes, indicating that the grinding process controls the size features.

2. The slow-dried EtOH V process leaves particles that are also irregular in shape and have a wide distribution of sizes, again indicating that grinding controls the sizes. The material containing 17.3 wt % MoO_3, corresponding to 1.5 monolayers, contains molybdena crystallites up to $10-100$ μm in size (bright spots in the SEM pictures and identified as MoO_3 by EDX analysis). It is not known whether these are "seeded" by

TABLE 1.
Chemical Composition and Physical Characteristics of Molybdena
on Alumina Catalysts

Catalyst	MoO_3 (wt %)	Tap density (g/cm^3)
Alumina, as received	–	0.07
Alumina, wetted[a]	–	0.70
EtOH V	17.3	–
EtOH V	12.5	0.34
IW	12.5	0.56
CPSE	10.8	0.005
EtOH RV	11.9	0.29

[a] Immersed in water and then dried at 120°C in air.

the support or are grown in a purely homogeneous process. The tap density of the slowly and rapidly (RV) evaporated ethanol falls between the wetted and the *as*-received alumina.

3. The secondary particles left by the CPSE process are much more uniform in size and shape, which leads one to suspect the operation of a droplet phenomenon at the critical point. The tap density of this catalyst is an order of magnitude lower than the other impregnated alumina samples and is about the same as was obtained for the *as*-received alumina (Fig. 3). The grayish-black appearance of the particles following solvent evaporation and the restoration of a white color following calcination in air indicate that some decomposition of the alcohol occurs during the heating–evaporation process.

The data in Table 2 summarize the nitrogen adsorption data and the data for mercury penetration measurements are summarized in Table 3. The impact of wetting the alumina support followed by conventional drying in air in an oven is apparent from the isotherms shown in Fig. 4. For the *as*-received alumina the isotherm is a typical Type II with essentially no

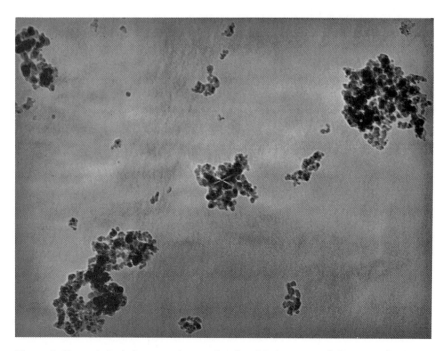

Figure 3. Transmission electron micrograph of a Mo-impregnated Degussa alumina (1 mm = 20 nm).

TABLE 2
Data from Nitrogen Adsorption Measurements

Catalyst	BET area (m²/g)	Heat of adsorption (J/mol)	ΣS_p adsorption (m²/g)	ΣV_p adsorption (cc/g)	R_p adsorption (Å)	ΣS_p desorption (m²/g)	ΣV_p desorption (cc/g)	R_p desorption (Å)
Alumina, *as-received*	98.3	3400	102.9	0.276	a	96.0	0.295	a
Alumina, wetted	111.7	2920	164 (183)[b]	0.768 (0.781)	91	161	0.762	112
EtOH V (17.3% MoO₃)	92.0	3050	94.3	0.421	150	87.3	0.391	175
EtOH V (12.5% MoO₃)	97.4	3120	108	0.568	158	100	0.589	150
IW	97.7	3090	120 (151)[b]	0.828 (0.852)[a]	115	120	0.819	150
CPSE (10.8% MoO₃)	95.1	3120	108	0.715	115	105	0.629	190
EtOH RV (11.9% MoO₃)	91.6	2830	101	0.264	a	89.0	0.311	a

[a] Not applicable, Type II isotherm.
[a] Sum excluding negatives.

TABLE 3
Data from Mercury Penetration Measurements

Catalyst	Total pore volume (cc/g)	ΣS_p (m²/g)	$\dot{R}_p$ (Å)
Alumina, *as*-received	3.88	186	118
Alumina, wetted	0.89	162	112
EtOH V (12.5% MoO₃)	2.60	160	162
IW (12.5% MoO₃)	1.33	132	188
CPSE (10.8% MoO₃)	7.30	23.8	148
EtOH RV (11.9% MoO₃)	2.17	141	148

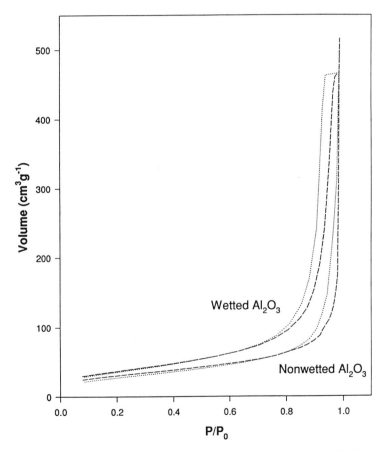

Figure 4. Adsorption isotherms (- - -) and desorption isotherms (•••) of *as*-received, nonwetted alumina and wetted alumina.

hysteresis between the adsorption and desorption isotherms. However, for the wetted–dried sample, the isotherm has a typical Type IV shape and there is a Type H1 hysteresis, which indicates the presence of pores with a narrow size distribution range. Our previous results indicate that the structure is as shown in Fig. 3 and that the particles agglomerate to produce a narrow size distribution of pore openings.

The BET surface area of this *as*-received alumina is 98 m²/g and that of the wetted–dried alumina is 117 m²/g. This difference plus the differences in the isotherms in the following description indicate that a large number of interparticle contacts are formed during the drying process. The surface area of the sample prepared by the critical pressure solvent evaporation is 95 m²/g; correction for the weight added by the molybdena gives a surface area of 107 m²/g, a value very close to that of the *as*-received alumina. The surface areas of the other materials that contain 12 wt % MoO₃ fall in the range of 95–98 m²/g (108–111 m²/g MoO₃ free basis). These values indicate that the "apparent surface area" gain is a result of nitrogen adsorption in the interparticle contact regions and this more than offsets the decrease that can be accounted for by the added weight of the molybdena.

The nitrogen adsorption–desorption isotherms for the sample prepared by the IW and the CPSE techniques are similar to those obtained for the wetted and the *as*-received alumina support isotherms, respectively. Plots of the pore size distribution obtained from the nitrogen desorption curves (Figs. 5a and 5b) show that the IW has well-defined pores with an average size of 15 nm, compared with 11.2 nm for the wetted–dried alumina sample. The CPSE sample does not produce a significant maximum in the dV/dR plot except for the very small pore size (maximum at about 3 nm); it is believed that this peak is an artifact of the Cohan cylindrical pore model. Interestingly, the surface area obtained from a summation of the areas associated with each pore size grouping equals the BET area; however, the reason for this is unexplained at present. The pore size distributions of the ethanol slurry catalysts, calculated from the nitrogen desorption curves, have a maximum at 16 and 19 nm for the EtOH V and EtOH RV samples, respectively. Of the group of catalysts, the IW material more closely resembles that of the wetted–dried alumina, and this is as expected. Moreover, the presence of the solute decreases the compacting effect of the water only slightly.

The total pore volume under 26 nm (from the nitrogen desorption data) falls in the order:

$$CPSE < EtOH\ V < EtOH\ RV < IW,$$

and this is the same order that one would place for the compaction effect of the process on the support. The mercury penetration data show that for the total pore volume:

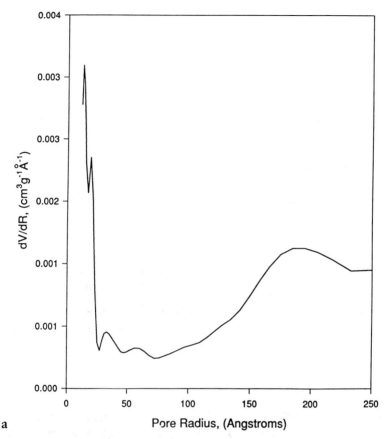

a   Pore Radius, (Angstroms)

Figure 5a. Pore volume versus pore size calculated from nitrogen desorption isotherm for *as*-received, nonwetted alumina.

$$CPSE > EtOH \, V > IW.$$

Thus decreasing the "compaction severity" of the process increases the pore volume, either through a lowering of the hydrogen bridging strength of the solvent or by increasing the rate of solvent removal and makes it available in larger pore sizes. This trend is also seen by comparing the data for the *as*-received and the wetted–dried alumina.

Mercury porosimetry has a tendency to crush loose structures and thus change the porosity that is being measured. For both CPSE and the *as*-received alumina the small dV/dR maxima indicated for pores in the 10–15 nm range is believed to be due to a compression of some of the ma-

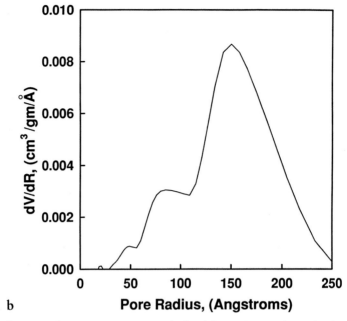

b

Figure 5b. Pore volume versus pore size calculated from nitrogen desorption isotherm for wetted alumina.

terial during the measurement. However, a comparison of the mercury penetration curves clearly shows that the step at higher penetration pressures is nearly absent for the CPSE sample but is very pronounced for the IW sample (Figs. 6a and 6b).

The samples were examined by transmission electron microscopy (TEM). The TEM micrographs showed the presence of MoO_3 crystallites in all but the IW catalysts. The size of the molybdena crystals ranged from 380 to 830 nm. X-ray diffraction data indicate that the molybdena is present predominately as the orthorhombic phase. The amount of crystalline MoO_3 was estimated from X-ray diffraction using the ratio of the MoO_3 (021) peak to that of the principal peak of the γ-Al_2O_3. Qualitatively, the conclusion is that the CPSE and the IW impregnations produce much better dispersions than the EtOH V method; furthermore, exceeding the monolayer coverage (about 12 wt %) causes the appearance of crystallites of MoO_3 for any of the preparative techniques.

Octene selectivity data depend upon the catalyst preparation procedure and upon the oxidation state of the molybdenum in the catalyst. The

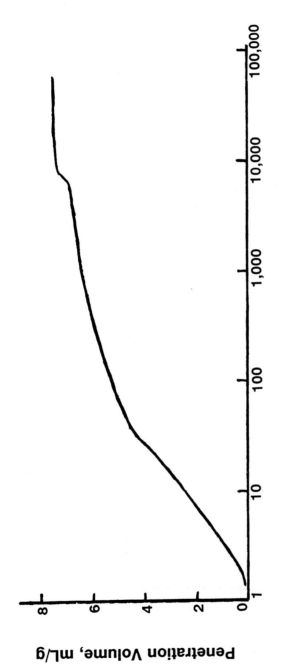

Figure 6a. Mercury penetration volume versus applied pressure for *as*-received, nonwetted alumina.

a

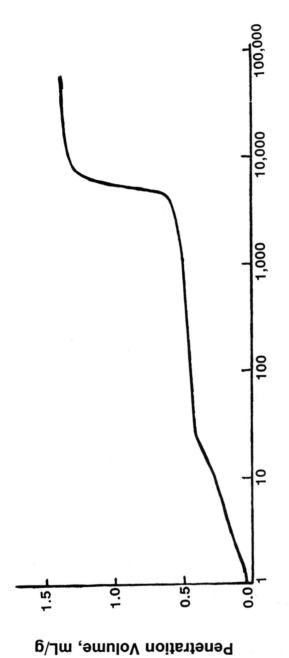

Figure 6b. Mercury penetration volume versus applied pressure for *as*-received, wetted alumina.

b

selectivity for the initial reaction period for oxidized samples is shown in Fig. 7. From this plot the coverage of the γ-Al$_2$O$_3$ by MoO$_3$ has the following ordering:

IW > EtOH V (17 wt % MoO$_3$) > EtOH V (12 wt % MoO$_3$) > EtOH RV > CPSE.

This indicates that IW produced a better dispersion than EtOH, which in turn is better than CPSE. The only way that CPSE could have a lower crystalline MoO$_3$ content and a poorer dispersion would be for the MoO$_3$ to be present in multilayers that are too small to be detected by the XRD technique. For any of the catalysts the octene product distribution shifts toward that obtained for alumina; reduction of any of the samples in hydrogen prior to the introduction of 2-octanol causes a similar shift in the octene distribution toward that of the alumina. It was demonstrated that the shift in octene distribution can be recycled during repeated periods of use and reoxidation. Thus a 12 wt % MoO$_3$ on alumina catalyst, prepared by the IW technique, showed a strong shift toward the alumina selectivity during an 8-h reaction period (Fig. 8).

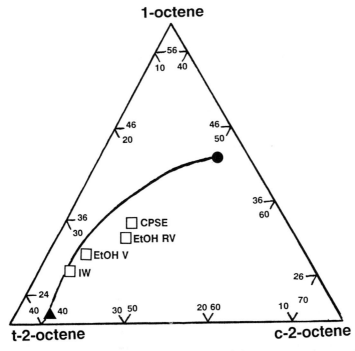

Figure 7. Octene distribution for initial sample collected during 2-octanol conversion (by MoO$_3$-Al$_2$O$_3$) catalysts (solid line is for varying Mo surface coverage, (•) is for pure Al$_2$O$_3$, and (▲) is for pure MoO$_3$).

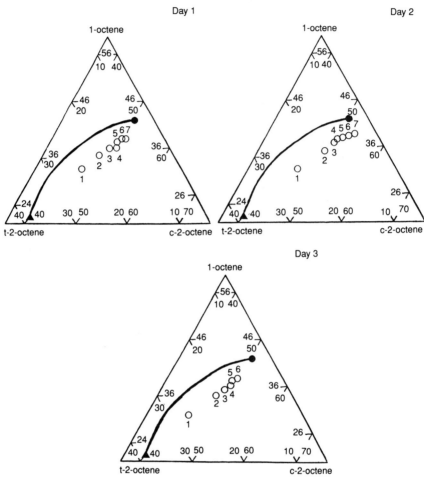

Figure 8. Octene distribution for initial sample collected during 2-octanol conversion (by IW catalyst during 8 h of operation) catalysts (solid line is for varying Mo surface coverage, (●) is for pure Al₂O₃, and (▲) is for pure MoO₃).

Treatment of the catalyst for 2 h at 300°C and then 2 h at 450°C in air returns the selectivity to that seen for the initial sample obtained on the first day, whereas the trend toward the alumina selectivity continues with catalyst use. This reversibility was repeated during a third oxidation and activity testing period. Thus it is believed that changes in the selectivity are due to the oxidation state of the molybdenum and not to a change in the dispersion of the Mo species during 2-octanol dehydration. This means that the surface coverage by MoO_3 should be based on the initial alcohol conversion data.

6A.2.1 Zirconia and Molybdena–Zirconia Catalysts

The room temperature X-ray diffraction study shows that dried hydrous zirconia powders, prepared by conventional air drying, are amorphous. However, high-temperature X-ray diffraction data show that these powders undergo a phase transition at about 450°C from an amorphous to a tetragonal phase [26, 27]. The differential thermal analysis (DTA) curve exhibits a sharp exotherm at 450°C (Fig. 9) which indicates that this phase transition is exothermic. This phase transition is also accompanied by a decrease in surface area (Fig. 10).

The surface area of the CPSE-dried zirconia was 325 m^2/g and that of the nominal 4 wt % MoO_3 was 434 m^2/g. Thus the presence of molybdena resulted in a significant increase in the surface area. The higher surface area of the Mo-containing material persisted following calcination at 500°C (111 versus 70 m^2/g) and 800°C (40 versus < 5 m^2/g). The nitrogen adsorption isotherms and the shape of the hysteresis for the samples prepared using the CPSE technique differ from those of a zirconia obtained by

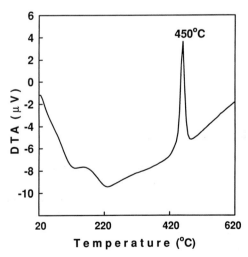

Figure 9. Differential thermal analysis curve for a sample of hydrous zirconia exhibiting an exotherm centered at 450°C.

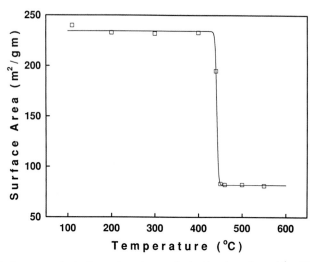

Figure 10. Surface area of a hydrous zirconia sample that has been heated for 15 min at various temperatures.

use of oven drying for water evaporation. Thus the sample prepared by the normal water evaporation exhibits a Type H2 hysteresis that is typical of a material with a narrow range of pores. The sample prepared by CPSE exhibits a hysteresis loop but it is one that is typical of a material that has a wide range of structures with a range of porosity (Fig. 11). However, as the samples prepared using the CPSE technique are calcined at higher temperatures, the shape of the isotherms and the hysteresis loop change to resemble one prepared using oven drying to remove water.

The thermal analysis data for the molybdena–zirconia sample is shown in Fig. 12. The TGA curve corresponds to 10.2% weight loss and is due to the desorption of water. A sample prepared using oven drying in air will lose 10–12 wt % and is therefore similar to that of the zirconia or molybdena–zirconia samples prepared by the CPSE technique. The DTA trace of the CPSE sample and the conventionally dried sample are very similar except for a decided endotherm that immediately precedes the exothermic event; furthermore, there is a very small but discernible weight loss that occurs during the temperature range of this small endotherm. Analysis of numerous unpromoted zirconia samples shows that an exothermic event occurs at $450 \pm 5°C$; however, when 4 wt % molybdena is incorporated into the sample, this exothermic event is shifted to a higher temperature and is centered at $482°C$. It has been shown that sulfate stabilizes the zirconia against the thermodynamically favorable transition from

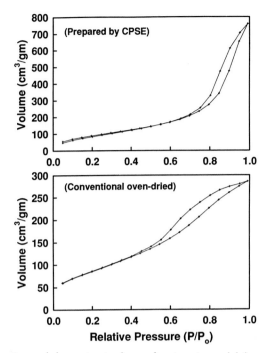

Figure 11. Adsorption and desorption isotherms for zirconia–molybdena prepared by conventional air drying and by CPSE.

the metastable tetragonal phase to the more stable monoclinic phase; furthermore, it has been demonstrated that the stabilization by the sulfate is the result of a surface phenomenon. Thus it appears that at least some of the molybdena must be present on the surface of the zirconia sample; otherwise the exothermic event would occur near 450°C.

The zirconia prepared by rapid precipitation and oven dried in air will produce predominantly the monoclinic phase of zirconia following calcination at 500°C. However, the molybdena-containing sample prepared using the CPSE technique, following calcination at either 500 or 800°C, exists predominantly (80% or greater) in the tetragonal form as shown by the XRD pattern (Fig. 13) [28, 29]. The XRD pattern of the material calcined at 500°C is similar to that of Fig. 13 except that the peaks are much broader, which indicates that much smaller crystal sizes are present in the material calcined at the lower temperature. The material that was obtained following the CPSE treatment at 205°C was amorphous to X rays.

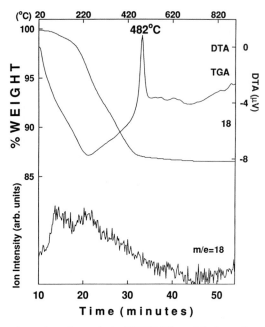

Figure 12. Thermal gravimetric analysis (TGA)/differential thermal analysis (DTA)/mass spectrometry (MS) characterization of a zirconia–molybdena sample prepared using the CPSE technique.

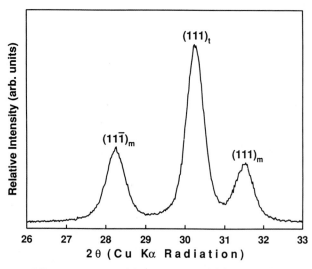

Figure 13. X-ray diffraction pattern of the zirconia–molybdena sample prepared using the CPSE technique following calcination in air for 4 h at 800°C.

The bulk density of the samples prepared using the CPSE technique were in the range of 0.25–0.3 g/mL whereas that of a sample prepared by conventional water washing followed by drying in air in an oven was about 1.0 g/mL. Thus the bulk density reflects the more compact nature of the material prepared by the conventional technique.

6A.3   Conclusions

The critical pressure solvent evaporation appears to provide a viable technique to prepare a catalytic material in which the ultrafine particles can be prevented from forming large agglomerates during the drying step. The technique has been shown useful for the preparation of a low-bulk-density, high-surface-area zirconia and for the preparation of a molybdena supported on a low-density alumina or zirconia. The low bulk density is a result of the elimination or dramatic reduction in the formation of secondary agglomerates by significantly reducing the solvent forces derived from the meniscus liquid at contact points of the ultrafine solid particles. The use of methanol appears to provide an advantage over ethanol, and presumably higher carbon number alcohols, because dehydration to produce olefins and subsequently carbon on the finished catalyst occurs to a much less extent with methanol. The technique certainly appears to have potential for the preparation of catalysts where the goal is to maintain a low bulk density of ultrafine particles that can subsequently be utilized in a highly dispersed system.

References

1. Donaldson, G. R., Pasik, L. F., and Haensel, V. (1955). *Ind. Eng. Chem. Res.*, **47**, 731.
2. Maselli, J. M., and Peters, A. (1984). *Catal. Rev.—Sci. Eng.*, **26**, 525.
3. Jager, B., Dry, M. E., Shingles, T., and Steynberg, A. P. (1990). *Catal. Lett.*, **7**, 293.
4. Beaton, W. I., and Bertolacini, R. J. (1991). *Catal. Rev. - Sci. Eng.*, **33**, 281.
5. Quarderer, G. J., and Moll, N. G. (1978). U.S. Patent 4 102 775.
6. Maatman, R. W., and Prater, C. D. (1957). *Ind. Eng. Chem.*, **49**, 253.
7. Kistler, S. S. (1932). *J. Phys. Chem.*, **36**, 52.
8. U.S. Patents 2 093 454, 2 188 007, and 2 249 767.
9. Kistler, S. S. (1931). *Nature*, **127**, 741.
10. Hunt, A. J. *Chemical Processing of Advanced Materials* (1992). (L. L. Hench and J. K. West, eds.) pp. 341–347, Wiley, New York.
11. Horn, R. G. (1990). *J. Am. Ceram. Soc.*, **73**, 1117–1135.
12. Adamson, A. W. (1982). *Physical Chemistry of Surfaces*, 4th ed., pp. 54–55, Wiley, New York.
13. Fisher, L. R., and Israelachivili, J. N. (1981). *Colloids Surf.*, **3**, 303–319.

14. Dell'Orco, P. C., Gloyna, E. F., and Buelow, S. (1993) *Super Fluid Engineering Science* (E. Kiran and J. F. Brennecke, eds.), pp. 314–325, 514.

15. Gallagher, J. S., and Haar, L. (1989) *Thermophysical Properties of Water*, National Institute of Standards and Technology (NIST) Standards, Reference Database 10.

16. Quist, A. S. (1970). *J. Phys. Chem.*, 74, 3396.

17. Marshall, W. L., and Frank, E. U. (1981). *J. Phys. Chem. Ref. Data.*, 10, 295.

18. Frank, E. U. (1970). *Pure Appl. Chem.*, 24, 13.

19. Japas, M. L., and Frank, E. U. (1985). *Ber. Bunsen-Ges. Phys. Chem.*, 89, 1268.

20. Japas, M. L., and Frank, E. U. (1985). *Ber. Bunsen-Ges. Phys. Chem.*, 89, 793.

21. Connolly, J. F. (1966). *AIChE J.*, 11, 13.

22. Marlynova, O. I., and Smirnov, O. K. (1964). *Teploenergetika*, 9, 145; from Reference 14.

23. Adkins, B. D., and Davis, B. H. (1986). *J. Phys. Chem.*, 90, 4866.

24. Adkins, B. D., and Davis, B. H. (1987). *Langmuir*, 3, 722.

25. Adkins, B. D., Reucroft, P. J., and Davis, B. H. (1986). *Ads. Sci. & Tech.*, 3, 123.

26. Srinivasan, R., Davis, B. H., Cavin, O. B., and Hubbard, C. R. (1992). *J. Am. Ceram. Soc.*, 75, 1217.

27. Srinivasan, R., Davis, B. H., Cavin, O. B., and Hubbard, C. R. (1993). *Chem. Mater.*, 5, 27.

28. Srinivasan, R., De Angelis, R. J., Ice, G., and Davis, B. H. (1991). *J. Mater. Chem.*, 6, 1287.

29. Simpson, S., and Davis, B. H. (1987). *J. Phys. Chem.*, 91, 5664.

CHAPTER 6B

Aerogel Synthesis as an Improved Method for the Preparation of Platinum-Promoted Zirconia–Sulfate Catalysts

G. Strukul,* M. Signoretto,* F. Pinna,* A. Benedetti,†
G. Cerrato,‡ and C. Morterra‡

*Department of Chemistry, University of Venice, 30123 Venice, Italy
†Department of Physical Chemistry, University of Venice, 30123 Venice, Italy
‡Department of Inorganic, Physical and Materials Chemistry, University of Turin, 10125 Turin, Italy

KEYWORDS: Zirconia sulfate, platinum catalysts, synthetic methods, sol-gel, butane isomerization

6B.1   Introduction

Sol–gel techniques for the preparation of materials with high surface area and tailored morphological properties have witnessed an impressive growth of applications over the past years in such diverse fields like optics, electronics, ceramics, etc. [1]. In a very schematic view sol–gel consists in the hydrolysis of a transition metal or a nontransition metal alcoxide (or salt) under suitable conditions to form the corresponding hydroxide (sol formation) and the subsequent polycondensation to form a thick gelatinous hydrous oxide (the gel). Both hydrolysis and condensation normally require the use of either acid or basic catalysis.

$$M(OR)_x + x\ H_2O \xrightarrow{H^+} M(OH)_x + x\ ROH. \tag{1}$$

$$n\ M(OH)_x \xrightarrow{H^+} H(O\text{-}M\text{-}O\text{-}M)_n OH + n\ H_2O. \tag{2}$$

Water and/or solvent evaporation can be performed in different manners: *in vacuo*, at atmospheric pressure, or under supercritical conditions, the last leading to the formation of the aerogel. In the last three decades the easy availability of a wide variety of alcoxide precursors to be used in Eq. 1 has greatly expanded the applicability of sol–gel methodology and of aerogel in particular. This has proved very fruitful for the preparation of oxides with large surface areas, wide-pore structure, and good mechanical stability that have been found to be very attractive for catalytic applications either as supports or as catalysts themselves. This subject has been reviewed by several authors [2–6], and some groups have begun exploring the preparation of supported transition metal catalysts via sol–gel [7–9]. In this work we shall survey in more detail our preliminary attempts [10] to apply the aerogel technique to the preparation of supported catalysts, discussing the specific case of Pt/ZrO_2–SO_4, its physical and chemical properties, and finally its catalytic activity in the isomerization of *n*-butane.

6B.2 Scope and Applications

The need to reformulate gasoline to meet a new standard that can be more compatible with environmental issues has led to an increase in the demand of branched, high-octane aliphatic hydrocarbons and of oxygenated compounds to be used as octane boosters such as MTBE or ETBE. This had led to an increasing fraction of *n*-butane being converted to the corresponding *iso*- compound (Eq. 3) and hence the necessity to carry out Eq. 3 in a selective fashion.

$$\diagup\!\!\!\diagdown\!\!\!\diagup \xrightarrow{\text{cat}} \diagup\!\!\!\diagdown\!\!\!\diagup \tag{3}$$

The current technology (e.g., the UOP or BP processes) to perform the Eq. 3 reaction [11] is based on the use of strongly acidic, chlorinated Al_2O_3 or Pt on chlorinated Al_2O_3 catalysts, with operation temperatures normally in the 250–300°C range, for which thermodynamics predicts an equilibrium more shifted to the left, thereby being intrinsically only moderately selective. Over the years the search for more-selective catalysts operating at low temperature has led to the formulation of both solid and liquid catalysts. The former, e.g., SbF_5 dispersed on either graphite or silica [12,13], are normally quite selective but rapidly deactivate under operating conditions. The latter, generally liquid superacids based on HF [14,15], are efficient and selective but difficult to separate from the reaction mixture, which leads to severe environmental problems.

Zirconia sulfate has emerged as the most promising contribution in the achievement of a new generation of catalysts for the isomerization of alkanes. Zirconia sulfate has been often referred to as a superacidic material, estimated $H_0 - 16$ [16,17], although this view has been questioned [18,19] and is particularly suited as catalyst for the Eq. 3 reaction. Its major advantages are the high selectivity towards *i*-butane formation and the ability to operate in the 100–150°C range. The disadvantage is the inability to withstand continuous flow operations for a sufficiently long time with consequent deactivation. The addition of a transition metal to improve the performance of ZrO_2-SO_4 by increasing its stability and lifetime has been practiced over the past few years. The use of Fe + Mn was first reported by Sun Company [20] and has been investigated by several authors [18,21,22]. On the other hand, the possible promotion effect of the precious metals (Ru, Os, Rh, Ir, Pd, Pt) has been also evaluated by Hino and Arata [23]. However, platinum has been by far the most widely studied promoter for the performance of ZrO_2-SO_4 following the early report of Hosoi *et al.* in 1988 [24].

Figure 1 summarizes as a flow sheet the traditional synthetic methodologies to Pt/ZrO_2-SO_4. The two methods consist of a series of similar operations and differ mainly in the order in which the various operations are performed.

Method A starts with the preparation of a hydrous zirconia gel, $Zr(OH)_4$, generally made with a conventional sol–gel technique starting from salts [25]. The first operation is a sulfation step that is generally performed by washing thoroughly at room temperature the hydrous zirconia with a sulfuric acid solution. This leads to adsorption of the sulfate anions on the surface of zirconia. Alternatively, some authors have reported the use of thermally labile sulfate salts, e.g., $(NH_4)_2SO_4$ [26], yielding final products with essentially the same features. Interestingly, Hino and Arata [27] have also tested a dry method that consists in kneading the hydrous zirconia with either ammonium sulfate or sulfur in the absence of water using a mortar and pestle. The washing procedure is accomplished either by packing a column with the zirconia and simply passing the sulfate solution through it or by immersing the zirconia in the solution, stirring, and filtering. The washing time and the concentration of the sulfate solution are variable from case to case, although the latter is always several times the amount necessary for the monolayer.

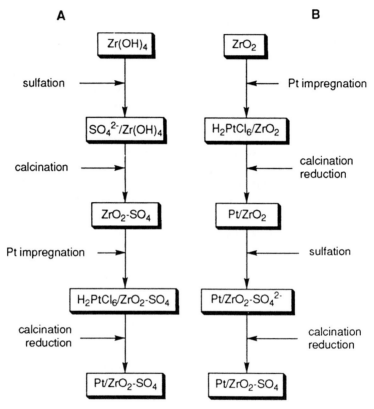

Figure 1. Flow sheet summarizing the traditional synthetic methods for the synthesis of Pt/ZrO_2-SO_4 catalysts.

The subsequent step is a calcination at temperatures normally higher than 500°C to induce acidic properties in the material. A slightly higher temperature is necessary when ammonium sulfate is employed as precursor [12]. In this step a large amount of the sulfate introduced is lost and only a minor fraction is grafted firmly on the surface. For example, a systematic study on the sulfate loss as a function of the calcination temperature (the higher the latter, the higher the former) has been reported by Nascimento *et al* [28]. So far the procedure is identical to the preparation of zirconia sulfate and a thorough review on this subject has appeared [29]. The nature of the acidic sites, especially in connection with the catalytic properties of these materials (Brønsted or Lewis type), has been the subject of much debate [e.g., 30–33]. However, because this is beyond the purpose of this review, it will not be summarized here. The most widely accepted view for the structure of surface sulfate groups is shown in Fig. 2 [29].

Once formed, zirconia sulfate is washed (Fig. 1) with a solution of a platinum complex, generally H_2PtCl_6, in a procedure that is quite similar to conventional wet impregnation. As in the preparation of traditional catalysts, the H_2PtCl_6 dispersed on zirconia sulfate is dried, calcined, and finally reduced to yield the catalyst. Calcination temperatures in the range 450–600°C and reduction temperatures around 300°C are the most common ones.

Method B (Fig. 1) is less common [28,34–37] and consists of the same elementary operations as method A, but the platination and sulfation steps are inverted. It starts with a conventional high-surface-area zirconia on which platinum is added by wet impregnation, followed by calcination and reduction to yield a traditional Pt/ZrO_2 catalyst. Then sulfate is introduced with the same methodology as in A. Finally the material is calcined to graft the sulfate and induce the acidic properties and eventually reduced to yield the Pt/ZrO_2-SO_4 catalyst.

A comparison between the two synthetic methods has been made by Baba *et al* [34]. It was found that using the same starting amounts of Pt, Zr, and SO_4^{2-}, either method (A or B) yielded catalysts with the same properties. Conversely, Hino and Arata [27] found that method B yielded catalysts with a much lower activity in butane isomerization.

The general procedure described entails the following series of criticism:

1. First of all, it is a complex, stepwise procedure. In principle this should not be a difficulty; however, it is common knowledge that in the preparation of heterogeneous catalysts the longer and more complex the procedure, the higher the number of variables that may escape a careful control and therefore alter the desired properties of the catalyst.

2. The sulfation carried out by washing with a large excess of sulfate may lead to an uncertain sulfur content. This point is critical, because it is

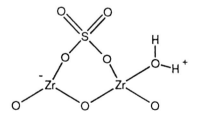

Figure 2. Structure of surface sulfate groups according to Ref. 29, evidencing the Brønsted and Lewis acid sites.

clear that most of the sulfur remains in solution. In this respect it is surprising that only a few authors [28,36,38–42] have reported the final sulfur content of their catalysts, although it is known that the catalytic properties of the catalysts depend critically on the nature and number of sulfate groups present on the surface. In most cases only the initial concentration of sulfate is given.

3. The calcination step, an essential requirement to induce the acidic properties and make an active catalyst, is carried out both without and with platinum present on the material. This will lead to different effects on sulfur. In fact, the calcination carried out without Pt results in the grafting on the surface of some of the sulfate present and in the decomposition of the remaining sulfate according to Eq. 4. On the other hand, Pt is a well-known oxidation catalyst and the oxidation of SO_2 to SO_3 on Pt is one of the oldest catalytic reactions known (the original patent by Phillips is dated 1831 [43]). The occurrence of this reaction has been reported by Dicko *et al* [44]. (Eq. 5–7) and may be responsible for a reassembly of the sulfate grafted on the surface with possible consequences on the catalytic properties.

$$SO_4^{2-} \longrightarrow SO_2 + O_2. \tag{4}$$

$$\begin{array}{c} Zr-O \\ \diagdown \\ \diagup \\ Zr-O \end{array} S \begin{array}{c} O \\ \diagdown \\ \diagup \\ O \end{array} \longrightarrow \begin{array}{c} Zr \\ \diagdown \\ \diagup \\ Zr \end{array} O + 1/2\ O_2 + SO_2. \tag{5}$$

$$PtO_2 + 2SO_2 \longrightarrow 2SO_3 + Pt^0. \tag{6}$$

$$2 \begin{array}{c} Zr \\ \diagdown \\ \diagup \\ Zr \end{array} O + PtO_2 + 2\ SO_2 \longrightarrow 2 \begin{array}{c} Zr-O \\ \diagdown \\ \diagup \\ Zr-O \end{array} S \begin{array}{c} O \\ \diagdown \\ \diagup \\ O \end{array} + Pt^0. \tag{7}$$

4. The final reduction step that in principle should lead to Pt metal may again affect the sulfate groups and Pt itself. In fact, Pt metal is also a hydrogenation catalyst, capable of reducing sulfate to hydrogen sulfide (Eq. 8). The latter has been actually observed by some authors [45,46]; it results in a further change in the number of sulfate-active sites and possibly in some contamination of Pt with H_2S.

$$SO_4^{2-} + 5\ H_2 \xrightarrow{\text{Pt}} H_2S + 4\ H_2O. \tag{8}$$

The consequence of the preceding observations is that the state and role of the S and Pt active sites in these catalysts are difficult to define and probably different from case to case. This may account for the differences (even strong ones) that have been found in catalysts that are similar, at least nominally, because made with the same synthetic procedure. Examples are found in the work of Ebitani et al [47]. and in the work of Sayari and Dicko [45]. Both groups analyzed the oxidation state of Pt in Pt/ZrO_2-SO_4 samples prepared with method A using the same spectroscopic technique (XPS). Although the catalysts were apparently similar, the conclusions were (consistently) opposite. According to Ebitani, platinum was essentially Pt(II) even after the reduction step, whereas Sayari concluded that platinum was already reduced to Pt metal after the calcination step. Analogous differences have been found also in the catalytic behavior, especially as far as the role of acidity (Brønsted/Lewis) is concerned [30–33].

These discrepancies, certainly related to a relatively uncertain preparation method, make the Pt/ZrO_2-SO_4 catalysts difficult to compare. In addition, because their mechanism of action is still far from confirmed, a rational synthetic approach aimed at tailoring the desired properties of the catalysts is difficult to design, but at least a systematic, careful control of the synthetic parameters would be highly desirable.

6B.3   Synthesis

We have applied [10] the use of the aerogel technique to the synthesis of Pt/ZrO_2-SO_4 because this reduces the number of synthetic steps and simplifies the procedure, which results in a reliable and reproducible synthetic method. Figure 3 summarizes the essential operations involved. The basic idea is to use the sulfuric acid that is necessary for the preparation of the final material also as the catalyst necessary to promote the gel formation (see Eqs. 1 and 2). A similar procedure has been reported by Ward and Ko for the preparation of zirconia sulfate [48].

Initially a solution is prepared by mixing $Zr(OPr)_4$ as precursor, an amount of sulfuric acid, a variable amount of H_2PtCl_6, and i-PrOH as the solvent. To this solution water dissolved in i-PrOH is added dropwise with stirring. This one-pot procedure leads to the formation of a zirconia gel that already contains the sulfate and Pt components of the final catalyst. As reference materials some zirconia sulfate catalysts containing different amounts of sulfur can be prepared using the same procedure.

Drying of the gel (Fig. 3) from either preparation (with Pt and without Pt) can be carried out either by evaporation in $vacuo$ at room temperature, yielding a xerogel that is subsequently calcined at 550°C (XZS

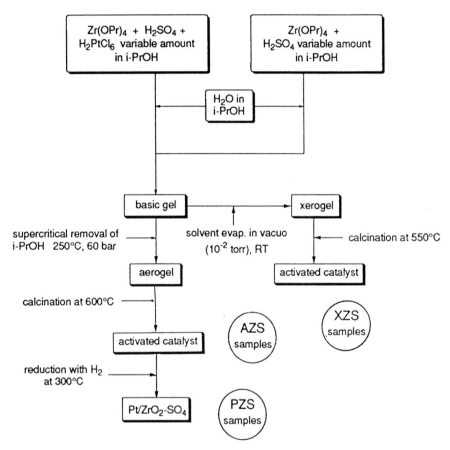

Figure 3. Flow sheet summarizing the operations involved in the synthesis of Pt/ZrO$_2$–SO$_4$ according to the new sol–gel method.

sample), or by removing *i*-PrOH under supercritical conditions (250°C, 60 bar) to give aerogels that are calcined at 600°C yielding the activated catalyst. If the material does not contain Pt (zirconia sulfate), this is termed AZS, whereas when Pt is present a subsequent reduction at 300°C is also performed to give the final Pt/ZrO$_2$–SO$_4$ catalysts (PZS samples). Aerogel formation needs no special equipment; an ordinary laboratory autoclave of sufficient capacity for the size of the experiment is perfectly adequate for this purpose.

6B.4 Analytical Properties

Some properties of the catalysts are reported in Table 1. The amount of sulfate retained can be determined by ion chromatography following dissolution of the materials [41]. Interestingly, this final sulfate content seems to be independent of the initial amount used (AZS samples), at least up to 17%, and also of both the presence and the amount of Pt of the sample (PZS samples). A parallel behavior is observed as far as the surface area is concerned that is always in the 120–140 m^2/g range. This corresponds to a surface density of about 3 SO$_4$ groups/nm^2 for most catalysts that is rather close to 4, i.e., the value estimated for the monolayer [28]. Interestingly, a surface density above the monolayer (4.5 SO$_4$ groups/nm^2 for AZS4) is observed only when loading the material with an initial amount of sulfate of 22% by weight.

TABLE 1
Analytical Data of the Samples

Sample	T calc (°C)	Nominal SO$_4$ (wt %)	Found SO$_4$ (wt %)	Pt (wt %)
AZS1	600	6.6	6.2	
AZS2	600	12.5	5.5	
AZS3	600	17	6.4	
AZS4	600	22.3	9.0	
XZS3	550	17	2.7	
PZS-A	600	12.3	6.1	0.65
PZS-B	600	12.3	6.2	1.06
PZS-C	600	11.7	5.8	5.76

A notable exception is the sample prepared by xerogel (XZS sample), where both the sulfate content and the surface area are much lower. This is typical of xerogels, where the collapse of the structure of the gel primary particles occurs when the solvent is evaporated under vacuum. The use of supercritical conditions in solvent evaporation leads to larger, porous primary particles that are also very stable during the calcination process, which results in a larger surface area in the final solid [49].

The data reported here seem to show, rather independently of the initial composition parameters, that there is a sort of "homogenizing" effect during the preparation procedure, which leads to samples with very similar surface areas and analytical features. This contrasts with some previous reports on samples where sulfate was introduced by washing with sulfuric acid [28,42,48]; in these materials the surface area is relatively constant but the final sulfate content is strongly dependent on the H_2SO_4 concentration. Interestingly, in our AZS samples a large amount of sulfate is lost during the aerogel formation. Analyses carried out prior to calcination indicated in AZS1, AZS2, and AZS4 samples a sulfate wt % of 6.4, 7.0, and 14.3, respectively.

6B.5   Structure

The crystal phase composition of the samples and some typical X-ray diffraction patterns are shown in Table 2 and Fig. 4, respectively. As is clear, with the exception of samples AZS1 and XZS3, the crystal phase composition of the catalysts seems unaffected by the calcination treatment independently on whether Pt is present or not, the only effect of calcination, if any, being a moderate increase in the crystalline particle size. Moreover, the increase in sulfuric acid amount, a factor that is known to increase the stability of the tetragonal phase [50] results (Table 2) in the presence of increasing monoclinic phase. These observations are in sharp contrast with the findings of Ward and Ko [48], whose zirconia sulfate samples were also prepared by aerogel technique following a procedure very similar to the present one. These authors found that independently on the initial amount of sulfuric acid used, all samples were amorphous at least up to a calcination temperature of 500°C and that the temperature of amorphous–tetragonal crystal phase transition increased with increasing amount of sulfuric acid used. The results of Ward and Ko are in all cases similar to those found by most authors utilizing the washing procedure to introduce sulfate on zirconia.

The differences here outlined are probably to be ascribed to the solvent evaporation conditions. The use of *i*-PrOH as supercritical fluid, very common in sol–gel technology, requires removal at relatively high temperature and pressure (Fig. 3), at variance with CO_2, the fluid used by

TABLE 2

Crystal phase composition of the samples and Pt particle size after calcination

Sample	Crystal phase		Pt particle size (Å)
	Before calcination	After calcination	
AZS1	A[a]	T[b]	
AZS2	T	T	
AZS3	T + 31% M[c,d]	T + 26% M[c,d]	
AZS4	T + 74% M[c,d]	T + 69% M[c,d]	
XZS3	A	T	
PZS-A	T	T	65
PZS-B	T	T	103
PZS-C	T	T	154

[a]Amorphous.
[b]Tetragonal.
[c]Monoclinic.
[d]Crystal phase composition calculated according to Toraya, H., Yoshimura, M., and Somiya, S. (1984). *J. Am. Ceram. Soc.*, **67**, C119.

Ward and Ko that can be removed supercritically at 70°C and atmospheric pressure [48]. In ZrO_2 when sulfuric acid is present, the amorphous–tetragonal transition generally occurs at temperatures above 500°C [48,51,52]. The pressure (60 bar) seems, therefore, to be the critical factor to induce crystallinity in the material during the aerogel formation in the present case, because one would normally expect higher pressure to favor the denser crystalline phase.

If we now consider the Pt region of the X-ray diffractogram (Fig. 5), Pt metal formation is already evident during aerogel preparation, particularly as the Pt content of the material increases (upper curves in the individual figures). Pt metal formation is complete after calcination (lower curves), and further reduction of the materials as indicated in the preparation procedure (Fig. 3) does not change the X-ray diffractograms. The complete pattern for PZS-C, including the peaks pertaining to zirconia, is shown in Fig. 5 and is particularly illustrative of the moderate structural changes occurring in the materials after calcination as compared with the aerogels themselves. The Pt particle size estimated by X-ray analysis according to Enzo *et al.* [53] after the calcination step is reported in Table 2.

The preceding observations are confirmed by TPR experiments. Figure 6 reports a comparison among AZS2, PZS-B, and an ordinary 1% Pt/ZrO_2 sample, all calcined at 600°C prior to the experiment. As can be seen, the

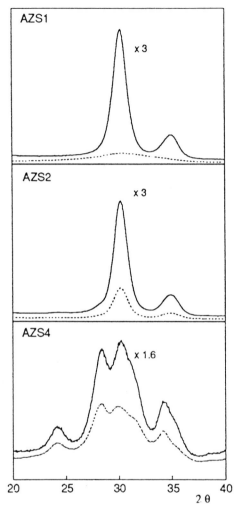

Figure 4. X-ray diffraction patterns relative to AZS samples evidencing the crystal phase composition. Dotted line: prior to calcination; solid line: after calcination. Scaling factors to avoid superimposition are indicated on the individual curves.

Pt/ZrO$_2$ sample exhibits only one reduction band of moderate intensity centered at 230°C typical of PtO [54]. On the other hand, AZS2 shows a sharp, intense band at 650°C, and an analysis of the off-gas with mass spectrometry reveals the presence of extensive amounts of SO$_2$ and traces of H$_2$S. Finally, the PZS-B sample exhibits neither of the previous bands but only one broad, strong band centered at 480°C consisting of H$_2$S as detected by mass spectrometry.

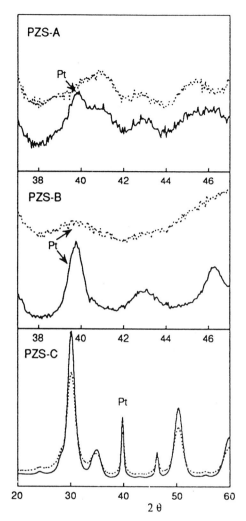

Figure 5. X-ray diffraction patterns relative to PZS samples evidencing the Pt 111 reflection (PZS-A and PZS-B). For PZS-C, the whole range is reported showing also the pattern of zirconia. Dotted line: prior to calcination; solid line: after calcination.

All the structural information reported so far can be accounted for in the following interpretation that accounts for the genesis of the catalysts starting from the gel formed in the one-pot procedure.

First of all, the gel is constituted by hydrous zirconia that contains sulfate both dissolved as H_2SO_4 in *i*-PrOH, and possibly, as suggested by Ward and Ko [48], as bulk sulfate trapped in the zirconia matrix by co-

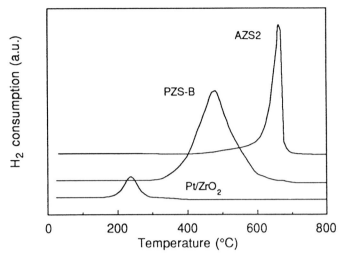

Figure 6. TPR profiles relative to AZS2, PZS-B, and a 1% Pt/ZrO$_2$ sample.

gelation. At this stage, due to the strongly acidic conditions used, Pt is still present as H$_2$PtCl$_6$.

Solvent evaporation under supercritical conditions (Scheme 1) leads to reorganization of the zirconia structure and (in most cases) crystallization in the appropriate crystal phase composition, possible expulsion on the surface of the sulfur oxide present in the bulk [48], decomposition of much of the excess sulfate, and partial grafting of the remaining sulfate on the zirconia surface. At the same time, at 250°C and under 60 bar of *i*-PrOH vapor, much of the H$_2$PtCl$_6$ present is reduced to Pt metal. In Scheme 1, *i*-PrOH is indicated as the reducing agent. One may think of water as doing the same role. However, the amount of H$_2$O introduced into the system during the preparation is only the stoichiometric one and is consumed during sol formation. Although some water is formed during the condensation reaction (Eq. 2), the amount of *i*-PrOH present as solvent is much higher.

The calcination at 600°C results (Scheme 1) in the definite grafting of sulfate on zirconia to cover approximately three-fourths of the surface available, the final decomposition of the excess sulfate, and transient formation of PtO both on the surface of the already formed Pt metal and as bulk oxide from the still-present H$_2$PtCl$_6$. Oxidation of sulfur dioxide (formed from the decomposition of sulfate) by PtO leads to the complete reduction of platinum to Pt metal as suggested also by Sayari and Dicko [45].

Finally, the reduction step (Scheme 1) consists merely in the catalytic hydrogenation of SO$_4$ to H$_2$S performed by Pt and is therefore unnecessary for the formation of the catalytically active material. Indeed, as will be evi-

solvent evaporation:

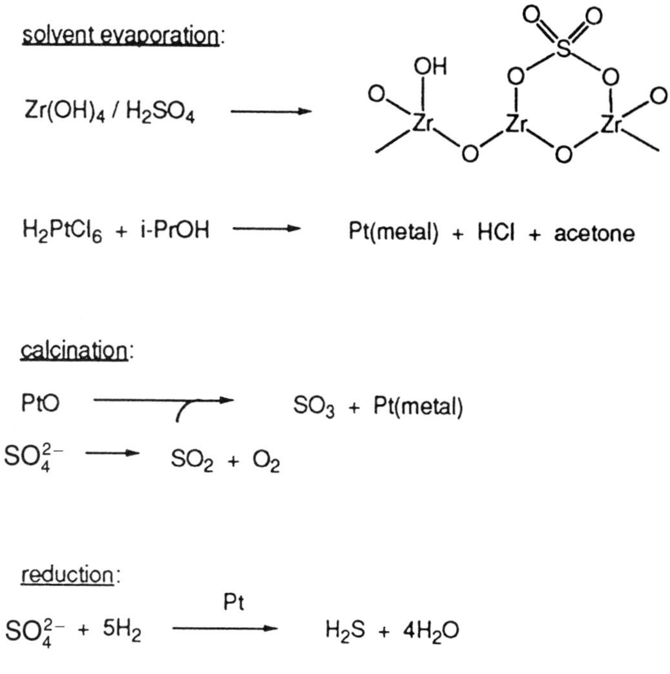

$Zr(OH)_4 / H_2SO_4 \longrightarrow$

$H_2PtCl_6 + i\text{-PrOH} \longrightarrow Pt(metal) + HCl + acetone$

calcination:

$PtO \longrightarrow SO_3 + Pt(metal)$

$SO_4^{2-} \longrightarrow SO_2 + O_2$

reduction:

$SO_4^{2-} + 5H_2 \xrightarrow{\;Pt\;} H_2S + 4H_2O$

Scheme 1

dent in the discussion of the catalytic activity (*vide infra*), the reduction step is detrimental because it results in a partial deactivation of the catalysts, probably due to surface contamination with H_2S.

6B.6 Nature of the Active Sites

An investigation of the nature of sulfate groups in the catalysts reported here has been carried out with IR spectroscopy. Some typical spectra are reported in Fig. 7, where the SO stretching region, the spectrum of adsorbed pyridine, and the CO stretching region after addition of carbon monoxide are shown. The spectra reported in Figs. 7(A) and (B) are very similar to those reported for zirconia sulfate made with the traditional synthetic method [55,56], and lead to similar interpretation. They indicate the presence of surface sulfates in a highly covalent form [Fig. 7(A)], similarly to organic sulfonates and sulfates [57]. The addition of pyridine [Fig. 7(B)] evidences a high concentration of strong (aprotic) Lewis acid sites and a fairly low (though never null) concentration of (protonic) Brønsted acid sites.

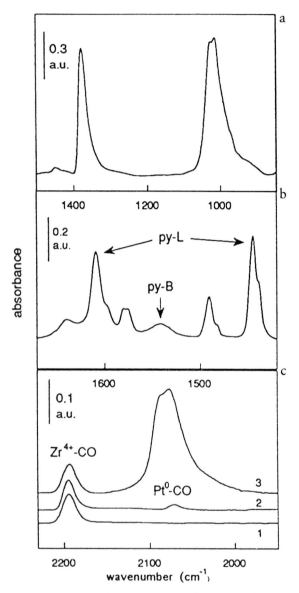

Figure 7. Typical IR spectra of PZS samples. (a) sulfate region, (b) spectrum of adsorbed pyridine showing the interactions with Brønsted and Lewis acid sites; (c) spectrum of adsorbed CO on (1) catalyst activated *in vacuo*; (2) catalyst activated in air; (3) 1% Pt/ZrO$_2$.

Interestingly, the CO stretching region, after the addition of carbon monoxide on the catalysts activated *in vacuo* [Fig. 7(C, 1)], indicates a weak adsorption on Zr^{4+} centers but no adsorption on Pt, as is evidenced by the comparison with an ordinary Pt/ZrO_2 sample containing the same amount of platinum [Fig. 7(C, 3)]. Treatment with air prior to admission of CO shows only a very weak adsorption capacity of the Pt surface [Fig. 7(C, 2)].

These results are surprising, because the PZS samples show regular hydrogen chemisorption capacity. In other words, the Pt surface is hindered by species that inhibit the adsorption of CO but do not inhibit the adsorption of H_2. Because under the conditions used for the experiments the presence of H_2S on the Pt surface can be excluded, it seems reasonable to conclude that the species hindering the Pt surface is sulfate which, as shown by the analytical data in Table 1, covers about 3/4 of the available ZrO_2 surface. This view is supported by the observation that when the hydrogen chemisorbed on Pt is titrated with oxygen, the amount of O_2 consumed is much less than would be required, which indicates that much of the H_2 adsorbed by the catalyst is not located on Pt and is most likely spilled on zirconia sulfate [58].

6B.7 Catalytic Activity

The catalytic activity of the PZS samples was tested in the isomerization of butane by using a continuous flow reactor and by adding hydrogen to the feed. The beneficial use of hydrogen in this reaction both with and without Pt present in the catalyst was first observed by Garin *et al.* [59] and is now com-

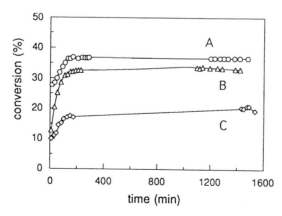

Figure 8. Conversion versus time plot for the isomerization of butane at 250°C with PZS-2 catalyst. Curves refer to different activation conditions. (A) activation in air at 450°C for 2 h; (B) activation in air at 450°C for 2 h, then in H_2 for 2h; (C) activation in H_2 for 2 h.

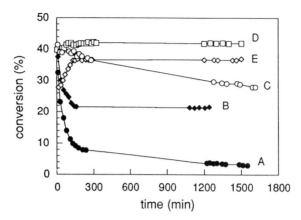

Figure 9. Effect of hydrogen on the activity of AZS2 and PZS-B in the isomerization of butane at 250°C. Catalysts activated in air for 2 h AZS2: (A) H_2/butane 1:1; (B) H_2/butane 6:1; PZS-B: (C) H_2/butane 1:1; (D) H_2/butane 3:1; (E) H_2/butane 6:1.

mon practice in reactions involving this class of catalysts. In Fig. 8 a conversion versus time plot is reported for PZS-B activated under different conditions prior to reaction. Clearly the use of hydrogen in the activation procedure results in a loss of catalytic activity, in agreement with the observations reported in this chapter that the catalyst is already completely formed after calcination and that the reduction operation may be detrimental, leading to formation of hydrogen sulfide with possible surface contamination.

A comparison between AZS2 and PZS-B is shown in Fig. 9. The two catalysts were tested in the presence of hydrogen and both were activated in air prior to the reaction. As can be seen, they display a sharp difference in activity, particularly at a low H_2/butane ratio (1:1), although for both catalysts a slow decline in conversion is observed for long reaction times (curves A and C). The deactivation can be stopped by increasing the H_2/butane ratio in the feed (6:1) and even in this case (curves B and E) the Pt-containing sample is significantly more active.

As is clear from Fig. 9, the effect of the hydrogen pressure in the feed for PZS-B is not a simple one, as the optimum activity is observed for intermediate values (3:1) of the H_2/butane ratio. Under these conditions a conversion higher than 40% is observed and the activity of the catalyst is constant for at least 2 days.

It is interesting to notice that PZS catalysts display the same steady state catalytic activity independently of the Pt content [58]. This seems to demonstrate that platinum is mainly a promoter in maximizing the performance of the catalysts (the role of Pt in this respect is largely unknown), whereas the activity depends essentially on sulfate (the same for all PZS samples) and hence on the acidity induced on zirconia. The fundamental role of Lewis centers on the activity of zirconia sulfate catalysts has been already demonstrated by us [31,56] and later confirmed by other authors [60] also for platinum-containing samples.

6B.8   Conclusions

The example reported here of application of sol–gel technique to the synthesis of catalysts suggests a few final considerations, some of which are specific to the particular case chosen whereas others seem relevant for a more general applicability of the method.

1. The aerogel synthetic method reported seems to represent a simple, easily accessible synthetic procedure for the preparation of Pt/ZrO_2-SO_4 catalysts, because everything can be done in one pot without the use of any special equipment. This is indeed a simplification with respect to the more traditional synthetic method and allows the overcoming of many of the drawbacks we have outlined.

2. The specific conditions used for solvent evaporation, according to the aerogel synthetic procedure, are very stringent with respect to the determination of the analytical properties of the materials. This leads to a "homogenizing" effect with the removal of the "excess" sulfate and yields a constant surface concentration. In fact, the materials prepared have essentially the same amount of sulfate and the same surface area independently of the presence of platinum and the starting amount of sulfate.

3. The morphology of the materials is also predetermined by the solvent evaporation procedure. The crystal phase composition, the surface area, and the particle size of zirconia are essentially the same both before and after calcination and independently of the presence of platinum.

4. Platinum is partly reduced during the solvent evaporation step, due to the reducing properties of *i*-PrOH, and completely reduced after calcination so that the reduction step is not only unnecessary but even detrimental for the catalytic activity, probably because of hydrogen sulfide formation.

5. Indeed, as they come off the autoclave, the materials are almost ready to use and their final properties in the catalytic isomerization of butane are excellent both in terms of activity and stability.

Some of the points mentioned probably may be generalized—in particular, the possibility of preparing supported metallic catalysts in a single stage through the use of a solvent that may act as reductant. In this respect the combination of high pressure and high temperature necessary for the aerogel formation seems effective when *i*-PrOH is used, at least with easily reducible metals.

Acknowledgments

Thanks are expressed to MURST (Rome) and CNR (Rome) for financial support of this work.

References

1. Brinker, C. J., Scherer, G. W. (1990). *Sol–Gel Science: The Physics and Chemistry of Sol–Gel Processing*, Academic Press, New York.
2. Gesser, H. D., and Goswami, P. C. (1989). *Chem. Rev.*, **89**, 765.
3. Fricke, J. (1992). *J. Non-Cryst. Solids*, **147–148**, 356.
4. Pajonk, G. M. (1991). *Appl. Catal.*, **72**, 217.
5. Ko, E. I. (1993). *Chem–tech*, 31.
6. Teichner, S. J. (1989). *Rev. Phys. Appl.*, **24**, C4-1.
7. Bosch, P., Lopez, T., Lara, V.-H., and Gomez, R. (1993). *J. Mol. Catal.*, **80**, 299.
8. Azomoza, M., Lopez, T., Gomez, R., and Gonzalez, R. D. (1992). *Catal. Today*, **15**, 547; Balakrishnan, K., and Gonzalez, R. D. (1993). *J. Catal.*, **144**, 395; Zau, W., and Gonzalez, R. D. (1993). *Appl. Catal., A*, **102**, 181; Zou, W., and Gonzalez, R. D. (1995). *Appl. Catal., A*, **126**, 351.
9. Cauqui, M. A., and Rodriguez-Izquierdo, J. M. (1992). *J. Non-Cryst. Solids*, **147–148**, 724 and references therein.
10. Signoretto, M., Pinna, F., Strukul, G., Cerrato, G., and Morterra, C. (1996). *Catal Lett.*, **36**, 129.
11. Ware, K. J., and Richardson, A. H. (1972). *Hydrocarbon Process.*, **51**, 161.
12. Tanabe, K., Hattori, H., and Yamaguchi, T. (1990). *Critical Rev. Surf. Chem.*, **1**, 1.
13. Hattori, H., Tahakashi, O., Takagi, M., and Tanabe, K. (1981). *J. Catal.*, **68**, 132.
14. Olah, G. A., Prakash, G. K. S., and Sommer, J. (1985). *Superacids*, p. 53, Wiley-Interscience, New York.
15. Olah, G. A., Farooq, O., Husain, A., Ding, N., Trivedi, N. J., and Olah, J. A. (1991). *Catal. Lett.*, **10**, 239.
16. Hino, M., Kobayashi, S., and Arata, K. (1979). *J. Am Chem Soc.*, **101**, 6439.
17. Hino, M., and Arata, K. (1980). *J. Chem. Soc. Chem. Commun.*, 851.
18. Adeeva, V., de Haan, J. W., Jänchen, J., Lei, G. D., Schünemann, V., van de Ven, L. J. M., Sachtler, W. M. H., and van Santen, R. A. (1995). *J. Catal.*, **151**, 364.
19. Kustov, L. M., Kazansky, V. B., Figueras, F., and Tichit, D. (1994). *J. Catal.*, **150**, 143.
20. Hollstein, E. J., Wei, J. T., Hsu, C.-Y. (1990). U. S. Patent, 4 918 041.
21. Hsu, C.-Y., Heimbruch, C. R., Armes, C. T., and Gates, B. C. (1992). *J. Chem. Soc. Chem. Commun.*, 1645; Cheung, T.-K., d'Itri, J. L., and Gates, B. C. (1995). *J. Catal.*, **151**, 464.
22. Adeeva, V., Lei, G. D., and Sachtler, W. M. H. (1994). *Appl. Catal., A*, **118**, L-11.
23. Hino, M., and Arata, K. (1995). *Catal. Lett.*, **30**, 25.
24. Hosoi, T., Shimidsu, T., Itoh, S., Baba, S., Takaoka, H., Imai, T., and Yokoyama, N. (1988). *ACS Div. Petr. Chem. Prepr.*, **33**, 562; Yamaguchi, T. (1990). *Appl. Catal.*, **61**, 1; Arata, K. (1990). *Adv. Catal.*, **37**, 165.
25. See for example: Mercera, P. D. L., Van Ommen, J. G., Doesburg, E. B. M., Burggraaf, A. J., and Ross, J. R. H. (1991). *Appl. Catal.*, **78**, 79; Benedetti, A., Fagherazzi, G., Pinna, F., and Polizzi, S. (1990). *J. Mater Sci.*, **25**, 1473; Srinivasan, R., Hubbard, C. R., Cavin, O. B., and Davis, B. H. (1993). *Chem. Mater.*, **5**, 27; Srinivasan, R., and Davis, B. H. (1992). *Catal. Lett.*, **14**, 165.
26. See for example: Batamack, P., Bucsi, I., Molnar, A., and Olah, G. A. (1994). *Catal. Lett.* **25**, 11; Waqif, M., Bachelier, J., Saur, O., and Lavalley, J.-C. (1992). *J. Mol. Catal.*, **72**, 127; Yamaguchi, T. (1990). *Appl. Catal.*, **61**,. 1.

27. Hino, M., and Arata, K. (1995). *J. Chem. Soc. Chem. Commun.*, 789.
28. Nascimento, P., Akratopoulou, C., Oszagyani, M., Coudourier, G., Travers, C., Joly, J.-F., and Vedrine, J. C. (1993). *New Frontiers in Catalysis*, Part B (L. Guczi, F. Solymosi, and P. Tetenyi, (eds.), p. 1185, Akademiai Kiadò, Budapest.
29. Davis, B. H., Keogh, R. A., and Srinivasan, R. (1994). *Catal. Today*, **20**, 219.
30. Chen, F. R., Coudourier, G., Joly, J.-F., and Vedrine, J. C. (1993). *J. Catal.*, **143**, 616.
31. Pinna, F., Signoretto, M., Strukul, G., Cerrato, G., and Morterra, C. (1994). *Catal. Lett.*, **26**, 339.
32. Comelli, R. A., Vera, C. R., and Parera, J. M. (1995). *J. Catal.*, **151**, 96.
33. Babou, F., Bigot, B., and Sautet, P. (1993). *J. Phys. Chem.*, **97**, 11501.
34. Baba, S., Shibata, Y., Takaoka, H., Kimura, T., and Takasaka, K. (1986). Jap. Patent, 61 153 140.
35. Wen, M. Y., Wender, I., and Tierney, J. W. (1990). *Energy Fuels*, **4**, 372.
36. Iglesia, E., Soled, S., and Kramer, G. M. (1993). *J. Catal.*, **144**, 238.
37. Soled, S., Iglesia, E., and Kramer, G. M. (1994). *Stud. Surf. Sci. Catal.*, Vol. 90, p. 507, Elsevier, Amsterdam.
38. Corma, A., Martinez, A., and Martinez, C. (1994). *J. Catal.*, **149**, 52.
39. Bensitel, M., Saur, O., Lavalley, J. -C., and Mabilon, G. (1987). *Mater. Chem. Phys.*, **17**, 249.
40. Ebitani, K., Konishi, J., and Hattori, H. (1991). *J. Catal.*, **130**, 257.
41. Sarzanini, C., Sacchero, G., Pinna, F., Signoretto, M., Cerrato, G., and Morterra, C. (1995). *J. Mater Chem.*, **5**, 353.
42. Chokkaram, S., Srinivasan, R., Milburn, D. R., and Davis, B. H. (1994). *J. Colloid Interface Sci.*, **165**, 160.
43. Donovan, J. R., and Salamone, J. M. (1983). *Kirk-Othmer Encyclopedia of Chemical Technology*, 3rd ed., Vol. 22, p. 190, Wiley-Interscience, New York.
44. Dicko, A., Song, X., Adnot, A., and Sayari, A. (1994). *J. Catal.*, **150**, 254.
45. Sayari, A., and Dicko, A. (1994). *J. Catal.*, **145**, 561.
46. Ng, F. T. T., and Horvat, N. (1995). *Appl. Catal.*, *A*, **123**, L197.
47. Ebitani, K., Konno, H., Tanaka, T., and Hattori, H. (1992). *J. Catal.*, **135**, 60.
48. Ward, D. A., and Ko, E. I. (1994). *J. Catal.*, **150**, 18.
49. Brinker, C. J., and Scherer, G. W. (1990). *Sol–Gel Science: The Physics and Chemistry of Sol–Gel Processing*, Academic Press, New York.
50. Yamaguchi, T., Tanabe, K., and Kung, Y. C. (1986). *Mater. Chem. Phys.*, **16**, 67; Srinivasan, R., Taulbee, D., and Davis, B. H. (1991). *Catal. Lett.*, **9**, 1; Parera, J. M. (1992). *Catal. Today*, **15**, 481 and references therein.
51. Norman, C. J., Goulding, P. A., and McAlpine, I. (1994). *Catal. Today*, **20**, 313.
52. Srinivasan, R., Keogh, R. A., Milburn, D. R., and Davis, B. H. (1995). *J. Catal.*, **153**, 123.
53. Enzo, S., Polizzi, S., and Benedetti, A. (1985). *Z. Kristallogr.*, **170**, 275.
54. Hurst, N. W., Gentry, S. J., Jones, A., and McNicol, B. D. (1982). *Catal. Rev.—Sci. Eng.*, **24**, 233.
55. Bensitel, M., Saur, O., Lavalley, J. C., and Morrow, B. A. (1988). *Mater. Chem. Phys.*, **19**, 147.
56. Morterra, C., Cerrato, G., Pinna, F., Signoretto, M., and Strukul, G. (1994). *J. Catal.*, **149**, 181.
57. Bellamy, L. J. (1980). *The Infrared Spectra of Complex Molecules*, Vol. 2, p. 225, Chapman and Hall, London.
58. Chies, P., Signoretto, M., Pinna, F., Strukul, G., Cerrato, G., and Morterra, C. *J. Catal.*, submitted.
59. Garin, F., Andriamasinoro, D., Abdulsamad, A., and Sommer, J. (1991). *J. Catal.*, **131**, 199.
60. Keogh, R. A., Srinivasan, R., and Davis, B. H. (1995). *J. Catal.*, **151**, 292.

CHAPTER 7

Surfactant-Stabilized Nanosized Colloidal Metals and Alloys as Catalyst Precursors

Helmut Bönnemann and Werner Brijoux
Max-Planck-Institut für Kohlenforschung
45466 Mülheim an der Ruhr, Germany

KEYWORDS: Catalyst precursor, colloidal alloy, colloidal metal, hydrosol, lipophilic surfactant, nanometal powder, organosol

7.1 Introduction

The chemistry of colloidal metal particles began with Michael Faraday's synthesis of the ruby-red gold colloids, gained momentum with Wilhelm Ostwald who in 1915 coined the term of "the world of neglected dimensions," and obtained its present character in the early 1980s when chemists rediscovered the field and physicists provided a plethora of the most powerful analytical tools that for the first time allowed the comprehensive characterization of these nanostructured materials. The present state of the art is reviewed in Günter Schmid's book *Clusters and Colloids* [1]. The object of this chapter is to summarize some new routes toward the preparation of nanoscale metal and metal–alloy colloids via chemical reduction utilizing either surfactants or solvents as the stabilizing agents. In addition, the application of these nanostructured materials as precursors for a new type of catalyst will be discussed. The reduction of transition metal salts and oxides by using alkali hydrotriorganoborates in organic media in the absence of stabilizers at ambient temperature leads to X-ray amorphous nanopowders of metals and alloys [2]. The reduction of ether and thioether adducts of early transition metal halides, with $K(BEt_3H)$, gives isolatable organosols of zero-valent Ti, Zr, V, Nb, and Mn stabilized by ether or thioether molecules, respectively. The use of tetraalkylammonium hydrotriorganoborates as reducing agents leads to mono- and bimetallic organosols. Pretreatment of transition metal halides with various hydrophilic surfactant types prior to the reduction using conventional agents (e.g., H_2 or HCOOH) opens an easy access to nanostructured mono- and bimetallic hydrosols of Group 6–11 metals. This synthesis can be performed even in water. Mono- and bimetallic hydrosols of this type serve as effective precursors for heterogeneous metal catalysts for the selective hydrogenation and oxidation of organic substrates.

7.2 Survey and Key to the Literature

Due to their catalytic potential, colloidal mono- and bimetallic materials have, after a period of hibernation, attracted more attention, notably through the contributions of Boutonnet et al. [3,4], Bradley et al. [5–7], Braunstein [8], Esumi et al. [9–12], Evans et al. [14–16], Heaton [19], Henglein [22], Klabunde et al. [24–27], Knözinger [31], Larpent and Patin [32], Lewis et al. [33–36], Moiseev et al. [37,38], Schmid et al. [44–47,52], and Toshima et al. [58–62]. Some key publications of the major contributors to the field have been collected here [3–62]. An important progress consists of the preparation of water soluble nanoclusters using hydrophilic P- or N-donors as stabilizer [63–67]. Further miscellaneous agents have been used for this purpose [68–85]. The first nanosized metals stabilized by surfactants were reported in 1976 by Lisichkin et al. [103] and in 1979 by Kiwi and Grätzel [97]. In the course of our research in this area, we have developed new methods for the preparation of stable and very soluble metal organo- and hydrosols having a narrow particle size distribution. The metallic core derived from elements of the periodic table Group 6–11 is protected by lipophilic and hydrophilic surfactant molecules [85,88–96]. Reetz and co-workers have found an alternative electrochemical access to this class of nanoparticles [106,107,110]. In the early transition metal series (e.g., Ti, Zr, Nb, and Mn) the stabilization is achieved by THF (tetrahydrofurane) or the corresponding thioether to give very soluble organosols [87]. Colloidal metal systems stabilized either by surfactants or by solvents are very monodisperse and have been applied successfully as precursors to

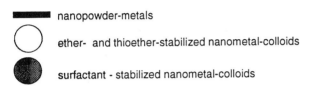

Ti	V	Cr	Mn	Fe	Co	Ni	Cu
Zr	Nb	Mo		Ru	Rh	Pd	Ag
Hf	Ta	W	Re	Os	Ir	Pt	Au

■■■ nanopowder-metals

◯ ether- and thioether-stabilized nanometal-colloids

⬤ surfactant - stabilized nanometal-colloids

Figure 1. Survey of powderous or colloidal nanometals prepared via Eqs. 2–6.

heterogeneous catalysts [86,88,89,92]. Electron microscopy has confirmed that the discrete metal particles may be deposited onto supports without any unwanted agglomeration. This is a major prerequisite for size-selective studies in heterogeneous catalysis (Fig. 1).

7.3   Results and Discussion

7.3.1 X-ray Amorphous Metal Nanopowders

Metal salts of Groups 6–12 and $SnCl_2$ may be reduced using alkali hydrotriorganoborates in hydrocarbons between -20 and $80°C$ to give boron-free powderous metals. According to X-ray diffraction the particles are nearly amorphous. The particle size of the powders according to transmission electron microscopy (TEM) is between 1 and 100 nm, depending on the metal. By simple co-reduction of suspended metal salts, binary or ternary alloys and intermetallic compounds were obtained [62] (Fig. 2).

$$u\,MX_v + v\,M'(BEt_3H)_u \xrightarrow{\text{THF}} u\,M\downarrow + v\,M'X_u + uv\,BEt_3 + uv/2\,H_2\uparrow, \quad (1)$$

where X = halogen.

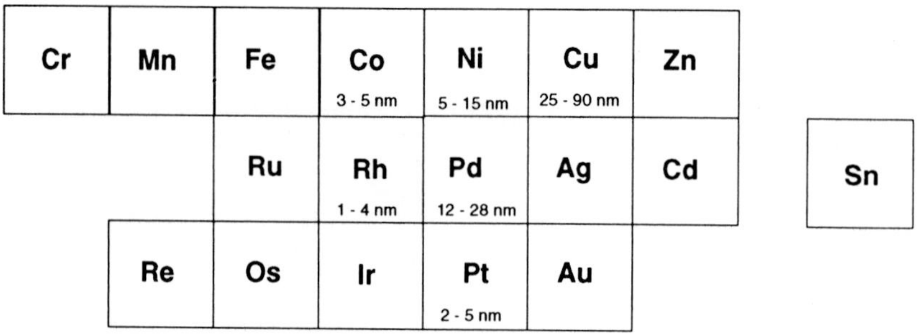

Cr	Mn	Fe	Co	Ni	Cu	Zn		
			3 - 5 nm	5 - 15 nm	25 - 90 nm			
		Ru	Rh	Pd	Ag	Cd		Sn
			1 - 4 nm	12 - 28 nm				
	Re	Os	Ir	Pt	Au			
				2 - 5 nm				

Figure 2. X-ray amorphous nanopowders and alloys.

7.3.2 Ether-Stabilized Organosols of Ti(0), Zr(0), V(0), Nb(0), and Mn(0)

The reduction of $TiCl_4 \cdot 2THF$ or $TiCl_3 \cdot 3THF$ in THF with $K(BEt_3H)$ gives colloidal titanium stabilized by THF. After evaporating the solvent and BEt_3 and thoroughly drying *in vacuo*, a black pyrophoric powder (containing small amounts of KCl) is obtained. Protonolysis and cross experiments using $K(BEt_3D)$ as the reducing agent indicated the presence of residual hydrogen in the resulting Ti colloid [87]. In contrast, the reduction of $TiBr_4 \cdot 2$ THF **1** according to Eq. 2 yields a hydrogen-free colloidal $(Ti \cdot 0.5$ THF$)_x$ **2** consisting of 44% Ti and 1% KBr.

$$x \cdot (TiBr_4 \cdot 2\ THF) + x \cdot 4\ K(BEt_3H) \xrightarrow[\text{2h , 20°C}]{THF}$$

1

$$(Ti \cdot 0.5\ THF)_x + x \cdot 4\ BEt_3 + x \cdot 4\ KBr \downarrow + x \cdot 2\ H_2 \uparrow . \qquad (2)$$

2

No particles were detected by HRTEM (high resolution TEM) indicating that the size is <0.8 nm. The careful analytical examination of the electronic and geometric structure of this extremely oxophilic Ti organosol **2** by means of XRD (X-ray diffraction), XPS (X-ray photoelectron spectroscopy), XANES (X-ray absorption near edge structure), and EXAFS (extended X-ray absorption fine structure) [87] revealed that **2** consists of very small Ti particles in the zero-valent state stabilized by intact THF molecules. The EXAFS spectra showed signals at 1.6 and 2.4 Å; no signals from backscatterers >3 Å were observed. The Ti–Ti distance found (2.804 Å) was smaller than in the bulk Ti metal, and the Ti–O distance was 1.964 Å. On the basis of these findings **2** is best described as a regular Ti cluster consisting of thirteen or less Ti atoms and having six or less THF–O atoms possibly in an octahedral configuration (Fig. 3a).

Analogously, the corresponding Zr colloid may be isolated; on the

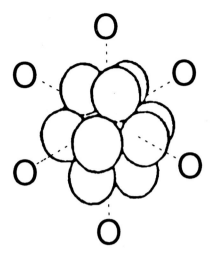

Figure 3a. Regular $Ti(0)_{13}$ cluster having six THF–O atoms in an octahedral configuration.

slow addition of the filtered THF solution to pentane (removal of residual KCl) the Zr colloid precipitates. The workup of the V and Nb colloids is performed similarly. The reduction of the THF adduct to $MnBr_2$ at 40°C yielded a stable, isolated $(Mn \cdot 0.3 \ THF)_x$ colloid containing typically 60% Mn (Eq. 3a).

$$x \cdot (MnBr_2 \cdot 2 \ THF) + x \cdot 2 \ K(BEt_3H) \xrightarrow[\text{2h , 40°C}]{\text{THF}}$$
$$(Mn \cdot 0.3 \ THF)_x + x \cdot 2 \ BEt_3 + x \cdot 2 \ KBr + x \cdot H_2 \uparrow . \tag{3a}$$

IR and NMR data of the colloid show intact THF coordinated to Mn. HRTEM shows the fringes of Mn particles of the size 1–2.5 nm. No bromine was detectable by EDX analysis of the Mn nanoparticles, and EXAFS confirmed that colloidal Mn(0) was formed according to Eq. 3a [93]. Table 1 summarizes the results achieved so far with THF-stabilized organosols of early transition metals [111,130].

The colloidal stabilization of zero-valent nanometals may be also achieved by using tetrahydrothiophene as the donor molecule in a synthesis similar to Eq. 3a. In the case of Mn, Pd, and Pt, stable organosols were isolated; however, sols of Ti and V quickly decomposed (Fig. 3b).

7.3.3 Organosols via the Stabilization of Nanometals by Lipophilic Surfactants

Lipophilic surfactants of the cationic type such as tetraalkylammonium halides have been used by different workers [e.g., 97,103] as highly efficient protecting groups for nanometals giving very stable organosols,

TABLE 1
THF-Stabolized Organosols of Early Transition Metals

Products	Starting material	Reducing agent	T (°C)	t (h)	Metal content	Size (nm)
(Ti·0.5 THF)$_x$	TiBr$_4$·2 THF	K(BEt$_3$H)	rt[a]	6	43.5%	(<0.8)
(Zr·0.4 THF)$_x$	ZrBr$_4$·2 THF	K(BEt$_3$H)	rt	6	42%	—
(V·0.3 THF)$_x$	VBr$_3$·3 THF	K(BEt$_3$H)	rt	2	51%	—
(Nb·0.3 THF)$_x$	NbCl$_4$·2 THF	K(BEt$_3$H)	rt	4	48%	—
(Mn·0.3 THF)$_x$	MnBr$_2$·2 THF	K(BEt$_3$H)	50	3	70%	1–2.5

[a]rt = room temperature.

isolatable in the form of dry powders containing up to 85 wt % of metal [86,88,92,94]. The reduction of suspended metal salts using tetraalkylammonium hydrotriorganoborates in organic solvents (Eq. 3b) occurs very smoothly, giving metal organosols stabilized by NR$_4^+$ [86], which is present at the reduction center in a high local concentration. The metal particles (1–10 nm) are well protected from agglomeration by the long-chain alkyl groups so that very little metal (if any) precipitates. This very lipophilic type of a surfactant makes the resulting organosols very soluble in organic phases so that up to 1 M metal solutions are easily obtainable.

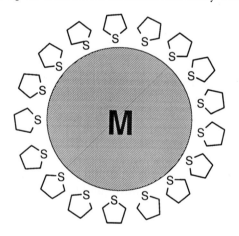

M = Ti, V (decomposition)
M = Mn, Pd, Pt stable organosols

Figure 3b. Organosols stabilized by tetrahydrothiophene. M = Ti, V (decomposition); S = Mn, Pd, Pt stable organosols.

$$MX_v + v\,NR_4(BEt_3H) \xrightarrow{\text{THF}} M_{colloid} + v\,NR_4X + v\,BEt_3 + v/2\,H_2 \uparrow, \quad (3b)$$

where M = metals of Groups 6–11; X = Cl, Br; v = 1, 2, 3, and R = alkyl, C_6–C_{20}.

The necessary pre-preparation of the alkylammonium organoborate (Eq. 3b) can be avoided by coupling the NR_4X agent to the metal salt prior to the reduction step. This again provides a high local concentration of the proctecting agent right at the reduction center so that the reduction itself may now be performed using a large variety of conventional inorganic or organic reducing agents (Eq. 4a).

$$(NR_4)_w\,MX_vY_w + v\,Red \rightarrow M_{colloid} + v\,RedX + w\,NR_4Y, \quad (4a)$$

where M = metals; Red = H_2, HCOOH, K, Zn, LiH, $LiBEt_3H$, $NaBEt_3H$, $KBEt_3H$; X, Y = Cl, Br; v, w = 1–3; and R = alkyl, C_6–C_{12}.

In the case of noble metals, the simple reduction of the THF-suspended metal salts by using H_2 in the presence of trialkylamines proved to be an effective and clean synthetic alternative. This pathway avoids the formation of alkaline salt byproducts completely (Eq. 4b).

$$MX_n + n\,N(R^1)_2R^2 + n/2\,H_2 \xrightarrow{\text{THF}} M_{colloid} + n[(R^1)_2R^2N^+H]X^-, \quad (4b)$$

where M = Ru, Rh, Pt; X = Cl, Br; $R^1 = C_8H_{17}$; $R^2 = C_3, C_8H_{17}$. Conditions were Ru: 60°C, 1–50 bar H_2, 24 h; Rh: 60°C, 1–50 bar H_2, 16 h; Pt: 20°C, 1 bar H_2, 1 h.

Comparing the utility of the three synthetic alternatives given in Eqs. 3b, 4a, and 4b, it can be stated that the "ammonio borate method" (Eq. 3b), which uses a powerful reducing agent, provides the most general access to mono- and bimetallic organosols via the reduction or coreduction of metal salts between Groups 6 and 11. The "double-salt method" (Eq. 4a), in contrast, provides an easy access to these materials; however, Group 6 metal organosols and colloidal Fe(0) cannot be produced this way. The amine variation described in (Eq. 4b), where the proctective ammonio group is formed via "self-construction" at the particle surface during the reduction step, is certainly the cleanest access to metal organosols; however, its application appears to be limited to noble metal salts that are easy to reduce.

Interestingly, Fe(0) organosols are the most difficult example to prepare in the Group 6–11 transition metal series. In using the "ammonio borate method" (Eq. 3b) no reduction of, for example, $FeBr_2$ occurs at 25°C even after 10 days of stirring. At 60°C, however, the salt suspended in THF dissolves completely within 24 h; under the evolution of hydrogen about 50% of the Fe precipitates as a magnetic powder, which is removed by filtration (Eq. 5).

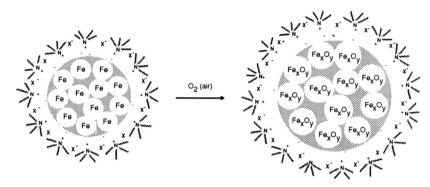

Figure 4a. Oxidation of the NR_4^+-stabilized Fe colloid.

$$FeX_2 + 2\,N(Octyl)_4BEt_3H \xrightarrow{\text{THF , 60°C}}$$

$$(1 - u)\,Fe_{colloid} + u\,Fe\downarrow + 2\,N(Octyl)_4X + 2\,BEt_3 + H_2\uparrow, \qquad (5)$$

where X = Br, I.

Subsequent workup of the reddish-black colloidal Fe solution and further purification of the raw product by precipitation of the redispersed material in THF by slow addition of pentane to the solution gave the dry organosol (particle size 3 nm) containing 14% Fe(0) (oxidation state evident from Mössbauer spectra), which proved to be fully redispersible in

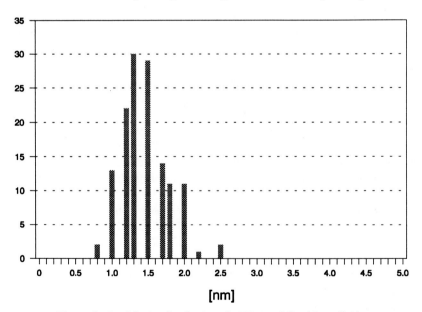

Figure 4b. Particle size distribution of a NR_4^+-stabilized Ru colloid.

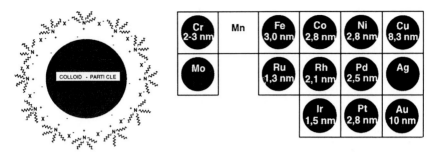

Figure 4c. NR_4^+-stabilized organosols and diameter of the metal core.

THF and toluene [111]. This nanosized colloidal Fe is extremely oxophilic both in solution and in powder form. On the carefully controlled addition of air (16 h), however, the dissolved Fe(0) N(octyl)$_4$ Br organosol slowly oxidizes giving a rusty-brown solution of colloidal iron oxide, which is stable in air for several days and may be kept under argon even for months without any sedimentation [112] (Fig. 4a).

Because the synthesis pathways (Eqs. 3b, 4a, and 4b) provide a high local concentration of the protecting agent right at the reduction center, the resulting colloid particles consequently are rather small and monodisperse. A typical particle size distribution histogram, derived from the electron micrograph, of a Ru organosol is shown in Fig. 4b.

Figure 4c summarizes the results obtained in the preparation of transition metal organosols stabilized by lipophilic NR_4^+ surfactants as the protecting agent [94].

7.3.4 Hydrosols via the Stabilization of Nanometals by Hydrophilic Surfactant

For the effective synthesis of stable nanometal hydrosols via chemical reduction we first have tried to transfer the principle of combining the reduction agent with the protective group (used in Eq. 3b for the synthesis of organosols) to hydrophilic surfactants such as betaines. For example, 3-(N,N-dimethyldodecylammonio)-propanesulfonate (Sulfobetaine 12) reacts in THF suspension with lithium triorganoborohydride to form a strongly reducing THF soluble 1:1 adduct [113], the structure of which is currently under investigation. This adduct was used to reduce a number of transition metals salts of Groups 8–11 in THF to give mono- and bimetallic hydrosols, which precipitate from THF. The resulting nanometallic materials (particle size 1–6 nm) proved to be highly soluble in water. An example of this new synthesis of transition metal hydrosols is given in Eq. 6a.

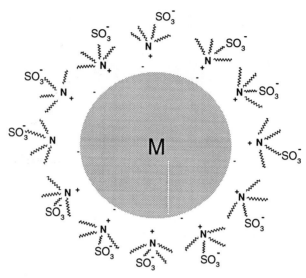

Figure 5. Sulfobetaine-stabilized colloidal metals and alloys. M = Cu, Ru, Rh, Pd, Ag, Ir, Pt, Ru/Fe, Ru/Co, Ru/Ni, Ru/Cu, Pt/Co, Pt/Cu, Ru/Rh, Ru/Ir, Ru/Pt, Rh/Ir, Rh/Pt, Pd/Pt, Ir/Pt.

$$C_{12}H_{25}N(CH_3)_2(C_3H_6-SO_3)\cdot LiBEt_3H + PtCl_2 \xrightarrow{\text{THF}}$$
THF soluble

$$Pt_{colloid}\cdot C_{12}H_{25}N(CH_3)_2(C_3H_6-SO_3). \qquad (6a)$$
water soluble

The zero-valent status of the metal core was checked by XAS (X-ray absorption spectroscopy) and XPS [114]. All analytical data obtained of the various sulfobetaine-stabilized nanometallic hydrosols are in full agreement with the schematic structure depicted in Fig. 5.

The method of coupling the stabilizing surfactant to the metal salt prior to the reduction (analogous to Eq. 4a) allows the metal hydrosol synthesis even in water. This variation (eq. 6b) led us to highly water soluble colloidal nanometals (mono- and bimetallic), stabilized by hydrophilic surfactants of the anionic, nonionic, or amphiphilic types using, for example, hydrogen, Li formates, or alkali borohydrides as reducing agents in aqueous solution.

$$PtCl_2 + 4\ C_{12}H_{25}N(CH_3)_2(C_3H_6-SO_3) + Li_2CO_3 + H_2 \xrightarrow{\text{water, 20°C}}$$
(SB12)

$$Pt_{colloid} + 4\ C_{12}H_{25}N(CH_3)_2(C_3H_6-SO_3) + 2\ LiCl + H_2O + CO_2 \uparrow. \qquad (6b)$$
$$(SB12)$$

Our consequent application of both synthetic pathways (described in Eqs. 6a and 6b) has made a plethora of mono- and bimetallic hydrosols accessible (Table 2).

Because both the synthesis methods (Eqs. 6a and 6b) favor a high local concentration of the protecting agent at the reduction center, the resulting hydrosol particles are found by TEM to be rather small (1–10 nm). The electron micrographs further show a very narrow particle size distribution (comparable with Fig. 4b) to be present also in the hydrosols.

7.4   Nanometal Powders from Organosols

After the extraction of the protecting shell from the metal core of the colloid using, for example, ethanol nanometal powders, under full conservation of the particle size of the colloidal starting material can be obtained. For example, the extraction of colloidal Pt (average particle size is 2.8 nm) with ethanol yields a gray Pt powder, again of 2.8 nm size according to TEM [88].

During the physical characterization of the colloidal materials through electron microscopy (specifically EDX) and XPS, both of which employed ultrahigh vacuum conditions (10^{-7} to 10^{-8} Pa), neither the halogen nor the nitrogen of the NR_4X groups could be detected. This clearly indicates that the protecting tetraalkylammonium halide can be removed under certain conditions, leaving the bare metal core behind. This observation prompted us to develop a chemical procedure for the extraction of the protecting shell from the colloids at room temperature to produce nanoscale metal powders (Table 3). The preparation of nanoscale platinum powder from the corresponding platinum colloid represents a typical example. The gray-brown colloid powder is treated with an excess of ethanol, after which the supernatant solution containing the tetraalkylammonium halide is siphoned off. This procedure is repeated several times during which a continuous darkening of the product color is observed. The resulting black pyrophoric metal powder (93 wt % Pt) is no longer redispersible in THF (Table 3, No. 5). The TEM image of this product shows that the mean particle size of the platinum powder after the extraction corresponds exactly to that of the initial colloidal platinum sample. As determined by EDX analysis the extracted sample still contains traces of the protecting shell, which can be completely removed by heating the metal powder at 700°C under vacuum (0.1 Pa). According to the TEM image of the heated product the mean particle size is 2.8 nm, the same value found for the

TABLE 2
Stabilization of Water Soluble Colloids by Various Surfactant Types

Surfactants	Trademark	Metals
Cationic Type		
$\begin{array}{ccc} CH_3 & OH & Cl \\ \mid & \mid & \mid \\ C_{18}H_{37}-N^+-CH_2-CH-CH_2 & & Cl^- \\ \mid & \\ CH_3 \end{array}$	QUAB 426 (Degussa AG)	Pt
$\begin{array}{c} O \\ \parallel \\ (R\ COCH_2CH_2)_n\ \overset{+}{N}\ \ (CH_2CH_2OH)_{3-n} \\ \mid \\ CH_3 \end{array}$		
R = partially hydrogenated C_{15} group	ESTERQUAT AU35 (Henkel KG a A)	Pt
Anionic Type		
Li-dodecylsulfate		Ru
Nonionic Type		
Polyoxyethylenelaurylether	BRIJ 35 (Atlas Powder)	Co, Ni, Ru, Rh, Pt, Pt/Ru, Pt/Pd
Polyoxyethylenesorbitanmonolaurate	TWEEN 20 (Atlas Chem. Ind. Inc., ICI America Inc.	Pt
Alkylpolyglycosides	APG 600 (Henkel KG a A)	Pt

Zwitterionic Type

3-(*N,N*-Dimethylnonylammonio)propanesulfonate (SB9)		Pt
3-(*N,N*-Dimethyldodecylammonio)-propanesulfonate (SB12)		Ni, Cu, Ru, Rh, Pd, Ag, Ir, Pt, Ru/Fe, Ru/Co, Ru/Ni, Pt/Co, Pt/Cu, Ru/Rh, Ru/Ir, Ru/Pt, Rh/Ir, Rh/Pt, Pd/Pt, Pd/Ag, Ir/Pt
3-(*N,N*-Dimethyloctadecylammonio)-propanesulfonate (SB18)		Pd[a]
2-(*N,N*-Dimethyldodecylammonio)acetate	REWOTERIC AM DML (*Witco Surfactants GmbH*)	Pd, Pt, Pt/Co, Pt/Ni, Pt/Cu, Pt/Ir, Pt/Ru, Rh, Ru
Cocoamidopropylbetaine	DEHYTON K (*Henkel KG a A*) AMPHOLYT JB 130 (*Hüls AG*)	Pt, Rh

$$
\underset{\displaystyle R = \text{Cocoalkyl}}{\overset{\displaystyle \underset{\|}{\text{O}}}{R\ \text{CNH(CH}_2)_3\text{N}-\text{CH}_2\text{CO}_2\text{Na}}}
$$
$$
\text{CH}_2\text{CH}_2\text{OH}
$$

TABLE 2 *continued*

Surfactants	Trademark	Metals
Na-cocoamidoethyl-N-hydroxyethylglycinate	DEHYTON G (*Henkel KG a A*)	Pt
$CH_3(CH_2)_{11} - O - \overset{\overset{O}{\|\|}}{P} - OCH_2CH_2 - \overset{\overset{CH_3}{\|}}{\underset{\underset{CH_3}{\|}}{N^+}} - CH_3$ $\overset{\|}{O_-}$		
Dodecyl-2-(trimethylammonio)ethylphosphate		Pt
$CH_3(CH_2)_{11} - \overset{\overset{CH_3}{\|}}{\underset{\underset{CH_3}{\|}}{N^+}} - CH_2CH_2PO_3Na^-$		
2-(Dimethyldodecylammonio)phosphonic acid, Na-salt		Pt
$^-O_3SCH_2CH_2CH_2 - \overset{\overset{CH_3}{\|}}{\underset{\underset{CH_3}{\|}}{^+N}} - (CH_2)_{12} - \overset{\overset{CH_3}{\|}}{\underset{\underset{CH_3}{\|}}{N^+}} - CH_2CH_2CH_2SO_3$		
1,12-bis[Dimethyl(3-propanesulfonate)ammonio]dodecane		Pt

[a]Water soluble above 50°C.

TABLE 3
Preparation of Nanoscale Metal Powders via Metal Colloids

No.	Starting material		Solvent for extraction of NR_4X	Product metal content after extraction (%)	Mean particle size	
	Colloid	Mean particle size (nm)			After extraction (nm)	After heat treatment (700°C, 4 h, 0.1 Pa) (nm)
1	Co	2.8	Ethanol	82.84	3.8	—
2	Ni	2.8	Ethanol	88.18	3.0	—
3	Rh	2.1	Ethanol	82.76	2.7	2.9
4	Pd	2.5	Ethanol	98.13	5.8	6.0
5	Pt	2.8	Ethanol	92.90	2.8	2.8
6	Rh/Pt	2.3	Ether/ethanol 1:10	—	2.7	3.0
7	Pd/Pt	2.8	Ether/ethanol 1:10	—	2.8	—

original colloidal platinum. Hence, an agglomeration of the metallic nanoparticles as a consequence of heat treatment is not observed.

This finding holds basically true in almost every preparation of metallic or intermetallic powder from the corresponding colloidal material (compared with the particle sizes listed in Table 3). The only exception found so far involves palladium (Table 3, No. 4), where a particle growth from 2.5 to 5.8 nm is seen (via TEM) upon removing the protecting shell. After the thermal treatment no further particle augmentation is detected.

Precipitated nanopowders obtained by chemical reduction may be transferred into soluble metal colloids by subsequent reaction with NR_4X. For example, a sample of magnetic cobalt powder (particle size 4 nm), precipitated from $CoBr_2$ by reduction in THF, was reacted with an excess of $N(octyl)_4Br$ to give a clear, dark-red solution of the corresponding cobalt organosol.

7.5 Colloidal Alloyed Metals

The coreduction of a mixture of different tetraalkylammonium metalates yields colloidal metal alloys. An HRTEM of the product of the coreduction of $(NR_4)_3RhCl_6$ and $(NR_4)_2PtCl_4$ (magnification $6.3 \cdot 10^5$) showed particles of an average size of 2.3 nm and a net–plane distance of 0.25 nm. Under the microscope 70 particles were analyzed by EDX with a point resolution of 1 nm. In every particle examined, both Rh and Pt were found to be present, which indicates that, in fact, a colloidal Pt/Rh alloy is generated during coreduction. An EXAFS study of a $Pt_{56}Rh_{44}$ colloid at both the Rh K edge and the Pt L(III)-edge verified the formation of a bimetallic alloy nanoparticulate system [95]. In subsequent X-ray absorption spectroscopy studies, the structural characterization of a series of Pt/Rh colloids with varying stoichiometries was achieved. The results of these investigations together with electrochemical studies, XPS, and X-ray diffraction measurements will be the subject of forthcoming articles [96].

7.6 Catalytic Applications

The goal behind the preparation of novel metal and alloy colloids includes, among possible applications in the field of electronic "nanodevices" [115], the development of promising new homogeneous and heterogeneous catalysts. The catalytic potential of the colloidal materials was explored both in the homogeneous phase and in the supported state. Regarding the former, the activities of a series of palladium colloids were measured for the hydrogenation of cyclohexene in THF under normalized conditions [89]. The catalytic activity as well as selectivity and long-time stability of noble metal colloids adsorbed on charcoal, 5 wt % Pd [90], 5 wt % Rh [88], and 5 wt % Pt [88], were tested in the hydrogenation of cinnamic acid, butyronitrile, and crotonic acid, respectively. Further, the selective oxidation of glucose giving gluconic acid using a bimetallic Pd/Pt colloid as the active component has been included in these studies. The experimental apparatus and the test conditions employed in the investigations of the heterogeneous charcoal-supported catalysts are described in detail in Reference 89.

7.6.1 The Precursor Concept

Pre-prepared nanometals stabilized by surfactants may be used as easily accessible precursors for a new type of heterogeneous catayst. These precursors may be optimized independent of the support by varying the particle size, composition, and structure of bimetallic systems. Further, the coverage of the metal surface by various protective shells and intermediate layers, for example, oxygen or sulfur, may be used for modifications of the active component. This is visualized in Fig. 6.

Because modern analytical tools such as HRTEM, EDX, XPS, XANES, and EXAFS provide detailed physical data about the size, composition, oxidation state, and structure of the precursor particles, all prerequisites of the "molecular design" of catalyst precursors are given. The perfect protection of the nanometallic hydrosols by the various hydrophilic surfactants listed in Table 2 allows the handling of the precursors even in concentrated aqueous solution. Further, the use of surfactants as the protective shell around the metal core enables the efficient adsorptive fixation

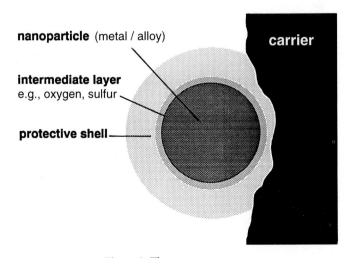

Figure 6. The precursor concept.

of the metal particles even on single-crystal oxides, such as quartz, sapphire, and highly oriented pyrolitic graphite (HOPG), simply by dipping these substrates into aqueous solutions of the nanometal hydrosols at 20°C. A combined AFM, STM, and XPS study by Behm *et al.* [116] reveals that the supported metal particles are very resistant to agglomeration even under extreme conditions. For example, atomic force microscopy of an alumina-supported Pt-SB12-hydrosol after annealing the sample *in vacuo* at 500°C still showed individual, sphere-shaped Pt species randomly distributed over the support surface.

7.6.2 Heterogeneous Catalysts on the Basis of Surfactant-Stabilized Precursors

To prepare heterogeneous catalysts, the precursors may be adsorbed from aqueous solutions on the supports simply by dipping at ambient temperature. According to TEM no agglomeration of the particles occurs [88]. The advantage of supported colloid catalysts is demonstrated, for example,

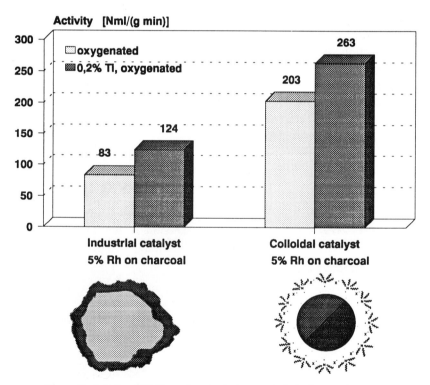

Figure 7. Activity of Rh/C catalysts in the butyronitrile hydrogenation test.

by comparison of the activity of two Rh catalysts (5% Rh on charcoal) in the butyronitrile hydrogenation test (hydrogenation of butyronitrile to butylamine) (Fig. 7).

The electron micrograph of the conventional catalyst (left) made by salt impregnation and subsequent calcination shows large metal agglomerations on the surface besides a minor part of Rh particles of 1–5 nm size. With an Rh colloid precursor of 1.2- to 2.2-nm particle size on the same support, the activity observed in the test reaction is virtually doubled. Both catalyst types show a significant increase of activity when doped with 0.2% of the Ti(0) colloid **2**.

In addition, the lifetime of the colloid catalyst is superior to that generated by the conventional precipitation systems. Whereas the activity of a conventional Pd/C catalyst tested in the hydrogenation of cyclooctene to cyclooctane expires completely after the performance of 38×10^3 catalytic cycles per Pd atom, the Pd colloid/C catalyst still shows a residual activity after 96×10^3 catalytic turnovers (Fig. 8).

Selectivity control may be brought about by doping and by adding a second metal. For example, doping of Rh colloid catalysts with Sn has a strong effect on the selective $C = O$ group hydrogenation of α,β-unsaturated aldehydes.

An Rh colloid/C catalyst doped with Sn (Rh:Sn = 1.5:1) exhibits 86% selectivity in the hydrogenation of cinnamic acid to cinnamic alcohol [117]. Using the catalytic hydrogenation of crotonic acid to butanoic acid as a test reaction, we were able to observe the synergistic effect of bimetal-

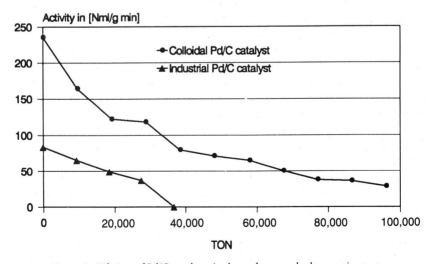

Figure 8. Lifetime of Pd/C catalysts in the cyclooctene hydrogenation test.

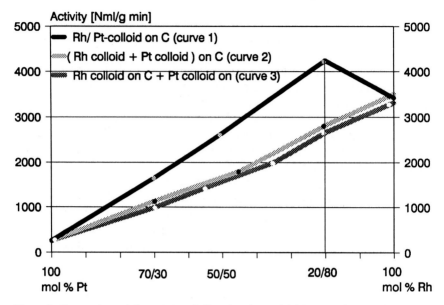

Figure 9. Comparison of the activity of alloyed and mixed Rh/Pt/C catalysts in the crotonic acid test.

lic colloidal precursors (Fig. 9) [88]. The catalysts prepared by mixing Pt colloid/C and Rh colloid/C powders or by consecutive adsorption of Pt and Rh colloids on charcoal show a linear increase of activity with increasing content of Rh (additive effect). In contrast, the corresponding activity plot of the bimetallic Pt/Rh colloid systems [95,96] clearly shows a maximum at $Pt_{20}Rh_{80}$. Because this maximum exceeds the activity found for Rh colloid/C alone, this finding is significant for a synergistic effect.

The superior properties of the new type of bimetallic colloid catalysts formed by the adsorptive fixation of surfactant-stabilized bimetallic precursor particles to the support over conventional "bimetallic" catalysts (prepared via coimpregnation of two metal salts followed by calcination) were confirmed in the catalytic oxidation of, for example, D-glucose by molecular oxygen, this gives D-gluconic acid using a bismuth-promoted Pd/Pt colloid as the active component on a charcoal support (Fig. 10) [118].

The carbohydrate oxidation reaction shown in Fig. 10 is currently under intensive investigation both in academia [119] and in industry [120,121]. In our study we first compared the activity of a conventionally prepared catalyst from industry (atomic ratio Pd:Pt = 4:1, supported on charcoal) and the activity of a Pd_{88}/Pt_{12} colloid supported on an identical sample of charcoal. Both catalysts were promoted by bismuth. The results are shown in Fig. 11. Without affecting the high selectivity of the reaction

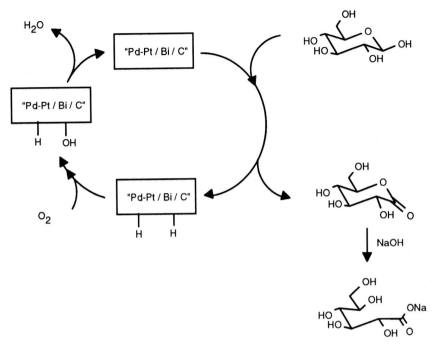

Figure 10. Catalytic oxidation of D-glucose on bismuth-promoted Pd/Pt/C colloid catalysts.

(98%), the catalytic activity of the supported bimetallic Pd/Pt colloid was found to be virtually doubled.

The improved activity of the Pd/Pt colloid catalyst may be explained—in addition to the true bimetallic composition of the active component—by the fact that the adsorbed precursor particles sit at the most exposed sites of the support and consequently are more accessible for the substrates than in the case of the conventional type of catalyst made via salt impregenation, where part of the metal is inevitably "buried" under the support surface. The practical disadvantage of highly active catalyst components sitting on exposed positions of the support often lies in a rather insufficient long-time stability, because the sensitivity to poisons is drastically increased. Figure 12 compares the stability of an optimized conventional catalyst with the colloidal Pd/Pt system. Surprisingly, the expected decay in activity after recycling the catalysts was found to be much less in the colloidal than in the conventional system [118].

Because chemisorption measurements indicate that metal colloid precursors adsorbed on the surface of the support are still partly covered by surfactant molecules [122], it seems reasonable to assume that the catalytically active nanometal particles are protected by a "coat"—however per-

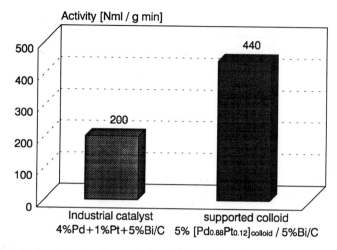

Figure 11. Initial activity in the catalytic oxidation of D-glucose on bismuth-promoted Pd/Pt/C colloid catalysts. Conditions: O₂ pressure = 1 bar; T = °C; glucose = 0.88 M; glucose/noble metal = 2065.

meable for small molecules such as H_2 or O_2—that prevents the direct contact with poisons.

To check the possible influence of stabilizing groups on the catalytic properties of colloidal nanometals, we have investigated the stereoselectivity of colloidal platinum (particle size 2 nm) stabilized by a chiral ammo-

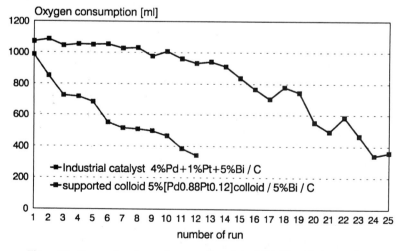

Figure 12. Conversion decay after recycling of the Pd/Pt/C oxidation catalysts.

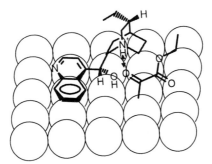

Figure 13. Dihydrocinchonidine-stabilized Pt colloids (2 nm) for the enantio-selective hydrogenation of ethyl pyruvate (Eq. 7) [128,129].

nium group derived from dihydrocinchonidine (Fig. 13) in the enantioselective hydrogenation of ethylpyruvate (Eq. 7) [123].

The enantioselectivity control of conventional heterogeneous Pt catalysts in the reaction (Eq. 7) using modifiers derived from cinchona alkaloids is well established in the literature, and enantiomeric excess (ee) values above 90% have been reported [124,125,126]. Further, the influence of conversion and bulk diffusion limitations on the stereoselectivity was studied in detail [127].

$$ \text{(7)} $$

R-ethyl lactate

Unsupported colloidal Pt of 2.0-nm particle size stabilized by dihydrocinchonidonium acetate in homogeneous acetic acid solution gave, under the conditions quoted in Table 4, the (R)-ethyl lactate in 70% ee. This clearly accounts for a strong selectivity control of the catalytic reaction induced by the chiral ammonium group present at the surface of the colloid particle. Our current investigations have further revealed that the alkaloid derivative acts as an "accelerating ligand" in favor of the formation of the R enantiomer. Consequently, the ee values increase when higher concentrations of H_2 are applied in solution. In homogeneous phase, however, the Pt colloid tends to precipitate when pressurized hydrogen is applied. Therefore, we have adsorbed the Pt colloid to SiO_2 and Al_2O_3 supports. The resulting chirally modified heterogeneous Pt colloid catalysts performed the

TABLE 4
Dihydrocinchonidine-Stabilized Pt Colloids[a]

Particle size ⌀ (nm)	Support	Pressure [p (bar)]	Conversion (%)	ee (%)
2.0	In solution	1	100	69
2.0	SiO$_2$, 5 wt %	100	100	81
2.0	Al$_2$O$_3$, 5 wt %	100	100	85

[a]Conditions: 25°C, 5 ml substrate; in solution: 100 ml CH$_3$COOH/H$_2$O (2:1), 22 mg Pt colloid (10 mg Pt); supported: 10 ml CH$_3$COOH, 100 mg catalyst (5 mg Pt), 20 mg cinchonidine.

reaction (Eq. 7) under 100 bar of hydrogen giving the *R* enantiomer in 81 and 85% ee, respectively.

Work is in progress to further explore the scope and limitations of nanosized metal colloids as precursors for a new type of highly active, stable, and selective catalyst.

7.7   Experimental

All reductions using metals or hydrides were conducted under argon in dry organic solvents. Using hydrogen for reductions in aqueous solution, we applied desoxigenated water.

7.7.1 NR$_4$Cl-Protected Ir Organosol

Procedure: A solution of N(C$_8$H$_{17}$)$_4$(BEt$_3$H) in THF (50 ml, 0.37 M) was added within 1 h at room temperature to a stirred suspension of anhydrous IrCl$_3$ (1.84 g, 6.16 mmol) in 200 ml of THF. Almost complete dissolution of the IrCl$_3$ occurred after 16 h at 60°C. After filtration the clear, dark black-brown solution was concentrated *in vacuo,* and the black-brown, waxy residue was dried for 3 h at room temperature and 0.1 Pa. The product (7.38 g) was soluble in THF, ether, toluene, and acetone; was insoluble in ethanol; and contained 10.85% Ir. The residue was suspended in 200 ml of technical quality ethanol without protective gas and addition of 20 ml of technical quality ether caused a gray-black precipitate to form. It was left to stand for 1 h, before the clear supernatant solution was removed by inert gas pressure (argon) on the liquid surface. The precipitate was washed once with a mixture of 40 ml of ethanol and 4 ml of ether. Drying in vacuum (0.1 Pa, 1 h, room temperature) yielded a gray iridium colloid powder (0.36 g), which was very soluble in THF; soluble in acetone; and insoluble in ether, ethanol, and toluene. Elemental analysis was Ir, 65.55%; N, 0.27%; C, 19.97%; H, 3.49%. Mean particle size was 1.5 nm.

7.7.2 NR₄Br-Protected Pd Organosol

Procedure: A mixture of 0.5 g (2.23 mmol) of $Pd(CH_3CO_2)_2$ and 1 g (1.30 mmol) of $N(dodecy)_4Br$ was dissolved in 110 ml of THF. The flask was evacuated several times, and then H_2 was introduced via a gas burette under normal pressure. After a while the solution turned black and after 16 h, 60.4 Nml (120%) of H_2 was taken up. THF was added to give a total volume of 110 ml, and on addition of 55 ml of H_2O/Ar, a brown-black precipitate resulted. After being allowed to stand for 16 h, the supernatant liquid was siphoned off and the solid was dried for 3 h in vacuum (0.1 Pa). The black powder isolated in this manner was redispersible in THF. Yield was 0.27 g (87%). Elemental analysis was Pd, 76.73%; C, 13.19%; H, 2.47%; N, 1.64%; Br, 3.06%. XPS was Pd(0). Mean particle size was 1.8 nm.

7.7.3 3-(*N,N*-Dimethyldodecylammonio)-Propanesulfonate (SB12)-Protected Pt Colloid in Water

Procedure: A mixture of 1.4 g (5.3 mmol) of $PtCl_2$, 7.2 g (21.2 mmol) of 3-(*N,N*-dimethyldodecylammonio)-propanesulfonate (SB12), and 0.4 g (5.3 mmol) of Li_2CO_3 was stirred in 100 ml of H_2O; and H_2 was introduced under normal pressure for 3 h at 20°C. After approximately 30 min a clear, black solution was formed and all volatile compounds were evaporated in vacuum (0.1 Pa, 40°C). The resulting black powder was redispersible in water. Yield was 8.4 g. Metal content was Pt 10.7%. Mean particle size was 2.2 nm.

7.7.4 Polyoxyethylenelaurylether-Protected Pt/Pd Colloid in Water

Procedure: A mixture of 1.35 g (2.65 mmol) of $H_2PtCl_6 \times 6H_2O$ and 0.7 g (2.65 mmol) of Pd $(NO_3)_2 \times H_2O$ is dissolved together with 7 g of polyoxyethylenelaurylether and 1.0 g of (13.25 mmol) Li_2CO_3 under argon in 100 ml of H_2O, and within 4 h H_2 gas is passed through it at 20°C. The resultant deep-black reaction mixture is filtered over a D4 glass frit, and the deep-black, clear solution is concentrated in high vacuum (0.1 Pa, 40°C) to dryness. Pt/Pd colloid (11.2 g) is obtained in the form of a black solid having a metal content of 4.3% Pt and 2.3% Pd.

7.7.5 Water Soluble Pt Colloid Stabilized by Dihydrocinchonidine

Procedure: In a 100-ml two-neck flask, provided with a reflux condenser and a septum, 0.104 g (0.31 mmol) of $PtCl_4$ g is dissolved in 83 ml of distilled water and is heated to reflux temperature in an oil bath. The temperature of the oil bath is 140°C (± 5°C) during the synthesis. A solution of 0.092 g of dihydrocinchonidine (0.31 mmol) in 7 ml of 0.1 N formic acid is rapidly injected through the septum. In the beginning, the reaction mixture becomes turbid and begins to become black after some minutes. The reaction is finished approximately 10 min after the beginning of the black coloration. The reaction mixture is frozen in liquid nitrogen and is liberated of water and forming hydrochloric acid by freeze-drying. A black powder is obtained that can be completely dispersed in water. If the formed platinum colloids should be applied on carrier materials, the aqueous product dispersion can be used without isolation of the metal particles before the fixing on the carrier. The yield is 0.18 g (103% of the theory) in this reaction. The elemental analysis shows 24.5% Pt, 16% Cl, 39.5% C, 5% H, and 5% N. TEM examinations show an average particle size of 2 nm.

7.7.6 Polyoxyethylenelaurylether-Protected Pt/Rh Colloid in Water

Procedure: A mixture of 1.35 g (2.65 mmol) of $H_2PtCl_6 \times 6H_2O$ and 0.7 g (2.65 mmol) of $RhCl_3 \times H_2O$ is dissolved together with 7 g of polyoxyethylenelaurylether and 1.0 g (13.25 mmol) of Li_2CO_3 under argon in 150 ml of H_2O; and within 20 h a solution of 2.86 g (55.0 mmol) of Li formate in 50 ml of H_2O is added at 60°C. The resultant deep-black reaction mixture is filtered over a D4 glass frit, and the deep-black, clear solution is concentrated in high vacuum (0.1 Pa, 40°C) to dryness. Obtained is 12.5 g of Pt/Rh colloid in the form of a black solid having a metal content of 4.0% Pt and 2.0% Rh.

7.7.7 Preparation of a Pd (SB12)/Activated Carbon Catalyst for the Partial Oxidation of Carbohydrates (5% per Weight of Pd on C)

Procedure: A microporous powdery active carbon (1.254 g) having a grain size of 20 μm is suspended in 50 ml of deoxygenated H_2O; and 64.7 ml of a solution of Pd (SB12) colloid in deoxygenated water (1.02 mg Pd per ml) is added within 16 h under stirring. The covered active carbon is separated over a glass frit, yielding a colorless filtrate. It is washed twice with 25 ml of deoxygenated water, respectively, and dried during 16 h in vacuum (0.1 Pa) at room temperature. Subsequently, the catalyst is oxygenated within 16 h at 10 Pa (approximately 0.2% O_2). The obtained catalyst can be handled in air.

7.7.8 Preparation of a Heterogeneous Pt Catalyst by Adsorption of Pt/Dihydrocinchonidine Colloid on SiO_2 and Activated Carbon

Procedure: The colloid solution (100 ml) described in Sect. 7.7.5 is directly taken up after the synthesis in 100 ml of cold, distilled water and is dropped within 1 h to 100 ml of the carrier suspension. Either the highly dispersed silicon dioxide Aerosil P 25 (Degussa) or the active carbon carrier 196 (Degussa), which was oxidized with NaOCl before the fixing of the colloid to the carrier, can be used as a carrier. The obtained suspensions are stirred with a magnetic stirrer at a low rotational speed for 2 days, and they are subsequently filtrated. The filtrate is completely discolored, from which can be concluded that the metal colloids were quantitatively absorbed on the carrier. The heterogeneous Pt catalysts obtained were dried in a drying oven, and they can be used subsequently as hydrogenation catalysts without any further intermediary step. A uniform and agglomeration-free distribution of the colloids on the carrier materials could be proved by TEM examinations.

7.7.9 Use of the Pd (SB12)/Activated Carbon Catalyst for the Oxidation of Glucose to Gluconic Acid

Procedure: An aqueous solution of glucose (100 ml) with 16 g of glucose (99% per weight) (88 mmol) and 0.24 g of the catalyst described in Sect. 7.7.6 (1.5% per weight in relation to the amount of glucose) are transferred to a 250-ml stirring reactor equipped with gassing stirrer, thermometer, alkali metering, pH electrode, and oxygen feeding. The oxygen is distributed at normal pressure by means of the gassing stirrer in the solution at a reaction temperature of 56°C. The resulting gluconic acid is neutralized by dropping 10% per weight of caustic soda thereto. Thereby the pH value of the suspension is 10.0. The catalyst is filtered off, and the filtrate is analyzed by means of ion chromatography and HPLC (high pressure liquid chromatography). Conversion (120 min) is 49%; selectivity (120 min) is 92%; activity (120 min) is 327 g (gluconic acid)/g (Pd) × h.

7.7.10 Enantioselective Hydrogenation of 2-Keto-propane Acid Ethylester to 2-Hydroxy-propane Acid Ethylester

Procedure: A 100-ml autoclave is charged with 100 mg of the catalyst described in Sect. 7.7.8 (Pt on SiO_2; metal content 5%), 5 ml of 2-keto-propane acid ethylester (45 mmol), 20 ml of dihydrocinchonidine (0.1 mmol), 10 ml of acetic acid, and a magnetic stirrer nucleus having a size of 3 cm. The pressure vessel is degassed after being closed, and subsequently 100 bar hydrogen is pressed on under vigorous stirring. The reaction takes place at 25°C, and it is terminated approximately after 15 min. Following the expansion of the pressure vessel, the product mixture is liberated of the catalyst by filtration, the clear filtrate is taken up in 180 ml of saturated sodium bicarbonate solution and subsequently extracted three times with each 20 ml of diethyl ether. The combined organic phases are concentrated on a rotary evaporator, the remaining clear solution is examined by NMR spectroscopy and mass spectroscopy, and it is identified as 2-hydroxy-propane acid ethylester. The yield was determined by gas chromatography to 90%. The optical yield of the reaction was examined by gas chromatography on a chiral column and yields an excess of enantiomer of 81%.

Acknowledgments

The authors gratefully acknowledge the valuable support of the following scientists, companies, and institutions: Dr. B. Tesche, Fritz-Haber-Institut der MPG, Berlin (Germany) now at MPI f. Kohlenforschung, Mülheim (Germany), for numerous TEM images taken with a Siemens Elmiskop 102 and DEEKO 100 at 100 kV and for many helpful discussions; Mr. T. Kamino, Hitachi Instruments Engineering Co. Ltd., 882 Ichige, Katsuta-shi (Japan), and Dipl. Ing. B. Spliethoff, MPI für Kohlenforschung, Mülheim (Germany), for HRTEM images performed with a Hitachi HF 2000 at 200 kV including EDX point analyses; Professor Dr. R. Courths and Dipl.-Phys. B. Heise, Universität-Gesamthochschule Duisburg (Germany), for XPS spectra obtained using an ESCALAB Mark II; Professor Dr. W. Keune and Dipl.-Ing. U. von Hörsten, Universität-Gesamthochschule Duisburg (Germany), for the measurement and interpretation of the Mössbauer spectum; Professor Dr. J. Rozière and Dr. D. J. Jones, Université Montpellier (France), for the X-ray absorption measurements recorded on the Ti K edge (77 K) with an EXAFS 3 spectrometer in DCL (French Synchrotron Facility in Lure); Professor Dr. J. Hormes and Dipl.-Phys. J. Rothe and Cand.-Phys. R. Becker, Universität Bonn (Germany), for the XANES and EXAFS measurements on the Pt/Rh colloid samples using synchroton radiation at beamline BN3 of the storage ring ELSA at Bonn University. Further we are indebted to Professor P. Kleinschmit, and Dr. P. Panster, and Dr. A. Freund, Degussa AG, ZN Wolfgang, Hanau (Germany), for a gift of commercial noble metal catalysts and the test procedures for noble metal charcoal catalysts. The support of this work by Fonds der Chemischen Industrie, Frankfurt (Germany); and a grant from the German Ministry of Education, Science, Research, and Technology (BMBF FKZ 03 D 0007 A2) are also gratefully acknowledged.

References

1. Schmid, G. (ed.) (1994). *Clusters and Colloids*, VCH Publishers, Weinheim, Germany.
2. Bönnemann, H., Brijoux, W., and Joussen, Th. (1990). *Angew. Chem. Int. Ed. Engl.*, 29, 273.
3. Boutonnet, M., Kizling, J., Stenius, P., and Maire, G. (1982). *Colloids Surf*, 5, 209.
4. Boutonnet, M., Kizling, J., Touroude, R., Maire, G., and Stenius, P. (1986). *Appl. Catal.* 20, 163.
5. Bradley, J. S., (1983). *Adv. Organomet.*, 6, 687.
6. Bradley, J. S., Hill, E., Leonowicz, M. E., and Witzke, H. J. (1987). *J. Mol. Catal.*, 41, 59.
7. Bradley, J. S., Millar, J. M., Hill, E. W., Behal, S., Chaudret, B., and Duteil, A. (1991). *Faraday Discuss.*, 92, 225.
8. Braunstein, P. (1986). *Nouv. J. Chim.*, 10(7), 365.
9. Esumi, K., Shiratori, M., Ishizuka, Tano, T., Torigoe, K., and Meguro, K. (1991). *Langmuir*, 7, 457.
10. Meguro, K., Torizuka, M., and Esumi, K. (1988). *Bull. Chem. Soc. Jpn.*, 61, 341.
11. Meguro, K., Tano, T., Toigoe, K., Nakamura, H., and Esumi, K. (1988). *Colloids Surf.*, 34, 381.
12. Tano, T., Esumi, K., and Meguro, K. (1989). *J. Colloid Interface Sci.*, 133, 530.
13. Duff, D. G., Edwards, P. P., Evans, J., Gauntlett, J. T., and Jefferson, D. A. (1989). *Angew. Chem.*, 101, 610.
14. Evans, J. (1988). *NATO AS1 Ser.*, Ser. C (Surf. Organomet. Chem., Mol. 31 Approaches Surf. Catal.), 231, 47.
15. Evans, J., Hayden, B., Mosselmans, and Murray, A. (1992). *J. Am. Chem. Soc.*, 114, 6912.
16. Evans, J., Hayden, B., Mosselmans, F. and Murray, A. (1992). *Surf. Sci.*, 279 (1–2), 159.
17. Lamb, H. H., Gates, B. C., and Knözinger, H. (1988). *Angew. Chem.*, 100, 1162.
18. Gates, B. C. (1993). *Catalytic Chemistry*, Wiley, New York.
19. Heaton, B. T. (1988). *Pure Appl. Chem.*, 60(12), 1757.
20. Brown, D., Heaton, B. T., and Iggo, J. A. (1991). *Catal. Met. Complexes*, 12, 329.
21. Devenish, R. W., Mulley, S., Heaton, B. T., and Longoni, G. (1992). *J. Mater. Res.*, 7(10), 2810.
22. Henglein, A. (1993). *J. Phys. Chem.*, 97, 5457.
23. Henglein, A., Mulvaney, P., Holzwarth, A., Sosebee, T. E., and Fojtik, A. (1992). *Ber. Bunsenges. Phys. Chem.*, 98, 754.
24. Klabunde, K. J., Li, Y.-X., and Tan, B.-J. (1991). *Chem. Mater.*, 3, 30.
25. Klabunde, K. J. (1984). *Science*, 224, 1329.
26. Klabunde, K. J., and Imizu, Y. (1984). *J. Am. Chem. Soc.*, 106, 2721.
27. Klabunde, K. J., and Tanaka, Y. (1983). *J. Mol., Catal.*, 21, 57.
28. Matosuo, K., and Klabunde, K. J. (1982). *J. Org. Chem.*, 47, 843.
29. Klabunde, K. J., Davis, S. C., Hattori, H., and Tanaka, Y. (1978). *J. Catal.*, 54, 254.
30. Klabunde, K. J., Efner, H. F., Murdock, T. O., and Ropple, R., (1976). *J. Am. Chem. Soc.*, 98, 1021.
31. Knözinger, H. (1992). *Cluster Models for Surface and Bulk Phenomena (G. Paccioni, P. S. Bagus, and F. Parmigiami, eds.)*, Plenum Press, New York.
32. Larpent, C., and Patin, H. (1988). *J. Mol. Catal.*, 44, 191.
33. Lewis, L. N. (1993). *Chem. Rev.*, 93, 2693.

34. Lewis, L. N., and Lewis, N. (1989). *Chem. Mater.*, **1**, 106.
35. Lewis, L. N., and Lewis, N. (1986). *J. Am. Chem. Soc.*, **108**, 7228.
36. Lewis, L. N., Uriarte, R., and Lewis, N. (1991). *J. Catal.*, **127**, 67.
37. Stromnova, T. A., Busygina, I. N., Vargaftik, M. N., and Moiseev, I. I. (1990). *Metalloorg. Khim.*, **3**, 803.
38. Moiseev, I. I. (1989). *Pure Appl. Chem.*, **61**, 1755.
39. Berenbyum, A. S., Knizhnik, A. G., Mund, S. L., and Moiseev, I. I. (1982). *J. Organomet. Chem.*, **234**, 219.
40. Vargaftik, M. N., Zagorodnikiv, V. P., Stolyarov, I. P., Moiseev, I. I., Likholobov, V. A., Kochubey, D. J., Churilin, A. L., Zaikovsky, V. I., Zamaraev, K. I., and Timofeeva, G. I. (1985). *J. Chem. Commun.*, 937.
41. Moiseev, I. I., Stromnova, T. A., and Vargaftik, M. N. (1994). *J. Mol. Catal.*, **86**, 71.
42. Moiseev, I. I. (1984). *Mekh. Katal. Novosibirsk*, **21**, 172; from Ref. *Zh, Khim.* (1985). Abstr. No. 4B4121.
43. Moiseev, I. I. (1984). *Itogi Nauki Tekh. VINITI. Kinet. Katal.* **13**, 47; from Ref. *Zh. Khim.* (1984). Abstr. No. 16B4121.
44. Schmid, G. (1990). *Endeavour*, **14**, 172.
45. Schmid, G. (1990). *Aspects Homogen. Catal.*, **7**, 1.
46. Schmid, G., Smit, H. A., van Staveren, M. P. J., and Thiel, R. C. (1990). *New. J. Chem.*, **14**, 559.
47. Schmid, G., Klein, N., Morun, B., Lehnert, A., and Malm, J. O. (1990). *Pure Appl. Chem.*, **62**, 559.
48. Schmid, G. (1988). *Polyhedron*, **7**, 2321.
49. Schmid, G. (1988). *Chem. Unserer Zeit*, **22**, 85.
50. Schmid, G. (1987). *Nachr. Chem., Tech. Lab.*, **35**(249), 252.
51. Schmid, G. (1978). *Angew. Chem.*, **90**, 417.
52. De Jongh, L. J., De Aguiar, J. A. O., Brom, H. B., Longoni, G., van Ruitenbeek, J. M., Schmid. G., Smit, H. A., van Staveren, M. P. J., and Thiel, R. C. (1989). *Z. Phys. D: At., Mol., Clusters*, **12**, 455.
53. Schmid, G. (1985). *Struct. Bonding (Berlin)*, **62**, 51.
54. Hirai, H., Komatsuzaki, S., and Toshima, N. (1984). *Bull. Chem. Soc. Jpn.*, **57**(2), 488.
55. Hirai, H., Chawanya, H., and Toshima, N. (1985). *Bull. Chem. Soc. Jpn.*, **58**(2), 682.
56. Hirai, H., Komatsuzaki, S., and Toshima, N. (1986). *J. Macromol. Sci., [Part A] Chem.*, **23**(8), 933.
57. Toshima, N., Teranishi, T., and Saito, Y. (1992). *Macromol. Chem., Macromol. Symp.* **59**, 327.
58. Toshima. N., and Takahashi, T. (1992). *Bull. Chem. Soc. Jpn.*, **65**(4), 400.
59. Toshima. N., Harada, M., Yonezawa, T., Kushihashi, K., and Asakura, K. (1991). *J. Phys. Chem.*, **95**, 7448.
60. Toshima. N., Takahashi, T., and Hirai, H. (1988). *J. Macromol. Sci. [Part A] Chem.*, **25**(5-7), 669.
61. Toshima. N. (1990). *J. Macromol. Sci. [Part A] Chem.*, **27** (9-11), 1225.
62. Zhao, B., and Toshima. N. (1990). *Chem. Express*, **5**(10), 721.
63. Paal, C., and Amberger, C. (1904). *Berichte*, **37**, 124.
64. Paal, C., and Amberger, C. (1905). *Berichte*, **38**, 1398.
65. Bradley, J. S. (1994). *Clusters and Colloids* (G. Schmid, ed.), VCH Publishers, Weinheim, Germany.
66. Schmid, G. (1992). *Chem. Rev.*, **92**, 1709.
67. Liu, H., and Toshima, N. (1992). *J. Chem. Soc., Chem. Commun.*, 1095.
68. Bradley, J. S. (1994). *Clusters and Colloids* (G. Schmid, ed.), VCH Publishers, Weinheim, Germany.

69. Bradley, J. S. (1993). *Chem. Mater.*, 5, 254.
70. Torigoe, K., and Esumi, K. (1993). *Langmuir*, 9, 1664.
71. Fendler, J. H. (1994). *Membrane-Mimetic Approach to Advanced Materials*, Springer-Verlag, Berlin.
72. Larpent, C., Brisse-Le Menn, F., and Patin, H. (1991). *Mol. Catal.*, 65, L35.
73. Sato, T., Kuroda, S., Takami, A., Yonezawa, Y., and Hada, H. (1991). *Appl. Organomet. Chem.*, 5, 261.
74. Sato, T. (1990). *J. Appl. Phys.*, 68, 1297.
75. Sato, T. (1987). *J. Chem. Soc., Faraday Trans. 1*, 83, 1559.
76. Sato, T., Kuroda, S., Takami, A., Yonezawa, Y., and Hada, H. (1991). *Appl. Organomet. Chem.*, 5, 261.
77. Hirai, H., Nakao, Y., and Toshima, N. (1978). *Chem. Lett.*, 5, 545.
78. Ohtaki, M., Komiyama, M., Hirai, H., and Toshima, N. (1991). *Macromolecules*, 24, 5567.
79. Toshima, N. (1991). *J. Phys. Chem.*, 95, 7448.
80. N. Toshima, and T. Yonezawa (1992). *Macromol. Chem., Macromol. Symp.*, 59, 287.
81. Toshima, N. (1992). *J. Phys. Chem.*, 96, 9927.
82. Hirai, H., Nakao, Y., and Toshima, N. (1976). *Chem. Lett.*, 9, 905.
83. Ohtaki, M., Komiyama, M., Hirai, H., and Toshima, N. (1991). *Macromolecules*, 24, 5567.
84. Toshima, N., Ohtaki, M., and Teranishi, T. (1991). *Reactive Polym.*, 15, 135.
85. Toshima, N., and Takahashi, T. (1992). *Bull. Chem. Soc. Jpn.*, 65, 400.
86. Bönnemann, H., Brijoux, W., Brinkmann, R., Dinjus, E., Joussen, Th., and Korall, B. (1991). *Angew. Chem.*, 103, 1344; (1991) *Angew. Chem., Int. Ed. Engl.*, 30, 1312.
87. Bönnemann, H., and Korall, B. (1992). *Angew. Chem.*, 104, 1506.
88. Bönnemann, H., Brijoux, W., Brinkmann, R., Fretzen, R., Joussen, Th., Köppler, R., Korall, B., Neiteler, P., and Richter, J. (1994). *J. Mol. Catal.*, 86, 129.
89. Bönnemann, H., Brijoux, W., Brinkmann, R., Dinjus, E., Fretzen, R., Joussen, Th., and Korall, B. (1992). *J. Mol. Catal.*, 74, 323.
90. Bönnemann, H., Brinkmann, R., Köppler, R., Neiteler, P., and Richter, J. (1992). *J. Adv. Mater.*, 4, 804.
91. Bönnemann, H., Brijoux, W., Brinkmann, R., Dinjus, E., Fretzen, R., and Korall, B. (1991). German Patent DE 4 111 719.
92. Bönnemann, H., Brinkmann, R., and Neiteler, P. (1994). *Appl. Organomet. Chem.*, 8, 361.
93. Bönnemann, H., and Brijoux, W. (1994). *Molecularly Designed Ultrafine Nanostructured Materials*, 351 (K. E. Gonsalves, G. M. Chow, T. D. Xiao, and R. C. Cammarata, eds.), from the *MRS Symposium Proceedings Series*, MRS, Pittsburgh, Pennsylvania.
94. Bönnemann, H., and Brijoux, W. (1995). *Active Metals* (A. Fürstner, ed.), VCH Publishers, Weinheim, Germany (1995).
95. Aleandri, L. E., Bönnemann, H., Jones, D. J., Richter, J., and Rozière, J., *J. Mater. Chem.*, submitted.
96. Bönnemann, H., Brijoux, W., Richter, J., Becker, R., Hormes, J., and Rothe, J. (1995). *Z. Naturforsch. Teil B*, 50, 333.
97. Kiwi, J., and Grätzel, M. (1979). *J. Am. Chem. Soc.*, 101, 7214.
98. Deshpande, V. M., Singh, P., and Narasimhan, C. S. (1989). *J. Mol. Catal.*, 53, L21.
99. Deshpande, V. M., Singh, P., and Narasimhan, C. S. (1990). *J. Mol. Catal.*, 63, L5.
100. Deshpande, V. M., Singh, P., and Narasimhan, C. S., (1990). *J. Chem. Soc., Chem. Commun.*, 1181.
101. Esumi, K., Shiratori, M., Ihshizuka, H., Tano, T., Torigoe, K., and Meguro, K., (1991). *Langmuir*, 7, 457.

102. Larpent, C., Brisse-Le Menn, F., and Patin, H. (1991). *New J. Chem.,* 15, 361.
103. Lisichkin, G. V., Yuffa, A. Ya., and Khinchagashvii, V. Yu. (1976). *Russ. J. Phys. Chem.,* 50, 1285.
104. Nakao, Y., Kaeriyama, K. (1986). *J. Colloids Surf. Sci.,* 110(1), 82.
105. Petrow, H. G., and Allen, R. J. (1977). U. S. Patent 4 044 193.
106. Reetz, M. T., and Helbig, W. (1994). *J. Am. Chem. Soc,* 116, 7401.
107. Reetz, M. T., Helbig, W., Quaiser, S., Stimmig, U., Breuer, N., and Vogel, R. (1995). *Science,* 267, 367.
108. Toshima, N., Takahashi, T., and Hirai, H. (1985). *Chem. Lett.,* 1245.
109. Toshima, N., and Takahashi, T. (1988). *Chem. Lett.,* 573.
110. Reetz, M. T., and Quaiser, S. (1995). *Angew. Chem.,* 107, 2461.
111. Köppler, R. (1995). Ph.D. thesis, University of Aachen, 58–64.
112. Köppler, R. (1995). Ph.D. thesis, University of Aachen, 65–67.
113. Bönnemann, H., Brijoux, W., and Richter, J. (1992). unpublished results.
114. Bönnemann, H., Brijoux, W., Richter, J., Siepen, K., Franke, R., Hormes, J., Pollmann, J., and Rothe, J., *Fresenius Z. Anal. Chem.,* submitted.
115. Simon, U., Schön, G., and Schmid, G. (1993). *Angew. Chem., Int. Ed.,* 250.
116. Behm, R. J., Witek, G., Noeske, M., Mestl, G., and Shaikhutdinov, Sh., *Catal. Lett.,* submitted.
117. Egeler, N. (1992). Ph.D. thesis, University of Aachen.
118. Schulze Tilling, A., Ph.D. thesis, University of Aachen, in preparation.
119. Brönnimann, C., Mallat, T., and Baiker, A. (1995). *J. Chem. Soc., Chem. Commun.,* 1377.
120. EP 0 142 725 (Prior. 24.10.1983) to Kao Corporation.
121. EP 0 350 741 B 1 (Prior. 09.07.1988) to Degussa AG.
122. Siepen, K., Ph.D. thesis, University of Aachen, in preparation.
123. Braun, G., Ph.D. thesis, University of Aachen, in preparation.
124. Orito, Y., Imai, S., and Niwa, S. (1979). *J. Chem. Soc. Jpn.,* 8, 1118.
125. Blaser, H. U., Jalett, H. P., Monti, D. M., Reber, J. F., and Welnli, J. T. (1988). *Stud. Surf. Sci, Catal.,* 59, 153.
126. Welnli, J. T., Baiker, A., Monti, D. M., and Blaser, H. U. (1990). *J. Mol. Catal.,* 61, 207.
127. Singh, U. K., Landau, R. N., Sun, Y., LeBlond, C., Blackmond, D. G., Tanielyan, S. K., and Augustine, R. L. (1995). *J. Catal.,* 154, 91.
128. Sutherland, I. M. (1990). *J. Catal.,* 125, 77–88.
129. Wells, P. B. (1994). *Recl. Trav. Chim. Pays-Bas,* 113, 465–474.
130. Hindenburg, Th. (1995). Ph.D. thesis, University of Aachen.

CHAPTER 8

Sonochemical Preparation of Nanostructured Catalysts

Kenneth S. Suslick, Taeghwan Hyeon, Mingming Fang, and
Andrzej A. Cichowlas
School of Chemical Sciences
University of Illinois at Urbana–Champaign
Urbana, Illinois 61801

KEYWORDS: acoustic cavitation, molybdenum carbide, nanostructured materials, silica-supported Fe, sonochemical decomposition, sonochemistry, ultrasound

8.1 Introduction

Sonochemistry arises from acoustic cavitation: the formation, growth, and implosive collapse of bubbles in a liquid [1]. The collapse of bubbles generates localized hot spots through adiabatic compression or shock wave formation within the gas of the collapsing bubble. This local heating produces a wide range of high-energy chemistry and also induces the emission of light, sonoluminescence. The conditions formed in these hot spots have been experimentally determined, with transient temperatures of ~5000 K, pressures of ~1800 atm, cooling rates in excess of 10^{10} K/s [2,3]. Using these extreme conditions, we have explored a variety of applications of ultrasound to materials chemistry [4].

Among these applications, we have developed a new synthetic technique for the synthesis of nanostructured inorganic materials. When solutions of volatile organometallic compounds are irradiated with high-

197

intensity ultrasound, high-surface-area solids are produced that consist of agglomerates of nanometer clusters. For Fe and Co, nanostructured metals and alloys are formed; for Mo and W, the metal carbides (e.g., Mo_2C) are produced. These sonochemically produced nanostructured solids are active heterogeneous catalysts for hydrocarbon reforming and CO hydrogenation. When polymeric ligands (e.g., polyvinylpyrrolidone) or oxide supports (alumina or silica) are used, the initially formed nanometer clusters can be trapped as colloids or supported catalysts, respectively.

8.2   Literature Survey

The preparation of nanostructured materials and catalysts is the focus of intense study in materials science [5,6]. A variety of chemical and physical preparative methods have been developed to produce materials with nanometer structure [7], including metal atom evaporation [8], thermal decomposition of organometallic compounds [9], and reduction of metal salts [10,11]. Sonochemical routes to nanostructured catalysts are also being developed and will be the focus of this chapter.

Ultrasonic irradiation can affect the reactivity of heterogeneous catalysts in various ways: by altering the formation of heterogeneous catalysts, by perturbing the properties of previously formed catalysts, or by affecting the reactivity during catalysis. There are review articles on sonochemical application to heterogeneous catalysis [12,13]. Reports of modest effects of sonication on heterogeneous catalysis include the hydrogenation of alkenes and benzylethers with Pd on carbon [14,15,16]. More impressive accelerations have also been reported. For instance, the hydrogenation of alkenes by ordinary Ni powder was enormously enhanced by ultrasonic irradiation [17]. The high enhancement of catalytic reaction rate was due to the surface morphology change and removal of passivating oxide layer [18].

More directly relevant to this chapter, ultrasound has also been used in the preparation of high-surface-area, nanostructured, amorphous iron [19–21]. The preparation of amorphous metals requires high cooling rates to prevent crystallization, and acoustic cavitation provides a convenient laboratory process with which to do this. Given their high-surface areas and nanometer substructure, one might anticipate good catalytic activity from such materials; this proved to be the case for Fischer–Tropsch hydrogenation of CO [22] and alkane dehydrogenation [23,24].

8.3   Results and Discussion

Sonochemical decomposition rates for volatile organometallic compounds depend on a variety of experimental parameters. To achieve good sonochemical yields, precursors should be highly volatile because the primary sonochemical reaction site is the vapor inside the cavitating bubbles [25]. Good thermal stability is also important, so that decomposition takes place only during cavitation. In addition, the solvent vapor pressure should be low at the sonication temperature, because significant solvent vapor inside the bubble reduces bubble collapse efficiency [1].

Our sonochemical synthesis of nanostructured materials is extremely versatile: various forms of nanophase materials can be generated simply by changing the reaction medium (Scheme 1). When precursors are sonicated by high-boiling alkanes, nanostructured metal powders are formed. Sonication in the presence of a polymeric ligand, for example, polyvinylpyrrolidone (PVP), stable nanophase metal colloids can be isolated and characterized. A transmission electron micrograph (TEM) of the nanocolloid Fe/PVP is shown in Fig. 1. Additionally, sonication of the precursor in the presence of an inorganic support (e.g., silica or alumina) provides an alternative means of trapping the nanometer clusters formed during cavitation and produces active supported heterogeneous catalysts. Why we get nanometer clusters sono-

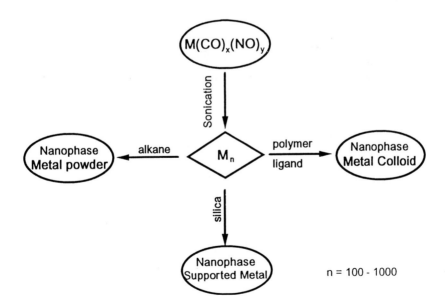

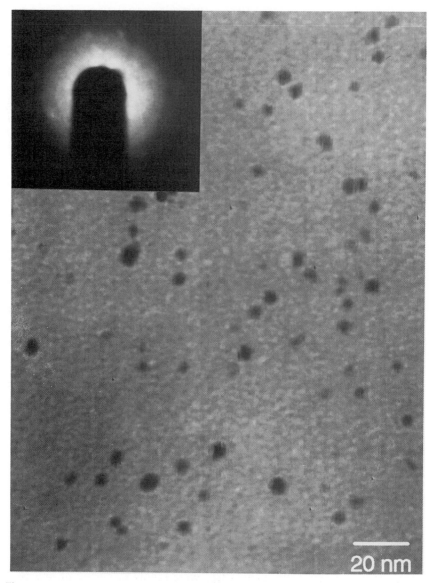

Figure 1. Transmission electron micrograph of nanostructured Fe colloid sonochemically prepared with polyvinylpyrrolidone, obtained on Phillips 420 electron microscope. Reprinted from *Materials Science & Engineering*, A204, 186–192, 1995, with kind permission from Elsevier Science S.A., Lausanne, Switzerland.

chemically is not yet clear. At 20 kHz, the cavitation bubble will be roughly 120 μm in diameter. If the metal carbonyl inside a single bubble were to give rise to a single metal particle, it would be micrometer sized. Further studies are underway to explain these observations.

To show by example the utility of sonochemistry as an advanced technique in catalyst preparation, we will examine here the sonochemical synthesis and heterogeneous catalytic studies of nanostructured Fe on silica, nanophase Fe–Co alloys, and nanostructured Mo_2C.

8.3.1 Synthesis and Catalytic Studies of Nanostructured Silica-Supported Fe

Ultrasonic irradiation of decane solutions of iron pentacarbonyl, $Fe(CO)_5$, in the presence of silica gel produces a silica-supported amorphous nanostructured iron. The iron loading on the SiO_2 can be easily varied by changing the initial concentration of the $Fe(CO)_5$ solution. Elemental analysis reveals Fe, Si, O, and a trace amount of carbon (<1%) to be present. The origin of carbon most likely arises from the decomposition of CO or the alkane solvent during ultrasonic irradiation.

The amorphous nature of these supported iron particles has been confirmed by several different techniques, including differential scanning calorimetry (DSC), powder X-ray diffraction (XRD), and electron-beam microdiffraction. Differential scanning calorimetry shows one irreversible exothermic transition at 335°C corresponding to a disorder–order transition (i.e., crystallization) of the amorphous iron. X-ray powder diffraction shows no diffraction peaks from the material as initially prepared; after heat treatment under He at 400°C for 4 h, which is sufficient to induce crystallization, only the lines characteristic of α–Fe metal (d-spacings of 2.03, 1.43, 1.17, and 1.04 Å) are observed. After crystallization, the X-ray powder diffraction pattern contains no peaks attributable to iron oxide, iron carbide, or other iron-based phases. Electron microdiffraction with a transmission electron microscope confirms these observations and shows only a diffuse ring characteristic of amorphous iron particles.

Transmission electron micrographs (TEM) showed that the iron particles produced by sonolysis of $Fe(CO)_5$ were highly dispersed on the SiO_2 surface. The iron particles range in size from 3 to 8 nm. Chemisorption of CO permits measurement of the dispersion and the average particle size of iron supported on silica surfaces [26]. CO chemisorption measurement data at −78°C on our samples give an average iron particle size of 7 nm, in good agreement with the TEM data.

The catalytic activity of the silica-supported nanostructured iron was probed with the commercially important Fischer–Tropsch synthesis reaction (i.e., hydrogenation of CO). Figure 2 compares the activity (in terms of turnover frequency of CO molecules converted per catalytic site per second) of silica-supported nanophase iron and conventional silica-supported iron, prepared by the incipient wetness method, as a function of temperature. These catalytic data were obtained at high iron loading and low dispersion to minimize the effects of support and dispersion. The sonochemically produced iron-on-silica catalyst is an order of magnitude more active than the conventionally supported iron. Moreover, the silica-supported nanostructured iron catalyst exhibits high activity at low temperatures (<250°C), where the silica-supported conventional iron catalyst has no activity. We suggest that the dramatic difference in activity between the two samples below 300°C may be due to the amorphous nature of iron and the inherently highly defected surface formed during sonolysis of $Fe(CO)_5$ when the amorphous state of iron is preserved. At higher temperature, activity decreases, which may be due to iron crystallization, surface annealing, or catalyst deactivation from surface carbon deposition.

Differences between the catalytic properties of the nanostructured iron and of conventionally supported catalysts are also observed in selectivities of hydrocarbon synthesis. Under our conditions, the major reaction products for both catalysts are short-chain C_1 to C_4 hydrocarbons and CO_2.

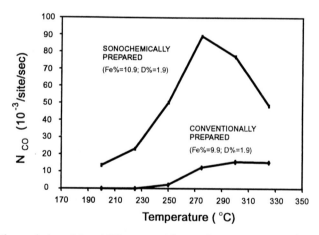

Figure 2. The catalytic activity of SiO_2-supported amorphous nanostructured iron sonochemically prepared from $Fe(CO)_5$ and SiO_2 slurry (iron loading = 10.94 wt % and dispersion, D = 1.85%) and SiO_2-supported crystalline iron prepared by the incipient wetness method (iron loading = 9.91 wt %, D = 1.86%) as a function of temperature for Fischer–Tropsch synthesis (H_2/CO = 3.48, 1 atm, 25°C). Reprinted from *Materials Science & Engineering*, A204, 186–192, 1995, with kind permission from Elsevier Science S.A., Lausanne, Switzerland.

Product distribution of hydrocarbons showed that at temperatures lower than 275°C, the silica-supported nanostructured iron catalyst shows higher selectivity toward long-chain hydrocarbons (C_{5+}), whereas the conventionally supported iron shows no activity at these temperatures. At temperatures higher than 275°C, the reaction product distributions are similar for both types of catalysts.

8.3.2 Synthesis and Catalytic Studies of Nanostructured Fe–Co Alloys

$Fe(CO)_5$ and $Co(CO)_3(NO)$ were chosen as precursors because of their high vapor pressures at modest bulk solution temperatures where they are still thermally stable. The composition of the Fe–Co alloys can be controlled simply by changing the ratio of solution concentrations of the precursors; alloy compositions ranging from pure Fe to pure Co are readily obtained.

The solid–solution nature of the alloys was confirmed by TEM-EDX (energy dispersive X-ray spectroscopy) results, which were made on different spots of the polycrystalline alloy powders. The EDX results show that the alloys are homogeneous on a nanometer scale: electron-beam size was 1-nm diameter at 100 kV on a VG-HB5 electron microscope. The original Fe, Co, and Fe–Co alloys produced by ultrasound are amorphous, as determined by XRD, electron-beam microdiffraction, and DSC. After heat treatment under H_2 gas flow at 400°C for 2 h, all samples underwent crystallization. The XRD results show no peaks attributable to iron/cobalt oxide, iron/cobalt carbide, or other iron/cobalt impurity phases. Pure Fe crystallizes to body-centered cubic (bcc) structure; pure Co crystallizes to face-centered cubic (fcc) and hexagonal close-packed (hcp) mixed structures. All the alloys that we have tested so far crystallize in the bcc structure; this is consistent with the known Fe–Co equilibrium phase diagram that strongly favors the bcc structure [27]. Elemental analysis results show that nearly pure metal and alloys are produced after H_2 treatment. A scanning electron micrograph (SEM) at high magnification indicates that these materials are porous aggregates of small clusters of 10 to 20-nm particles. Surface electronic structures and surface compositions of the sonochemically prepared Fe–Co alloys were also examined by using X-ray photoelectron spectroscopy (XPS). The XPS measurements have been performed on heat-treated samples before catalytic reactions. The electronic structures of the surfaces of these samples appear to be the same as the pure metals. The surface compositions of the alloys demonstrate some small enrichment of Fe over Co. Similar trends toward an iron-enriched surface have been reported by other researchers with other preparations using co-precipitation methods [28].

Catalytic studies of the sonochemically prepared Fe–Co alloys were made for cyclohexane dehydrogenation and hydrogenolysis reactions. All catalysts were treated under H_2 gas flow at 400°C for 2 h before the catalytic studies. While this does not alter the nanostructure of the material significantly, it does cause crystallization of the nanometer clusters. H_2 treatment is necessary, however, to provide a reproducible catalytic surface. The catalytic activity (in terms of turnover frequency of cyclohexane molecules converted to benzene per surface Fe or Co atom per second) as a function of temperature is shown in Fig. 3. Two kinds of products were formed during the cyclohexane reaction: benzene was the only dehydrogenation reaction product and aliphatic hydrocarbons (mostly methane) were the hydrogenolysis reaction products. The catalytic selectivity (in terms of the percentage of benzene among all the reaction products) as a function of temperature is shown in Fig. 4. The catalytic properties of the sonochemically prepared Fe, Co and Fe–Co alloys in the cyclohexane reaction exhibit interesting trends. First, they are all active catalysts for cyclohexane conversion: pure Co has the highest activity (although primarily for hydrogenolysis), pure Fe has the lowest activity, and Fe–Co alloys have intermediate activity between pure Fe and Co. Second, Fe–Co alloys generate much more dehydrogenation product (benzene) than pure Fe or Co does. Third, the 1:1 Fe–Co alloy has both much higher dehydrogenation activities and selectivities at all reaction temperatures (250–300°C) than the other alloys or pure metals do. In the best cases the selectivity for dehydrogenation approaches 100%.

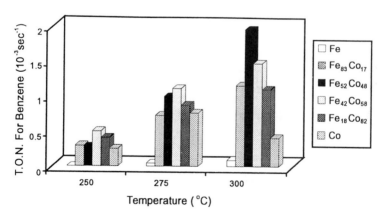

Figure 3. The catalytic activity of Fe, Co, and Fe–Co alloys for dehydrogenation of cyclohexane to benzene as a function of temperature. Reprinted from *Materials Science & Engineering*, A204, 186–192, 1995, with kind permission from Elsevier Science S.A., Lausanne, Switzerland.

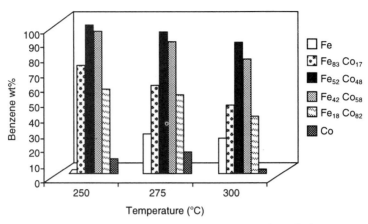

Figure 4. The catalytic selectivity of Fe, Co, and Fe–Co alloys for dehydrogenation versus hydrogenolysis of cyclohexane as a function of temperature. Reprinted from *Materials Science & Engineering*, A204, 186–192, 1995, with kind permission from Elsevier Science S.A., Lausanne, Switzerland.

8.3.3 Synthesis and Catalytic Studies of Nanostructured Molybdenum Carbide

Molybdenum and tungsten carbides have been explored as heterogeneous catalysts because of the similarity in activity that these carbides share with platinum group metals [29–31]. For catalytic applications, high-surface-area materials are generally needed. The preparation of interstitial carbides of molybdenum and tungsten with high-surface areas, however, is very difficult. Volpe and Boudart [32] prepared carbides of molybdenum and tungsten with high-surface areas by the temperature-programmed carburization of the corresponding nitrides. We present here a simple sonochemical synthesis of nanophase molybdenum carbide from the ultrasonic irradiation of molybdenum hexacarbonyl.

Sonochemical decomposition of molybdenum hexacarbonyl in hexadecane produced a black powder. Powder XRD showed extremely broad peaks centered at d-spacings of 2.4, 1.5, and 1.3 Å, which do not match bcc lines of molybdenum metal. After the heat treatment at 450°C under helium flow for 12 h, sharper peaks in the XRD were observed at d-spacing values of 2.39, 1.49, and 1.27 Å, which match very well with fcc molybdenum carbide, Mo_2C (Fig. 5). Elemental analysis also confirmed the stoichiometry of 2Mo/C, but with some oxygen as discussed in the following text. The formation of molybdenum carbide can be explained by

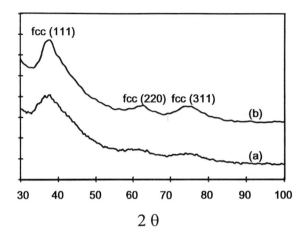

Figure 5. Powder X-ray diffraction patterns of sonochemically produced Mo$_2$C (a) after synthesis and (b) after heat treatment under He at 450°C for 12 h. Reprinted from *Materials Science & Engineering*, A204, 186–192, 1995, with kind permission from Elsevier Science S.A., Lausanne, Switzerland.

the disproportion of carbon monoxide on the active metal surface to form carbon and carbon dioxide [33].

The SEM confirms that the solid is an extremely porous aggregate of nanometer clusters. The high-resolution TEM showed that the solids were made up of aggregates of 2-nm-sized particles (Fig. 6). The particle size calculated from the line broadening of powder XRD was 1.6 nm. The surface are determined by Brunauer–Emmett–Teller (BET) gas adsorption isotherms was 188 m^2/g.

Even after heat treatment at 450°C under helium, the sample still contained about 4 wt % of oxygen; this is also a problem for the conventional synthesis of Mo$_2$C catalysts [35]. Because the presence of oxygen could poison the catalytic activity, it was removed before catalytic studies by heating in a flow of 1:1 CH$_4$/H$_2$ mixture at 300°C for 1 h, then at 400°C for 1 h, and finally at 500°C for 48 h. The flow rate of the CH$_4$/H$_2$ mixture was 27.5 cm^3 (STP)/min. After the heat treatment, excess carbon, hydrogen, and oxygen had been largely removed. The elemental analysis results showed the sample had a stoichiometry of Mo$_2$C$_{1.02}$ with less than 0.09 wt % oxygen (by difference) and <0.02 wt % hydrogen. Electron micrographs showed that the material was still porous and was composed of particles of 3 nm in diameter. The BET surface area decreased slightly to 130 m^2/g. After carburization, care must be taken to avoid sample exposure to oxygen.

The catalytic activity of the sonochemically produced molybdenum carbide was tested for dehydrogenation of cyclohexane. Figure 7 shows the

10 nm

Figure 6. Transmission electron micrograph of sonochemically produced Mo₂C, obtained on Phillips CM-12 electron microscope. Reprinted from *Materials Science & Engineering*, A204, 186–192, 1995, with kind permission from Elsevier Science S.A., Lausanne, Switzerland.

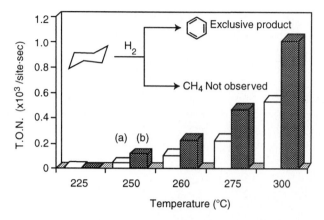

Figure 7. Catalytic activity of sonochemically produced Mo$_2$C for dehydrogenation of cyclo-hexane (a) sample heat-treated under helium at 450°C for 12 h and (b) sample heat-treated under CH$_4$/H$_2$ at 500°C for 48 h. Reprinted from *Materials Science & Engineering*, A204, 186–192, 1995, with kind permission from Elsevier Science S.A., Lausanne, Switzerland.

catalytic activity (in terms of turnover frequency of cyclohexane molecules converted per site per second, as determined by CO chemisorption on fully carburized samples) as a function of temperature for the sample pretreated under CH$_4$/H$_2$ at 500°C for 48 h and for the sample pretreated under helium at 450°C for 12 h. For comparison, the catalytic dehydrogenation activity of Pt under similar conditions was roughly 20-fold higher, on a BET surface area comparison of turnover.

At all the reaction temperatures, benzene was the *only* product formed for either sample. No hydrogenolysis products were detected. Samples heat-treated under helium (i.e., with oxide impurity) had the same selectivity (benzene is still the only product), but with somewhat lower activity. These results demonstrate that the sonochemically prepared molybdenum carbide is an excellent dehydrogenation catalyst and a rather poor hydrogenolysis catalyst. This is also confirmed by the complete lack of activity for ethane hydrogenolysis to methane.

The catalytic properties of Mo$_2$C (fcc) and Mo$_2$C (hcp) have been studied intensively since 1985 [29–36]; in all cases oxygen contamination remains a serious problem. CO hydrogenation, alkene hydrogenation, and hydrocarbon isomerization and hydrogenolysis have been investigated. In spite of the importance of the reaction, however, few reports mention Mo$_2$C as an active catalyst for alkane dehydrogenation. As a precedent for our studies, we note that carburization of Mo is known to temper Mo metal for dehydrogenation of hydrocarbons [37]. In contrast with our high selectivity against hydrogenolysis of cyclo-hexane, Ranhotra *et al.* [29] reported that there is some catalytic activity for ethane hydrogenolysis using conventional Mo_2C (fcc) prepared by reduction and carburization of MoO_3. It should be noted that there are substantial amounts of oxygen left in the conventionally prepared nanophase Mo_2C: as initially formed the C:O ratio is 2:7, and even after extensive carburization the C:O ratio is 0.74. In sonochemically prepared samples, we have been able to reduce this ratio to <0.09. This may contribute to the highly selective catalytic behavior of sonochemically generated molybdenum.

8.4   Experimental Details

8.4.1   General Procedures

All manipulations for the preparation of samples were performed using Schlenk vacuum line and inert atmosphere box (Vacuum Atmospheres, <1 ppm O_2) techniques. Pentane was distilled over sodium/benzophenone. Decane and hexadecane were distilled over sodium. Ultrasonic irradiation was accomplished with a high-intensity ultrasonic probe (Sonic and Materials, model VC-600, 1-cm diameter Ti horn, 20 kHz, 100 Wcm^{-2}).

Powder X-ray diffraction data were collected on a Rigaku D-max diffractometer using Cu Kα radiation (λ = 1.5418 Å). Scanning electron micrographs were taken on a Hitachi S800 electron microscope. Transmission electron micrographs were taken on a Phillips CM-12 electron microscope. Samples for elemental analysis were submitted in sealed vials without exposure to air.

Hydrogen (99.99%, Linde), methane (99.97%, Matheson), and CO (99.0+%, Linde) were further purified through 4A molecular sieves and oxy-traps (Alltech). Cyclohexane (99+%, Fisher) was dried over molecular sieves prior to use. In cyclohexane reaction, a digital mass flow controller maintained the flow of hydrogen at 27.5 cm^3 (STP)/min to carry the cyclohexane vapor at a constant partial pressure of 0.1 bar through the catalyst. A quartz reactor was used for both adsorption and gas–solid catalytic studies. The catalysts were transferred from an inert atmosphere box to the catalytic rig without exposure to air. Surface areas were calculated by applying the BET equation to the N_2 adsorption isotherm measured at 77 K. The gas products obtained during the temperature-programmed desorption (TPD) and temperature-programmed reduction (TPR) experiments were analyzed by a quadruple mass spectrometer (Spectra Instruments). The catalytic reaction products were analyzed by gas chromatography (Hewlett-Packard 5730A) on a N-octane–Porasil C column with flame ionization detector.

8.4.2 Synthesis of Nanostructured Catalysts by Sonochemical Decomposition

For the preparation of nanostructured Fe/SiO_2 catalysts, silica gel (Universal Scientific Incorporated chemicals, 63 – 100 mesh) was pretreated at 450°C under vacuum (1×10^{-5} torr) for 10 h before use. To this, a 0.1 M solution of $Fe(CO)_5$ in dry decane was added, in which the vapor pressure of $Fe(CO)_5$ is roughly 3 torr at 20°C. The slurry was then irradiated at 20°C with a high-intensity ultrasonic probe for 3 h under argon. After irradiation, the black powder was filtered and washed with dry pentane in an inert atmosphere box.

Conventional silica-supported crystalline iron catalysts were prepared using the incipient wetness impregnation method by dissolving $Fe(NO_3)_3 \cdot 9H_2O$ in an aqueous solution which was added to silica gel [38]. These samples were dried at 220°C for 12 h and calcined at 450°C under an O_2 flow for 1 h. Reduction of iron supported on silica was conducted in a flow of hydrogen at 200°C for 1 hr, at 300°C for 1 h, and finally at 450°C for 2 h.

The synthesis of nanostructured Fe–Co alloy catalysts followed a procedure similar to that developed for the sonochemical synthesis of amorphous iron [19,22]. Solutions containing various relative concentrations of $Fe(CO)_5$ and $Co(CO)_3(NO)$ were irradiated in dry decane at 0°C with a high-intensity ultrasonic probe for 3 h under argon. After irradiation, black powders were formed, which were filtered and washed with dry pentane in the glovebox.

Nanostructured molybdenum carbide catalysts were prepared by ultrasonic irradiation of a slurry of molybdenum hexacarbonyl, 1 g of $Mo(CO)_6$ in 50 mL of hexadecane; and were sonicated at 90°C for 3 h under argon. Hexadecane was chosen as solvent because its vapor pressure is low at the sonication temperature. The black powder was filtered inside a glovebox and washed several times with purified, degassed pentane.

8.5 Conclusions

Sonochemical decomposition of volatile organometallic precursors in high-boiling solvents produces nanostructured materials in various forms with high catalytic activities. Nanometer colloids, nanoporous high-surface-area aggregates, and nanostructured oxide-supported catalysts can all be prepared by this general route. For example, sonication of iron pentacarbonyl with silica generated supported amorphous nanostructured Fe/SiO_2 catalyst. The nanostructured Fe/SiO_2 catalyst showed higher catalytic activity for Fischer–Tropsch synthesis than the conventional Fe/silica catalyst prepared by the incipient wetness method. Sonochemical synthesis of high-surface-area alloys can be accomplished by the sonolysis of $Fe(CO)_5$ and $Co(CO)_3(NO)$ to make nanostructured Fe and Co metals and Fe–Co alloys. The sonochemically prepared Fe–Co alloys have large surface areas relative to bulk metal even after heat treatment. We find very high catalytic activity for these Fe, Co, and Fe–Co powders for the dehydrogenation and hydrogenolysis of cyclohexane. Surprisingly, the Fe–Co alloys show very high selectivities for dehydrogenation. Other bimetallic catalysts that have been examined for dehydrogenation of cyclohexane often involved Pt; in these cases, too, selectivity is improved in the alloys [39–40]. Finally, ultrasonic irradiation of molybdenum hexacarbonyl produces aggregates of nanometer-sized clusters of face-centered cubic molybdenum carbide. The material was extremely porous with a high-surface area and consisted of aggregates of ~2-nm-sized particles. The catalytic properties showed that the molybdenum carbide generated by ultrasound is an active and highly selective dehydrogenation catalyst.

Acknowledgments

This work was supported by National Science Foundation. We thank Peggy Mochel; Vania Petrova; and the UIUC Center for Microanalysis of Materials, which is supported by the U. S. Department of Energy, for their assistance in the electron microscopic studies.

References

1. Suslick, K. S. (1988). *Ultrasound: Its Chemical, Physical, and Biological Effects,* (K. S. Suslick, ed.), p123. VCH Publishers, New York.
2. Suslick, K. S. (1990). *Science, 247,* 1439.
3. Flint, E. B., and Suslick, K. S. (1991). *Science, 253,* 1397.
4. Suslick, K. S. (1995). *MRS Bull., 20,* 29.
5. Weller, H. (1993). *Adv. Mater., 5,* 88.
6. Ozin, G. A. (1992). *Adv. Mater., 4,* 612.
7. Gonsalves, K. E., Chow, G. M., Xiao, T. O., and Cammarata, R. C., (eds.), (1994). *Molecularly Designed Nanostructured Materials,* MRS Symposium Proceedings, Vol. 351, Materials Research Society, Pittsburgh.
8. Davis, S. C., and Klabunde, K. J. (1982). *Chem. Rev., 82,* 152.
9. Lisitsyn, A. S., Golovin, A. V., Chuvilin, A. L., Kuznetsov, V. L., Romanenko, A. V., Danilyuk, A. F., and Yermakov, Y. I. (1989). *Appl. Catal., 55,* 235.
10. Bönnemann, H., Brijoux, W., Brinkmann, R., and Joussen, T. (1990). *Angew. Chem. Int. Ed. Engl., 129,* 273.

212 / Kenneth S. Suslick et al.

11. Tsai, K. -L., and Dye, J. L. (1991). *J. Am. Chem. Soc.*, **113**, 1650.
12. Suslick, K. S. (1993). *Encyclopedia of Materials Science and Engineering*, 3rd Suppl., (R. W. Cahn ed.), p. 2093, Pergamon Press, Oxford.
13. Suslick, K. S. (1994). *Encyclopedia of Inorganic Chemistry* Vol. 7, (King, R. B., ed.) pp. 3890–3905, Wiley, New York.
14. Boudjouk, P., and Han, B. H. (1983). *J. Catal.*, **79**, 489.
15. Townsend, C. A., and Nguyen, L. T. (1981). *J. Am. Chem. Soc.*, **103**, 4582.
16. Han, B. H., and Boudjouk, P. (1983). *Organometallics* **2**, 769.
17. Suslick, K. S., and Casadonte, D. J. (1987). *J. Am. Chem. Soc.*, **109**, 3459.
18. Suslick, K. S., Casadonte, D. J., and Doktycz, S. J. (1989). *Solid State Ionics*, **32/33**, 444.
19. Suslick, K. S., Choe, S. B., Cichowlas, A. A., and Grinstaff, M. W. (1991). *Nature (London)*, **353**, 414.
20. Grinstaff, M. W., Salamon, M. B., and Suslick, K. S. (1993). *Phys. Rev. B*, **48**, 269.
21. Bellissent, R., Galli, G., Grinstaff, M. W., Migliardo, P., and Suslick, K. S. (1993). *Phys. Rev. B*, **48**, 15797.
22. Grinstaff, M. W., Cichowlas, A. A., Choe, S. B., and Suslick, K. S. (1992). *Ultrasonics*, **30**, 168.
23. Suslick, K. S., Hyeon, T., Fang, M., and Cichowlas, A. A. (1994). *Molecularly Designed Nanostructured Materials*, MRS Symposium Proceedings, Vol. 351 (K. E. Gonsalves, G. M. Chow, T. O. Xiao, and R. C. Cammarata, eds.), pp. 201–206, Materials Research Society, Pittsburgh.
24. Suslick, K. S. Fang, M., Hyeon, T., and Cichowlas, A. A. (1994). *Molecularly Designed Nanostructured Materials*, MRS Symposium Proceedings, Vol. 351, (K. E. Gonsalves, G. M. Chow, T. O. Xiao, and R. C. Cammarata, eds.), pp. 443–448, Materials Research Society, Pittsburgh.
25. Suslick, K. S., Cline, R. E., Jr., and Hammerton, D. A. (1986). *J. Am. Chem. Soc.*, **106**, 5641.
26. Dumesic, J. A., Topsφe, H., and Boudart, M. (1975). *J. Catal.*, **37**, 513.
27. Nishizawa, T., and Ishida, K. (1984). *Bull. Alloy Phase Diagrams*, **5**, 250.
28. Nakamura, M., Wood, B. J., Hou, P. Y., and Wise, H. (1981). *Proceedings 4th International Congress on Catalysis, Tokyo*, Kodansha Ltd., Tokyo, p. 432.
29. Ranhotra, G. S., Haddix, G. W., Bell, A. T., and Reimer, J. A. (1987). *J. Catal.*, **108**, 40.
30. Lee, J. S., Oyama, S. T., and Boudart, M. (1990). *J. Catal.*, **125**, 157.
31. Ledoux, M. J., Pham-Huu, C., Guille, J., and Dunlop, H. (1992). *J. Catal.*, **134**, 383.
32. Volpe, L., and Boudart, M. (1985). *J. Solid State Chem.*, **59**, 332.
33. Rodriguez, N. M., Kim, M. S., and Baker, R. T. K. (1993). *J. Catal.*, **144**, 93.
34. Lee, J. S., Volpe, L., Ribeiro, F. H., and Boudart, M. (1988). *J. Catal.*, **112**, 44.
35. Ranhotra, G. S., Haddix, G. W., Bell, A. T., and Reimer, J. A. (1987). *J. Catal.*, **108**, 24.
36. Pham-Huu, C., Ledoux, M., and Guille, J. (1993). *J. Catal.*, **143**, 249.
37. Ko, E. E., and Madix, R. J. (1980). *Surf. Sci.*, **100**, L449, L505.
38. Bianchi, D., Tan, L. M., Borcar, S., and Bennett, C. O. (1983). *J. Catal.*, **84**, 358.
39. Sinfelt, J. H. (1983). *Bimetallic Catalysts: Discoveries, Concepts, and Applications*, pp. 18–31, Wiley, New York.
40. Klabunde, K. J., and Li, Y. -X. (1993). *Selectivity in Catalysis*, (M. E. Davis, and S. L. Suib, eds.), pp. 88–108, American Chemical Society, Washington, DC, and references therein.

CHAPTER 9

Preparation and Characterization of Polymer-Stabilized Rhodium Particles

G. W. Busser, J. G. van Ommen, and J. A. Lercher
Catalytic Processes and Materials
Faculty of Chemical Technology
University of Twente
7500 AE Enschede, The Netherlands

KEYWORDS: Rhodium catalyst, polyvinyl-2-pyrrolidone-stabilized colloidal suspension, rhodium colloid, nanoscale rhodium particle, cyclohexene hydrogenation

9.1   Introduction

The preparation of small metal particles on inorganic carriers has been investigated thoroughly and is well documented [1,2]. Generally accepted preparation techniques involve ion exchange or impregnation of an inorganic oxide surface (e.g., alumina, silica) with metal salt solutions, followed by calcination and subsequent reduction with hydrogen. However, the surface of the carrier on a molecular scale, and therefore that surrounding the metal cluster, is not well defined; and often it is difficult to obtain monodispersed metals on oxides, because the interaction of the reduced metal with the carrier is low compared with the interaction of the individual metal atoms.

Polymer-stabilized nanoscale particles are employed in a variety of applications ranging from precursors for metal deposition [3,4] to catalysis [5–9]. The stabilization of the small particles (d = 0.5–5 nm) by the polymer offers the means to influence the electronic properties of the metal clusters (via interaction with the functional groups of the polymer) and the accessibility of the metal (through constraints imposed by the polymer). The preparation of such well-defined metal clusters is quite complex, is difficult to reproduce, and even for a given metal–polymer combination is found to depend subtly on a variety of preparation parameters.

To be able to synthesize colloids with a uniform size distribution, one needs to acquire an understanding of the parameters that influence the resulting particle size on a molecular–atomic level. These parameters include the action of the reducing agent, the temperature, the stabilizing agents, and the nature of the metal complexes [10,11].

For polymer-stabilized colloidal suspensions, the polymer usually acts as a steric stabilizer by balancing the van der Waals forces that cause coagulation of the particles [12]. The adsorption of the polymer on the metal cluster–particle is considered to be irreversible in the sense that simultaneous desorption of all polymer segments is statistically unlikely [13]. When these materials are used as catalysts, it can be expected that both the strength and way of interaction of the polymer with the metal surface will influence the accessibility and the chemical reactivity of the metal toward reactants and products, and thus will also direct its activity and selectivity. The majority of research on the fundamental properties of polymer-stabilized metal particles consists of model studies. Electronic properties of metal particles were compared with quantum chemical predictions and spectroscopic techniques such as nuclear magnetic resonance (NMR), Fourier transform infrared (FTIR), and x-ray absorption fine structure (XAFS) spectroscopy were used for structural characterization. Estiu and Zerner [14], for example, performed theoretical calculations for rhodium particles consisting of 13 and 19 atoms and suggest that these clusters have electron-deficient surface atoms.

Bradley et al. [15] reported on the characterization of polyvinyl-pyrrolidone-stabilized palladium and palladium/copper particles by IR and ^{13}C NMR spectroscopy of adsorbed CO in the liquid phase. Bradley et al. [16] find an increase in bridged-bound CO relative to the linearly bound form when particle size increases from 2.5 to 6 nm. This was also reported for silica-supported Pd crystallites [17], which underlined the similarities of the catalysts. It is speculated that the N or O donor atoms of the polymer are attached to metal ions on the metal particle surface, but preliminary XAFS experiments show such contacts only after intentional oxidation of the metal surface. Thus, it was concluded that for fully reduced particles the polymer rather loosely interacts with the metal particles.

Harada et al. [18] observed that bimetallic clusters prepared under air are more active than those prepared under nitrogen in 1,3-cyclooctadiene hydrogenation. When XAFS was used, it was possible to conclude that this is primarily the result of a high concentration of defects on the surface of clusters prepared in air.

This contribution addresses the preparation of small polyvinyl-2-pyrrolidone, stabilized metal particles well defined in size and size distribution. The relationship between the average particle size and the catalytic properties for cyclohexene hydrogenation is discussed.

9.2   Experimental

9.2.1 Materials

Polyvinyl-2-pyrrolidone (MW = 40,000), methanol, ethanol, 1-propanol, and 1-butanol were all obtained from Merck and $RhCl_3 \cdot 3H_2O$ was obtained from Aldrich. All materials were of pro analysis (p. a.) quality and used without further purification.

9.2.2 Preparation

9.2.2.1 Method I

Rhodium chloride (15 mg) dissolved in water (20 ml) was added to a solution of polyvinyl-2-pyrrolidone (225 or 450 mg) in alcohol (methanol, ethanol, 1-propanol, or 1-butanol) (130 ml) at reflux temperature. After 48 h of refluxing and mixing, the solution was cooled in liquid nitrogen and dried under vacuum. Finally, the material was redissolved in 1-butanol.

9.2.2.2 Method II

Rhodium chloride (15 mg) and polyvinyl-2-pyrrolidone (225 mg) were dissolved in water (20 ml). This mixture was heated at reflux 373 K for 2 h. Then it was rapidly mixed with alcohol (methanol, ethanol, 1-propanol, or 1-butanol) (130 ml) at reflux temperature. After 48 h of refluxing and mixing, the solution was cooled in liquid nitrogen and dried under vacuum. Finally, the material was redissolved in 1-butanol.

9.2.3 Characterization

Particle size distributions were determined by placing a drop of the colloidal solution on a carbon-covered copper grid (Balzers) and analyzing with a high-resolution transmission electron microscope (model JEOL 200 CX). Particle size distributions were determined by optical inspection of the photographs. The metal areas of the catalysts were estimated assuming spherical particle shape. ^{13}C NMR spectra were recorded on a Bruker spectrometer operating at 62.9 MHz. UV-VIS spectra were recorded on a Philips UV-VIS spectrophotometer (PU 8700 Series).

9.2.4 Catalytic Testing

For catalytic cyclohexene hydrogenation, the reactor was filled with 25 ml of the colloidal solution and 50 ml of dry 1-butanol. This mixture was treated with hydrogen at 7 bar for 1 h at 343 K. Subsequently the pressure was released and the temperature was decreased to 323 K. Then 6.5 ml of cyclohexene was added to the reactor. The hydrogen pressure was again increased to 7 bar, and with an injection needle samples were taken during reaction. These samples were analyzed by injecting 1 μl into a Varian 3700 gas chromatograph equipped with a flame ionization detector (FID) detector and a 2.5 m · 1/8 ss column packed with 5% SP 1200/1.75% Bentone 34 on Supelcoport, 100–120 mesh.

9.3   Results

9.3.1 Catalyst Preparation and Characterization

With method I, rhodium colloids were produced by the simultaneous contact of rhodium chloride in water with polyvinyl-2-pyrrolidone and the reducing agent (methanol, ethanol, 1-propanol, or 1-butanol). During the reduction process, part of the alcohol was converted to the corresponding aldehyde [19,20].

The resulting particle size distributions, determined from TEM photographs, are compiled in Fig. 1a; and TEM photographs are displayed in Fig. 1b. Except for the methanol-reduced sample, the particle size distribution observed was quite narrow and the particle size increased with increasing molecular weight of the alcohol.

With method II, the metal ions were first allowed to interact with the polymer by refluxing the metal salt with the polymer in water for 2 h. Subsequently this mixture was added to the alcohol at reflux temperature, followed by the same steps that were applied in method I. The TEM photograph of the methanol-reduced sample does not show metal particles. After the material was dried and redissolved in 1-butanol and treated with hydrogen (7 bar, T = 343 K, 60 min), very small particles were observed. Figures 1a and 2a show size distributions, and Figs. 1b and 2b display TEM photographs of the materials prepared by different reducing agents.

When the particle size distributions obtained by the different methods are compared (see Figs. 1a and 2a), it is striking that the methanol-reduced sample contains much smaller particles (< 1 nm) if prepared via method II. The difference for the other alcohols is less dramatic.

In order to support the results obtained by TEM, UV-VIS spectroscopy of colloid solutions (as described by Furlong *et al.* [21]) was used as the

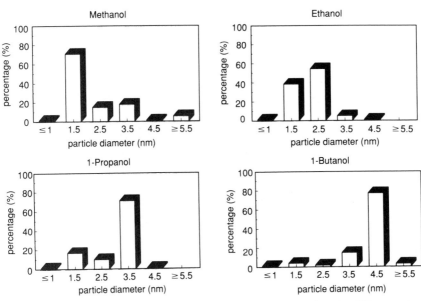

Figure 1a. Particle size distributions of materials prepared by method I using different reducing agents.

means of parallel characterization of the average particle size. The spectra show a monotonous increase in absorption with decreasing wavelength between 300 and 600 nm. In general, the UV-VIS spectrum is sensitive to concentration, particle size, particle shape, and coagulation state. The flatness of the spectrum is characterized by the parameter S:

$$S_{\lambda_1} = -\,d\log A_{\lambda_1}/d\log\lambda_1. \tag{1}$$

S can be approximated by S':

$$S' = -\,(\log A_{\lambda_1} - \log A_{\lambda_2})/(\log\lambda_1 - \log\lambda_2). \tag{2}$$

For platinum colloids, S has been found to decrease with increasing particle size up to 4 nm [21]. In Table 1 S' (calculated from the absorption values at 400 and 600 nm) was compiled for the materials prepared by the two methods, before and after use as a catalyst. There is a correlation between S' and the metal particle size distribution derived by TEM for samples prepared by method II, in contrast with those prepared by method I. For the former samples S', as expected [21], decreases with increasing particle size.

The interaction of the metal salt with the polymer (i.e., the catalyst precursor) was characterized by ^{13}C NMR spectra of a series of polymer samples with varying concentrations of rhodium chloride in water (see Fig. 3). From similar samples UV-VIS spectra were recorded (Fig. 4).

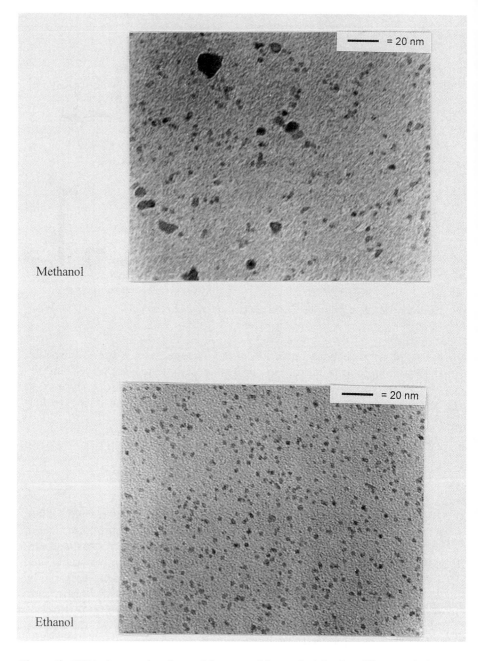

Figure 1b. TEM photographs of materials prepared by method I using different reducing agents.

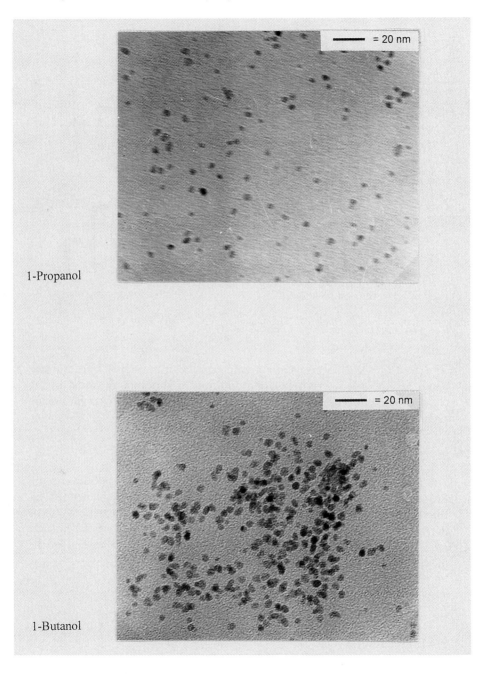

1-Propanol

1-Butanol

Figure 1b. *continued*

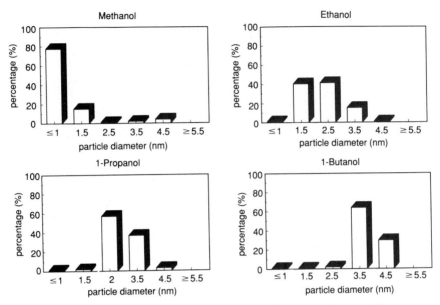

Figure 2a. Particle size distributions of materials prepared by method II using different reducing agents (methanol sample after treatment with H_2).

The ^{13}C NMR spectra show a gradual shift of the signal from the carbonyl group to a lower field upon addition of rhodium chloride. In the UV-VIS spectra a shift of the maxima to lower wavelengths with increasing polymer concentration is observed.

9.3.2 Catalytic Testing

The results of using these catalysts for liquid-phase cyclohexene hydrogenation are compiled in Figs. 5 and 6. The turnover frequency (TOF), that is, the activity normalized to the concentration of surface metal atoms derived from TEM measurements, is shown as a function of the particle size. With decreasing particle size, the TOF increased, independent of the method of preparation.

9.4   Discussion

9.4.1 Preparation

Figure 1 suggests that a direct correlation between the molecular weight of the alcohol used for reduction and the rhodium particle size exists, that is, that the particle size increases with increasing molecular

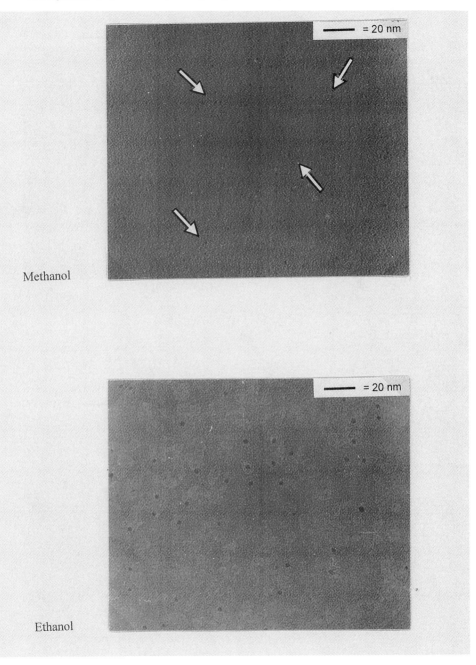

Methanol

Ethanol

Figure 2b. TEM photographs of materials prepared by method II using different reducing agents.

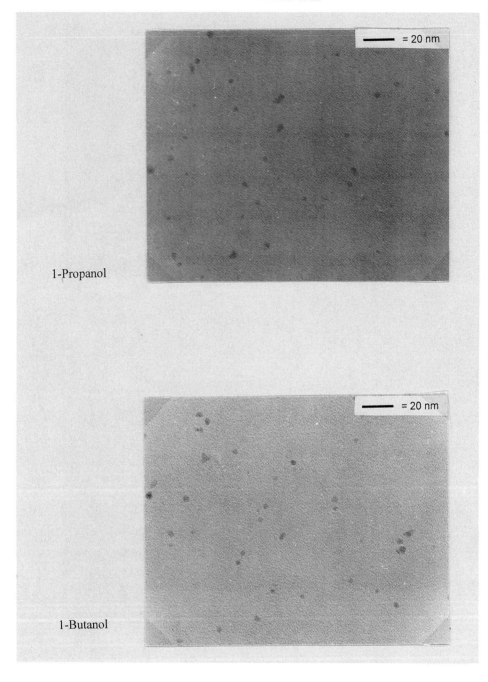

Figure 2b. *continued*

TABLE 1
Particle Size Characterization of Catalysts Prepared by Method I and Method II

		Method I		Method II	
Reducing agent		S'	Average particle size (nm)[a]	S'	Average particle size (nm)[a]
Methanol	Before use	2.6	1.5–5	15.8	<1
	After use	2.5	1.5–5	5.4	1
Ethanol	Before use	2.6	2.5	2.9	2
	After use	2.8	2.5	2.8	2.5
1-Propanol	Before use	2.8	3.5	2.9	2
	After use	3.0	3.5	2.8	2.5
1-Butanol	Before use	2.7	4.5	2.7	3.5
	After use	2.8	4.5	2.5	4

[a]Particle sizes were determined from TEM photographs.

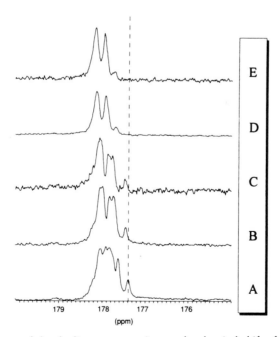

Figure 3. Influence of the rhodium concentration on the chemical shift of the carbonyl C atom (polyvinylpyrrolidone concentration = 4.1 M; loading of the functional groups: A = 0 mol % Rh, B = 7.5 mol % Rh, C = 20 mol % Rh, D = 40 mol %, Rh, E = 60% Rh).

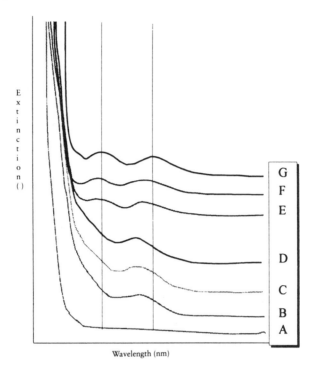

Figure 4. Influence of the polymer concentration on ligand surroundings of Rh(III) (concentration rhodium = 3 mM; loading of the functional groups: A = 0 mol % Rh, B = 2.8 mol % Rh, C = 5.6 mol % Rh, D = 12.5 mol % Rh, E = 25 mol % Rh, F = rhodium chloride refluxed in water, G = rhodium chloride in water).

weight of the alcohol. Following the arguments of Kirkland *et al.* [10] (Fig. 7) we think that a high concentration of thermodynamically stable nuclei leads to a small particle size, provided that these particles are sufficiently stabilized (i.e., prevented from clustering). The solution is depleted with unreduced material that can potentially cause growth of the metal particles. To obtain a high concentration of nuclei, the reduction process should be faster than the growth process. Because the reducing agent has to interact with the metal complex during the reduction process, the process will be influenced by the ligands surrounding the metal ion and the reducing agent. Bulky ligands and reducing agents, as well as a strong interaction of the ligands with the metal ion, can make interaction more difficult and hence slow down reduction, which in turn increases the particle size. Using method I, we find a broader distribution for the methanol-reduced sample than for the other alcohols. We attribute this to formation of a high initial concentration of nuclei that are not sufficiently stabilized in the solution, re-

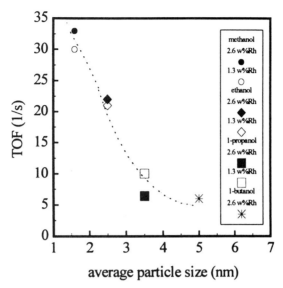

Figure 5. Activity of cyclohexene hydrogenation as a function of particle size (catalysts prepared by method I).

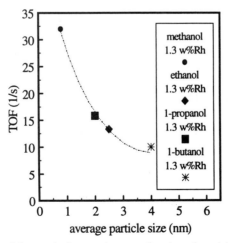

Figure 6. Activity of cyclohexene hydrogenation as a function of particle size (catalysts prepared by method II).

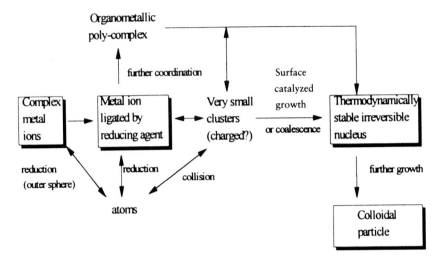

Figure 7. Processes involved in the formation of colloidal metal particles.

sulting in particle agglomeration. For the samples reduced by the higher alcohols, the concentration of nuclei is lower than with methanol; and therefore initial stabilization is easier to achieve and is less important.

There is also a marked difference between the methanol-reduced samples (Figs. 1 and 2). Particles with diameters below 1 nm are only formed when prepared with method II. Differences in particle size resulting from reduction with ethanol, propanol, and butanol are less dramatic. This is in accordance with the presence of a lower concentration of nuclei, which makes coordination of the polymer with the metal salt prior to reduction less important.

Note that the results for methanol are dramatically improved in comparison with Hirai [20] who also prepared rhodium sols stabilized by polyvinylpyrrolidone, but found average particles sizes of 4 nm. The fact that he found a bigger particle size for the methanol-reduced sample could be explained by a slower initial reduction rate (compared with preparation method II) during the heating of the materials. This again underlines the importance of a fast reduction rate. From the results presented in Table 1, it follows that for materials prepared by method II there is a correlation between particle size and the S' value before and after reaction. However, for materials prepared by method I the propanol and butanol-reduced samples show a high S' value, which is not in accordance with the TEM results. The unexpectedly high values of S' may be caused by two effects. One possibility could be a high concentration of rhodium chloride, because rhodium chloride has a relatively high absorption at 400 nm and no absorption at 600 nm. However, we do not have an indication of the presence of a large fraction of rhodium chloride. The other possibility is related to the roughness of the metal particles formed. According to Furlong *et al.* [21], rougher particles are more effective scatterers resulting in lower S values. Although the TEM photographs do not give a clear indication of a difference in surface structure, we speculate that the differences are related to the roughness of the particles. This implies that with method I we prepared smoother particles, at least for the samples reduced with propanol and butanol. The smoother particles are assumed to be the result of a gradual particle growth by the reduction of rhodium chloride on the surface of the metal nucleus, while the rougher particles are formed by coagulation of small metal crystallites.

9.4.2 Interactions of the Catalyst Precursor with the Polymer

The ^{13}C NMR signal of the carbonyl group consists of several peaks, which can be attributed to carbonyl groups with different surroundings as a result of the "tacticity" of the polymer containing an asymmetric carbon atom [22,23]. There is a gradual shift of these carbonyl peaks to a lower field upon increasing the rhodium concentration. This indicates coordination of rhodium cations and the establishing of a sorption equilibrium of rhodium chloride.

In principle, the oxygen and the nitrogen group can be involved in coordination. Conceptually, one expects the carbonyl group to be the more important group, because it is less sterically hindered than the nitrogen. Indeed, this seems to be the case, because the carbonyl C atom signal displays a significant change in the NMR spectrum upon addition of rhodium chloride, while it hardly affects the other peaks.

The mode of interaction of rhodium ions with the polymer is also well reflected in the UV-VIS spectra. A spectrum of rhodium chloride in water displays two maxima that can be attributed to the transfer of d-electrons and is influenced by ligand surroundings.

Refluxing rhodium chloride in water results in a shift of the maxima to lower wavelengths, which is attributed to the exchange of chlorine ligands for water ligands [24] that induces an increase in crystal field splitting. Addition of polyvinyl-2-pyrrolidone induces a further shift that indicates the coordination of rhodium chloride to the polymer by exchanging one or more water ligands for carbonyl ligands (Fig. 8).

Thus both NMR and UV-VIS spectroscopy clearly indicate that a well-defined interaction between the polymer and the rhodium chloride exists. The interaction between the metal particles and the polymer is more difficult to prove directly. Indirectly, the stability of the obtained colloids indi-

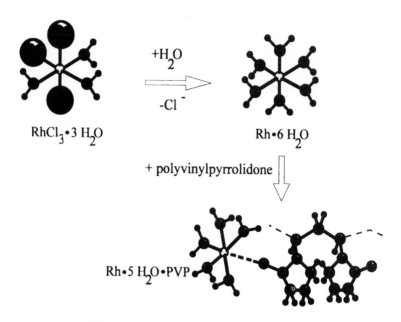

$RhCl_3 \cdot 3\ H_2O$

$+H_2O$

$-Cl^-$

$Rh \cdot 6\ H_2O$

$+$ polyvinylpyrrolidone

$Rh \cdot 5\ H_2O \cdot PVP$

Figure 8. Ligand exchange during preparation.

cates that the interaction between the polymer and the nanoscale rhodium particles is strong enough to stabilize very small clusters.

9.4.3 Catalytic Testing

Catalytic results show that there is an increase in TOF with decreasing particle size (see Figs. 5 and 6). This is quite surprising because cyclohexene hydrogenation is reported to be a structure-insensitive reaction. Boudart and Sajkowski [25], for example, found liquid-phase cyclohexene hydrogenation over Rh/Al_2O_3 to be independent of the size of rhodium particles (ranging from approximately 12 to more than 150 atoms). The same structure insensitivity has been reported for platinum, palladium, and nickel catalysts [26–28]. However, Augustine et al. [29] indicated that there exist at least three different catalytic sites on Pt with different activities toward cyclohexene hydrogenation, which suggests the possibility of structure sensitivity. In this context it is interesting to note that Hub et al. [30] found butene hydrogenation to be structure sensitive. They attributed this to an increase in adsorption coefficient with decreasing particle size, the smaller particles containing an electron-deficient surface structure.

Finally, it should be pointed out that for homogeneous catalysts [31] the ligand concentration is able to affect cyclohexene hydrogenation markedly. These examples indicate that the fact that we find an increase in TOF with decreasing particle size might be attributed to variations in the sorption constant, different concentrations of active sites, or ligand effects. Experiments to distinguish these possibilities are being currently performed.

9.5 Conclusions

Methods for the preparation of metal particles of desired size stabilized by polyvinyl-2-pyrrolidone have been developed. The metal particle size is governed by the interaction of the polymer with the catalyst precursor and the reduction rate. Stronger interaction of the metal salt precursor with the polymer and higher reduction rates result in smaller particle sizes on the final catalyst.

Acknowledgments

The research was supported by the Onderzoeks Stimulerings Fonds of the University of Twente. We are grateful to Ing. Marc Smithers and Dr. Thomasz Kachlicki of the Centrum voor Materiaalkundig Onderzoek for the TEM analysis, and to Dr. John van Duynhoven for the discussions involving the NMR interpretation.

References

1. Gonzalez, R. D., and Miura, H. (1994). *Catal. Rev. Sci. Eng.*, **36**(1), 145–177.
2. Bond, G. C., and Wells, P. B. (1985). *Appl. Catal.*, **18**, 221–224.
3. van der Putten, A., de Bakker, J. W., and Fokkink, L. G. J. (1992). *Electrochem. Soc.*, **139**(12), 3475–3480.
4. Boonekamp, E. P., Kelly, J. J., and Fokkink, L. G. J. (1994). *Langmuir*, **10**(11), 4089–4094.
5. Harrison, D. P., and Rase, H. F. (1967). *Ind. Eng. Chem. Fundam.*, **6**(2), May.
6. Hoang-Van, C., Tournier, G., and Teichner, S. J. (1984). *J. Catal.*, **86**, 210–214.
7. Michel, C., Hoang-Van, C., and Teichner, S. J. (1978). *J. Chim. Phys.*, **75**(9), 819–825.
8. Dini, P., Dones, D., Montelatici, S., and Giordano, N. (1973). *J. Catal.*, **30**, 1–12.
9. Hirai, H., Chawanya, H., and Toshima, N. (1981). *Makromol. Chem. Rapid Commun.*, **2**, 99–103.
10. Kirkland, A. I., Edwards, P. P., Jefferson, D. A., and Duff, D. G. (1990, Publ. 1991). The structure, characterization and evolution of colloidal metals, *Annu. Rep. Prog. Chem.*, *Sect. C*, **87**, 247–304.

230 / *G. W. Busser* et al.

11. van Rheenen, P. R., McKelvy, M. J., and Glaunsinger, W. S. (1987). *J. Solid State Chem.*, **67**, 151–169.
12. Napper, D. H. (1983). *Polymeric Stabilization of Colloidal Dispersions*, Academic Press, London.
13. Ohtaki, M., Komiyama, M., Hirai, H., and Toshima, N. (1991). *Macromolecules*, **24**, 5567–5572.
14. Estiu, G. L., and Zerner, M. C. (1993). *Quantum Chem. Symp.*, **27**, 195–211.
15. Bradley, J. S., Millar, J. M., Hill, E. W., Klein, C., Chaudret, B., and Duteuil, A. (1993). *New Frontiers in Catalysis* (L. Guczi et al., eds.), *Proceedings 10th International Congress Catalysis*, July 19–24, 1992, Budapest, Hungary, Elsevier Science Publisher.
16. Bradley, J. S., Hill, E. W., Behal, S., Klein, C., Chaudret, B., and Duteil, A. (1992). *Chem. Mater.*, **4**, 1234–1239.
17. Shien, L. L., Karpinski, Z., and Sachtler, W. M. H. (1989). *J. Phys. Chem.*, **93**, 4890.
18. Harada, M., Asakura, K., Ueki, Y., and Toshima, N. (1992). *J. Phys. Chem.*, **96**, 9730–9738.
19. Toshima, N., (1990). *J. Macromol. Sci. Chem.*, **A27**(9–11), 1225–1238.
20. Hirai, H., (1979). *J. Macromol. Sci. Chem.*, **A13**(5), 633–649.
21. Furlong, D. N., Launikonis, A., Sasse, W. H. F., Sanders, J. V. (1984). *J. Chem. Soc. Faraday Trans.*, **1**(80), 571–588.
22. Ebdon, J. R., Huckerby, T. N., and Senogles, E. (1983). *Polymer*, **24**, 339–343, March.
23. Cheng, H. N., Smith, T. E., and Vitus, D. M. (1981). *J. Polym. Sci. Polym. Lett. Ed.*, **19**, 29–31.
24. *Gmelin Handbook of Inorganic Chemistry*, (1982). Rhodium, Compounds, Suppl. Vol. B1, Springer-Verlag, New York.
25. Boudart, M. and Sajkowski, D. J., (1991). *Faraday Discuss.*, **92**, 57–67.
26. Davis, S. M., and Somorjai, G. A. *J. Catal.*, **65**, 78.
27. Gonzo, E. E. and Boudart, M. (1978). *J. Catal.*, **52**, 462.
28. McConica, C. M., and Boudart, M. (1989). *J. Catal.*, **117**, 33.
29. Augustine, R. L., Thompson, M. M., and Doran, M. A. (1987). *J. Chem. Soc. Chem. Commun.*, 1173–1174.
30. Hub, S., Hilaire, L., and Touroude, R. (1988). *Appl. Catal.*, **36**, 307–322.
31. Hostetler, M. J., Butts, M. D., and Bergman, R. G. (1993). *J. Am. Chem. Soc.*, **115**, 2743–2752.

CHAPTER 10

Gas-Phase Synthesis of Nonstoichiometric Nanocrystalline Catalysts

Jackie Y. Ying and Andreas Tschöpe
Department of Chemical Engineering
Massachusetts Institute of Technology
Cambridge, Massachusetts 02139-4307

KEYWORDS: catalytic oxidation, gas-phase synthesis, heterogeneous catalysts, Joule-heating evaporation, magnetron sputtering, multicomponent oxides, nanocrystalline CeO_2, surface chemistry

10.1 Introduction

The activity of catalytic materials can be strongly affected by their preparation method. Commonly, oxide catalysts are synthesized by chemical precipitation. To prepare multicomponent systems, coprecipitation or impregnation are often employed. These approaches are limited in the following aspects in the derivation of the ideal catalyst structure [1]:

- It is difficult to produce molecularly homogeneous mixed oxide systems by coprecipitation or impregnation.
- Oxygen deficiency in materials may be desirable for defect-related catalytic reactivity. However, highly nonstoichiometric oxides cannot be generated easily by wet chemical approaches.
- In supported catalysts, catalytic activity is affected by the interaction between an active component and its support. The microstructure of

supported catalysts and the dispersion of active components are strongly dependent on the preparation method and pretreatment history. Coprecipitation or impregnation may not readily produce highly dispersed supported metal catalysts.

This chapter describes a gas-phase synthesis technique that enables us to generate unique nanocrystalline materials for heterogeneous catalysis. The objective is to design advanced catalysts with the following desirable characteristics: (1) ultrahigh surface-to-volume ratio, (2) homogeneous multicomponent systems, (3) ultrahigh active component dispersion, and (4) controlled oxygen vacancy concentrations. The importance of the first three attributes has been widely recognized. The fourth feature is related to oxide stoichiometry, which directly affects the material properties in a variety of applications [2]. The defect concentrations in oxides are typically changed by altervalent dopant and oxygen partial pressure. However, a high degree of nonstoichiometry cannot be achieved easily or stabilized without going to high temperature and reducing atmosphere. The ability to control defect concentration would allow us to tailor oxygen conductivity in applications such as sensors, fuel cells, ceramic membrane reactors, and oxidation catalysis. In the latter, the catalytically active sites are often associated with oxygen vacancies and surface adsorbates [3,4]. The availability of a high concentration of such defects and surface species in a catalyst could facilitate catalytic activity at lower reaction temperatures. In this chapter, synthesis and applications of nanocrystalline catalysts with ultrahigh dispersion and nonstoichiometry are reviewed.

10.2 Gas-Phase Synthesis of Nanocrystalline Materials

Nanocrystalline materials are solid-state systems constituting of crystallites 5–20 nm in dimension. Various techniques have been adopted for their synthesis [5–7], many of which rely on solution chemistry to precipitate fine particles from chemical precursors [8,9]. While the chemical routes are capable of producing large quantities of materials, the resultant powder usually needs to be heat-treated to burn off the organic residue before the sample becomes purified for further use. Such treatment may promote grain growth, and may not enable the derivation of nonequilibrium structures and defects that can be desirable in catalytic applications. To allow for high-purity, low-temperature processing of nanocrystalline materials, a gas-phase synthesis, known as inert gas condensation, was developed.

10.2.1 Joule-Heating Evaporation and Gas Condensation

In the inert gas condensation method, a volatilized monomer population is established by evaporating materials commonly with oven or Joule-heating sources. The monomers are cooled by collisions with "cold" inert gas atoms, and are aggregated into clusters from collisions between monomers. Near the source, small clusters of fairly uniform sizes are observed [10]. Farther from the source the clusters become larger with a broader size distribution. The particle size produced can be decreased by reducing either the gas pressure in the chamber or the evaporation rate, and by using lighter gases (such as He) instead of heavier ones (such as Xe) [11–13]. Clusters with mean diameters as small as 3–4 nm have been made by this process, and they exhibit a log-normal size distribution. Such a distribution is characteristic of cluster–cluster aggregation and is probably due to the relatively slow rate at which particles are transported away from the crucible or oven [14].

Gleiter and co-workers [15,16] adapted the above technique to produce nanocrystals and assembled them into three-dimensional nanostructured compacts with an ultrahigh volume fraction of grain boundaries. The synthesis is performed in an ultrahigh vacuum (UHV) chamber in a setup illustrated in Fig. 1 [17]. The chamber is baked to $<10^{-6}$ Pa before it is filled with an inert gas. Precursors of the nanostructured materials are placed in a refractory metal boat which is resistively heated (Fig. 1 (a)). The precursor atoms or molecules effused from the source rapidly lose their kinetic energy by colliding with the inert gas molecules. The short collision mean free path results in efficient cooling of the precursor vapor and produces a high supersaturation of vapor locally that leads to a homogeneous nucleation, followed by cluster and particle growth through a coalescence mechanism [12].

The aerosol of particles generated by evaporation and inert gas condensation is transported via natural gas convection to a liquid nitrogen cooled rotating cold-finger, where it is collected via thermophoresis [18]. After evaporation, a vacuum of $<10^{-6}$ Pa is restored, and the particles are scraped from the cold-finger and can be compacted *in situ* into a nanostructured pellet in a piston and anvil device. Generally, in producing nanostructured materials, the evaporation parameters are adjusted to produce clusters having the smallest particle size while maintaining a fairly fast evaporation rate, so that sufficient quantities of materials can be produced in a reasonable time. The conditions that are usually used are a few hundreds pascal of He, and an evaporation temperature that corresponds to a vapor pressure of ~10 Pa for the precursor [19]. The average cluster diameters presently produced range between 5 and 15 nm.

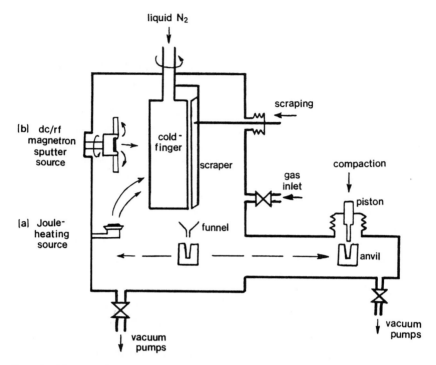

Figure 1. Schematic drawing of a gas condensation chamber for the synthesis of nanometer-sized clusters by (a) Joule heating and (b) dc- and rf-magnetron sputtering. Clusters are transported to the liquid nitrogen cooled substrates in an inert or reactive gas. The powders are subsequently scraped from the cold-finger or the modified ground shield in vacuum, and collected via the funnel in the anvil device. The anvil can be moved by a manipulator to situate right under the piston for *in situ* compaction. Reprinted from *J. Aerosol Sci.*, 24, J. Y. Ying, 315, Copyright (1993), with kind permission from Elsevier Science Ltd., The Boulevard, Langford Lane, Kidlington 0X5 1GB, UK.

This physical vapor condensation method for nanocrystalline synthesis has many inherent advantages: (1) processing in a prebaked UHV chamber minimizes sample contamination during material evaporation and condensation. The as-synthesized material is already nanocrystalline in nature and does not require further treatment for burnout of precursor complexes or crystallization. (2) Isolated clusters can be effectively collected and easily assembled into a bulk material by compaction. (3) During *in situ* compaction, the free particle surfaces are converted into internal interfaces or grain boundaries with a high degree of cleanliness. The large fraction of atoms residing in the grain boundaries of these materials (almost 50% in the case of 5-nm grains, assuming a 1-nm grain boundary thickness) may allow for new atomic arrangements to be formed in a significant volume fraction of the material, and novel material properties may result. (4) This processing technique allows for versatility with regard to both the variety of materials that can be produced and the control of different particle sizes [12]. Ceramic materials can be generated via evaporation in a reactive atmosphere or through *in situ* posttreatment. Oxides, for example, can be derived by postoxidation of the metallic clusters on the substrate [20,21]. By varying the nature and pressure of the reactive gas, the degree of oxidation can be manipulated to yield highly nonstoichiometric materials. (5) This approach also allows composite materials with ultrafine grain size and controllable composition to be synthesized with great flexibility in designing advanced materials [22]. Composites or alloys can be produced by using two or more evaporation sources. The rotation of the cold-finger facilitates good mixing of different constituents during the deposition of a multicomponent system. It has been demonstrated that the intermixing of particles of two metals (Cu and Er) in the convective He gas flow and on the rotating cold-finger is sufficient to almost totally react the powder by compaction alone [21].

Conventionally, Joule heating of metallic crucibles has been principally used for vaporizing the source materials. Several process limitations are imposed by Joule heating: (1) chemical reaction between most metals with the refractory metal crucibles causes continuous changes in the evaporation conditions. Often, the reactions result in the failure of crucibles: (2) nonhomogeneous temperature distributions in the molten metal may lead to unsatisfactory control and reproducibility in evaporation; (3) the powder production of alloys, intermetallics, or composites is restricted since the thermodynamic activities and vapor pressures might be quite different for the constituents, resulting in difficulties in controlling the composition and homogeneity of multicomponent systems; and (4) most ceramic materials have very high melting points and are, therefore, not good candidates for evaporation by resistive heating. Vaporization of many oxides in oxygen-containing atmosphere is limited by oxidation of the refractory boats (mostly tungsten), causing WO_3 contamination of the powder or boat failure. Consequently, direct preparation of nanostructured ceramics has been limited to materials with low-melting points or high-vapor pressures, such as CaF_2 and MgO [23]. Postoxidation of metallic clusters typically leads to materials with oxygen deficiency, and further heat treatment may be necessary where stoichiometric materials are needed.

10.2.2 Magnetron Sputtering

To enable the use of refractory or reactive precursors and the synthesis of more complex multicomponent systems, sputtering [23], electron-beam [24], plasma [25], laser ablation [26], or other potentially well-controlled evaporation methods have been employed as alternatives to Joule heating. Several publications reported the use of dc and rf-magnetron sputtering for synthesizing nanocrystalline materials [1,17,23,27–30]. Sputtering has a number of distinct advantages [1,23]: (1) low contamination, because crucibles are not used; (2) greater flexibility in the choice of source materials because high-melting-temperature metals, ceramics, and intermetallics can be sputtered; (3) preservation of the source composition for alloys and composites; (4) stable and easily controllable process parameters (e.g., plasma current); and (5) low heat load on the chamber walls, reducing outgassing and subsequent impurity incorporation into the samples.

During sputtering, ions of a suitable substance (such as Ar or Kr) accelerated to high energies are directed toward a surface, from which atoms and clusters, both neutral and ionic, are ejected. In thin-film deposition [31,32], dc- and rf-magnetron sputtering have become standard procedures; and they can be applied to the deposition of metals, alloys, semiconductors, and ceramics. However, the normal operating pressure of sputter sources (10^{-1} to 10^{-2} Pa) is several orders of magnitude lower than the pressure range required for particle formation (10^1 to 10^3 Pa). It had been shown that simple diode sputtering can be used to produce ultrafine particles in the desired range [33]. Sputtering at such high pressures is possible because the width of the dark space in the plasma is inversely proportional to the gas pressure, so that the number of collisions between accelerating ions and gas atoms in the dark space is independent of the gas pressure.

Typically, the sputter source is positioned normal to the axis of the substrate, such as a cold-finger. The low gas pressure in thin-film synthesis assures line-of-sight deposition of sputtered material. However, as the gas pressure is increased to >1 Pa, the mean free path decreases resulting in a thermalization distance of less than 2 cm [34]. Under these conditions, material transport is governed by diffusion. This effect can be demonstrated by measuring the steady-state deposition rate of dc-sputtered Cu with a film thickness monitor at different locations [1] (see Fig. 2). The deposition rate in a line-of-sight position decreases above an argon pressure of 1 Pa (Fig. 2 (a)), whereas the deposition in the plane of the sputter gun above the ground shield increases with increasing argon pressure until 3 Pa (Fig. 2 (b)). The drop in the deposition rate above 3 Pa in Fig. 2 (b) indicates that substantial material deposition occurs at locations closer to the opening of the sputter gun as thermalization length becomes further reduced. To utilize the diffuse transport to effectively collect the nanocrystalline particles at high argon pressures, we have modified the ground shield of the sputter gun [1] (Fig. 1 (b)). The ground shield was designed with the following features: (1) an extended diameter for use as a collection substrate,

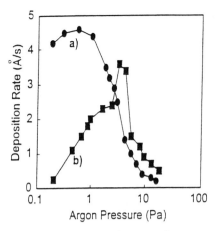

Figure 2. Deposition rates of sputtered Cu as a function of argon pressure, measured with a quartz crystal film thickness monitor at different locations: (a) 10 cm in front of the target facing the sputter gun, and (b) on the ground shield—in the same plane as the sputter gun. Reprinted from *Nanostr. Mater.*, 4, A. S. Tschöpe and J. Y. Ying, 617, Copyright (1994), with kind permission from Elsevier Science Ltd., The Boulevard, Langford Lane, Kidlington 0X5 1GB, UK.

(2) the capability of being cooled by liquid nitrogen, and (3) a mechanism for rotating in the plane of the sputter gun such that a scraper may be engaged for *in situ* scraping of the powder samples.

The pure metals and alloys can be prepared by dc sputtering using Ar gas. Particles might be formed with He gas, but with very low sputtering yield. Oxides can be synthesized by sputtering in different ways: (1) post-oxidation of the metallic clusters prepared by dc sputtering in an Ar atmosphere; (2) reactive rf sputtering of metals in an $Ar-O_2$ mixture; and (3) direct rf sputtering with an oxide target [17,23].

10.3 Nonstoichiometric Nanocrystalline Oxides for Catalytic Oxidation

Catalytic oxidation reactions, such as partial oxidation, complete combustion, and oxidative dehydrogenation, are extensively employed in industrial chemicals manufacturing and environmental pollution control. Typical oxidation catalysts are metal oxides and pure metals on which chemisorbed oxygen is the oxidizing agent. In the former, the nature of the reacting oxygen species is less obvious. In addition to adsorbed surface species, lattice oxygen can participate in the reaction if the oxide is highly reducible.

Catalytic oxidation reactions on metal oxides are often described in terms of a general redox mechanism:

$$Cat\text{-}O + Red \rightarrow Cat + Red\text{-}O, \tag{1}$$

$$Cat + Ox\text{-}O \rightarrow Cat\text{-}O + Ox. \tag{2}$$

Here, the oxide catalyst surface (Cat-O) is reduced by a reductant (Red) and reoxidized by an oxidant (Ox-O) to its initial state. The net result of this two-step reaction is the transfer of oxygen from one species to another. Based on isotope exchange experiments, this general scheme has been divided more specifically into (1) extrafacial reactions, in which only adsorbed surface oxygen reacts and lattice oxygen does not participate in the reaction; and (2) interfacial reactions, where lattice oxygen is extracted and oxygen vacancies are created [35]. Stability of the oxide and the type of active oxygen species might also determine the selectivity of the oxidation reaction. It has been proposed that redox reactions involving lattice oxygen result in partial oxidation of hydrocarbons (nucleophilic reactions), whereas active surface species such as superoxides lead to complete combustion products (electrophilic reactions) [35]. This hypothesis suggests that designing the oxidative properties of a catalyst would require control of the nature of active oxygen species. Oxidation characteristics become even more complex in the case of oxide-supported catalysts. A variety of metal–support interactions can lead to synergistic effects and enhanced catalytic properties [36].

Cerium oxide is well known for its large deviation from stoichiometry at low oxygen partial pressures and temperatures above 500°C [37,38]. We have selected it as the catalyst for oxidation studies because of its easy reducibility and the high oxygen vacancy mobility associated with its fluorite crystal structure. The defect chemistry studies of bulk CeO_2 indicate that upon reduction, oxygen vacancies are created in this material which are initially doubly ionized and then singly ionized with greater deviation from stoichiometry [39]. Electrons that are released from the vacancy sites are back-donated to localized Ce-4f states and are mobile by a small polaron-hopping process [40]. Reduced CeO_{2-x} is therefore a mixed ionic–electronic conductor. The availability of quasi-free electrons is assumed to be responsible for the formation of surface oxygen species such as superoxide O_2^- and peroxide O_2^{2-} on reduced CeO_{2-x} [41,42]. In contrast to bulk defect chemistry, information on defect thermodynamics at CeO_2 surfaces is very limited. Recent computer simulation shows that the enthalpy of vacancy formation at the surface is considerably lowered compared to the bulk and depends on the crystallography of the surface [43].

We have synthesized nonstoichiometric nanocrystalline CeO_{2-x} catalysts for two reactions: selective SO_2 reduction by CO (Eq. 3) and CO oxidation by O_2 (Eq. 4) [3,4,44],

$$SO_2 + 2CO \rightarrow S + 2CO_2, \tag{3}$$

$$O_2 + 2CO \rightarrow 2CO_2, \tag{4}$$

Both reactions are important for environmental pollution control. In our study, cerium oxide was investigated for potentially enhanced catalytic activity and selectivity, as well as for poisoning resistance. We have derived nanocrystalline nonstoichiometric CeO_{2-x} by inert gas condensation [1]. The catalytic activities of this material were compared to those of the chemically precipitated stoichiometric catalysts. Attempts to reduce unsupported stoichiometric CeO_2 by hydrogen are not successful because the reduced material will reoxidize readily when exposed to an oxidizing atmosphere. By comparing the nanocrystalline CeO_{2-x} and precipitated CeO_2, two possible changes in the catalytic materials due to nonstoichiometry are explored: (1) presence of oxygen vacancies, and (2) availability of quasi-free electrons. The effect of these changes will be illustrated by the different activities of the pure nanocrystalline CeO_{2-x} and the precipitated CeO_2. La doping is well known for generating oxygen vacancies without liberation of electrons; thus stoichiometric La-doped CeO_2 is a pure ionic conductor [45]. In nonstoichiometric $La-CeO_{2-x}$, additional oxygen vacancies and free electrons exists simultaneously. Hence, precipitated $La-CeO_2$ will elucidate the effect of oxygen vacancies in the absence of free electrons, whereas nanocrystalline $La-CeO_{2-x}$ represents a material with essentially the same chemical composition but mixed ionic–electronic conductivity.

Supported metal catalysts such as Pt or Pd on Al_2O_3 are typical catalysts for CO oxidation. Due to the high cost of the precious metals, continuous research efforts have been in progress to investigate base metal catalysts for oxidation reactions. Copper is one of the most studied base metals exhibiting high activity in methanol synthesis [46] and CO oxidation [47]. The effective dispersion and the promoting effect of Cu on nanocrystalline CeO_{2-x} will be discussed in this chapter.

10.3.1 Synthesis and Structural Characterization of Nanocrystalline CeO_{2-x}-Based Materials

10.3.1.1 Synthesis

Nanocrystalline Ce clusters were generated by dc-magnetron sputtering from a metallic Ce target in Ar atmosphere (30 Pa) [1,3]. These clusters are collected on the liquid nitrogen cooled modified ground shield described in Sec. 10.2.2 [1]. The UHV chamber was evacuated after deposition and slowly backfilled with O_2 to a final pressure of 100 Pa. The color of the deposit on the substrate changed from black to brownish-yellow during oxidation. The oxidized material was then scraped off and studied in this powder form for catalytic properties. For X-ray diffractometry (XRD), X-ray photoelectron spectroscopy (XPS), and electrical conductivity studies, the clusters were consolidated to porous solid discs of 5-mm diameter and 1-mm thickness with an uniaxial pressure of 0.5 GPa. For doped nanocrystalline CeO_{2-x}, sputtering from mixed metal targets (10 atomic [at.] % La/90 at. % Ce and 15 at. % Cu/85 at. % Ce) was undertaken, followed by postoxidation [1]. Sputtering has the advantage of reproducing the target composition in the samples and generating the most intimate mixture possible [1,48]. Stoichiometric CeO_2 was prepared by pyrolysis of cerium hydroxy-carbonates that was precipitated from an aqueous solution of $Ce(NO_3)_3$ and $(NH_4)_2CO_3$ [49,50]. The corresponding doped CeO_2 samples were coprecipitated from mixed nitrate solutions. The precipitate was washed in water and calcined at 650°C in air, and used as the stoichiometric reference materials.

10.3.1.2 Microstructure and Surface Chemistry

The specific surface areas of nanocrystalline and precipitated samples were obtained from five-point Brunauer–Emmett–Teller (BET) analysis of nitrogen adsorption (Micromeritics ASAP 2000). They were 50–65 m²/g for the materials generated by inert gas condensation, and 25–55 m²/g for the coprecipitated–calcined catalysts. The materials were also characterized by powder XRD (Rigaku rotating anode diffractometer). They all gave the cubic XRD pattern of the fluorite crystal structure of CeO_2. The widths of the cerium oxide (111) and (222) XRD peaks were determined by a least-squares fit of a Cauchy function. Deconvolution of the instrumental line broadening yielded the physical peak width. Kochendörfer analysis [51] and the Scherrer equation [52] were then applied to calculate the mean crystal size of the CeO_{2-x} clusters, corrected for internal strain contributions. The grain sizes of nanocrystalline and precipitated samples averaged ~8 nm and 8–14 nm, respectively. Grain growth and sintering of the nanocrystalline materials were investigated as a function of heat treatments (Fig. 3) [1]. They were initiated above 500°C, and after annealing at 650°C the nanocrystalline samples became comparable in surface area to the precipitated samples that have been calcined at the similar temperature. Annealing of nanocrystalline La-doped CeO_2 in water vapor saturated oxygen showed that sintering was not affected by the presence of moisture.

The surface structure of nanocrystalline CeO_{2-x} was examined with photoacoustic Fourier-transform infrared spectroscopy (PA-FTIR) [53]. This specially developed technique is particularly suitable for studying

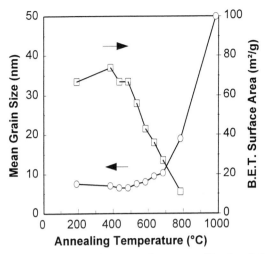

Figure 3. Mean grain size and BET surface area of nanocrystalline La–CeO_{2-x} after anneal-ing for 10 h in 15% O_2/He at the indicated temperatures. Reprinted from *Nanostr. Mater.*, 4, A. S. Tschöpe and J. Y. Ying, 617, Copyright (1994), with kind permission from Elsevier Sci-ence Ltd., The Boulevard, Langford Lane, Kidlington 0X5 1GB, UK.

both the surface and bulk phonon structure of solid samples [54,55]. The PA-FTIR spectrum of the as-prepared CeO_{2-x} nanoclusters (Fig. 4 (a)) in-dicated the presence of a great deal of adsorbed species associated with the large surface area of the sample, including hydrogen-bonded water and hy-droxyl groups, adsorbed hydrocarbon and water from atmosphere, and carbonate species. These surface adsorbates were gradually removed with heat treatment due to reduction of surface area. They were completely eliminated when the sample was sintered at 1400°C in O_2.

The spectra of the samples heated to 950 and 1400°C consisted of CeO_2 phonon vibrations. The as-prepared sample possessed additional phonon and adsorbate vibrations uniquely associated with its nonstoi-chiometry. It has phonon bands identified with substoichiometric CeO_{2-x}, and peaks which might be attributed to adsorbed species that formed on the surface defects of cerium oxides. Electrons were supplied to these adsor-bates by the coordinately unsaturated Ce^{3+} ions, which would then stabilize the formed species [42]. The presence of unique surface adsorbed species and CeO_{2-x} vibrations in the as-prepared nanoclusters illustrates the high concentration of surface defects and oxygen vacancies in this sample.

X-ray photoelectron spectroscopy (XPS) (Perkin Elmer PHI-5500) is employed to investigate the oxidation state of the nanocrystalline mate-rials [56,57]. The Ce-3d core level spectrum of CeO_2 exhibits six lines, de-

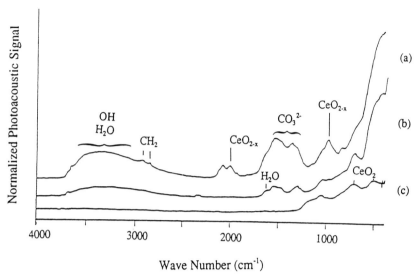

Figure 4. PA-FTIR spectra of CeO_{2-x} nanoclusters: as-prepared (a); and after annealing in O_2 at 950°C (b) and 1400°C (c).

noted v, v″, v‴, u, u″, and u‴ [58]. As-prepared nanocrystalline CeO_{2-x} showed two additional lines, v′ and u′ (Fig. 5 (a)) [56]. These peaks were also found in hydrogen-reduced CeO_2 and were attributed to Ce^{3+} cations. They were no longer observed after the nanocrystalline samples were oxidized at 500°C. To study the oxidation process, nanocrystalline CeO_{2-x} was annealed in 1 kPa of 1% CO_2/He in a reaction chamber connected to the XPS system. Deconvolution of the spectrum into the different components was used to determine the percentage of Ce^{3+} [56,57]. Figure 6 illustrates that the Ce^{3+} component decreased gradually with thermally activated oxidation from ~22% in the as-prepared sample to ~3% after treated at 475°C [56]. The degree of nonstoichiometry in nanocrystalline CeO_{2-x} is unusually high and cannot be eliminated by extensive exposure to oxidizing atmosphere below 500°C.

10.3.1.3 Dispersion in Multicomponent Oxides

Inert gas condensation synthesis and posttreatment of nanocrystalline $Cu-CeO_{2-x}$ were thought to provide unique control on the structural evolution of this system. XRD and scanning transmission electron microscopy (STEM) were employed to follow the phase transformation, morphological changes, and chemical dispersion in this nanocomposite [48,59]. XRD characterization of the as-prepared $Cu-CeO_{2-x}$ gave only the pattern corresponding to cerium oxide. No changes in XRD pattern and peak widths

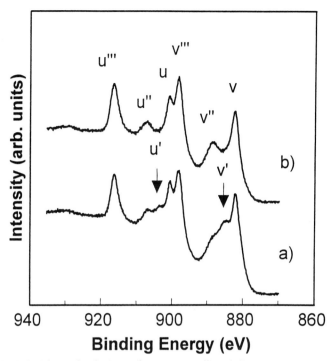

Figure 5. XPS Ce-3d core level spectra for nanocrystalline CeO_{2-x}: (a) as-prepared, and (b) after oxidation in 1 kPa of 1% CO_2/He at 475°C. Reprinted with permission from Tschöpe, A., Liu, W., Flytzani-Stephanopoulos, M., and Ying, J. Y. (1995). *J. Catal.,* **157**, 42, with permission from Academic Press.

were noted in annealing to 500°C under 15% O_2/He, indicating only the cerium oxide phase with ~5-nm grains was present (Fig. 7) [1]. CuO was found as a second phase only for samples heated at 550°C or above. The formation of a separate CuO phase coincided with the commencement of CeO_{2-x} grain growth (indicated by the sharpening of the diffraction peaks).

To investigate the dispersion of Cu in CeO_{2-x}, high-resolution STEM studies were performed using the VG HB5 system on powdered Cu–CeO_{2-x}. For a sample heated to 450°C, STEM revealed that the Cu concentration at the edge of the cluster was about two times higher than that at the center of the grain. This indicates that prior to the formation of a bulk CuO crystalline phase at 550°C (see XRD results), surface segregation of Cu in the Cu–CeO_{2-x} was present from annealing at ~450°C [59].

To investigate the thermal stability of the Cu dispersion in the Cu–CeO_{2-x} sample, HB603 STEM and elemental mapping were also em-

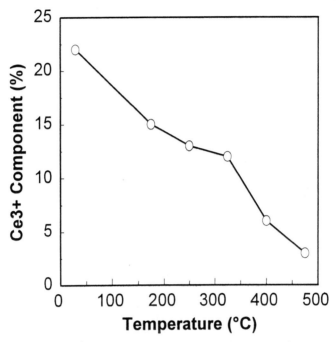

Figure 6. Percentage of Ce^{3+} component determined by deconvolution of the XPS Ce-3d core level spectrum after annealing in 1 kPa of 1% CO_2/He at the temperatures indicated. Reprinted from Tschöpe, A., Liu, W., Flytzani-Stephanopoulos, M., and Ying, J. Y. (1995). *J. Catal.*, **157**, 42, with permission from Academic Press.

ployed to examine a sample annealed for 10 h at 650°C (Fig. 8) [59]. XRD indicated that such treatment would have produced grain growth and a separate CuO phase. The STEM image and the elemental maps for Ce and Cu confirmed that 650°C oxidation did give rise to large Cu-rich clusters. Nevertheless, a substantial portion of Cu remained highly dispersed in the sample (Fig. 8 (c)).

Our preparation of Cu–CeO$_{2-x}$ catalyst is unique in that inert gas condensation was employed to generate this material first as $Cu_{0.15}Ce_{0.85}$ alloy nanoclusters through magnetron sputtering from a mixed metal target. The phase diagram of metallic Cu and Ce [60] shows a series of intermetallic phases, indicating chemical affinity and ionic radius mismatch between Cu and Ce. Inert gas condensation does not yield such intermetallic phases readily because of the high quenching rate during thermalization. Therefore, vapor-phase synthesis resulted in nanoclusters of a homogeneously dispersed Cu–Ce solid solution. The as-prepared samples were derived from controlled postoxidation of these Cu–Ce nanoclusters. Unfor-

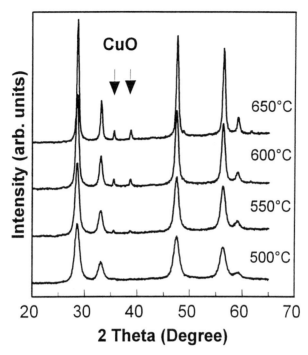

Figure 7. XRD pattern of nanocrystalline $Cu-CeO_{2-x}$ after oxidation in 15% O_2/He at the indicated temperature. Reprinted from *Nanostr. Mater.*, 4, A. S. Tschöpe and J. Y. Ying, 617, Copyright (1994), with kind permission from Elsevier Science Ltd., The Boulevard, Langford Lane, Kidlington 0X5 1GB, UK.

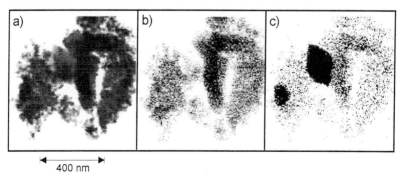

Figure 8. STEM image (a), Ce elemental map (b), and Cu elemental map (c) of nanocrystalline $Cu-CeO_{2-x}$ after annealing at 650°C.

tunately, the miscibility limits of the Cu–Ce–O system were not described in any existing phase diagram. We note, however, that Cu exhibits very low solubility in ZrO_2 [61], which shares the common crystal structure as CeO_2. Because Ce is an even larger cation than Zr, one might expect that the same low solubility would be true for Cu in CeO_2. However, recent electron paramagnetic resonance spectroscopy (EPR) studies indicate that Cu possesses a small solubility in CeO_2 in the form of Cu(II) ion pairs [62]. Cu might also be present in eightfold-coordinated interstitial sites as reported in the case of Cu–ThO_2 [63], or be highly dispersed on surface sites. Considering these thermodynamic constraints, one might anticipate that the $Cu_{0.15}Ce_{0.85}$ nanoclusters underwent the following structural evolution in postoxidation and thermal treatments: (1) a solid solution of metallic Cu and Ce prior to significant oxidation; (2) a supersatured solid solution of Cu in CeO_{2-x} after postoxidation; (3) segregation of Cu to the surface of CeO_{2-x} nanocrystals upon low-temperature annealing; and (4) formation of a separate copper oxide phase after high-temperature thermal treatment [59].

Based on these structural evolutions, we anticipate that at low temperatures, Cu and CeO_{2-x} might have existed as a solid solution, with the probable interstitial Cu ions in the CeO_{2-x} phase providing the dominant catalytic-promoting effects. By treatment at ~450°C, surface Cu ions became available from solute segregation in addition to interstitial Cu ions. The increase in Cu concentration at the surface would provide more active adsorption sites for CO. At 550°C and above, a bulk CuO phase became distinct in addition to the cerium oxide XRD peaks. Substantial grain growth was also noticed by 600°C. Both of these factors suggested a reduction of Cu dispersion in cerium oxide. However, the STEM experiment on samples treated at 650°C illustrated that significant amount of Cu remained dispersed within the CeO_{2-x} clusters or on their surfaces despite the formation of large CuO particles. It is unusual to see such high dispersion of a base metal on an oxide support after this high-temperature calcination. It may be attributed to the unique interaction between Cu and CeO_{2-x} and to the stable support CeO_{2-x} nanocrystals were able to provide. Such features would be very desirable to the preservation of thermally stable catalytic activities associated with high Cu dispersion.

10.3.2 Catalytic Properties of Nanocrystalline CeO_{2-x}-Based Materials

Catalytic testing was performed in a packed-bed microreactor under continuous flow [44,64]. Reactant gases for SO_2 reduction by CO were mixed in stoichiometric composition (1% SO_2/2% CO/He). Air was used as the oxidant for CO oxidation. The gas mixture was passed over the catalyst at a flow rate of 75–100 cm³/min at a constant temperature until steady state was reached. The composition of the effluent gas was measured by an HP-5880A gas chromatograph equipped with a thermal conductivity detector. The catalytic activities were compared based on the light-off temperature at which 50% conversion was achieved.

10.3.2.1 SO_2 Reduction by CO

The two possible reactions that can be anticipated are the reduction of SO_2 to elemental S and the formation of COS from elemental S and CO. Because COS is a highly toxic compound, high selectivity toward elemental S is critical.

A typical activity profile for this reaction is shown in Fig. 9 (a) for co-precipitated 10 at. % La-doped CeO_2 [4]. The S yield, which was the ratio of elemental S produced to the initial SO_2 feed, increased rapidly above 550°C to more than 95%. High selectivity toward elemental S and low COS formation were obtained for all CeO_2-based catalysts. After activation, the reaction temperature could be reduced to 500°C without loss of activity until the S yield fell off below 500°C. This process gave rise to a

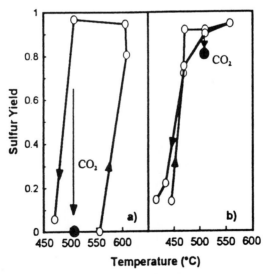

Figure 9. Catalytic activity profile for SO_2 reduction by CO over (a) precipitated La–CeO_2 and (b) nanocrystalline La–CeO_{2-x} with light-off and fall-off behavior. The effect of CO_2 poisoning is also indicated. Reprinted from *Nanostr. Mater.*, 6, A. Tschöpe and J. Y. Ying, 1005–1008, Copyright (1995), with kind permission from Elsevier Science Ltd., The Boulevard, Langford Lane, Kidlington 0X5 1GB, UK.

hysteresis loop in the activity profile. It was found that the fall-off temperature in this reaction was independent of contact time above a certain critical value [49]. Therefore, this temperature is a measure for thermodynamic limitations in the catalytic process.

Catalytic activities of nonstoichiometric and precipitated catalysts were compared in terms of their light-off and fall-off temperatures corresponding to 50% conversion (Fig. 10) [44]. The two main results were that (1) the nonstoichiometric catalysts were more active than the respective precipitated materials, and (2) the nonstoichiometric materials exhibited much less hysteresis [4]. The light-off temperature of nanocrystalline CeO_{2-x} was at 460°C, which was 120°C lower than for precipitated CeO_2, and hysteresis was reduced from 80 to 25°C.

The effect of excess CO_2 in the feed gas is also depicted in Fig. 9 [65]. The precipitated La-doped CeO_2 catalyst that was active at 510°C became completely deactivated by the addition of 1% CO_2 (Fig. 9 (a)). CO could no longer reduce the catalyst surface at this temperature, and the catalyst has to be heated to the light-off temperature to be reactivated. For nanocrystalline La-doped CeO_{2-x}, rather than complete deactivation, the sulfur yield decreased only slightly to 80% (Fig. 9 (b)) and remained high even at a CO_2 concentration of 6.5% in the feed gas containing 1% SO_2 and 2% CO. We found this remarkable stability against CO_2 poisoning for all nanocrystalline catalysts.

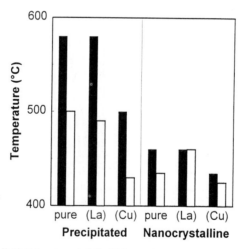

Figure 10. Light-off (full bars) and fall-off (open bars) temperatures of precipitated and nanocrystalline cerium oxide-based catalysts for SO_2 reduction by CO at 50% conversion. Reprinted from Tschöpe, A., Liu, W., Flytzani-Stephanopoulos, M., and Ying, J. Y. (1995). J. Catal., 157, 42, with permission from Academic Press.

Cerium oxide is well known for the ease in its reduction compared to other fluorite-type oxides. Temperature-programmed reaction (TPR) studies show that surface reduction of precipitated CeO_2 occurs with a maximum rate at about 500°C [66], which is close to the fall-off temperature of this sample in SO_2 reduction. Variation of the nature of dopant cations also reveals a correlation between catalytic activity and oxygen vacancy mobility induced by the dopant [49]. These results are interpreted by assuming a redox reaction mechanism for SO_2 reduction by CO as suggested by Happel *et al.* [67] for the same reaction over prereduced $LaTiO_3$ catalysts. The surface of the CeO_2 catalyst is continuously reduced by CO such that oxygen vacancies are created. SO_2 reoxidizes the reduced catalyst surface by donating its oxygen atoms to available vacancies. It is assumed that the reduction process involving several carbonate species is the rate-determining step.

The hysteresis found in SO_2 reduction with stoichiometric catalysts indicated the need for activation. Furthermore, the active state of the surface was lost in the presence of excess CO_2. Here, the drop in S yield was not caused by a simple competition between the two oxidizing species, SO_2 and CO_2. Under an overall oxidizing condition, the surface reduction process ceased completely and no further CO was oxidized to CO_2. These results could be consistently explained if the active state required a partially reduced surface. The hysteresis was caused by the lack of reduced surface sites on a fully oxidized surface. The initial reduction step required higher thermal activation, but once the surface was partially reduced, continuous reduction at a lower activation energy was possible. Under stoichiometric feed of reactants, the degree of surface reduction was given by the dynamic equilibrium between the rates of the two reactions. However, as excess CO_2 was introduced, all available vacancies were rapidly filled and the surface became oxidized to its initial state.

Absence of hysteresis in the case of nonstoichiometric nanocrystalline materials suggested that the activation process was not necessary or was much easier on a bulk-reduced catalyst. This could be due to the effect of a change in surface oxygen species. In addition to capping oxygen on a CeO_2 surface, a second surface species O_2^- has been identified by electron spin resonance (ESR) spectroscopy [41] and Fourier-transform infrared spectroscopy (FTIR) [53,68] after oxygen adsorption on prereduced or nanocrystalline nonstoichiometric CeO_{2-x}. The existence of this superoxide species is related to the availability of quasi-free electrons in reduced CeO_{2-x}. TPR measurements show that these species are less stable than regular surface-capping oxygen, and are readily available for oxidation reactions at temperatures below 500°C [66]. These active species were not part of the catalytic reaction between SO_2 and CO. However, the reaction of O_2^- with CO during activation was important to create the partially reduced surface before the catalytic reaction could begin. Continuous CO_2 admixture in the feed gas during SO_2 reduction oxidized the surface and deactivated the precipitated catalysts, but it could not oxidize the bulk-reduced nanocrystalline materials to full stoichiometry at temperatures below 500°C. Oxidation of nanocrystalline CeO_{2-x} in the XPS reaction chamber showed that oxidation did not occur spontaneously, but was thermally activated (see Fig. 6) [56]. Therefore, the effect of CeO_{2-x} nonstoichiometry on catalyzing SO_2 reduction by CO arises from a change of surface oxygen species and the resulting lower activation energy for initial surface reduction.

Doping of the cerium oxide with La did not increase catalytic activity in terms of the light-off temperatures for both types of catalytic materials. However, kinetic studies show that La doping permits full conversion at shorter contact times [49]. Furthermore, variation of the nature of dopants shows that catalytic activity correlates with oxygen ionic conductivity. However, increased concentration of oxygen vacancies through doping did not change the light-off temperature. In contrast, nonstoichiometric CeO_{2-x} catalysts contained oxygen vacancies and quasi-free electrons. The lower light-off and fall-off temperatures and the negligible hysteresis in the nanocrystalline La-doped CeO_{2-x} must be related to the change in the electronic structure of the material, compared to that of the precipitated sample.

The most active catalyst for SO_2 reduction by CO was the Cu-doped cerium oxide, which gave minor differences in activity in the precipitated and nanocrystalline forms. It appeared that the promoting effect of Cu dominated the effect of nonstoichiometry, except in the hysteresis behavior. Possible interactions for the promoting effect of Cu include [44]: (1) the presence of surface Cu^+ as preferred CO adsorption site, (2) interstitial Cu^+ in CeO_2 acting as electron donor and allowing easier CO adsorption even on the CeO_2 surface, and (3) change in the enthalpy of vacancy formation due to electronic interaction between Cu and CeO_2 [36]. In the first two cases, CO adsorption would be promoted by the availability of electrons, and oxygen vacancies could be created on the CeO_2 surface. These vacancies either were oxidized at its originated site by adsorbed SO_2 species or diffused from the CO adsorption site and became oxidized elsewhere. The interaction (3) involves the reducibility of CeO_2. The electronic structure of CeO_2 could be altered by the presence of Cu. Because oxygen vacancy formation involves electron transfer from the valence band to the next available state, thermodynamics of vacancy formation might be changed significantly. A metal oxide containing doping cations in lower oxidation states has new acceptor levels available. Electronic interaction in the discussions of supported catalysts usually are limited to metal–semiconductor systems. However, there might also be significant electron transfer at semiconductor–semiconductor junctions resulting in band-bending within CeO_2 (5.5-eV bandgap) and Cu_2O (1.5-eV bandgap) [69].

10.3.2.2 CO Oxidation by O_2

Neither nanocrystalline nor precipitated CeO_2-based catalysts exhibited hysteresis behavior in the activity profile for this reaction. The light-off temperature of 400°C for precipitated stoichiometric CeO_2 was considerably lowered to ~200°C with the use of nanocrystalline pure and La-doped CeO_{2-x} (Fig. 11) [44]. The most active catalyst in this study was again the Cu-doped cerium oxide with a light-off temperature of 80°C for both the nanocrystalline and the coprecipitated materials.

Compared to SO_2 reduction, the CO oxidation reaction is different in that there is: (1) a lowering in the light-off temperatures to the range below 200°C for all the nanocrystalline materials and the precipitated $Cu–CeO_2$, and (2) the absence of any hysteresis effects. This can be explained if CO oxidation proceeds by an extrafacial mechanism between adsorbed oxygen molecules and CO at low temperatures, as reported for V_2O_5 catalysts [70]. Formation of oxygen vacancies by CO would no longer be the rate-limiting step if CO could react with adsorbed oxygen instead of lattice oxygen. Variation of the nature of dopants shows no correlation between oxygen vacancy conductivity and catalytic activity, as

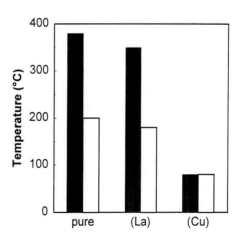

Figure 11. Temperatures for 50% conversion in CO oxidation for precipitated (full bars) and nanocrystalline (open bars) cerium oxide-based catalysts. Reprinted with permission from Tschöpe, A., Liu, W., Flytzani-Stephanopoulos, M., and Ying, J. Y. (1995). *J. Catal.*, **157**, 42, with permission from Academic Press.

found in SO_2 reduction [49]. This might also indicate the minor role of lattice oxygen in CO oxidation. In this case, the higher activity of nanocrystalline catalysts was caused by the rapid formation of activated ionized surface oxygen species O_2^- due to the availability of quasi-free electrons. CO oxidation catalytic activities of nanocrystalline and precipitated Cu-doped catalysts were identical. The promoting effect of Cu on the enthalpy of vacancy formation was not significant in CO oxidation because oxygen vacancies were not involved in this reaction. Therefore, the enhanced catalytic activity of Cu-doped catalyst must be related to (1) surface Cu that could be preferred adsorption sites for CO; or (2) interstitial Cu^+ that could donate free electrons toward chemisorption of oxygen as O_2^-, even on stoichiometric precipitated catalysts [44].

The presence of activated surface oxygen in the form of superoxide O_2^- species is important to both SO_2 reduction and CO oxidation. This was implied from the enhanced catalytic activity of nanocrystalline non-stoichiometric catalysts as compared with the corresponding stoichiometric precipitated materials. For both reactions, the Cu-doped catalyst was significantly more active than the La-doped catalyst. These two reactions are similar in that CO is the reducing species. In both cases, chemisorption of CO and formation of carbonate surface species that finally desorb as CO_2 is the rate-limiting step. We may therefore relate the promoting effect of Cu in SO_2 reduction and CO oxidation to the adsorption and surface reaction of CO.

By means of these studies, it was shown that the nature and extent of the effects of nonstoichiometry and dopants in CeO_2-based catalysts could be understood in terms of: (1) the presence of superoxide species O_2^- on nonstoichiometric CeO_{2-x}, and (2) preferred CO adsorption on Cu-doped cerium oxide [44]. The thermally stable Cu dispersion on cerium oxide was also notable, providing excellent catalytic activity in SO_2 reduction and CO oxidation even at temperatures above 600°C for this supported base metal system. This study demonstrates the potential of generating strong synergistic effects with nanocomposite processing for optimizing catalytic performance.

10.3.3 Electrical Conductivity of Nanocrystalline CeO_2

The electrical conductivity is determined by the product of charge carrier density, mobility, and electrical charge. Both the effective carrier density (which is controlled by defect thermodynamics) and the carrier mobility may be affected by the microstructure of the material. We have examined the electrical properties of nonstoichiometric nanocrystalline CeO_{2-x} to relate the catalytic oxidation activity to the formation of charge carriers.

The electrical conductivity of nanocrystalline CeO_{2-x} was analyzed by an HP4192A impedance meter [71]. The sample cell was a hydraulic compaction unit with electrically isolated pistons and external heating, and was sealed with Viton O-rings. The sample pellet was held under a pressure of 100 MPa between two platinum foils that acted as electrical contacts. The total conductivity for nanocrystalline CeO_{2-x} is presented in Fig. 12 as a function of temperature in oxidizing, curve (a), and reducing, curve (b), atmospheres [71]. The formation of mobile carriers upon reduction in 1% H_2/He gave rise to an increase in conductivity in this material. The slope of the plotted curves corresponds to the respective apparent activation energy, which is composed of a migration and carrier formation enthalpy. The value of the slope did not change upon reduction. The activation energy of electrical conductivity found for nanocrystalline CeO_{2-x} (0.7 eV) is less than half the value reported for single-crystal ceria (1.97 eV [39]). Because the difference between these values is much larger than the value of hopping mobility energy (0.4 eV [72]), it can be assumed from these results that the enthalpy of carrier formation was lower in nanocrystalline CeO_{2-x}. The formation of charge carriers can be caused by: (1) the removal of chemisorbed oxygen and the corresponding release of trapped

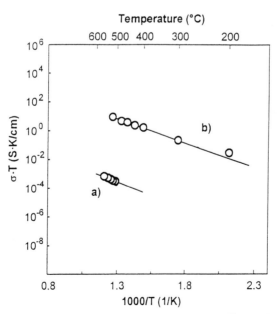

Figure 12. Temperature dependence of conductivity for nanocrystalline CeO_{2-x} in (a) oxidizing atmosphere (10% O_2/He) and (b) reducing atmosphere (1% H_2/He).

electrons, (2) the formation of new electronic carriers upon vacancy formation at the surface or at the grain boundaries, and (3) the bulk reduction of cerium oxide. The first two mechanisms are expected to exhibit lower enthalpies for carrier formation in comparison to bulk reduction in cerium oxide [66]. The low enthalpy of carrier formation in nanocrystalline CeO_{2-x} would explain its lower temperature requirements for catalytic oxidation. We may further use the light-off reaction temperatures of nanocrystalline CeO_{2-x} for CO oxidation by O_2 (200°C) and SO_2 reduction by CO (460°C) as a probe to distinguish the two types of charge carrier formation of interest: (1) the surface reaction between CO and chemisorbed oxygen vs (2) the surface reduction with vacancy formation [71]. Tying the electrical conductivity discussion with the catalytic study suggests that mechanism (1) that involves chemisorbed oxygen removal for charge carrier formation should be dominant at ~200°C, enabling CO oxidation by O_2 in nanocrystalline CeO_{2-x} at this low temperature. Mechanism (2) that is associated with vacancy formation may become significant at ~460°C, by which temperature nanocrystalline CeO_{2-x} was able to derive high catalytic activity for SO_2 reduction by CO.

10.4 Summary

Modified magnetron sputtering and inert gas condensation were combined as a new technique to produce nanocrystalline and nanocomposite catalysts. Via controlled postoxidation, nonstoichiometric CeO_{2-x}-based catalysts were generated for SO_2 reduction by CO and CO oxidation by O_2. These new materials exhibited outstanding catalytic activities in both reactions compared with the conventional catalysts. The findings could be explained by assuming that the active state of the catalyst involved a partially reduced surface. The higher activity might be attributed to a change in the nature of surface oxygen species for the nonstoichiometric CeO_{2-x}. Superoxide species could be formed easily on bulk-reduced CeO_{2-x} and were readily available for catalytic reactions. This difference translated into easy partial surface reduction, absence of hysteresis, and CO_2 poisoning resistance for the nonstoichiometric nanocrystalline materials. The oxidation catalytic activity could be further improved by doping CeO_{2-x} with Cu. A thermally stable dispersion of Cu on CeO_{2-x} could be achieved by the gas-phase synthesis and controlled posttreatment, and it provided synergistic chemical and electronic effects in this nanocomposite catalyst. This study demonstrates the uniqueness and applications of gas-phase nanocrystalline synthesis in advanced catalyst preparation. In particular, the importance of connecting microstructural control, surface chemistry, component dispersion, and electrical properties in designing a successful oxidation catalyst is emphasized. Such an approach may be further extended to the tailoring of other catalytic materials.

Acknowledgments

This work was supported by the National Science Foundation (CTS-9257223, DMR-9400334) and Sloan Fund/M.I.T. A. Tschöpe acknowledges fellowship support from the German National Scholarship Foundation (BASF program). The authors thank M. L. Trudeau of Hydro-Quebec, as well as W. Liu, M. Flytzani-Stephanopoulos, H. L. Tuller, and Y.-M. Chiang of M.I.T. for their collaborative efforts and contribution to this study.

References

1. Tschöpe, A. S., and Ying, J. Y. (1994). *Nanostr. Mater,* **4,** 617.
2. Ying, J. Y., Nanocrystalline Processing of Surface Reactive Materials: Effects of Non-stoichiometry and Dispersion, *Reviews of Naval Research,* Office of the Chief of Naval Research, Virginia, in press.
3. Tschöpe, A. S., and Ying, J. Y. (1994). *Nanophase Materials: Synthesis-Properties-Applications,* (G. C. Hadjipanayis and R. W. Siegel, eds.), p. 781, Kluwer, The Netherlands.
4. Tschöpe, A. S., Ying, J. Y., Liu, W., and Flytzani-Stephanopoulos, M. (1994). *Materials Research Society Symposium Proceedings,* Vol. 344, p. 133, Materials Research Society, Pittsburgh.
5. Gleiter, H. (1989). *Prog. Mater. Sci.,* **33,** 223.
6. Ying, J. Y. (1994). *Nanophase Materials: Synthesis-Properties-Applications,* (G. C. Hadjipanayis and R. W. Siegel, eds.) p. 37, Kluwer, The Netherlands.
7. Trudeau, M. L., and Ying, J. Y. Nanocrystalline Materials in Catalysis and Electrocatalysis: Structure Tailoring and Surface Reactivity, *Nanostr. Mater.,* in press.
8. MacKenzie, J. D. (1984). *Ultrastructure Processing of Ceramics, Glasses and Composites,* (L. L. Hench and D. R. Ulrich, eds.), p. 15, Wiley, New York.
9. Barringer, E. A., and Bowen, H. K. (1985). *Langmuir,* **1,** 414.
10. Uyeda, R. (1974). *J. Cryst. Growth,* **24,** 69.
11. Kimoto, K., Kamiya, Y., Nonoyama, M., and Uyeda, R. (1963). *Jpn. J. Appl. Phys.,* **2,** 702.
12. Granqvist, C. G., and Buhrman, R. A. (1976). *J. Appl. Phys,* **47,** 2200.
13. Thölen, A. R. (1979). *Acta Metall.,* **27,** 1765.
14. Andres, R. P., Averback, R. S., Brown, W. L., Brus, L. E., Goddard, W. A., III, Kaldor, A., Louie, S. G., Moscovits, M., Peercy, P. S., Riley, S. J., Siegel, R. W., Spaepen, F., and Wang, Y. (1989). *J. Mater. Res.,* **4,** 704.
15. Gleiter, H. (1981). Deformation of Polycrystals: Mechanisms and Microstructures, (N. Hansen, A. Horsewell, T. Leffers, and H. Lilholt, eds.), p. 15, Risø National Laboratory, Denmark.
16. Birringer, R., Gleiter, H., Klein, H. P., and Marquardt, P. (1984). *Phys. Lett.,* **A102,** 365.
17. Ying, J. Y. (1993). *J. Aerosol Sci.,* **24,** 315.
18. Siegel, R. W. (1991). *Annu. Rev. Mater. Sci.,* **21,** 559.
19. Siegel, R. W., and Eastman, J. A. (1989). *Materials Research Society Symposium Proceedings,* Vol. 132, p. 3, Materials Research Society, Pittsburgh.

20. Karch, J., Birringer, R., and Gleiter, H. (1987). *Nature (London)*, **330**, 556.
21. Siegel., R. W., and Hahn, H. (1987). *Current Trends in the Physics of Materials*, (M. Yussouff, ed.), p. 403, World Scientific, Singapore.
22. Averback, R. S., Hahn, H., Höfler, H. J., Logas, J. L., and Shen, T. C. (1989). *Materials Research Society Symposium Proceedings*, Vol. 153, p. 3, Materials Research Society, Pittsburgh.
23. Hahn, H., and Averback, R. S. (1990). *J. Appl. Phys.*, **67**, 1113.
24. Iwama, S., Hayakawa, K., and Arizumi, T. (1982). *J. Cryst. Growth*, **56**, 265.
25. Baba, K., Shohata, N., and Yonezawa, M. (1989). *Appl. Phys. Lett.*, **54**, 2309.
26. Matsunawa, A., and Katayama, S. (1985). Laser Welding, Machining, and Materials Processing, Proceedings, ICALEO '85, (C. Albright, ed.), IFS Publ.
27. Haas, V., Gleiter, H., and Birringer, R. (1993). *Scripta Metall. Mater.*, **28**, 721.
28. Haas, V., and Birringer, R. (1992). *Nanostr. Mater.*, **1**, 491.
29. Chow, G. M., and Edelstein, A. S. (1992). *Nanostr. Mater.*, **1**, 107.
30. Ying, J. Y., Benziger, J. B., and Gleiter, H. (1993). *Phys. Rev. B*, **48**, 1830.
31. Maissel, L. I., and Glang, R. (eds.) (1970). *Handbook of Thin Film Technology*, McGraw-Hill, New York.
32. Reichelt, K., and Jiang, X. (1990). *Thin Solid Films*, **191**, 91.
33. Yatsuya, S., Yamauchi, K. Kamakura, T., Yanagida, A., Wakayama, H., and Mihama, K., (1985). *Surf. Sci.*, **156**, 1011.
34. Westwood, W. D. (1988). *MRS Bull.*, Dec. 46.
35. Bielanski, A., and Haber, J. (1991). *Oxygen in Catalysis*, Marcel Dekker, New York.
36. Boudart, M., and Djega-Maliadassou, G. (1984). *Kinetics of Heterogeneous Catalytic Reactions*, Princeton University Press, New Jersey.
37. Bevan, D. J. M., and Kordis, J. (1964). *J. Inorg. Nucl. Chem.*, **26**, 1509.
38. Sørensen, O. T. (1976). *J. Solid State Chem.*, **18**, 217.
39. Tuller, H. L., and Nowick, A. S. (1979). *J. Electrochem. Soc.*, **126**, 209.
40. Tuller, H. L. (1981). *Nonstoichiometric Oxides*, (O. T. Sørensen, ed.), p. 271, Academic Press, New York.
41. Che, M., Kibblewhite, J. F. J., Tench, A. J., Dufaux, M., and Naccache, C. (1973). *J. Chem. Soc. Faraday Trans.*, **69**, 857.
42. Li, C., Domen, K., Maruya, K., and Onishi, T. (1989). *J. Am. Chem. Soc.*, **111**, 7683.
43. Sayle, T. X. T., Parker, S. C., and Catlow, C. R. A. (1992). *J. Chem. Soc. Chem. Commun.*, 977.
44. Tschöpe, A., Liu, W., Flytzani-Stephanopoulos, M., and Ying, J. Y. (1995). *J. Catal.*, **157**, 42.
45. Tuller, H. L., and Nowick, A. S. (1975). *J. Electrochem. Soc.*, **122**, 255.
46. Herman, R. G., Klier, K., Simmons, G. W., Finn, B. P., Bulko, J. B., and Kobylinski, T. P. (1979). *J. Catal.*, **56**, 407.
47. Choi, K. I., and Vannice, M. A. (1991). *J. Catal.*, **131**, 22.
48. Ying, J. Y. (1994). *Ceramic Transactions*, Vol. 44, p. 67, The American Ceramic Society, Columbus, OH.
49. Liu, W., Sarofim, A. F., and Flytzani-Stephanopoulos, M. (1994). *Appl. Catal.*, B4, 1967.
50. Liu, W., and Flytzani-Stephanopoulos, M. *J. Catal.*, **153**, in press.
51. Kochendörfer, A. (1944). *Z. Kristallogr*, **105**, 393.
52. Cullity, B. D. (1978). *Elements of X-Ray Analysis*, 2nd ed., p. 284, Addison-Wesley, Reading, MA.
53. Ying, J. Y. (1994). *Nanophase Materials: Synthesis-Properties-Applications*, (G. C. Hadjipanayis and R. W. Siegel, eds.), p. 197, Kluwer, The Netherlands.

54. Ying, J. Y. (1991). *Structural Evolution of Sol-Gel Derived Ceramics During Sintering*, Ph.D. thesis, Princeton University, New Jersey.
55. McGovern, S. J., Royce, B. S. H., and Benziger, J. B. (1984). *Appl. Surf. Sci*, 18, 401.
56. Tschöpe, A. S., Ying, J. Y., Amonlirdviman, K., and Trudeau, M. L. (1994). *Materials Research Society Symposium Proceedings*, Vol. 351, p. 251, Materials Research Society, Pittsburgh.
57. Trudeau, M. L., Tschöpe, A., and Ying, J. Y. (1995). *Surf. Interface Anal.*, 23, 219.
58. Laachir, A., Perrichon, V., Badri, A., Lamotte, J., Catherine, E., Lavalley, J. C., El Fallah, J., Hilaire, L., Le Normand, F., Quemere, E., Sauvion, G. N., and Touret, O. (1991). *J. Chem. Soc. Faraday Trans.*, 87, 1601.
59. Tschöpe, A., Ying, J. Y., and Chiang, Y.-M. (1995). *Mater. Sci. Eng. A*, 204, 267.
60. Subramanian, P. R., and Laughlin, D. E. (1990). *Binary Alloy Phase Diagrams*, 2nd ed., Vol. 2, (T. B. Massalski, ed.), p. 1051, ASM International.
61. Levin, E. M., Robbins, C. R., and McMurdie, H. F. (1969). *Phase Diagrams for Ceramists, 1969 Supplement*, (M. K. Reser, ed.), The American Ceramic Society, Columbus, OH.
62. Aboukais, A., Bennani, A., Aissi, C. F., Guelton, M., and Vedrine, J. C. (1992). *Chem. Mater.*, 4, 977.
63. Bechara, R., Wrobel, G., Aissi, C. G., Guelton, M., Bonnelle, J. P., and Aboukais, A. (1990). *Chem. Mater*, 2, 518.
64. Ying, J. Y., Tschöpe, A., and Levin, D. (1995). *Nanostr. Mater.*, 6, 237.
65. Tschöpe, A., and Ying, J. Y. (1995). *Nanostr. Mater.*, 6, 1005.
66. Yao, H. C., and Yu Yao, Y. F. (1984). *J. Catal.*, 86, 254.
67. Happel, J., Leon, A. L., Hnatow, M. A., and Bajars, L. (1977). *Ind. Eng. Chem. Prod. Res. Dev.*, 16, 150.
68. Li, C., Domen, K., Maruya, K., and Onishi, T. (1989). *J. Am. Chem. Soc.*, 111, 7683.
69. Strehlow, W. H., and Cook, E. L. (1973). *J. Phys. Chem. Ref. Data*, 2, 163.
70. Boreskov, G. K., (1973). *Kinet. Katal.*, 14, 7.
71. Tschöpe, A., Ying, J. Y., and Tuller, H. L. (1996). *Sensors Actuators*, B31, 111.
72. Tuller, H. L., and Nowick, A. S. (1977). *J. Phys. Chem. Sol.*, 38, 859.

C H A P T E R   1 1

A Flow-Through Hydrothermal Method for the Synthesis of Active Nanocrystalline Catalysts

D. W. Matson, J. C. Linehan, J. G. Darab, M. F. Buehler,
M. R. Phelps, and G. G. Neuenschwander
Pacific Northwest National Laboratory
Richland, Washington 99352

KEYWORDS: active nanocrystalline catalyst, iron-based catalytic system, iron-based oxides, iron-based oxyhydroxides, powder synthesis method, titanium oxide powder

11.1   Introduction

Many commercial chemical processes are aided by the heterogeneous addition of catalytic materials designed to enhance reaction rates, reduce process temperatures, modify selectivities, or otherwise improve the process economics. The use of catalytic materials is especially critical in the petrochemical industry, where hydrocarbon dissociation or rearrangement reactions must be achieved efficiently to ensure market competitiveness. In addition, increased environmental concerns have driven the mandated regulation of emissions from a wide variety of waste streams, as well as the cleanup of historical solid and liquid waste repositories. Processes enhanced through the use of heterogeneous catalytic materials are essential to achieve the goals of these regulations. Environmental considerations also increasingly play a role in determining catalyst selection for specific

259

processes, both in terms of catalyst preparation and in how spent catalysts are disposed or recycled. As a result, there is considerable interest in developing new synthesis methods capable of producing novel catalysts having unique properties or catalytic capabilities, catalysts having enhanced activities or extended lifetimes, or more environmentally compatible (e.g., disposable) catalysts to replace existing materials. [1]

Among the approaches being taken to achieve the preceding goals is the development of methods for producing very fine grained materials having catalytic properties. By producing grain sizes into the nanometer regime, specific surface areas can be maximized for catalysts applied as dispersed or supported powders, as microporous aggregates, or in the form of suspended slurries. Consequently, the volume of catalyst required for a specific application can be reduced through the use of nanometer-scale powders, providing both the economic and environmental benefits desired. There are currently few economically feasible methods suitable for producing large quantities of nanometer-sized materials for use as catalysts. The development of new synthesis methods for high-volume production of nanometer-sized materials is currently of high interest for the production of catalysts as well as for other uses that may benefit from powders having very fine crystallite sizes.

Workers at the Pacific Northwest National Laboratory (PNNL) have developed several general methods for producing nanometer-sized materials having a variety of potential applications, including the development of fuel cells, high-strength and specialty ceramics, and next-generation catalysts. Among these techniques are a glycine nitrate combustion method [2], a modified reverse micelle technique [3], and the rapid thermal decomposition of precursors in solution (RTDS) method [4,5]. We take this opportunity to provide an overview of the RTDS process and to discuss its use for generating nanocrystalline powders having demonstrated and potential catalytic properties.

RTDS is a continuous flow-through powder synthesis method having considerable flexibility in its ability to produce ultrafine nanocrystalline particulate products [4–7]. As used here, the term "nanocrystalline particulates" is defined as those particles having individual crystallites less than 50 nm in diameter, although those crystallites may be aggregated into high-surface-area particles (generally less than 10 μm in diameter). The RTDS process was originally developed to generate nanocrystalline iron oxides and oxyhydroxides for use as "disposable" first-stage direct coal liquefaction catalysts [6]. We and others have shown that the process is indeed capable of producing iron-based oxide and oxyhydroxide powders having high catalytic activities toward cleavage of C—C bonds in coal [9–12] and coal model compounds [7,13]. Many other applications, both catalytic and noncatalytic, exist that could benefit from the use of products having the ultrafine crystallite sizes produced by the RTDS method. As we show in the following text, the RTDS powder synthesis technique has the flexibility to produce a broad range of both pure and doped single-phase powders as well as mixed-phase nanocrystalline materials. This technique also has the potential for high-capacity powder production, making RTDS a commercially viable ultrafine powder synthesis method.

11.2 The RTDS Powder Synthesis Method and Apparatus

The RTDS method involves the rapid heating of a homogeneous precursor-bearing solution as the solution is continuously forced under pressure through a low-volume linear reactor. At the conditions present in the reactor, thermally induced reactions involving the dissolved precursor occur, causing the formation of a finely divided suspension of solids in the liquid solvent. Typical reactions may include thermal degradation of the precursor species, reactions between solute species present in the feed solution, or interactions between the solute precursor and the solvent. After only a very brief exposure to the high temperatures during passage through the reactor vessel, the solution pressure is abruptly dropped and the product suspension is thermally quenched by its passage out of the heated region through a pressure let-down device. At that point, the conditions promoting particle formation and growth are terminated. A simplified schematic of an RTDS apparatus is shown in Fig. 1, and a more detailed description of the process follows.

As a first step in RTDS processing, homogeneous feed solutions containing particle-forming precursor species dissolved in an appropriate solvent are prepared. To produce oxide or oxyhydroxide powders, water is an appropriate RTDS solvent and its use is desirable for a number of reasons. Aqueous solutions are easily prepared and handled; and they are not subject to many of the toxicity, flammability, volatility, or other potential problems associated with alternative (non-aqueous) solvents. The majority of RTDS studies to date were performed using aqueous solutions containing dissolved ionic salts or neutral water soluble complexes. However, RTDS is not limited to aqueous systems; and other solvents such as hydrocarbons, carbon dioxide, or ammonia may be used with the process if deemed desirable or necessary to produce a specific product. Some of our work using CO_2 as the solvent to produce metal particles will also be discussed briefly.

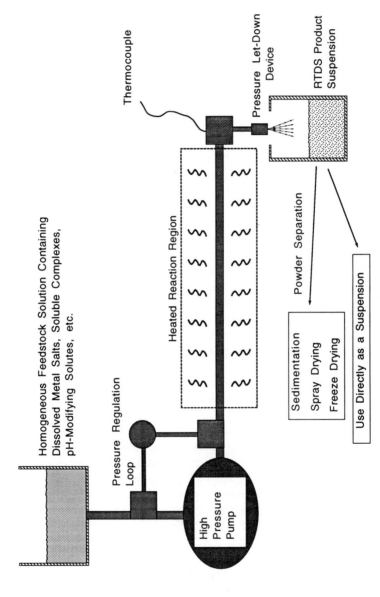

Figure 1. Simplified diagram of the RTDS nanocrystalline powder synthesis apparatus.

Concentrations of particle-forming precursor species in RTDS feed solutions are generally dilute (≤ 0.5 M with respect to metal content; more typically on the order of 0.1 M). These concentrations tend to minimize system plugging problems associated with the accumulation of solids on the reactor walls or buildup of precipitate at the entrance to the pressure let-down device, yet are sufficiently concentrated to provide a reasonable solid content in the product suspension. Halide-containing salts are generally avoided as RTDS particulate-forming precursors because aqueous solutions containing these salts are highly corrosive under RTDS conditions. However, corrosion-resistant tubing and fittings of nickel alloy (e.g., Hastelloy C-276) have been successfully used in place of the more commonly used stainless steel components when processing chloride solutions.

The RTDS feed solutions are pressurized using a high-pressure–high-capacity pump appropriate for the solvent and solute being used. At PNNL, aqueous RTDS precursor solutions are pressurized using either air or electrically driven reciprocating piston pumps having all wetted surfaces of type 316 stainless steel. A pressure regulation loop is incorporated downstream from the pump and consists of a backpressure regulator whose output is returned to the pump feed. The system pressure has not been found to be particularly important regarding its effect on the characteristics of the RTDS products. However, sufficient solution pressure must be maintained in the heated reactor region to ensure that a single liquid phase is maintained in the reactor. Typical RTDS operating system pressures are on the order of 4000–8000 psi, with 1000-psi fluctuations resulting from normal pump and backpressure-relief operations.

As used at PNNL, RTDS reactor vessels generally consist of lengths of high pressure type 316 stainless steel tubing, although (as noted previously) other materials can be substituted if increased corrosion resistance is required. Fittings used in the system are of the same material as that used for the reactor and are standard, commercially available high-pressure components. The RTDS reactor tube may be heated in any of a variety of ways. We have found that for small-diameter reactor tubes, resistive heating of the tube itself by applying DC current along its length is an effective method. For larger diameter reactors, passage through a tube furnace or series of tube furnaces provides sufficient heat to drive the RTDS reactions.

At the downstream end of the RTDS reaction tube, the final fluid temperature is monitored by a sheathed thermocouple mounted in the fluid flow. Power being applied to the devices providing heat to the reactor tube (i.e., the DC power supply or tube furnace) is controlled by a feedback loop tied into the sensed fluid temperature at the thermocouple. Just beyond the thermocouple the reaction fluid pressure is dropped across a let-down device (i.e., a nozzle). Crystallite nucleation and growth processes, resulting from exposure of the precursor species to the elevated temperatures and pressures in the RTDS reactor, are effectively terminated at the point at which the solution passes through the let-down device. It is important to recognize that crystallite formation occurs in the RTDS process before the solution–suspension is transferred through the pressure let-down device. Consequently, it should not be confused with spray nucleation [14], spray pyrolysis [15], spray-drying [16], or any other technique in which particle formation occurs as a result of, or subsequent to, the spraying process.

Unless the spray produced by the pressure drop across the let-down device is directed onto a substrate (e.g., for direct application of a catalyst powder onto a support), it is condensed and collected as a suspension in a cooled, enclosed vessel. Depending on the characteristics of the particles and other components present in water-based systems, rapid aggregation leading to the settling of solids may occur or the particles may remain suspended for weeks or months. If solid separation is desirable, sedimentation can be accelerated by centrifugation or by adjusting the product pH, or the suspensions can be spray- or freeze-dried.

As commonly used with an aqueous solvent, the RTDS process can be described as a continuous-flow hydrothermal powder synthesis method in which crystallite formation is first induced and then terminated over a very short time. Residence times of the precursor-bearing solutions in the reactors typically range from 1 to 30 s, depending on the reactor volume and the fluid flow rate, which in turn is determined by the diameter and configuration of the pressure let-down device and the reactor temperature. Thus, exposure of the precipitated solid to crystallite growth-promoting conditions is limited. Furthermore, essentially all of the precursor solution passing through the RTDS reactor experiences the same conditions at any given point in the reactor, ensuring a highly uniform powder product, both in terms of composition and particle characteristics for a given solvent–solute combination and set of RTDS processing parameters.

In aqueous feeds in which polyvalent metal-containing precursors are present in the form of dissolved ionic salts, solid particle formation can be initiated by thermally induced hydrolysis of the metal ions according to the following generalized reactions:

$$X^{m+} = nH_2O \rightarrow X(OH)_n^{(m-n)+} + nH^+, \tag{1}$$

and

$$X^{m+} + nH_2O \rightarrow XO_{n/2}^{(m-n/2)+} + 2nH^+. \tag{2}$$

In acidic media such reactions can yield either insoluble hydrous metal oxides or full oxides, depending on the reactor conditions and the metal ion in question and the presence of other species in solution. The rates of these reactions are accelerated dramatically under the elevated temperature and pressure conditions present in the RTDS reactor. If precursors are present in the form of dissolved nonionic metal complexes, thermal dissociation reactions likely occur under the hydrothermal conditions, followed by hydrolysis of the metal.

Aqueous RTDS feedstocks may also be prepared with additional solutes such as urea, $CO(NH_2)_2$, that decompose under hydrothermal conditions to yield a reactive species. Urea is commonly added to increase the pH of the solution in the hydrothermal region by thermal dissociation and reaction with water to form ammonia and carbon dioxide in the high-temperature solution. Consequently, if the urea decomposition occurs at a lower temperature than is necessary to promote significant hydrolysis of the metal ion, an amorphous hydroxide gel is produced. Further heating of the precipitated gel as it continues to flow through the RTDS system before its removal through the pressure let-down device can result in the production of a crystalline or partially crystalline product. Caution must, however, be exercised when using urea with solutions containing metal ions that readily form a stable carbonate phase (e.g., Ba^{2+}) in preference to the hydroxide.

The reactions just described are commonly used in other, more conventional, hydrothermal powder processing methods [17–19]. However, because those reactions typically take place in either batch reactors or large volume linear autoclaves in which growth processes are allowed to occur for much longer times, crystallites produced using conventional hydrothermal processing are generally significantly larger than those produced by RTDS.

PNNL currently has two operational RTDS units. One of these is referred to as a "bench-scale" unit and is used primarily for quick response feasibility studies and small batch (1–10 liters of precursor solution) sample processing. The second, called "engineering-scale" unit, has a larger capacity and is used exclusively for processes involving greater than 10 liters of precursor solution. Specifications of these units in typical operating configurations are presented in Table 1.

The production capacity of the engineering-scale RTDS unit currently varies from 2 to 3 lb of nanocrystalline powder per day, depending on the material produced. We envision the next step in scaleup of the process to a pilot-scale unit that uses a shell and tube heat exchanger with multitube bundles, each having an inlet and outlet manifold. The tubes comprising the individual bundles would each represent an individual small-volume linear reactor. Bundles would be valved independently, so that they could be individually removed for cleaning. The precursor solutions could be

TABLE 1

Configurations of Operational RTDS Units at PNNL

	Bench-scale unit	Engineering-scale unit
Pump type	Reciprocating piston (air driven)	Reciprocating piston (electric)
Pump materials	All wetted parts Type 316 stainless steel (ss)	All wetted parts Type 316 ss
Reactor vessel	1 m × 1/8 in. o.d. × .035 in. wall	2.25 m × 9/16 in. o.d. × 1/8 in. wall
Reactor material	Type 316 ss or nickel alloy	Type 316 ss
Heating mechanism	Resistive/dc current	Multistage tube furnace
Pressure let-down device	5 mm × 90 μm i.d. Pt/Ir capillary tube	125-μm Sapphire orifice, tapered inlet
Operating temperature range	<100–400°C	<100–400°C
Operating pressure range	4–8 kpsi	4–8 kpsi
Solution residence time in reactor	1–5 s	5–30 s
Solution feed rate	50–75 cc/min	150–250 cc/min[a]

[a]Operating at ~ 1/3 of total pump capacity.

preheated in linear autoclaves positioned between the primary pump and the multitube reactor bundles. This will reduce the heat load required to raise the feed solution to reaction temperature and enable precise control of residence time at that temperature. A multihead pump could be used to ensure a uniform flow distribution in the inlet and outlet manifolds. This proposed pilot-scale design is expected to satisfy catalyst manufacturer needs by increasing the RTDS production capacity by several orders of magnitude relative to the production capability of existing equipment.

11.3   RTDS Products

A major benefit of the RTDS nanocrystalline powder-generating process is that the flow-through hydrothermal design allows a broad range of different oxide and oxyhydroxide materials to be produced. These include single-phase and doped single-phase materials, intimately dispersed multiphase materials, and combinations thereof. Table 2 provides a partial list of oxide and oxyhydroxide materials processed by RTDS.

In the discussion that follows, we describe three metal oxyhydroxy systems that have been most extensively studied for the production of nanocrystalline powders by the RTDS method and that have considerable potential for use in catalytic applications. We will also briefly discuss the use of RTDS technology for processing metal carbonyl–CO_2 systems.

RTDS powder products are commonly characterized by crystalline phase, as determined using powder X-ray diffraction (XRD). Crystallite size is determined by analysis of XRD line-broadening effects that occur as the size of crystallites in the powder is decreased [20]. Crystallite sizes in many RTDS powders are less than 10 nm in diameter, and the line-broadening effect may be sufficiently large to complicate or prevent definitive identification of a specific product phase. Other techniques such as transmission electron spectroscopy (TEM), Raman spectroscopy, or Mossbauer spectroscopy have sometimes been used to help identify or verify the existence of a specific phase or crystallite diameter in the RTDS product.

11.3.1 Iron-Based Oxides and Oxyhydroxides

Iron-based materials are typically low in both cost and toxicity compared to conventional (e.g., Mo- or Co-based) catalysts used in cracking or other processing operations on hydrocarbon feedstocks, yet they are well known to have some degree of catalytic activity for those types of operations [1,8]. Consequently, iron-based catalysts have received attention for applications in which catalyst recovery or regeneration after reaction is excessively difficult or expensive, such as situations where the use of highly dispersed catalysts is desirable. Optimizing the properties of iron-based

TABLE 2
Selected Nanocrystalline Power Products Produced Using the RTDS Process

Product	Precursor	Crystallite size (nm)[a]
$5Fe_2O_3\cdot9H_2O/Fe_2O_3$ (6-line ferrihydrite/hematite)	$Fe(NO_3)_3$/urea	<10
α-Fe_2O_3 (hematite)	$Fe(NO_3)_3$	6–20
Fe_3O_4 (magnetite)	$FeSO_4$/urea	12–19
Ferric oxyhydroxysulfate	$Fe(NO_3)_3$/Na_2SO_4	<10 (4)
$NiFe_2O_4$ (trevorite)	$Fe(NO_3)_3$/$Ni(NO_3)_2$/urea	<10 (8)
NiO (bunsenite)	$Ni(NO_3)_2$	12
$NiO/Ni_2Cr_3O_5(OH)_4$ (bunsenite/brindleyite)	$Ni(NO_3)_2$/$Cr(NO_3)$/urea	<10
NiO/ZrO_2 (Ni-doped ZrO_2)	$Ni(NO_3)_2$/$ZrO(NO_3)_2$	<10
$ZrO_2/(SO_4)$ (sulfated ZrO_2 superacid)	$Zr(SO_4)$/urea	<10
TiO_2 (anatase)	$K_2TiO(C_2O_4)_2$	<10 (3)
TiO_2/Pd (Pd-doped anatase)	$K_2TiO(C_2O_4)_2$/$PdCl_2$	<10 (3)
ZnO (zincite)	$Zn(NO_3)_2$/urea	34

[a]Crystallite size estimated by XRD line-broadening analysis. Number in parentheses are estimates based on dark-field TEM images.

materials is extremely important for such applications to obtain the best of both catalytic and physical properties. Among the variables that can be addressed for iron oxyhydroxy catalysts are the crystallite phase, for which there are a number of possibilities; and the crystallite size, a factor affecting the number of catalytically active surface sites available to substrate molecules. The latter effect is also influenced by the degree of crystallite aggregation and by the aggregate size [21]. In addition, the use of biphasic or doped iron-based products may enhance catalytic properties without significantly degrading the environmental advantages of these materials.

There are at least nine crystallographically distinct common iron oxide and oxyhydroxide phases [22]. Several of these phases having inherent crystallite sizes that range into the nanometer regime can be produced directly in an aqueous RTDS system by choosing the appropriate iron-bearing precursor material, by using additives such as urea, or by adjusting other process conditions under which the particle-forming reactions occur (e.g., RTDS reaction temperature or residence time in the RTDS reaction vessel). Additional phases having nanocrystalline characteristics can be produced by subsequent processing (e.g., by calcination) of RTDS-generated products (Table 3). An example of the effect of increasing RTDS pro-

TABLE 3
Common Iron Oxide and Oxyhydroxide Phases and Their Synthesis by RTDS[a]

Phase	Formula	RTDS conditions used to produce iron-based powders[a]	
		Precursor solution components	Reaction temperature (°C)
Hematite	$\alpha\text{-Fe}_2\text{O}_3$	$\text{FE(NO}_3)_3$	>300
2-Line ferrihydrite[b]	$\text{Fe}_5\text{HO}_8\cdot4\text{H}_2\text{O}$	$\text{Fe(NO}_3)_3$	<250
6-Line ferrihydrite[b]	$\text{Fe}_5\text{HO}_8\cdot4\text{H}_2\text{O}$	$\text{Fe(NO}_3)_3/\text{urea}$	>300
Magnetite	Fe_3O_4	$\text{Fe(II)SO}_4/\text{urea}$	>250
Ferric oxyhydroxysulfate[c]	$\text{Fe}_8\text{O}_8(\text{OH})_6\text{SO}_4$	$\text{Fe(NH}_4)(\text{SO}_4)_2/\text{Na}_2\text{SO}_4$	80–200
Maghemite[d]	$\gamma\text{-Fe}_2\text{O}_3$	—	—
Goethite[e]	$\alpha\text{-FeOOH}$	—	—
Akaganeite	$\beta\text{-FeOOH}$	—	—
Lepidocrocite	$\gamma\text{-FeOOH}$	—	—
Feroxyhyte	$\delta'\text{-FeOOH}$	—	—

[a]Conditions described are for the bench-scale RTDS unit, 0.1 M iron content in solution, 5–8 kpsi pressure, 75 cc/min feed rate.

[b]The two ferrihydrites are distinguished by the number of broad peaks present in their XRD patterns.

[c]Formation of solid ferric oxyhydroxy sulfate requires a brief (1–7 day) dialysis step after RTDS processing.

[d]Nanocrystalline maghemite is produced by air oxidation of RTDS-generated magnetite.

[e]Goethite can be produced by reaction of the six-line or two-line ferrihydrite suspensions with ammonium or sodium sulfide in air.

270 / D. W. Matson et al.

cessing temperature on the product phase distribution produced from a 0.1 M Fe(NO₃)₃ solution is shown in Fig. 2.

Postprocessing of RTDS-generated powders can also be used to modify phase characteristics of some of the iron-based products (Fig. 3). In addition to the simple oxides and oxyhydroxides, synthesis of a sulfated oxyhydroxide phase, ferric oxyhydroxysulfate (OHS), can be accelerated using the RTDS method relative to conventional bench-top synthesis. A single pass of solution containing ferric and sulfate ions through the RTDS reactor, followed by a relatively brief dialysis of the product solution (1 to 7 days) and centrifugation, yielded the OHS product shown in Fig. 4. The bench-top synthesis procedure for this material recommends strict adherence to a 2-liter batch size and a 30-day dialysis time [22].

Iron-based materials are of interest as precursors to catalysts used for the direct conversion of coal-to-liquid products suitable for processing into transportation fuels. It is commonly assumed that the iron oxide or oxyhydroxide phases do not directly act as catalysts for the carbon-carbon bond cleavage reactions. Rather, the iron-based phases are themselves converted at reaction conditions into a poorly defined iron sulfide, which is the active catalyst [23]. With iron-based powders produced using bulk synthesis

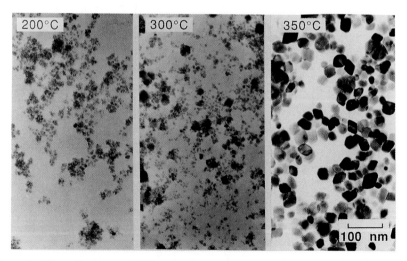

Figure 2. Effect of increasing RTDS processing temperature on the product phase characteristics of a powder produced from 0.1 M Fe(NO₃)₃. Powder XRD analysis indicates that the samples produced at 200 and 350°C consist exclusively of ferrihydrite and hematite, respectively. The 300°C product is a mixture of the two end-member phases. Reprinted with permission from Energy and Fuels, 8, 14, 1996, American Chemical Society.

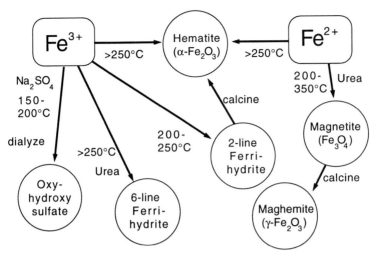

Figure 3. Mechanisms for producing various nanocrystalline iron oxide and oxyhydroxide phases either directly or indirectly by the RTDS process.

Figure 4. TEM micrograph of a ferric oxyhydroxysulfate (OHS) aggregate produced using the RTDS process.

methods available in the literature [22], we have shown that at a uniform set of test conditions with a model compound substrate, the iron precursor phase is an important factor in determining the measured catalytic activity [13,24]. We speculate that certain iron-based precursor phases are more readily converted to the active catalyst phase by virtue of their crystallographic structures and thermodynamic free energies relative to the active catalyst phase. The flexibility of the RTDS process to produce a variety of distinct iron-based phases is then an important consideration in establishing the usefulness of the process as a method for generating iron-based coal liquefaction catalysts. In fact, two iron-based phases that were shown to exhibit some of the highest activity as carbon-carbon bond scission catalyst precursors—six-line ferrihydrite and the sulfated oxyhydroxide (OHS)—can be generated using the RTDS method. The application of nanocrystalline iron-based powders in other hydrocracking processes may also increasingly prove to be feasible.

The iron-based catalytic systems also demonstrate another capability of the RTDS method. Using the engineering-scale RTDS unit, we were able to successfully increase the output of six-line ferrihydrite to pound per hour levels using the parameters developed with the bench-scale RTDS unit. The increased production rate was accomplished primarily through increasing the concentration of the ferric nitrate in the feed from 0.1 to 0.5 M. The recovery of the iron in the feed solution as the iron oxyhydroxide remained close to 100% even for the highest production rates. The XRD patterns of the iron oxyhydroxides produced at the different rates all showed six-line ferrihydrite as the only phase present. The catalytic activity of six-line ferrihydrite powders showed essentially no effect of variations in production rate, demonstrating the scalability of the RTDS process.

The formation of ultrafine magnetic powder, identified using XRD as nanocrystalline magnetite, was accomplished by using an RTDS feed solution containing Fe^{2+} ions and urea. In the absence of urea, Fe^{2+} ions are fully oxidized to Fe^{3+}. Under those conditions hematite (α-Fe_2O_3) is the primary product phase, with the remainder consisting of highly hydrated species (ferrihydrites).

A further advantage of the RTDS process is its ability to produce powders doped with a small percentage of a second metal ion. RTDS powder products are generated from homogeneous solutions and contain crystallites that are extremely small and rapidly formed. Consequently, foreign ions present in the feedstock, which are hydrolyzed under similar conditions as those comprising the bulk of the metal ions in solution, are readily incorporated into the resulting powder product. In some instances, the additional ions are incorporated into the crystalline iron oxyhydroxide structure by substitution for an iron atom, forming a single solid solution phase. Other species may be included at crystalline defect sites or at crystallite interfaces, or may form nanocrystalline secondary phases. Due to the short reaction time in the RTDS unit ($\leq$30 s,) the product phase may not be the one that is thermodynamically stable under those conditions.

In an effort to further enhance the already attractive carbon-carbon bond cleavage characteristics of six-line ferrihydrite, a number of samples were produced in which a small amount (1–10 atom (at.) % relative to Fe) of a second metal ion (including Ni, Cr, Mo, Zr, and Sn) was included in the RTDS feedstock. The solutions were processed under RTDS conditions that would normally produce a catalytically active six-line ferrihydrite powder with only Fe^{3+} in the feedstock. In all cases the only phase, other than a minor hematite component, that was detected by powder XRD in these doped samples was six-line ferrihydrite. Inductively coupled plasma–mass spectrometric (ICP-MS) analysis of the powders indicated a close correlation between feedstock metal content and the dopant metal concentration in the powder product (Table 4). While these analyses do not completely eliminate the possibility of compositional heterogeneities, they strongly suggest that the dopant metal was dispersed uniformly throughout the nanoscale powders.

TABLE 4

Ferrihydrite Powders Produced by Coprocessing Solutions of Ferric Nitrate and an Additional Metal Salt

Dopant salt	Nominal mol percent dopant metal in feed	Measured mol percent dopant metal in isolated powder[a]	Phases detected by XRD
Cobalt acetate	10.0	9.3	6-Line ferrihydrite
Cobalt acetate	1.0	1.1	6-Line ferrihydrite
$Ni(NO_3)_2$	10.0	9.8	6-Line ferrihydrite
$Ni(NO_3)_2$	1.0	1.2	6-Line ferrihydrite
$SnCl_2$	1.0	1.1	6-Line ferrihydrite/ hematite
$Cr(NO_3)_3$	10.0	9.9	6-Line ferrihydrite
$Cr(NO_3)_3$	1.0	1.2	nd
$ZrO(NO_3)_2$	10.0	9.4	6-Line ferrihydrite
$(NH_4)_2MoO_4$	1.0	1.0	6-Line ferrihydrite

[a]Analysis method: ICP/MS.

11.3.2 Titanium Oxide Powders

Nanocrystalline TiO_2 and metal-doped TiO_2 have considerable potential as semiconductor materials suitable for photocatalytic applications. TiO_2 is particularly attractive for catalytic reactions in aqueous systems because it is resistant to photocorrosion and has a bandgap that allows electronic excitation by UV radiation. Electrons and the associated holes produced by bandgap excitation of TiO_2 migrate to particle surfaces where they can participate in redox reactions. The use of well-dispersed nanocrystalline TiO_2 for these types of applications is conceptually attractive because of the short distances required for electron and hole migration before encountering a particle surface. Among the reactions for which TiO_2 has been investigated as a photocatalytic agent are the dissociation of water, the oxidation of dissolved organics, and the reduction of hazardous metal ions in solutions [25–27].

Titanium dioxide is a particularly amenable material for processing into nanocrystalline powders (anatase phase, <5-nm crystallite diameter) by RTDS. Among the water soluble materials that have been identified as precursors for RTDS processing of TiO_2 are potassium bis(oxalato)oxotitanate, $K_2TiO(C_2O_4)_2$, and its ammonium analog. Low-temperature processing of 0.1 M $K_2TiO(C_2O_4)_2$ solutions (100–200°C) using the bench-scale RTDS apparatus typically produces clear or slightly murky suspensions that gradually turn opaque white over a period of minutes to hours and eventually settle. This behavior suggests that initial primary particle formation takes place during the low-temperature heating, but that significant aggregation of the particles does not occur until the suspensions are allowed to set for some period of time. RTDS processing of the $K_2TiO(C_2O_4)_2$ solutions at temperatures of 250°C or higher tend to produce opaque suspensions that settle rapidly, indicating that significant aggregation occurred either in the RTDS reaction tube or immediately upon product collection. These results suggest that the TiO_2 product dispersion and degree of aggregation can be controlled, depending on the requirements of the application for which the material is being produced.

Another water soluble precursor, a commercially available water-based solution of the ammonium salt of a lactic acid titanium chelate (TYZOR LA, DuPont), also can be processed in dilute form using RTDS. The product obtained using this precursor is an indefinitely stable translucent suspension.

The RTDS process has been used to produce a 1% Pd-doped TiO_2 powder that was shown to have a high activity toward photocatalytic oxidation of organic species in aqueous solution. Using a 0.1 M $K_2TiO(C_2O_4)_2$ feedstock solution containing 10^{-3} M $PdCl_2$ and processing at a 250°C RTDS reactor temperature, a nanocrystalline TiO_2 product was obtained. After separating and drying the powder product, it was calcined at 250°C for 2 h. XRD analysis of this powder showed anatase (<10-nm diameter) as the only phase present. Energy dispersive X-ray analysis of the powder indicated that the palladium was uniformly distributed throughout the anatase product. We have found that this Pd-doped anatase powder has a higher activity than commercially available titania-based photocatalysts toward light-induced catalytic destruction of disodium ethylenediaminetetraacetic acid (Na_2EDTA) in aqueous solvents.

11.3.3 Zirconium Oxide Powders

Zirconium oxide is another material that has a number of catalytic applications and is amenable to nanocrystalline powder synthesis via the RTDS method. Zirconium oxide-based catalysts can be useful for a range of synthesis, dehydration, dehydrogenation, and isomerization reactions, although ZrO_2-based catalysts are very sensitive to preparation methods and activation treatments [28]. Bulk ZrO_2 is also often used as a catalyst support because of its attractive thermal and chemical stability characteristics. A major focus in the past few years has been to develop and characterize sulfate-promoted ZrO_2 catalysts exhibiting superacid properties. These solid superacids are being viewed as potential replacements for the liquid acid catalysts currently used in industrial hydrocarbon processing applications.

Water-soluble precursors that were evaluated and found to be suitable for formation of nanocrystalline ZrO_2 powders (≤5-nm crystallite diameter) by RTDS include $ZrO(NO_3)_2$, $Zr(SO_4)_2$, and a zirconium citrate ammonium complex (Noah Technologies Corporation). Precursor feedstock solutions (0.1 M) of $ZrO(NO_3)_2$ and $Zr(SO_4)_2$ are acidic (pH <2). Comparable solutions of the citrate ammonium complex are near neutral. In the absence of other pH-modifying components, the product pH of each of these solutions tends to be slightly lower than that of the starting solution because of the generation of excess hydrogen ions during the hydrolysis reactions leading to solid precipitation in the RTDS reactor. As will be discussed in further detail later, the specific product suspension characteristics produced from these solutions were found to be dependent on the precursor, on the RTDS processing temperature, and on the presence or absence of other components in the feed solution.

In general, a higher RTDS reaction temperature was required to produce changes in the appearance of the zirconium-bearing solutions (indicating the formation of a solid phase) than was necessary to produce similar changes in the appearance of either iron- or titanium-based feedstocks.

Using the bench-scale RTDS system (~2-s residence time in the RTDS reactor) and the zirconyl nitrate precursor, a processing temperature of at least 300°C was required to yield product suspensions that exhibited visible changes suggesting the formation of a suspended solid component. Solids contents in the product suspensions appeared to increase with increasing processing temperature to at least 350°C, although a temperature up to 400°C was evaluated. For every processing temperature above 300°C, the product was a translucent suspension that did not appreciably settle over a period of up to several months. Comparable feed solutions containing the precursor zirconium sulfate generated product suspensions having similar properties. RTDS processing of feed solutions containing the zirconium citrate ammonium complex behaved similarly in terms of the processing temperature requirements for initiating visible changes in the precursor-bearing solution. For this precursor, the product suspensions were typically pink-to-white, and again showed little tendency to settle over time.

By adding urea to the feed solution containing the dissolved zirconium precursor, the resulting product suspension characteristics were altered dramatically. When using the bench-scale RTDS unit with precursor solutions containing the nitrate or sulfate zirconium salts, an RTDS processing temperature of at least 300°C, and a urea content 2.5 times that of the zirconium ion concentration (e.g., 0.1 M zirconium sulfate, 0.25 M urea), the product pH ranged from 9 to 10. Under these conditions, the zirconium precipitated as a hydroxide gel and the product suspension was a white dispersed solid that began to settle within the first few minutes of collection. The liquid phase above the settling solid was clear, and the odor of ammonia generated by the decomposition and reaction of the urea in the RTDS reactor was detectable in the product suspension. There was little difference between the product pH of solutions processed at 300°C and of those processed at a higher temperature.

The RTDS-processing temperature significantly affected the characteristics of the solid products generated from zirconia precursor/urea solutions. For example, a 0.1 M $Zr(SO_4)_2$/0.25 M urea solution processed at increasing RTDS temperature from 300 to 400°C produced a series of products whose settling and crystallinity properties varied systematically as a function of the processing temperature. As collected, each of the five products had a similar white gel-like appearance, except for the 400°C sample that had a gray cast to the color of its suspension and separated solid. After settling overnight after collection, the proportion of product volume in the clear liquid phase above the settled solid increased from approximately 50% for the 300°C sample to greater than 70% for the 400°C sample. Samples produced at the intermediate temperatures settled at intermediate rates, and in direct proportion to their RTDS processing temperatures.

Settling rates of the solid component in these samples can be correlated with the degree of crystallinity exhibited by the separated solid product. X-ray diffraction analysis of the centrifuged and dried solids from these samples (Fig. 5) also suggests a systematic variation in the product characteristics. Higher RTDS processing temperature yielded as-collected powder products having a higher degree of crystallinity, as indicated by the development of broad, but distinct, crystalline features in the XRD patterns. Because of the amorphous character of the lowest temperature sample and increasing crystallinity with processing temperature, we suggest that gel formation occurred first as a result of urea decomposition and the formation of ammonium ions, followed by crystallization of ZrO_2 from the gel.

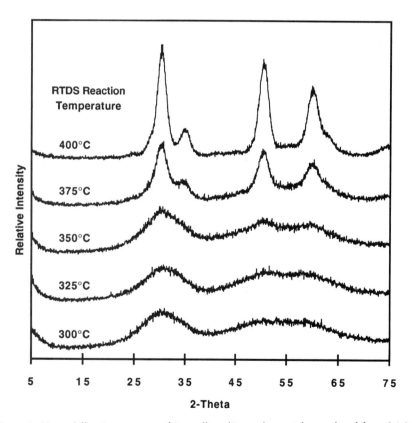

Figure 5. X-ray diffraction patterns of "as collected" powder samples produced from 0.1 M $Zr(SO_4)_2/0.25$ M urea solutions at RTDS temperatures from 300 to 400°C.

Variations in crystallization behavior of the five samples shown in Figure 5 persisted even when the powders were calcined at 500°C in air for an hour (Fig. 6). The sample generated at the lowest RTDS temperature (300°C) showed no increase in crystallinity as a result of the calcination treatment, while the 325 and 350°C samples both exhibited crystallite growth. Crystalline features shown by the XRD analysis of the 375°C sample became more pronounced as a result of the calcination. Samples of the as-collected powders similarly calcined at 600°C in air for an hour all exhibited a relatively high degree of crystallinity, although minor distinctions between the XRD patterns persisted to some extent. Even after calcination of the powders at 600°C, line-broadening analysis of the XRD patterns indicated a crystallite size of less than 10 nm in all of these samples. The effect of solution residence time on product crystallinity in a similar series of samples is shown in Fig. 7. This series of samples was produced

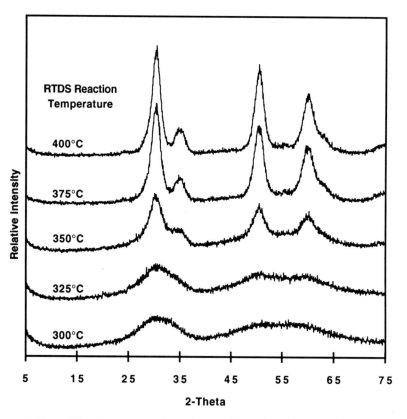

Figure 6. X-ray diffraction patterns of powder samples produced from 0.1 M Zr(SO$_4$)$_2$/0.25 M urea solutions at RTDS temperatures from 300 to 400°C after calcination in air at 500°C for 1 h.

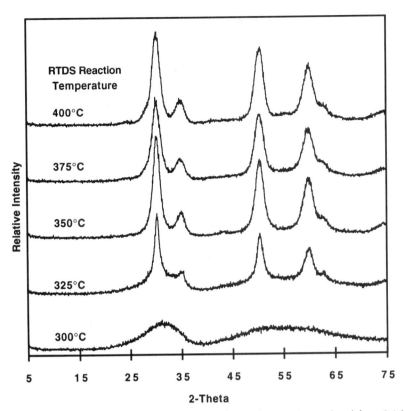

Figure 7. X-ray diffraction patterns of "as collected" powder samples produced from 0.1 M $Zr(SO_4)_2$/0.25 M urea solutions at RTDS temperatures from 300 to 400°C. Solution residence time in the RTDS reactor was approximately five times that used to produce the products shown in Fig. 1.5, and all other processing variables were constant.

from an equivalent feedstock solution and at comparable flow rates to those used to produce the samples in Fig. 6. However, the residence time of the solution at hydrothermal conditions was increased by a factor of approximately five. This was accomplished by increasing the length and inside diameter of the tube used as the reactor.

Properly prepared zirconium and other metal oxides containing adsorbed sulfate species have been found to have superacidic properties (i.e., acid strengths greater than that of 100% H_2SO_4) [28] and have been shown useful for promoting low-temperature hydrocarbon isomerization, alkylation, esterification, and cracking reactions [29–31]. These sulfated solid superacid catalysts have been proposed as environmentally acceptable substitutes for homogeneous liquid acids now widely used for those applications. In addition to environmental problems associated with their use, liquid acid catalysts suffer from a variety of disadvantages, including difficulty in separating them from product streams, inefficient utilization leading to excess acidic waste, and their inherent corrosive nature. Development of methods to produce highly active, long-lived solid superacid catalysts for hydrocarbon processing is an important technological goal for the near future.

Most attempts at producing sulfated zirconia superacid catalyst precursors have focused on use of a two-step process involving the production of a zirconia gel by adding a base to a dissolved zirconium salt, and then sulfating the gel surface by incipient wetness using a soluble sulfate source [32,33]. The resulting precursor must then be calcined at temperatures ranging from 400 to 650°C to produce the proper sulfate group coordination and water content to yield a catalytically active solid superacid species [34,35].

We have found that RTDS can be used as a one-step synthesis method for nanocrystalline sulfated zirconia precursor powder that is readily converted by calcination into a nanocrystalline solid having superacidic properties. By coprocessing a ZrO_2 precursor solution with a sulfate source (e.g., with sulfuric acid or a sulfate salt), a sulfated zirconia powder product was produced directly. Although active powders could be generated in acidic solutions (i.e., in the absence of added urea), as noted previously the solids were difficult to separate from the suspension under those conditions. Sulfated zirconium oxide powders that were easily separated from suspension and were catalytically active toward low-temperature isomerization of n-butane to isobutane were produced using a feedstock solution containing $Zr(SO_4)_2$ and urea. Specific activities of the sulfated ZrO_2 powders toward these reactions were found to vary as a function of RTDS processing conditions (e.g., the RTDS temperature) as well as variations in the calcination temperature used to activate the powders.

11.3.4 RTDS Processing in CO_2

Solvents other than water can also be used in the RTDS process. Some of the first experiments run using the RTDS configuration involved the processing of iron pentacarbonyl and molybdenum hexacarbonyl in a CO_2 solvent. In one example, $Fe(CO)_5$ (3.0 ml) was placed in a 50-ml autoclave and pressurized to 2500 psi with CO_2. The solution produced was then passed through a stainless steel RTDS reactor at 350°C, and the powder product was collected on a porous stainless steel frit. XRD analysis of the resulting black powder showed the powder to be extremely fine (<10-nm-diameter crystallite size) α-iron. XRD analysis of a powder produced under similar RTDS conditions at 250°C, but collected by bubbling the product into an octane bath, showed the major product to be magnetite.

11.4 Conclusions

The RTDS process is a novel powder synthesis method involving hydrothermal reactions of particle-forming precursor species in continuously flowing solutions. It uses short reaction times, on the order of seconds, to quickly form particulate species and then terminates the growth process by abruptly removing the resulting suspension from the hydrothermal environment through a pressure let-down device. This method is an attractive approach for the large-scale generation of active nanocrystalline oxide and oxyhydroxide powders for a variety of catalytic applications. Specific benefits of the process include the nanocrystalline character of the powder products, the capability to adjust product characteristics by varying the process conditions, a wide range of product materials, the ability to generate doped single phase or biphasic powders, and the potential to operate in a continuous mode for large-scale nanocrystalline powder generation.

Specific examples of RTDS-generated nanocrystalline powders that have been shown to have application as catalytic materials include iron oxyhydroxides, TiO_2-based products, and ZrO_2-based powders. RTDS-generated six-line ferrihydrite and ferric oxyhydroxysulfate powders were demonstrated to be highly active in C—C bond scission reactions specifically applicable to direct coal liquefaction, but may have additional applications in other petrochemical processes. Titanium oxide and ZrO_2 powders are two additional nanocrystalline materials that are readily produced by the RTDS process and have potential for catalytic applications. RTDS synthesized Pd-doped TiO_2 has high activity toward the photocatalytic destruction of EDTA in aqueous solutions. Sulfated ZrO_2 powders produced by one-step synthesis using the RTDS method have superacid characteristics, and variations in the catalytic activities of these materials toward butane isomerization are related to RTDS processing conditions. Evaluation of RTDS-generated nanocrystalline powders as catalysts for a number of applications is ongoing.

Acknowledgments

The Pacific Northwest National Laboratory (PNNL) is operated for the U.S. Department of Energy (DOE) by Battelle Memorial Institute under Contract DE-AC06-76RLO 1830. This work was sponsored by the U.S. Department of Energy, Office of Fossil Energy, and the Pacific Northwest National Laboratory Advanced Processing Technology Initiative. The authors gratefully acknowledge the technical and analytical assistance provided by T. D. Brewer, B. A. Armstrong, D. E. McCready, J. E. Coleman, and W. J. Shaw.

References

1. Rabo, J. A. (1994). *Advanced Heterogeneous Catalysts for Energy Applications*, Vol. 2, U.S. DOE Report DOE/ER-30201-H1, p. 1.1.
2. Chick, L. A., Pederson, L. R., Maupin, G. D., Bates, J. L., Thomas, L. E., and Exarhos, G. (1990). *Mater. Lett.*, **10**, 6.
3. Darab, J. G., Pfund, D. M., Fulton, J. L., Linehan, J. C., Capel, M., and Ma, Y. (1994). *Langmuir*, **10**, 135.
4. Matson, D. W., Linehan, J. C., and Geusic, M. E. (1992). *Part. Sci. Technol.*, **10**, 143.
5. Matson, D. W., Linehan, J. C., and Bean, R. M. (1992). *Mater. Lett.*, **14**, 222.
6. Darab, J. G., Buehler, M. F., Linehan, J. C., and Matson, D. W. (1994). *Better Ceramics Through Chemistry VI* (A. K. Cheetham, C. J. Brinker, M. L. Mecartney, and C. Sanchez, eds.), p. 499, Materials Research Society, Pittsburgh.
7. Matson, D. W., Linehan, J. C., Darab, J. G., and Buehler, M. F. (1994). *Energy Fuels*, **8**, 10.
8. Derbyshire, F. (1989). *Energy Fuels*, **3**, 273.
9. Linehan, J. C., Matson, D. W., and Darab, J. G. (1994). *Proceedings of the 1994 U.S. DOE Coal Liquefaction and Gas Conversion Contractors Conference*, p. 579, Pittsburgh, PA.
10. Stohl, F. V., Diegert, K. V., Goodnow, K. R., Rao, K. R. P. M., Huggins, F., and Huffman, G. P. (1994). *Proceedings of the 1994 U.S. DOE Coal Liquefaction and Gas Conversion Contractors Conference*, p. 605, Pittsburgh, PA.
11. Stohl, F. V., Diegert, K. V., and Goodnow, D. C. (1995). *Fuel Div. Preprints, ACS*, **40**, 335.
12. Darab, J. G., Buehler, M. F., Linehan, J. C., and Matson, D. W. (1994). *Better Ceramics Through Chemistry VI* (A. K. Cheetham, C. J. Brinker, M. L. Mecartney, and C. Sanchez, eds.), p. 505, Materials Research Society, Pittsburgh, PA.
13. Darab, J. G., Linehan, J. C., Matson, D. W., and Campbell, J. A. (1993). *Proceedings of the 7th International Conference on Coal Science*, p. 24, Banff, Alberta, Canada.
14. Matson, D. W., Petersen, R. C., and Smith, R. D. (1987). *J. Mater Sci.*, **22**, 1919.
15. Roy, D. M., Neurganonkar, R. R., O'Holleran, T. P., and Roy, R. (1977). *Ceram. Bull.*, **56**, 1023.
16. Armor, J. N., Fanelli, A. J., Marsh, G. M., and Zambri, P. M. (1988). *J. Am. Ceram. Soc.*, **71**, 938.
17. Rabenau, A. (1985). *Angew. Chem. Int. Ed. Engl.*, **24**, 1026.
18. Matijevic, E. (1985). *Annu. Rev. Mater. Sci.*, **15**, 483.
19. Dawson, W. J. (1988). *Ceram. Bull.*, **67**, 1673.
20. Cullity, B. D. (1978). *Elements of X-ray Diffraction*, Addison-Wesley, Reading, MA.
21. Darab, J. G., Linehan, J. C., and Matson, D. W. (1994). *Energy Fuels*, **8**, 1004.
22. Schwertmann, U., and Cornell, R. M. (1991). *Iron Oxides in the Laboratory*, VCH Publishers, Weinheim, New York.
23. Farcasiu, M., Smith, C., Pradhan, V. R., and Wender, I. (1991). *Fuel Process. Technol.*, **29**, 199.
24. Linehan, J. C., Matson, D. W., and Darab, J. G. (1994). *Energy Fuels*, **8**, 56.
25. Hoffmann, M. R., Martin, S. T., Choi, W., and Bahnemann, D. W. (1995). *Chem. Rev.*, **95**, 69.
26. Karakitsou, K. E., and Verykios, X. E. (1993). *J. Phys. Chem.*, **97**, 1184.
27. Prairie, M. R., Evans, L. R., Stange, B. M., and Martinez, S. L. (1993). *Environ. Sci. Technol.*, **27**, 1776.

28. Yamaguchi, T. (1990). *Appl. Catal.*, **61**, 1.
29. Scurrell, M. S. (1987). *Appl. Catal.*, **34**, 109.
30. Arata, K. (1990). *Adv. Catal.*, **37**, 165.
31. Yori, J. C., Luy, J. C., and Parera, J. M. (1989). *Catal. Today*, **5**, 493.
32. Chokkaram, S. C., Srinivasan, R., Milburn, D. R., and Davis, B. H. J. (1994). *Colloid Interface Sci.*, **165**, 160.
33. Kustov, L. M., Kazansky, V. B., Figueras, F., and Tichit, D. J. (1994). *J. Catal.*, **150**, 143.
34. Nascimento, P., Akratopulou, C., Oszagyan, M., Coudurier, G., Travers, C., Joly, J. F., and Vedrine, J. C. (1993). *Stud. Surf. Sci, Catal. (New Frontiers Catal.)*, **75**, 1185.
35. Ward, D. A., and Ko, E. I. (1994). *J. Catal.*, **150**, 18.

CHAPTER 12

The Synthesis of Nanostructured Pure-Phase Catalysts by Hydrodynamic Cavitation

William R. Moser, Joseph E. Sunstrom IV,
and Barbara Marshik-Guerts
Department of Chemical Engineering
Worcester Polytechnic Institute
Worcester Massachusetts 01609

KEYWORDS: Cabosil, catalyst synthesis, cavitation, cavitational synthesis, complex metal oxide, metal oxide catalyst, Microfluidizer, pure-phase catalysts, supported catalyst

12.1   Introduction

A new materials synthesis method, the Worchester Polytechnic Institute (WPI) cavitational materials process, was developed [1–4] for the fabrication of nanostructured materials in exceptionally high-phase purities. It is based on the use of mechanically generated, high shear hydrodynamic cavitational processing for advanced materials synthesis. Studies in our laboratories have examined a wide range of advanced catalyst syntheses as well as high-performance materials syntheses using this process. Characterization of these materials showed that the exceptional degree of mixing and *in situ* thermal treatment, afforded by cavitational processing, resulted in the formation of a wide variety of representative metal oxide catalysts and noble metal supported catalysts as phase pure materials. In addition, most syntheses afforded nanometer-sized grains of 1–5 nm without further processing. The cavitational effect was generated using a Microfluidizer [5]. It is a fluid processing system which combines a patented fixed geometry mixing chamber with a constant energy pumping system to uniquely process streams for liquid emulsion, dispersion, encapsulation, and cell rupture applications.

The method of generation of mechanical cavitation in these studies used laboratory models M-110Y and M-110EH Microfluidizers [5] which enables the processing of about 4 kg/h. Larger equipment is currently available with capacities which are 400–2000 kg/h depending on the desired intensity of the cavitational effect. Although the capabilities of the process have not yet been demonstrated at the previously mentioned commercial synthesis rates, the high-volume equipment is currently used in the processing of solids for applications in the food and cosmetics industries.

The synthesis of catalysts reported here will describe the equipment used to generate high shear hydrodynamic cavitation and the consequence of this type of processing on the eventual structure of the finished catalysts. The range of catalysts accessible using this technique will be described, and they include a wide range of typical catalysts that are currently used commercially.

12.1.1 Prior Studies Using Cavitational Techniques for the Preparation of Advanced Catalytic Materials

The method of generation of cavitation used in these studies was by mechanical means where the entire process stream is subjected to the cavitational effect. Our evidence indicates that this method generates only a modest cavitational effect, but one that is sufficiently powerful to cause the thermal decomposition of most metal salts. Other pioneering work in the development of high-powered ultrasound for the acoustic generation of cavitational effects during the processing of catalytic materials has been described by Suslick and co-workers [6–13]. The generation of cavitation by the hydrodynamic cavitation technique described in our research using hydrostatic pressures of 25,000 psi resulted in bulk stream heating to about 300–350°C. Suslick *et al.* [6] showed that local temperatures achievable by acoustic cavitational generation may be in the league of 5000°C. It is not yet clear how extensively this local heating may be useful for heating all components of the fluid stream, but the processing of a slurry of solids indicated that some particles were heated to the range of 3,000°C as evidenced by partial melting of micron-sized particles when processed under high-powered ultrasound.

The synthesis of catalysts using acoustic cavitation [7,8] demonstrated not only that nanostructured materials resulted, but also that the materials synthesized exhibited superior homogeneity and catalytic performance properties as compared to classical methods of synthesis of the same compositions. Suslick and co-workers [9,10] reported the synthesis of nanometer grains of iron particles when prepared by acoustic sonochemical means. These catalysts were prepared by sonolysis of $Fe(CO)_5$ in high-surface areas as amorphous metallic particles due to the high-cooling rates inherent in the cavitation effect. They were effective in syngas conversion [11] and hydrocarbon dehydrogenation [12]. The same type of experiment was used to sonicate $Fe(CO)_5$ in the presence of silica [13] resulting in supported metallic iron particles in the range of 3–8 nm. Sonochemistry led to the formation of nanostructured iron–cobalt alloys that were exceptionally active for cyclohexane dehydrogenation [13].

The use of the hydrodynamic cavitational method of processing in a Microfluidizer in our laboratories relies on the geometry of the interaction chamber for the generation of both high shear and cavitational heating to obtain nanostructured, polycrystalline, homogeneous metal oxide catalysts. The rapid thermal decomposition of solutions (RTDS) method developed by Matson and co-workers [14,15] uses a spray technique in which the feed solution is heated prior to passing through a nozzle that forms fine-grained materials. Given the velocities of fluids passing through the nozzle, it is likely that this process includes some cavitational heating and mixing during processing. These studies have resulted in the synthesis of nanostructured materials over a wide range of compositions [16]. The RTDS process was also found to be effective in the preparation of advanced catalytic materials [17] showing better catalytic performance properties as compared with conventional means for the synthesis of the same catalysts.

The WPI cavitational materials process was reported [3] to result in the formation of a wide range of nanostructured, high-phase purity materials including piezoelectrics, nanostructured zeolites, electronic and structural ceramics, and superconducting metal oxides. The data collected at this time suggest that the process is a general one with commercial application in the synthesis of high-purity, nanostructured metal oxides and metal-supported metal oxide materials.

12.2   High Shear Hydrodynamic Cavitation Equipment

The method of processing metal oxide streams used in these studies begins with a coprecipitation of the metal oxide components immediately before the high-pressure inlet to the Microfluidizer indicated as port B in Fig. 1. In some cases when the metal salt solutions are incompatible chemically, the dual-feed configuration shown in the figure is required; in most cases, a

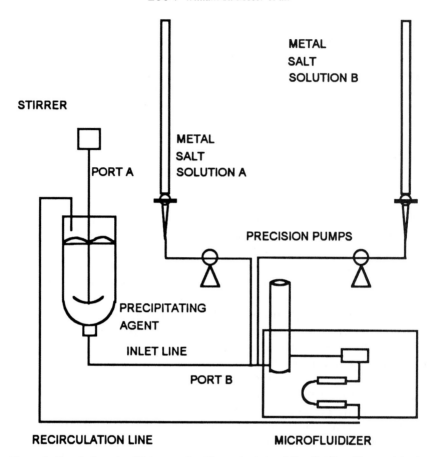

Figure 1. Two-feed system [3] for metal oxide synthesis in a Microfluidizer. The precipitating agent is recirculated through the Microfluidizer while the aqueous solutions of metal salts are metered in at Port B. (Reproduced with permission of Moser *et al., J. Mater. Res.*, 10, 2322 (1995).)

single-feed burette may be used. The precipitated slurry stream is then drawn into the device where it is elevated to high pressures within low internal volume tubing. The precipitated gel immediately passes into the interaction chamber where it experiences ultrashear forces and cavitational heating. These two effects are essential to the production of nanophase particles and high-phase purities in the synthesis of complex metal oxides.

The method used in the Microfluidizer for mechanically generated hydrodynamic cavitation and high shear is shown schematically in Fig. 2. At pressures up to 40,000 psi, the process stream is passed at high velocities through precisely defined microchannels resulting in highly focused sheer,

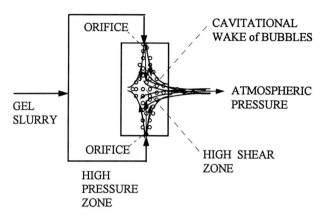

Figure 2. Mechanical generation of high shear and hydrodynamic cavitation [3]. This zone of the Microfluidizer operates at high pressures (usually 20,000 psi for catalyst synthesis).

impact, and transfer of cavitational energy to every element of the process stream. For materials synthesis, a process stream enters the device at atmospheric pressure as a precipitated slurry as shown in the Fig. 2. For most catalyst synthesis, the internal pressure of the liquid media is elevated from ambient to between 6.87 and 171.6 MPa (1,000–25000 psia). This stream is diverted into two equal parts which are redirected at one another, so that they impinge on each other at 180° in a low-volume interaction chamber while each traveling at velocities greater than 190 m/s. The inlet side of this chamber is at the high-stream pressures indicated above while the effluent stream is at ambient pressures. The two impinging streams result in exceptionally high shear. Additional effective shear and effective mixing occur due to the hydrodynamic wake of bubbles formed as each stream passes into the chamber through a separate orifice. The wake of bubbles arises due to the Bernoulli effect [18] resulting from a liquid passing through the orifice at high velocities where the effluent immediately experiences atmospheric pressure conditions. This causes the solvent to boil, forming a wake of bubbles. The rapid formation of this wake of bubbles followed by their rapid implosion results in local cavitational heating [6,18] and the application of severe shear on the precipitated slurry of the catalytic material. Figure 3 shows a schematic of the fluid dynamics when the stream at a high velocity passes through an orifice in the Microfluidizer. The wake of bubbles initially forms and then collapses causing shear and bulk heating of the media. Furthermore, the solids suspended in the slurry are projected at one another at very high velocities when the bubbles collapse; and when particles collide, additional heating of the particles themselves results [6].

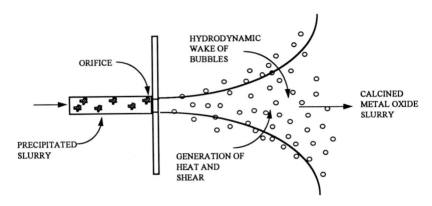

ORIFICE

HYDRODYNAMIC
WAKE OF
BUBBLES

CALCINED
METAL OXIDE
SLURRY

PRECIPITATED
SLURRY

GENERATION OF
HEAT AND
SHEAR

Figure 3. Schematic illustrating the formation of a wake of bubbles when a fluid stream of a catalyst slurry passes through an orifice at high velocities.

12.3   Catalyst Synthesis

This report describes our experimental studies on the synthesis and characterization of a wide range of catalytic materials commonly used in chemical, petrochemical, and fine chemical processes. The characterization data are centered on an examination of the grain and particle (agglomerate) sizes, phase purities of multimetallic metal oxides, thermal analyses, and surface areas. The principal method used for the determination of grain size was transmission electron microscopy (TEM). The phase identification of particles that were too small to be observed by X-ray diffraction (XRD) was done by indexing the rings resulting from a selected area electron diffraction (SAD) analysis of the materials that had been dried at 100°C immediately before analysis.

12.3.1   Nanostructured Catalytic Materials Synthesis

The cavitational process was used to synthesize several common metal oxides as nanostructured grains of 1–5 nm as determined by TEM analysis. These syntheses resulted in the formation of nanometer grains of several single metal oxide catalytic materials with well-defined SAD rings that were indexed for the following oxides: NiO, Co_3O_4, TiO_2, ZrO_2, Fe_2O_3, Cr_2O_3, CeO_2, and Y_2O_3. These syntheses were all conducted at 21,000 psi in the Microfluidizer. Following the synthesis, they were filtered, dried in air at 100°C, and stored. Thermal gravimetric analysis (TGA) on the materials after simple air-drying at room temperature showed a large weight loss due to water evolution appearing as a broad band centered at ca.

150°C. Drying the materials in the TGA at 100°C for 1–2 h, followed by remeasuring the TGA between 100 and 800°C, showed a very low weight loss and no other weight loss peaks due to any salt decomposition. This indicates that the nanostructured compounds as formed have surfaces that are readily hydrated. Indeed, SAD analysis on several materials after air-drying at ambient conditions resulted in no ring structures; however, after drying at 100°C for an hour and immediately mounting in the TEM spectrometer, all of the compounds resulted in smooth well-defined rings that were indexed solely as their metal oxides containing none of any known hydroxide component. Also, the SAD analysis resulted in rings with little or no amorphous halo indicating that a large fraction of the material was crystalline. The SAD data coupled with the TEM analysis indicate that after simple drying at 100°C the metal oxides are nanostructured materials and that the cavitational heating in the process resulted in the *in situ* decomposition of all salts (i.e., nitrates, chlorides, acetates, etc.) to the finished metal oxide.

The TEM analysis on these materials indicated that their grain sizes were in the range of 1–5 nm. Particle size determination by light-scattering analysis showed that the nanometer grains were clustered into larger particles, and that their sizes could be varied to a certain extent by process modifications. Particle size analyses were conducted on suspensions from titania synthesis, where it was shown that particles in the range of 100 nm resulted using one set of processing conditions in the Microfluidizer, while particles in the micron size resulted from another set of conditions. A series of titania preparations where the hydrostatic pressure was varied in four stages between 5,000 and 22,000 psi showed by SEM analysis that the size of the agglomerated grains varied from nanometer sizes at the highest pressures to several microns at the lowest process pressures.

The control of metal oxide agglomerate size through cavitational process modification is important due to the way the grains in this type of material grow upon mild calcination. The grain growth of TiO_2 was studied on a material that was prepared by cavitation at 21,000 psi. The grain sizes of the dried materials were determined by TEM to be between 2 and 8 nm, which were agglomerated into large particles ranging from 50 to 150 nm. The TEM of this material is shown in Fig. 4, and the electron diffraction pattern exhibited a ring pattern that was indexed for α-anatase. When this material was calcined in air at 300°C for 4 h, the grains grew just to the size of the agglomerate (i.e., 50–150 nm as shown in Fig. 5). The SAD analysis on the edge of the sintered particle is shown in Fig. 6, and the electron diffraction pattern was indexed for α-anatase. The SAD rings on the uncalcined material consisted of smooth rings consistent with the expectation for 2–8 nm grains of α-anatase, while the spots on the

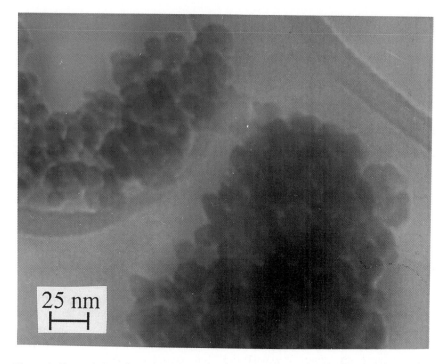

Figure 4. Transmission electron micrograph of a cavitation-produced TiO$_2$ sample at a hydrostatic pressure of 21,000 psi after crying at 100°C.

rings shown in Fig. 6 are indicative also of the grain growth of the material. The significance of this finding is that because the cavitational technique has the capability to synthesize different agglomerate sizes through process modifications and the grains grow only to the agglomerated size during calcination, the process has the potential capability of synthesizing a wide range of selected crystallites sizes where the grains are practically the same size as the agglomerate (particle) size. The property that makes this selectivity in particle sizes synthesis possible is the fact that the nanostructured metal oxide grains grow at a very low temperature. Other sintering studies in the nanostructured zirconia and β-Bi$_2$Mo$_2$O$_9$ systems resulted in the same grain growth behavior.

The zirconia synthesis resulted in a nanostructured material that demonstrated an SAD pattern with well-defined rings, but one could not distinguish between cubic and tetragonal phases due to the small differences in their diffraction rings. However, controlled calcination to 400°C resulted in sufficiently sharp reflections to make a structure assignment as the cubic phase [3]. This synthesis of cubic phase was unusual in that it was accomplished without modifying the zirconia with other metal ions.

Figure 5. Transmission electron micrograph of a cavitation-produced TiO_2 sample at a hydrostatic pressure of 21,000 psi after drying at 300°C for 4 h.

Because nanometer-sized grains of any metal or metal oxide particle lead to a very high fraction of the metal ions being located on the surface of the particle, this type of zirconia synthesis followed by surface modification with other ions should show different catalytic properties as compared with the same materials prepared by other techniques that do not result in nanometer-sized grains. To examine the relationship between the zirconia grain size after modification with sulfate ion for butane isomerization, a series of zirconia samples were synthesized at hydrodynamic pressures between 5,000 and 25,000 psi. These materials were modified with sulfate ion, and *n*-butane isomerization studies using these catalysts showed that the continuous modification of the surfaces of grains between 2 and 50 nm leads to substantially different catalytic properties.

12.3.2 Simple Metal Oxide Catalyst Synthesis

Aside from the simple metal oxides described in the preceding section, cavitational synthesis afforded the following materials: Fe_3O_4, PbO, Pb_3O_4, ZnO, MoO_3, Sb_2O_3, Sb_2O_4, CuO, η-alumina (380 m^2/g),

Figure 6. Selected area electron diffraction pattern for a cavitation-produced TiO_2 sample at a hydrostatic pressure of 21,000 psi after drying at 300°C for 4 h.

$Mg(OH)_2$, MgO, $Cu(OH)_2$, and La_2O_3. After these materials were synthesized and dried at 100°C, they were calcined in air in 100°C intervals between 100 and 500°C. An examination of the XRD after each stage of calcination showed the materials initially exhibiting broad reflections at low-temperature calcination stages that became progressively sharper at the highest temperature. As judged by the XRD line broadening, the materials calcined at lower temperature had grain sizes near 10 nm and those calcined at higher temperature resulted in 100–150 nm grains. Calcination at much higher temperature resulted in micron-size grains. These data suggest that this process enables the synthesis of particles of many important catalytic materials with the capability to regulate the grain sizes from a few nan-ometers to micron-size particles.

12.3.3 Complex Metal Oxide Synthesis in Pure Phases and Nanostructured Grains

12.3.3.1 Bismuth Molybdate Synthesis

Extensive catalyst syntheses and high-performance materials syntheses provided numerous examples to demonstrate that the cavitational process results in phase pure complex materials. The data in Fig. 7 describe the XRD analysis for $Bi_2Mo_3O_{12}$, $Bi_2Mo_2O_9$, and Bi_2MoO_6 (α-, β-, and γ-bismuth molybdate, respectively). The displayed diffraction patterns in Fig. 7 were obtained after calcining the cavitationally prepared materials between 300 and 400°C (temperature within the range commonly used in hydrocarbon partial oxidation). A comparison of all three diffraction patterns, after a 350°C calcination, to the calculated diffraction patterns demonstrated that the three samples contained diffraction lines for the specified phases and no others. A comparison of the XRD for the most important phase for hydrocarbon oxidation, β-bismuth molybdate, $Bi_2Mo_2O_9$, was made between the best available classical coprecipitation preparation [19,20] and the cavitational synthesis. A comparison of the two diffraction patterns of the materials calcined at the same temperature showed that the classical preparation led to the formation of several impure phases mixed

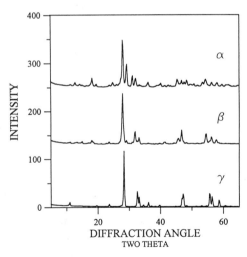

Figure 7. X-ray diffraction patterns for cavitational synthesis of $Bi_2Mo_3O_{12}$ (top curve, air-calcined at 325°C), $Bi_2Mo_2O_9$ (middle curve, air-calcined at 300°C), and Bi_2MoO_6 (bottom curve, air-calcined at 400°C).

in with $Bi_2Mo_2O_9$, while the cavitational preparation resulted in no extraneous phases. The TEM and SAD analyses of the cavitational-prepared particles after filtration and simple air-drying at 100°C showed that they were nanostructured grains, while the classically prepared catalysts were too thick to be analyzed by electron microscopy. We take these data along with other examples to illustrate that the cavitational process leads to pure phase complex metal oxides while classical processes generally do not.

Although dried samples of the bismuth molybdates were shown to be microcrystalline by TEM and SAD, their XRD patterns after drying between 100 and 120°C showed the broad set of reflections shown in Fig. 8. The interesting aspect to the data in Fig. 8, showing the XRD reflections after calcining in stages between 100 and 500°C, is that: (1) the grain sizes rapidly grew between 300 and 350°C; and (2) they grew very little after this point, even when calcined to 500°C. The particles after calcination to 350°C were still nanometer-size grains of around 20 nm as judged by XRD line broadening. This is interesting because these pure materials were fabricated as nanostructured materials and remained this size even at temperatures where they would normally be used in a hydrocarbon oxidation process. The fact that they are nanometer-size grains suggests that their activities and selectivities for hydrocarbon partial oxidation products would be higher [21,22] than a larger grain catalyst prepared by classical techniques because the fraction of edge sites is higher than basal plane sites as the grain size decreases to a few nanometers [23].

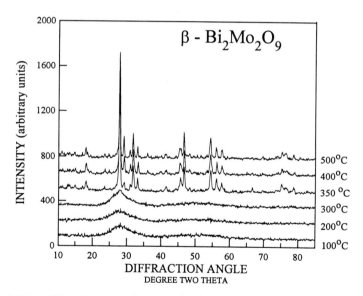

Figure 8. X-ray diffraction pattern for a sample of $Bi_2Mo_2O_9$ prepared by hydrodynamic cavitation and calcined in air between 100 and 500°C.

12.3.3.2 Oxygen- and Electron-Conducting Perovskites

This example illustrates the high-phase purities possible when well-known complex metal oxides containing several metal ions are formed under cavitation conditions. The synthesis of the ternary perovskite, $La_{0.6}Sr_{0.4}FeO_3$, was used as a model for this type of synthesis. This material was shown to be both an electron and oxygen solid-state conductor [24]. The XRD reflections at different levels of postsynthesis calcination are exhibited in Fig. 9. The diffraction patterns in the lower section of Fig. 9 show that the cavitational synthesis resulted in a pure perovskite phase after the material was calcined to 500°C. The top two diffraction traces show calcined materials that were prepared by classical coprecipitation means, using high-speed mechanical stirring, with the same reagents and concentrations. In comparison, the classical preparation resulted in a material that even after calcination to 600°C contained many impure, nonperovskite phases.

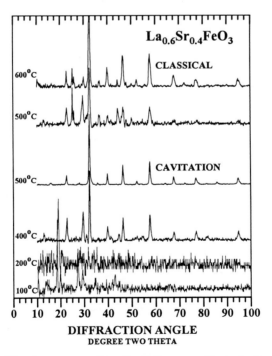

Figure 9. X-ray diffraction patterns of $La_{0.6}Sr_{0.4}FeO_3$ prepared by hydrodynamic cavitation at 96.1 MPa (14,000 psi) followed by calcination in air at 200, 400, and 500°C (lower curves), and conventional coprecipitation synthesis after calcination to 500 and 600°C (top two curves). (Reproduced with permission of Moser *et al.*, *J. Mater. Res.*, 10, 2322 (1995).)

Other complex metal oxide catalysts were synthesized and compared to classical synthesis. In these cases the grain sizes were much smaller and the phase purities were much higher than the classical preparations. Several examples of the catalysts synthesized were copper modified zinc oxides, iron chromium molybdates (e.g., $Fe_{1.8}Cr_{0.2}MoO_4$, iron molybdate), lanthanum aluminate, and nickel–copper–zinc oxide.

12.3.4 Supported Catalysts

12.3.4.1 Bismuth Molybdates on Silica

Supported catalysts were synthesized in two ways. The first procedure used either an aqueous or an alcoholic slurry of a support such as Cabosil initially contained in the reservoir of Fig. 12.1 with the precipitating agent. For example, an *n*-propanol slurry of Cabosil—as the precipitating agent—was recirculated through the Microfluidizer, while feed solutions of bismuth and molybdenum salts were metered in at two separate ports. The stoichiometry and flow rates for the additions into each port were adjusted, so that β-bismuth molybdate, $Bi_2Mo_2O_9$, was formed. When this material was calcined between 100 and 350°C, the formation of nano-structured $Bi_2Mo_2O_9$ resulted as shown in Fig. 10. Because each sample

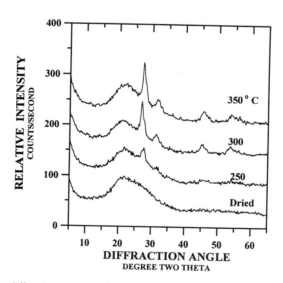

Figure 10. X-ray diffraction patterns of 10% w/w/ $Bi_2Mo_2O_9$ supported on Cabosil prepared by cavitational synthesis. The figure shows the XRD patterns of the materials which were air-calcined at 100°C (dried), 250, 300, and 350°C. (Reproduced with permission of Moser *et al., J. Mater. Res.*, 10, 2322 (1995).)

was calcined for 15 h, the broad diffraction lines indicate that the support appears to have stabilized $Bi_2Mo_2O_9$ against agglomeration. Several other concentrations of the same material were synthesized on Cabosil and calcined between 250 and 400°C. Figure 11 illustrates the XRD reflections resulting from the synthesis of $Bi_2Mo_2O_9$ supported on Cabosil where the concentration of β-bismuth molybdate on the support was varied from 10 to 30% w/w. The data shown in the figure resulted from catalysts that had been calcined at 250°C in air. The interesting aspect of the data is the observation that the $Bi_2Mo_2O_9$ reflections were progressively broadened as the concentration of the catalytically active phase decreased. This indicates that this technique has the capability to selectively fabricate catalysts of different sizes in the nanometer grain-size range where the grain sizes are stable at typical hydrocarbon oxidation temperature. Furthermore, by a simple adjustment of the hydrostatic pressure employed during the synthesis, the grain sizes of the supported catalysts can also be regulated. Several other supported catalyst series were conducted. By using the same technique, molybdena was synthesized on silica, and a multicomponent acrylonitrile catalyst was synthesized on silica.

12.3.4.2 Pt and Pd on Alumina

Noble metal catalysts on supports for environmental applications were conducted using a second method of synthesis. These syntheses used soluble salts in an aqueous solution containing both the noble metal component and the support metal component. For example, a series of five Pt on

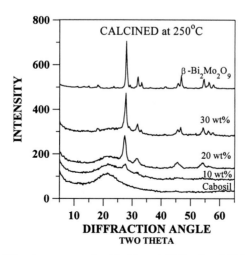

Figure 11. X-ray diffraction patterns for 10, 20, and 30% w/w of $Bi_2Mo_2O_9$ supported on Cabosil prepared by cavitational synthesis and calcined in air at 250°C.

alumina catalysts from 0.1 to 20% w/w were synthesized in which aluminum nitrate hydrate was dissolved in solution with chloroplatinic acid hydrate and precipitated in a single-step cavitational synthesis using ammonium hydroxide. The XRD reflections for the platinum on alumina catalysts after air calcination for 15 h at 400°C are shown in Fig. 12. It is interesting to note that the platinum is still well dispersed even at the highest concentrations as evidenced by the small, broad reflection for metallic platinum. A similar series was conducted using palladium where the metal weight-to-weight ratio was 1.5, 6.6, and 14.5%. After calcination of the samples to 400°C, no reflections for either metallic palladium of PdO could be observed in the XRD even for the catalyst of highest concentrations. The energy dispersive X-ray analysis indicated that the palladium was well dispersed throughout the support. Smooth, well-formed SAD diffraction lines were observed, which could be indexed for either γ- or η-alumina. In addition to the diffraction rings in the SAD for alumina, the major diffraction rings for Pd° were also observed. This indicates that the grain sizes of the alumina support, as well as the palladium particles even after calcination at 600°C, were a few nanometers. This conclusion was confirmed by the TEM analysis.

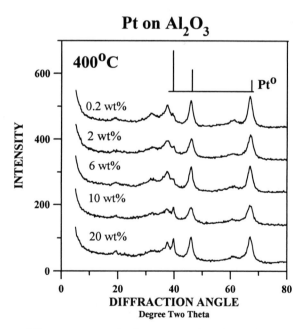

Figure 12. X-ray diffraction patterns for the cavitational synthesis of 0.2, 2, 6, 10, and 20% w/w of Pt on alumina after calcination in air at 400°C. The line drawing indicates the expected reflection positions for metallic platinum.

A second series of Pd-supported catalysts on a wide range of metal oxides was conducted where the Pd concentration was 1.0% w/w. This series included 1.0% Pd on MgO, Cr_2O_3, La_2O_3, Cr_2O_3, TiO_2, ZrO_2, and Al_2O_3. The synthesis conditions used in the Microfluidizer were set at those where the pure metal oxide preparations produced microcrystals of grain sizes in the range of 1–5 nm. The XRD analysis of the catalysts after air calcination at 600°C for 15 h demonstrated no palladium reflections for any of the catalysts; and the reflections for the metal oxide supports were broadened, indicating nanostructured grains. The XRD reflections for the 1% Pd on MgO calcined in air at 600°C is shown in Fig. 13. The absence of any trace reflections for Pd indicates that the Pd is well dispersed.

All of the previously mentioned Pd catalysts were examined for their activity in the conversion of a 1% methane in air stream to products of total combustion. These results indicated that the most active catalyst was the 1% Pd on magnesia catalyst, which resulted in 100% methane conversion at 450°C. Other catalysts in the series were evenly spread out in their conversion activity to the least active catalyst, which was 1% Pd on lanthana. Details of these catalytic studies will be reported separately.

12.3.4.3 Copper, Nickel, and Copper–Nickel Alloys Supported on Cabosil

A series of metal- and metal alloy-supported catalysts were synthesized where the metal oxides were deposited on Cabosil using the cavitational technique in concentrations ranging between 1 and 20% w/w. The individual metal oxides of Cu/Cabosil, Ni/Cabosil, and the bimetallic Cu–Ni/Cabosil catalysts were prepared. In general, the XRD analysis of this series showed that the peak line broadening increased as the metal concentration

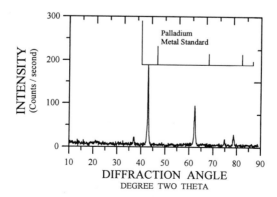

Figure 13. X-ray diffraction of cavitational synthesis of 1% w/w of Pd supported on MgO after calcination at 600°C in air. The line drawings indicate the diffraction peaks expected for metallic Pd.

on Cabosil decreased. Differential scanning calorimetric (DSC) studies on the dried materials showed characteristic exotherms for the individual metal oxides; however, the Cu–Ni/Cabosil catalyst resulted in a totally different set of thermal transitions in the DSC analysis. When the 5% Cu/Cabosil, 5% Ni/Cabosil, and 5% Cu–5Ni/Cabosil catalysts were first oxidized in air in the DSC, followed by cooling under helium, and then hydrogenated at 100 psig in the DSC starting from 25 to 500°C, the thermal transition observed for the alloy catalyst was quite different when compared to the same DSC analysis for the individually supported metal oxides. Figure 14 shows the DSC analysis of the three catalysts mentioned previously. The sharp transitions observed for the bimetallic catalyst and absence of transitions characteristic of the reduction of the single metal oxides of Cu and Ni indicate that the bimetallic catalyst has a structure of homogeneously mixed metal components.

This example illustrates the benefit of cavitational methods of supported metal synthesis and alloy metal synthesis to provide catalysts that not only are nanostructured compositions but also are homogeneously dispersed.

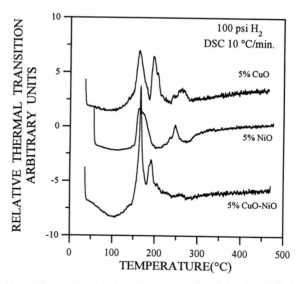

Figure 14. Differential scanning calorimetric (DSC) study of the 5% Cu/Cabosil (top), 5% Ni/Cabosil (middle), and 5% Cu–5 Ni/Cabosil (lower) catalysts which were first oxidized in air in the DSC followed by cooling under helium, and hydrogenated at 100 psig in the DSC starting from 25 to 500°C. The curves in the figure show the DSC analysis during the hydrogenation stage only.

12.3.5 Catalyst Support Materials

As indicated previously, the cavitation technique has resulted in our synthesis of a wide range of metal oxides that are commonly used as either inert or reactive supports for several types of heterogeneous catalysts. The advantages of fabricating support materials as nanostructured grains are threefold: (1) the surface-to-volume ratio of the nanostructured support material is much higher than large grain materials [25] and consequently the active catalyst may be distributed over the surface more evenly; (2) the superior dispersion should lead to a higher number of interactive, metal-to-support sites; and (3) due to the different thermodynamics of nanostructured grains, the interaction between the support metal oxide and the deposited active catalyst is likely to be much different.

12.3.6 Aluminum Oxide Synthesis

Alumina was synthesized using a solution of aluminum nitrate hydrate at 20,000 psi and dried at 100°C in the usual way. The XRD analysis on the dried product is illustrated in Fig. 15. The figure shows that both boehmite and bayerite were formed during the cavitation process. The standard XRD line positions for the two materials are displayed at the bottom of Fig. 15 for purposes of comparison to the synthesized aluminum oxide sample. It is seen from the X-ray line broadening that the boehmite portion of the sample is of a much smaller particle size as compared with the bayerite portion. When this material was calcined at the temperature

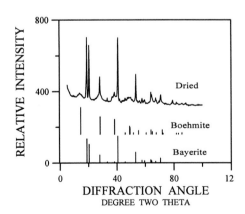

Figure 15. X-ray diffraction analysis of aluminum oxide samples prepared by hydrodynamic cavitation and dried at 100°C. The JDPDS reflections for boehmite and bayerite are shown as line drawings.

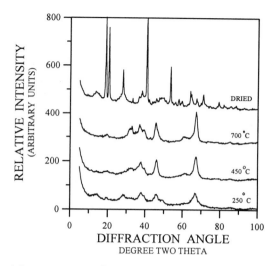

Figure 16. X-ray diffraction patterns for a cavitationally prepared aluminum oxide sample calcined in air at 100°C (dried), 250, 450, and 700°C.

indicated in Fig. 16, the XRD analysis shown in the figure indicates that the diffraction peaks were still quite broad indicating a fine grain material even after 700°C air calcination for 24 h. The diffraction pattern is consistent with a mixture of η- and γ-alumina. It is expected from the studies of Mardilovich and co-workers [26–28] that the boehmite fraction of the original sample would be converted on air calcination to γ-alumina, and the bayerite fraction would be converted to η-alumina. The Brunauer–Emmett–Teller (BET) surface area of the 500°C calcined material was 380 m^2/g.

12.4   Summary and Conclusions

The synthesis of materials by hydrodynamic cavitational processing in a Microfluidizer resulted in a large variety of catalysts in the form of nanometer-sized grains. The grains were agglomerated into particles of 100 nm to a few microns. The process provided higher purity than classical synthesis in the preparation of catalysts containing several metal ions. The technique offers a good route to the synthesis of homogeneous solid solutions of complex catalysts where one attempts to understand the relationship between ion modification of a host and catalytic performance.

We conclude that this process offers a route to the synthesis of a wide range of nanometer-structured catalysts, and it has the advantage of enabling one to probe the relationship of catalyst grain size and catalytic properties in a systemic way.

Although no commercial-scale quantities of catalysts have been fabricated using the WPI cavitational materials process to our knowledge, the process is capable of synthesizing a wide range of industrially important complex catalysts using currently manufactured equipment. An additional benefit from this type of commercial synthesis is the fact that the consistency of catalysts fabricated in this way is very high, resulting in very reliable and reproducible batches of catalysts. Thus, the fabrication of a variety of currently used commercial catalysts, where synthesis reproducibility is a problem, might be consistently conducted using the cavitational process reported here.

Acknowledgments

We wish to thank Microfluidics Corporation of Newton, Massachusetts, and Catalytica, Inc. of Mountain View, California for financial support of these studies.

References

1. Moser, W. R. (1995). U.S. Patent 5 417 956.
2. Moser, W. R., Marshik-Guerts, B. J., and Sunstrom, J. E. IV (1995). *Am. Chem. Soc., Pet. Div. Prepr.*, **40**, 100.
3. Moser, W. R., Marshik, B. J., Kingsley, J., Leemberger, M., Willette, R., Chan, A., Sunstrom, J. E. IV, and Boye, A. (1995). *J. Mater. Res.*, **10**, 2322.
4. Moser, W. R., Lemberger, M. J., Willett, R. D., Chan, A. M., Davis, P. J., and Boye, A. (1993). *Proceedings North American Catalysis Society*, March, DO7.
5. U.S. Patents 4 533 254 and 4 908 154, Manufactured by Microfluidics Corporation International.
6. Suslick, K. S. (1990). Sonochemistry, *Science*, **247**, 1439.
7. Suslick, K. S. (1988). *Ultrasound: Its Chemical, Physical, and Biological Effects* (K. S. Suslick, ed.), p. 123, VCH Publishers, New York.
8. Suslick, K. S. (1995). *Mater. Res. Soc. Bull.*, **20**, 29.
9. Suslick, K. S., Coe, S. B., Cichowlas, A. A., and Grinstaff, M. W. (1991). *Nature (London)*, **353**, 414.
10. Grinstaff, M. W., Salamon, M. B., and Suslick, K. S. (1993). *Phys. Rev. B*, **48**, 2269.
11. Grinstaff, M. W., Cichowlas, A. A., Choe, S. B., and Suslick, K. S. (1992). *Ultrasonics*, **30**, 168.
12. Suslick, K. S., Fang, M., Hyeon, T., Cichowlas, A. A. (1994). *Mat. Res. Soc. Symp. Proc.*, **351**, 443.
13. Fang, M., Cichowlas, A. A., Hyeon, T., and Suslick, K. (1995). *Am. Chem. Soc. Pet. Div. Prep.*, **40**, 67.

306 / *William R. Moser* et al.

14. Matson, D. W., Linehan, J. C., Geusic, M. E. (1992). *Part. Sci. Technol.*, **10**, 143.
15. Matson, D. W., Linehan, J. C., and Bean, R. M. (1992). *Mater. Lett.*, **14**, 222.
16. Matson, D. W., Linehan, J. C., Darab, J. G., and Buehler, M. F. (1994). *Energy Fuels*, **8**, 10.
17. Matson, D. W., Darab, J. G., Phelps, M. R., Linehan, J. C., Buehler, M. F., Neuenschwander, G. G. (1995). *Am. Chem. Soc. Div. Prep.*, **40**, 95.
18. Young, F. R. (1989). *Cavitation*, McGraw-Hill, New York.
19. La Lacono, M., Notermann, T., and Keulks, G. (1975). *J. Catal.*, **40**, 19–33.
20. Keulks, G. W., Hall, J. L., Daniel, C., and Suzuki, K. (1974). *J. Catal.*, **34**, 79.
21. Volta, J. C., and Portefaix, J. L. (1985). *Rev. Appl. Catal.*, **18**, 1.
22. Desikan, A. N., and Oyama, S. T. (1992). *ACS Symposium Series*, (D. J. Dwyer and F. M. Huffmann, eds.), 482, 260.
23. Van Hardeveld, R., and Hartog, F. (1972). *Adv. Catal.*, **22**, 75.
24. Battle, P. D., Gibb, T. C., and Lightfoot, P. (1990). *J. Solid State Chem.*, **84**, 271.
25. Siegel, R. W. (1991). *Annu. Rev. Mater. Sci.*, **21**, 559–578.
26. Mardilovich, P. P., Mukhurov, N. I., Rzhevskii, A. M., Govyadinov, A. N., and Paterson, R. (1995). *J. Membr. Sci.*, **98**, 131.
27. Mardilovich, P. P., Trokhimets, A. I., Zaretskii, M. V., and Kupchenko, G. G. (1985). *J. Appl. Spectrosc.*, **42**, 659.
28. Mardilovich, P. P., Trokhimets, A. I., and Zaretskii, M. V. (1984). *J. Appl. Spectrosc.*, **40**, 295.

CHAPTER 13

Nanocrystalline Zeolites
Synthesis, Characterization, and Applications with Emphasis on Zeolite L Nanoclusters

Mark C. Lovallo and Michael Tsapatsis
Department of Chemical Engineering
University of Massachusetts at Amherst
Amherst, Massachusetts 01003

KEYWORDS: aluminosilicate, catalyst synthesis, colloidal suspension, electron microscopy, film formation, light scattering, nanocrystals, seeded growth, zeolite

13.1   Introduction

Zeolites are crystalline aluminosilicates possessing a regular microporous channel network. Unique properties of zeolites such as the presence of framework cations, acid sites, and their well-defined porous structure — with pore sizes similar to those of small molecules — account for their traditional utilization in ion exchange, catalysis, and separations. In addition, zeolites are being investigated for novel emerging applications in a diversity of areas including optoelectronic devices and reactive membranes because they provide for discrimination, recognition, and organization of molecules with a precision of less than 1 Å [1].

Here, we present an overview of nanocrystalline (<100 nm) zeolite synthesis, characterization, and application with emphasis on our work re-

Advanced Catalysts
and Nanostructured Materials

307

garding the synthesis and characterization of zeolite L nanoparticles, and their use in controlled seeded growth and formation of asymmetric (porosity and grain-size gradient) films.

13.2 Synthesis and Characterization of Zeolite Nanoparticles

The synthesis of nanocrystalline zeolites was reported early in the literature. Breck in his book *Zeolite Molecular Sieves* showed a micrograph of cube-shaped zeolite A with sizes of 25–50 nm [2]. To date, zeolites synthesized in nanocrystalline form include sodalite and zeolites A, Y, ZSM-2, ZSM-5 [3–7], and L [8–11]. A wide variety of techniques have been employed to characterize these nanocrystalline systems. In the following section these characterization techniques are described emphasizing the need for their combined use, when possible, to provide a complete picture of crystallite characteristics.

13.2.1 Characterization

13.2.1.1 X-Ray Diffraction (XRD)

It is well known that an observed XRD pattern is a convolution of instrumental aberration functions with profiles that are indicative of the sample microstructure. Instrumental aberrations can readily be modeled using data from a suitable standard, and the information relative to sample broadening can be deconvoluted. Line broadening due to the microstructure of the sample contains contributions that are independent of the order of reflection (size broadening) and order-dependent contributions (strain broadening) [12]. The angular dependence of the profile width due to size broadening is described by the Scherrer equation: $l = K\lambda/((B^2 - \beta^2)^{1/2} \cos(2\Theta/2))$ where l is the average crystallite size, $K = 0.893$, $\lambda = 0.15405$ nm, B is the broadening of the peak of the sample in radians, and β is the instrumental peak broadening in radians determined using a standard exhibiting no strain or size broadening (such as lanthanum hexaboride, National Institutes of Standards and Technology [NIST] standard reference material 660). Size broadening depends on the distance over which diffraction is coherent. This distance is a weighted sum of crystal thickness in the corresponding direction, the reciprocal of dislocation density, and the mean distance of various other "mistakes." *The domain size (referred to as crystallite size) deduced from the breadth of a particular reflection is dominated by the smallest of the preceding dimensions.* Moreover, it is a volume average and thus tends to underestimate the contribution of the smaller domains. Size broadening is strongly influenced by the shape form of the domains. If the domain shape is spherical, the broadening is the same for all *hkl,* otherwise there can be a considerable variation in line breadth throughout the pattern for different *hkl* values. If such a variation is observed, the broadening is called anisotropic and some information of the domain size by other means (e.g., microscopy) is desirable in order to interpret the data.

13.2.1.2 Dynamic Light Scattering (DLS)

Due to the Brownian motion of colloidal particles, the light scattered from a dispersion fluctuates with time. The diffusion coefficient of the particles is determined by using DLS (or photon correlation spectroscopy) to detect and autocorrelate these fluctuations [13,14]. The Stokes diameter distribution (number or weight based) is then deduced from the diffusion coefficients. Evidently, DLS is not able to discriminate between amorphous and crystalline particles or crystalline particles coated with amorphous layers and provides information based on the agglomerate size. Therefore care should be taken in interpreting DLS results.

XRD and DLS (volume averaged) will provide comparable particle size estimates for nanocrystalline materials that consist of spherical nonagglomerated single (defect-free) crystallites.

13.2.1.3 Electron Microscopy

Direct observation of crystal morphology can be obtained by electron microscopy (EM). Conventional scanning electron microscopy (SEM) provides a 3-D view of particle surfaces, but resolution limitations (specific to the instrument employed) can often lead to erroneous interpretation of particle size and shape. Advances in SEM allow for observation of nanosized features. Using field emission-scanning electron microscopy (FE-SEM), it is possible to observe nanocrystalline zeolite surface shapes and sizes. Estimations can be drawn about grain and particle sizes, but positive identification of crystalline or amorphous regions is unattainable by this method. A powerful complementary tool for characterizing zeolite nanocrystals is transmission electron microscopy (TEM). High-resolution electron microscopy (HREM) imaging reveals crystallographic information and can probe the structure of grain boundaries and crystal defects. Also the presence of amorphous material can be readily assessed. Zeolite nanocrystals are ideally suited for such studies because, due to their small size, they can be examined in the TEM without any requirement for thinning.

In Figs. 1a–1f characterization data from XRD, EM, and DLS are presented for zeolite L ultrafine particles. Figure 1a shows the Stokes diameter

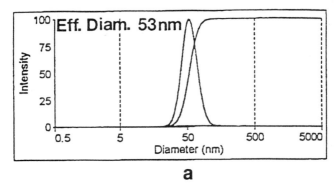

a

Figure 1a. Particle size distribution of zeolite L nanoclusters by DLS. Reprinted with permission from Tsapatsis, M., Lovallo, M., Okubo, T., Davis, M. E., and Sadakata, M., "Characterization of Zeolite L Nanoclusters," *Chemistry of Materials*, 7(9), 1734–1741, 1996, American Chemical Society.

distribution of zeolite L particles dispersed in water (pH ~7). An average (number-weighted) effective diameter of 53 nm is estimated.

In Fig. 1b the TEM photograph shows numerous nearly cylindrical zeolite domains that have approximate dimensions of ~15 nm in diameter and ~50 nm in length. These domains form larger clusters with dimensions comparable to those determined by DLS measurements (~60 nm). In the micrograph, most of the clusters are arranged on the Cu grid with their c-axis (channel direction) perpendicular to the electron beam. Less often clusters were found with their c-axis parallel to the electron beam allowing a

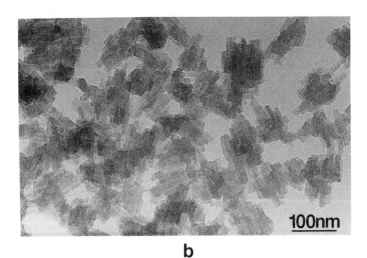

b

Figure 1b. TEM micrograph of zeolite L nanoclusters.

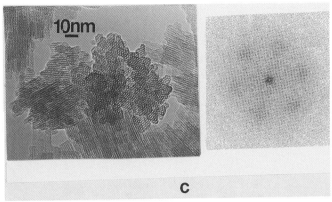

Figure 1c. TEM micrograph of zeolite L nanoclusters with view down c-axis with corresponding optical diffraction pattern. Reprinted with permission from Tsapatsis, M., Lovallo, M., Okubo, T., Davis, M. E., and Sadakata, M., "Characterization of Zeolite L Nanoclusters," *Chemistry of Materials*, 7(9), 1734–1741, 1996, American Chemical Society.

view down the zeolite channels. HREM images are shown in Figs. 1c and 1d. The crystalline domains are aligned with their c-axis nearly parallel to one another within a cluster. This allows the observation of the 12 T-atam channels in views down (001) as in Fig. 1c. Numerous small angle or 0° tilt boundaries are evident. Digital diffractograms of numerous such clusters show a six-fold symmetry with the characteristic spacing of zeolite L. In Fig. 1d the alignment is shown more clearly. Fringes corresponding to (001) planes are resolved and are shown to be aligned throughout the cluster.

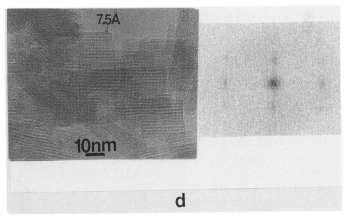

Figure 1d. TEM micrograph of zeolite L nanoclusters with view perpendicular to c-axis with corresponding optical diffraction pattern. Reprinted with permission from Tsapatsis, M., Lovallo, M., Okubo, T., Davis, M. E., and Sadakata, M., "Characterization of Zeolite L Nanoclusters," *Chemistry of Materials*, 7(9), 1734–1741, 1996, American Chemical Society.

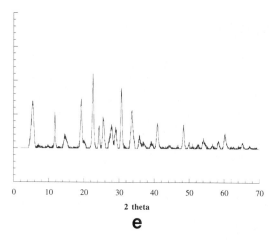

e

Figure 1e. XRD powder diffraction pattern of zeolite L nanoclusters.

An XRD pattern of the zeolite L nanocrystals is shown in Fig. 1e. Line-broadening calculations using the Scherrer equation (assigning all broadening to size effects) indicate anisotropic broadening. Analysis of the (001) peak gave an estimation of a particle size of ~30 nm while the (220) peak broadening corresponds to 15 nm. The anisotropic broadening in connection with the TEM observations suggests that the domains are prismatic with the prism height—indicated by the (001) peak broadening— greater than its diameter—indicated by the (220) peak broadening.

Figure 1f is an FE-SEM micrograph of the zeolite L crystallites. It provides a 3-D perspective of the particles at a lower resolution and is complementary to TEM observations (TEM provides 2-D projections of the 3-D particles, therefore TEM images cannot be used to readily assess the density of the clusters). The discrete nanocrystalline domains forming the clusters are evident in the SEM micrograph. Considerable porosity exists in each cluster formed by the void spaces between the nanocrystalline domains. The clusters have a nearly dentritic appearance with "fjords" having a length scale on the order of the cluster size. The TEM micrographs give the false impression of the clusters being dense because of overlapping of the projections of the nanocrystalline domains.

From the characterization results presented in Fig. 1 we can conclude that the zeolite L particles are clusters of aligned domains of nanoscale dimensions (nanoclusters). The nanoclusters have inhomogeneities ranging from 2 to 60 nm and therefore are expected to have a mass fractal character. An estimate of the fractal dimension of the nanoclusters is obtained from small-angle X-ray scattering (SAXS).

080808 15KV X100K 300nm

f

Figure 1f. FE-SEM micrograph of zeolite L nanoclusters.

13.2.1.4 SAXS

The log(I) vs log(h) plot from a zeolite L suspension is shown in Fig. 2 where I is the scattering intensity. The mass fractal dimension Df is obtained from the power law decay of the scattering intensity through the relationship: $I(h) \sim h^{-Df}$ and is ~2.2. The mass fractal dimension reflects the self-similar internal structure of the nanoclusters as a result of the distribution of voids with length scales ranging from ~60 nm (nanocluster size) to 2 nm (domain size). The end of the power law dependence provides an independent estimation of the domain size. It is obtained as ~2.5 nm in agreement with the XRD and TEM data.

13.2.1.5 Nitrogen Adsorption

The nitrogen adsorption isotherms for the zeolite L nanoclusters are shown in Fig. 3a along with the isotherm for micron-sized crystals. The adsorption results are consistent with the preceding results suggesting the

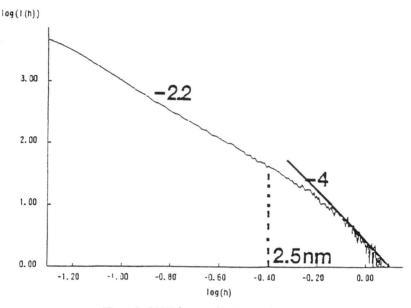

Figure 2. SAXS from zeolite L nanoclusters.

presence of inhomogeneities in the micro- and mesoscopic scales. In the pressure region where microporous adsorption occurs, the nanoclusters exhibit a greater adsorption capacity compared with the micron-sized crystals indicating the presence of additional microporosity between the nanocrystalline domains. At higher pressures the presence of mesoporosity accounts for the observed differences.

13.2.1.6 *Thermal Stability*

An issue of concern in nanocrystalline materials is thermal stability. The results shown in Fig. 3b provide information on the thermal stability of the zeolite L nanoclusters. The zeolite L nanoclusters are stable (no grain coarsening or phase transformation is observed by temperature-programmed XRD) up to 950°C. Larger, micron-sized crystals of zeolite L were found to be stable up to 1100°C. Thus, the nanocrystals are less stable than micron-sized zeolite L, but the differences are not large and do not prohibit their use.

The disordered, mosaic structure of the zeolite L nanoclusters does not seem to be an isolated case characteristic of zeolite L. Yamamura *et al.* [15] have employed XRD and FE-SEM to characterize zeolite nanoparticles

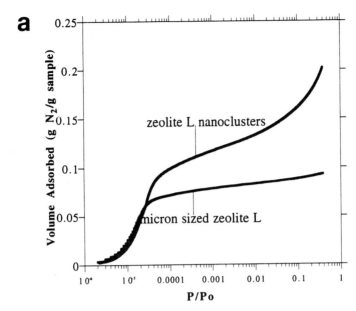

Figure 3a. Nitrogen adsorption @ 77 K, zeolite L nanoclusters, and micron-size zeolite L crystals.

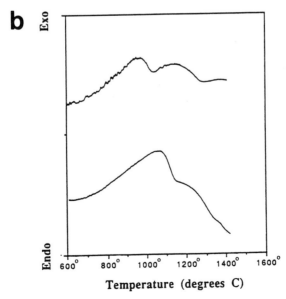

Figure 3b. Thermal analysis (DTA) of zeolite nanoclusters (top) and micron-size zeolite L crystals (bottom).

that they have synthesized. A striking similarity of particle morphology with the ZSM-5 nanocrystals synthesized by Yamamura *et al.* suggests that other ultrafine zeolites possibly exhibit similar characteristics and consist of aligned nanocrystalline domains. It is evident that no single characterization technique can provide a complete picture of the morphology of the zeolite nanoclusters, and a combination of techniques is necessary. Table 1 lists a number of zeolites synthesized as nanocrystals (some systems have been included with crystal sizes larger than 100 nm) along with their corresponding particle size as determined by the characterization techniques employed. It is our view that studies employing a combination of characterization techniques are needed in the other nanocrystalline systems reported.

13.2.2 Nucleation and Growth of Zeolite Nanocrystals

Zeolites are synthesized from alkaline aluminosilicate precursor gels or clear solutions. A mechanistic understanding of zeolite synthesis has remained elusive because of the complexity of these systems involving a multitude of species of various chemical composition and size, governed by

TABLE 1
Nanocrystalline Zeolite Syntheses

Zeolite type	Method of characteriazation	Average particle size (nm)	Grain size (nm)	Ref.
Zeolite L	XRD		30 × 15	30
	DLS	53		
	HRTEM	60	50 × 15	
	FE-SEM	60		
Zeolite Y	DLS	100		25
	SEM	100		
Sodalite	DLS	37		24
	SEM	40		
	TEM	40		
Silicalite	DLS	100		26
	SEM	100		
ZSM-5	SEM	130		28
	DLS	130		
ZSM-5	SEM	30		3
	XRD		13	
ZSM-2	SEM	<100		27
	DLS	50		

multiple equilibria. Studies on the mechanism of zeolite formation have led to several propositions regarding nucleation and growth. The proposed mechanisms of zeolite growth span from solution-mediated transport to the solid–hydrogel transformation. In the case of solution-mediated transport, it is proposed that species (monomeric, oligomeric, or extended structures) are transferred through the solution phase to the growing crystal. In contrast, in the solid–hydrogel transformation crystal growth proceeds through a rearrangement of the amorphous precursor gel. These rearrangements can be viewed as concerted bond rearrangements not involving release of species in the solution phase. Zeolite synthesis accomplished from clear homogeneous solutions provides evidence in support of the first mechanism [16], while zeolite formation from dried gels in nonaqueous solvents has provided evidence supporting the second mechanism [17]. There is a consensus that synthesis in an aqueous system should at some stage involve solution-mediated transport [18–21]. Of importance are the questions of the constitution of the species transported and the length scales involved. NMR studies [22] have shown a variety of species present in crystallizing mixtures; however, which, if any, of these species are directly involved in crystal growth is still a question of considerable debate [23].

Davis and Lobo [24] have drawn a parallel between zeolite crystallization and the formation of protein structures. Following the suggestion by Vaughan [23], they proposed the possibility that small structures (observed by NMR and Raman spectroscopy) may not be directly incorporated into a growing crystal, but form extended structures that assemble to form the longer range order of the crystals. To date, only indirect evidence supporting this proposition has been provided for low and pure silica zeolites [10,25,26], but such extended structures have not been isolated. The possibility exists that a variety of growth mechanisms may occur simultaneously, and it should be emphasized that general conclusions should not be drawn from one system to another.

Of an even more complicated nature is the issue of zeolite nucleation in these complex mixtures. Understanding the transformations occurring prior to and during nucleation is important for the production of materials with tailor-made properties. Experimental evidence has been provided in support of homogeneous, heterogeneous (in the solution phase or in the gel phase), secondary, and autocatalytic nucleation [27–30].

In the classical view homogeneous nucleation occurs in a sufficiently supersaturated solution through local energy fluctuations. Applying this view to zeolite synthesis requires the spontaneous formation of zeolite nuclei through the assembly of monomeric or oligomeric dissolved species. The aggregate of molecules will grow only if it exceeds a critical size, which depends on the supersaturation. Even in this view an issue of clarification is needed for defining what is meant by "zeolite nuclei". Strictly speaking, zeolite nuclei should possess the crystallographic structure of the zeolite being formed. Bronic and Subotic [31] have used classical nucleation theory to estimate rates of homogeneous nucleation. Their estimates indicated extremely large nucleation times (at least for low silica zeolites) leading to the proposition that homogeneous nucleation in the classical sense is not a viable nucleation mechanism. Such calculations should be viewed with care because they involve use of bulk properties for estimating properties of small particles, and this approach is known to be often subjected to errors [32]. However, in another view, zeolite nuclei can be considered to be homogeneously nucleated precursors not possessing (but closely related to) the structure of the zeolite. After reaching a certain size these precursors are transformed into the zeolite structure. Parallels from much simpler systems suggest that size-dependent spontaneous transformations thermodynamically or kinetically induced are possible [33]. For instance, Monte Carlo simulations on the growth of metal nanoparticles indicate such thermodynamically directed transitions, as particles grow from nuclei having structures different from the final crystal structure [34]. Despite matters of terminology, the issue of the initially nucleated particles having the zeolite structure or not is especially important for the case of zeolite nanoparticle synthesis in order to determine the smallest possible zeolite particle that can be discretely synthesized.

The issue of heterogeneous nucleation is also subject to considerable debate. Nucleation at the liquid–gel interface has been proposed and seems a likely possibility because at the interface the surrounding liquid provides for fast dissolution, while the gel can provide extended building entities. Autocatalytic nucleation, a variant of the heterogeneous mechanism, has been proposed in order to account for an increase in nucleation rate under constant supersaturation [30]. It involves release of zeolite nuclei (existing in the gel) into the solution by the dissolving gel. These nuclei are then assumed to grow. Simulations have demonstrated that the original autocatalytic mechanism does not conform to data in the literature [28]. Also, as in the case of homogeneous nucleation, a clarification of what constitutes a zeolite nuclei is needed.

In most of the preceding studies and simulations, a lumped precursor concentration was used and the multitude of species present in the crystallizing mixture was not explicitly taken into account. Lechert and Kacirek [35] have considered the issue of zeolite nucleation and growth in more detailed terms by employing polymerization kinetics and by using equilibria to estimate the relative abundance of species. In their more realistic formulation (applied to NaX synthesis) they have considered the presence of species with various Si and Al constitution and length. They have succeeded in capturing the nucleation rate dependence on OH concentration.

To elucidate some of the preceding aspects of zeolite synthesis, well-controlled experiments are needed. In particular, syntheses from clear solutions are desirable because they allow for *in situ* studies by light scattering to be combined with X-ray (wide and small angle *in situ*) and neutron scattering techniques (applicable also in gel synthesis) to provide information on length scales from 1 to >1000 Å. Dynamic light scattering was used to follow the zeolite growth from clear solution. Schoeman *et al.* [3] have followed the growth of sodalite particles by DLS and suggested homogeneous nucleation. They have further demonstrated that the midsynthesis addition of alumina to a sodalite mother liquor solution can result in further growth without the occurrence of any secondary nucleation. They also suggested, based on mathematical modeling, for the case of silicalite growth from clear solution that nucleation eventually ceased, while particle growth occurs [36]. However, Twomey *et al.* [37] have concluded that there is a continuous generation of nuclei during growth and explained the constant number of particles by proposing that the nuclei are incorporated on the growing particles. In the same study, crystal growth from sizes larger than 25 nm was found to follow a linear growth rate. The issue of which of these two mechanisms — nucleation followed by growth versus continuous particle nucleation and aggregation prevalent in the synthesis of zeolite nanoparticles — is important if size and particle size distribution are to be optimized. This is a question by no means limited to zeolite synthesis; and discussions, experiments, and mathematical modeling related to monodispersed particle formation in other systems can be found in numerous references where it is emphasized that the morphology of the particles as revealed by electron microscopy can assist in discriminating between the two mechanisms [38–44]. For the preceding zeolite synthesis, such information is not available.

Complementary to the work of Twomey *et al.* is a study by Dokter *et al.* [45] on a similar silicalite synthesis from clear solution. In this study SAXS and WAXS (wide angle X-ray scattering) were used to probe the reacting solution *in situ* to make it possible to examine smaller length scales than were possible with DLS. Zeolite formation, detectable by WAXS, was preceded by the formation of small microgel aggregates (~7 nm). It was suggested that they are composed of primary particles (<1.6 nm) that are preorganized tetrapropylammonium–silicate clusters as the ones observed, using NMR by Burkett and Davis [25] in the synthesis of ZSM-5.

The transformations reported by Dokter *et al.* [46] are similar to rearrangements observed by SANS (small angle neutron scattering) that occur in zeolite synthesis starting from a gel, and suggest that silicalite nucleation occurs through a heterogeneous mechanism (or at least not through direct homogeneous nucleation) involving the formation of a precursor phase even in syntheses from optically clear solutions. It can then be suggested that the formation of a gel having the composition and microstructure of the zeolite precursor phase may result in a spontaneous nucleation. That has not yet been demonstrated for silicalite or ZSM-5 but seems to be the case for the synthesis leading to zeolite L nanoclusters formation that is described in the following section.

13.2.2.1 Growth Curves and Crystallizing Mixture Evolution for Zeolite L

The synthesis of zeolite L starting from a homogeneous clear mixture of composition $10K_2O-1Al_2O_3-20SiO_2-400H_2O$ results in nanocluster suspensions[9]. The yields of solids in grams per gram of solution, the percentage of crystallinity from XRD, and the Si:Al and K:Al ratios are given in Table 2 (from Reference 9) for various times of crystallization at 175°C. The clear homogeneous mixture after 1 h of heating was transformed to a gelatinous precursor with a Si:Al:K ratio close to the one of the final zeolite product. Between 1 h 30 min and 2 h of heating, a transformation of the gel precursor to ~80% by weight zeolite is observed followed by a slower growth of zeolite to 100% crystallinity and the complete transformation of the gel to a zeolite L suspension. The yield of solid products (amorphous gel and zeolite) remains almost constant during these transformations. Similar evolution of the precursor solution is observed for crystallization temperatures ranging from 150 to 250°C. The percentage of

TABLE 2
Crystallinity, Yield, and Composition of Solid Products for Various Times
of Crystallization at 175°C Starting from a Clear Synthesis Mixture
of Composition $10K_2O-1Al_2O_3-20SiO_2-400H_2O$

Time	Crystallinity (%)	Si/Al ICP[a]	Si/Al EDAX[b]	K/Si ICP[a]	Yield (g/g of sol)
1 h	0	2.41	2.6	1.04	0.061
1 h 30 min	0	2.46	2.6	1.06	0.060
1 h 45 min	60	—	—	—	0.055
2 h	80	—	—	—	0.057
2 h 30 min	90	2.55	2.6	1.16	0.057
3 h	95	—	—	—	0.057
3 h 30 min	98	—	—	—	0.062
4 h	100	—	—	—	0.057
8 h	100	2.51	2.6	1.15	0.056

[a]Determined by ICP.
[b]Determined by energy dispersive X-ray analysis in the TEM.

crystallinity evolution for the two temperature extremes is shown in Fig. 4. It should be noted that the indicated temperature is that of the oven and not of the crystallizing mixture.

13.2.2.2 HREM Observations of Early Growth Events

As described in Sec. 13.2.2 on zeolite nucleation and growth, the use of SANS, SAXS, DLS, and WAXS supplied results pertinent to resolving the mechanism of zeolite formation. However, they cannot provide a description of the detailed morphology of zeolite crystals (crystal agglomerate shape, grain boundaries, defects) at early stages of growth and they lack the ability to directly discriminate between the zeolite and amorphous particles in a complex reaction mixture.

One way of directly observing nanometer-scale details is with the use of transmission electron microscopy (TEM). However, in a typical zeolite synthesis, because small zeolite crystals are scattered throughout the solution or gel at low concentrations, the chances of finding an area on a TEM sample that contains zeolite crystallites at early stages in development are low. Moreover, the possibility of introducing morphological and structural changes during the washing, drying, and thinning steps of sample preparation for TEM combined with the electron-beam sensitivity of zeolites further contributes to the difficulty of using TEM for observations of zeolite growth.

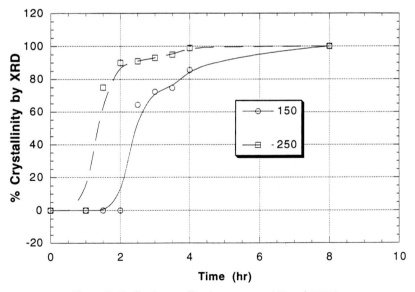

Figure 4. Zeolite L crystallization curves at 150 and 250°C.

The study of ultrafine crystal synthesis provides systems where nucleation rates are high, resulting in high yields of zeolite at early stages of growth, and making it possible to observe early growth events in the TEM. Also there is no need for sample thinning.

Zeolite L is relatively stable under the electron beam [47], and a considerable number of electron microscopy studies were reported as early as 1984 [23,47–49]. A most notable study is by De Gruyter *et al.* [49] who have used information on the grain boundaries of zeolite L to infer growth events that take place during crystallization.

Figures 5a–5f show a sequence of TEM micrographs of solid products during crystallization at 175°C. The morphology of the precursor gel formed after 1 h 30 min crystallization is shown in Fig. 5a. High resolution electron microscopy and microdiffraction fail to indicate the presence of crystalline material. EDAX analysis with a spot size ~100 Å gives a Si:Al ratio of 2.6 in agreement with the chemical analysis data and indicates a "uniform" distribution of Al and Si throughout the gel. Although sample preparation artifacts cannot be excluded, the electron micrographs suggest a finely divided hierarchial gel network structure with the typical macroporous-mesoporous-microporous structure of potassium silicate gels. The fact that the gel spans the whole sample and is translucent is consistent with the suggested microstructure [10]. From the preceding results

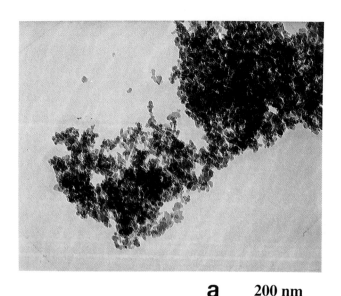

a 200 nm

Figure 5a. TEM micrograph of zeolite L precursor gel, (1 h 30 min @ 175°C).

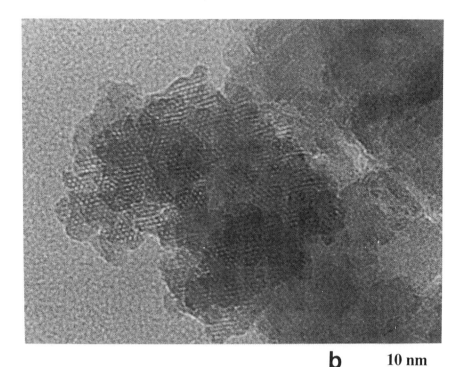

b   <u>10 nm</u>

Figure 5b. TEM micrograph of zeolite L in a sample with 60% crystallinity (1 h 45 min @ 175°C).

it is indicated that after 1 hr 30 min of heating, a gelatinous amorphous precursor is formed with an elemental composition very close to the one of the final zeolite product and a finely divided microstructure.

Between 1 h and 30 min and 2 h of crystallization, growth of zeolite L crystals to 80% crystallinity and the transformation of most of the gel to a suspension take place followed by a slower growth to 100% crystallinity and the complete disappearance of the gel within 4 h of heating. In the partially crystallized solid samples the presence of well-developed crystallites can be distinguished from an amorphous or partially transformed phase. Figure 5b shows HREM views of well-developed L crystals in a solid sample with 60% crystallinity. The HREM images reveal that the L crystals are actually clusters of aligned and connected nanometer domains similar to those of 100% crystalline samples shown in Figs. 1a–1f. Comparison of Fig. 5b with Fig. 5a indicates a correlation of the domain size with the particle size of the amorphous gel. This observation combined with the speed of gel to zeolite transformation and the constant weight and

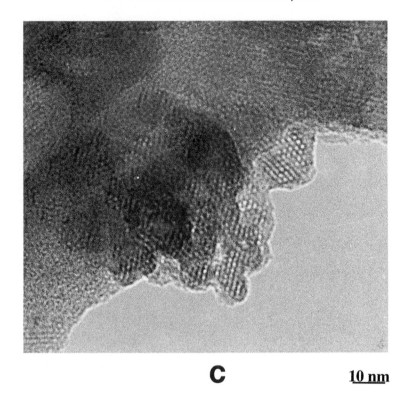

C   10 nm

Figure 5c. TEM micrograph of partially transformed gel in a sample with 60% crystallinity (1 h 45 min @ 175°C).

composition of the solid products during crystallization, strongly indicates a gel to zeolite transformation through local rearrangements (solid hydrogel or solution mediated) of the former. According to this hypothesis the small domain size and uniformity of the zeolite nanoclusters are due to the uniformity and finely divided microstructure of the precursor gel. In addition, the open and finely divided gel structure enhances the speed of rearrangements due to the large area of contact with the highly alkaline medium.

HREM observations of partially transformed areas of the solid products can be used to address the issue of morphological evolution. Figure 5c shows a typical micrograph of such partially transformed regions. It shows small zeolite L domains at early stages of development. Optical diffraction (Fig. 5d) indicates an alignment of these domains in an extended hexagonal lattice rather than in a random orientation. If the formation process were through agglomeration of individual nuclei, some misalignment should be expected. Analysis of micrographs from several such areas, as

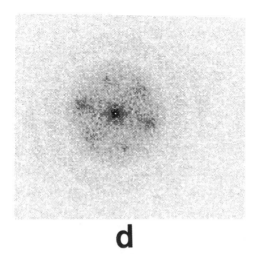

d

Figure 5d. Optical diffractogram of Figure 5c.

detailed elsewhere [10], fails to reveal the presence of misalignment. Therefore, it is indicated that images like the one in Fig. 5c represents a growing crystal from a single nucleation center.

It is, therefore, suggested that under the synthetic conditions employed here the morphology of the precursor gel plays a dominant role in determining the zeolite shape and size. With zeolite growth the solid shrinks due to the difference in density between the gel and zeolite. That leads to fractionation of the gel backbone, and the precursor gel is eventually transformed to a suspension of the zeolite L nanoclusters. The density of nucleation events within the gel, the fractionation process, and the gel connectivity determine the final size of the zeolite cluster.

Certainly, upon prolonged heating, dissolution and redeposition of the zeolite will lead to energetically more favorable dense and smooth particles. For example, an increase in temperature of crystallization results in increased domain size (estimated by TEM and XRD) as shown in Fig. 6.

It is evident that combination of sophisticated techniques is needed to provide a more complete view of nucleation and growth processes. Combinations of SAXS, WAXS, and DLS can provide information on the particle aggregate structure and morphology (primary particle, aggregate fractal dimension, gel and crystal interfacial area with the surrounding solution) evolution *in situ*. Nuclear magnetic resonance (NMR) complements these studies with information on the nature of bonds formed, while electron microscopy and especially HREM directly image crystal structure, defects, and morphology.

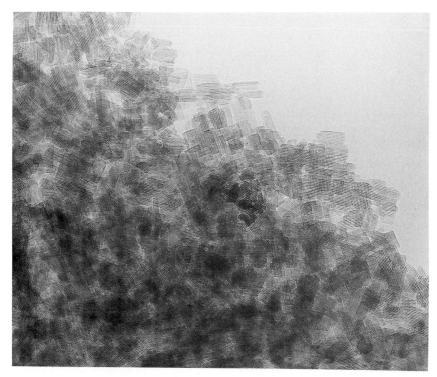

60 nm

Figure 6. TEM micrograph of zeolite L synthesized at 220°C.

The two examples: (1) silicalite synthesis from a clear dilute solution with slow crystal growth (and agglomeration) kinetics, in the absence of extended gel formation and (2) zeolite L with the formation of a precursor gel with fine microstructure and composition closely related to the zeolite composition, demonstrate two distinct approaches for zeolite synthesis leading to nanoparticle formation.

High nucleation rate with slow growth and slow agglomeration kinetics will generally favor monodisperse nanocrystal formation. General rules for selecting batch compositions leading to nanocrystal formation cannot be provided, but we believe that progress toward this direction can be achieved by studies and mathematical modeling of similar well-characterized systems. Moreover, some trends concerning synthesis composition and resulting particle size are apparent in the literature [50]. The most prevalent trend seems to be an increasing relative rate of nucleation and resulting smaller particle size with increasing alkalinity or, more specifically, increasing OH/SiO_2 content of the synthesis mixture. Lechert and Kacirek [35] successfully modeled that aspect for synthesis of zeolite X, indicating that the nucleation rate increases with a quadratic function of the OH concentration, and Yamamura et al. [15] used that trend to synthesize nanocrystals of ZSM-5.

As a final comment in this section, it should be mentioned that the batch crystallization method almost exclusively employed in zeolite synthesis is of limited flexibility for optimizing particle size and particle size distributions. Midsynthesis addition of reactants is one approach to circumvent this limitation and has been elegantly demonstrated by Schoeman et al. [4]. In the preparation of zeolites A and Y and mixtures of A and Y. Further developments in this area can be expected with crystallizers optimized for nanocrystal formation following developments in other zeolite syntheses [51,52].

13.3 Applications with Emphasis on Seeded Growth and Film Formation from Zeolite L Nanoclusters

For a given framework structure, the size and shape of zeolite crystals influence the characteristic diffusion time of molecules through the channels of the crystallites and can alter performance characteristics in catalytic applications.

Industrial use on nanocrystalline zeolites has been reported in Reference 19 of [53]. Several reports have addressed the role of external surface area and sites of zeolites. Methods on estimating the external surface area have been developed including the filled pore method [2,54,55] the BET method using large molecules [56]; the adsorption kinetics method, called $t_{1/2}$ plot or foot method [57]; and their combinations.

Theoretical analysis of the effects of the external surface sites of zeolites [53,58,59] indicates that external activity can be important, particularly for diffusion-limited reactions, for zeolite crystallites smaller than 100 nm. External versus internal intracrystalline activities have been investigated by Fourier transform infrared (FTIR) [60] and catalytic studies [61].

In a study Yamamura et al. [15] measured the activity and lifetime of ZSM-5 nanocrystals for the conversion of ethylene to larger hydrocarbons. The reported size of the ZSM-5 crystals was 13 nm by XRD and was 30 nm by FE-SEM. The zeolite nanocrystals showed excellent catalytic performance and exhibited the longest lifetime with low coke selectivity as compared with larger ZSM-5 crystals. A decrease in shape selectivity was observed as a result of the large external surface area. Another study, on nanocrystalline zeolite L, showed an increase in adsorptive capacity of heptane for nanocrystals (8-nm pore length by XRD) over larger crystals [62].

More indirect effects of crystal size and morphology have been speculated to occur [63]. For example, in addition to the short diffusion path, the more uniform distribution of aluminum in small ZSM-5 crystallites was shown to have an effect on p-xylene selectivity in toluene methylation and disproportionation [63].

In addition to the catalytic applications of zeolite ultrafine (nanocrystalline) particles, our interest in their synthesis stems from their potential use as precursors for thin-film formation through processing of colloidal suspensions [64]. The field of zeolites and molecular sieve thin films and membranes emerged in late 1980s and is under intensive investigation. Potential applications include catalytic processes [65], energy intensive separations [66], and advanced applications of zeolites for electronic and optical materials [67].

Several research groups have previously implemented post-zeolite synthesis procedures to assemble zeolite–silica [68] and zeolite–polymer [69] composite films. In these preparations zeolite crystals have been used as additives in silica and polymer films, resulting in moderate increases in selectivity for membrane separations [70] and selective sensing [71]. This method, however, does not fully exploit the shape-selective properties of the zeolite because the contribution from the matrix influences the transport characteristics.

In situ film formation has also been explored. This method consists of contacting the zeolite precursor gel or solution (mother liquor) with a suitable substrate. When in situ growth was used, the preparation of both freestanding [72,73] and supported films [74–80] of intergrown crystals have been demonstrated for certain zeolites, and promising membrane separation results have been reported for ZSM-5 membranes. For most of the zeolites and molecular sieves, because synthesis conditions involve elevated temperatures and high pH, the choice of substrate is limited and alumina and quartz are most widely used. The one-step trial and error in situ methods do not allow for efficient optimization of film microstructure and thickness. For example, typical film thicknesses for films synthesized by in situ preparation range from 5 to 100 μm, while it is recognized that films thinner than 100 nm are needed to enhance permeation through these materials [68].

We have proposed a versatile approach to zeolite film formation based on processing of zeolite ultrafine (nanoscale) particles [64]. This processing scheme for zeolite films decouples film casting–deposition and growth providing added flexibility for tailoring the film microstructure.

The processing scheme is outlined in Fig. 7. It includes the synthesis of zeolite nanoparticles and subsequent preparation of a suspension of these particles followed by film casting and postprocessing by secondary growth. The various steps of the process can be studied and optimized independently.

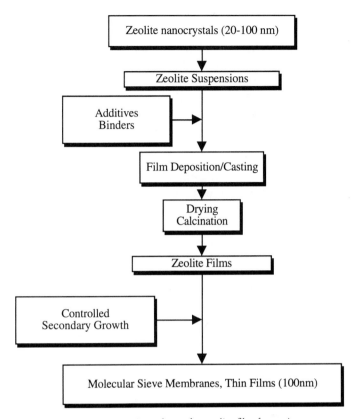

Figure 7. Processing scheme for zeolite film formation.

13.3.1 Preparations of Colloidal Suspensions in Water

The resulting suspensions of zeolite L nanoclusters in their mother liquid start settling when stored at room temperature, and a top layer of clear liquid is observed after 2 days. This is unlike the colloidal suspensions of ZSM-5 and sodalite that have been reported to be stable for months in their mother liquid [4]. DLS results provide evidence that aggregation of the nanoclusters takes place in the mother liquid [9]. However, stable suspensions of zeolite L nanoclusters can be prepared with repeated washing, centrifugation, and redispersion in water by sonication. Suspensions of the zeolite L clusters in water (pH ~7) with concentrations of up to 30–35 g/l do not show any change in particle size (as determined by DLS) for at least up to 3 months (~60 nm). The stability of the zeolite L suspensions in water (pH ~7) is due to the small particle size and the negative surface charge of the zeolite crystals at these conditions as indicated from mobility measurements. Mobility measurements were performed on a B1-Zeta Plus instrument. The samples were diluted to 8 mg/l with DI water, and the pH was adjusted using HNO_3 and NH_4OH. At pH 7.8 the mobility was recorded as -3.64 μm/s/V/cm and the corresponding zeta potential was -46.7 mV (Fig. 8). Mobility measurements are important in understanding colloidal stability and its effect on film processing.

13.3.2 Secondary Growth

The use of seed crystals in zeolite synthesis is often employed to increase the size of resulting crystals and to reduce the nucleation time. For the purposes of film formation [64] and size tailoring of zeolite L crystals, identification of conditions that lead to controlled growth of the nanocrystals is needed. We have performed experiments where weighted amounts of the colloidal suspensions of zeolite L nanoclusters were added to a homogeneous solution resulting in liquid compositions that do not nucleate zeolite L in the absence of the nanocluster seeds. Examples of such experiments that lead to growth of the zeolite particles are described in the following text. We refer to this process as secondary growth.

A zeolite suspension (10 g/l) in water (pH ~7) was added to a homogeneous, potassium aluminosilicate solution resulting in a solution with composition $10K_2O–1Al_2O_3–20SiO_2–2000H_2O$ with 5.5 g/l nanocluster

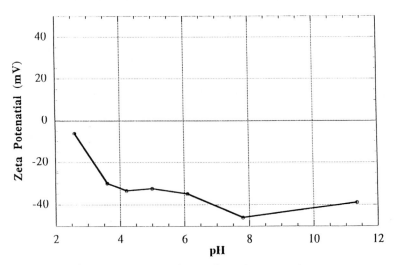

Figure 8. Zeta potential vs pH for zeolite L nanoclusters.

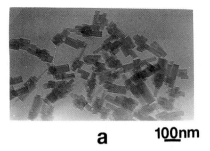

a **100nm**

Figure 9a. TEM micrograph of regrown zeolite L, composition $10K_2O-1Al_2O_3-20SiO_2-2000H_2O$ with 5.5 g/l nanocluster seeds.

seeds. A synthesis starting from a clear solution with this composition (no zeolite L added) results in an amorphous product when heated for up to 8 h at 175°C. With the zeolite nanoclusters present the crystals grow larger to almost perfect cylinders free of amorphous product after 6 h as shown in Fig. 9a. The zeolite nanoclusters due to their high-surface area can serve as seeds for growth without further nucleation. By varying solution compositions, seeded growth can result in different morphologies [9]. Figure 9b shows morphologies of regrown crystals in a solution with composition

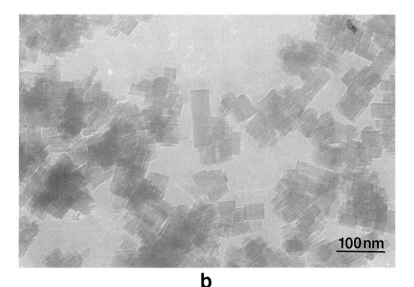

100nm

b

Figure 9b. TEM micrograph of regrown zeolite L, composition $10K_2O-1Al_2O_3-25SiO_2-1200H_2O$ with 54 g nanocluster seeds.

$10K_2O-1Al_2O_3-25SiO_2-1200H_2O$ with 54 g of nanocluster seeds (150°C for 6 h. In a control experiment, a solution with composition $10K_2O-1Al_2O_3-25SiO_2-1200H_2O$ with no added seeds was subjected to the same heating procedure. In the absence of zeolite L seed crystals, we observe the formation of an amorphous gel followed by the precipitation of an amorphous (by XRD and TEM) phase. With the zeolite nanoclusters present the amorphous gel formed encapsulates the zeolite nanoclusters and is gradually consumed for further crystal growth (Fig. 9c). No amorphous precipitate remains unconverted at the end. Table 3 shows an XRD study of the seeded zeolite L regrowth presented in Fig. 9c. After 1 h of heating, the seed crystals remain mostly unchanged and an amorphous gel has formed. On further heating, gel is consumed (note percentage of crystallinity) and seed particles begin to grow larger. The crystals grow larger along the (220) (perpendicular to the c-axis) and (221) directions while the (001) direction remains unchanged (parallel to the c-axis or channels). The

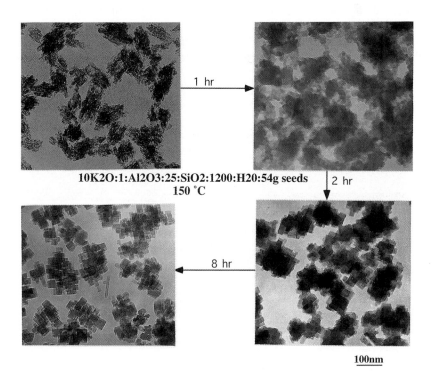

C

Figure 9c. TEM micrographs of solid products from zeolite L regrowth at various times.

TABLE 3
XRD Study of Seeded Zeolite L Crystal Regrowth[a]

Direction	Seed crystals L nanoclusters	1 h	2 h	8 h
(001)	30	30	30	30
(220)	15	15	25	25
(221)	20	20	25	25
% Crystallinity		40	82	100
Product(g)/solution(g)		0.030	0.034	0.034

[a]Solution was $10K_2O-1Al_2O_3-25SiO_2-1200H_2O$; product samples (size in nm) from various times heated @ 150°C; 0.01 g seed/g solution.

cylindrical grains of the nanoclusters grow mainly wider and not longer. The yield of solids recovered remains the same during regrowth and the final product is 100% crystalline, suggesting that the amorphous phase formed at early stages of heating is completely consumed for particle regrowth. Regrowth of other zeolite particles with similar results can be achieved. In Fig. 10 the morphology of silicalite crystals before and after secondary growth is shown.

13.3.3 Film Formation

The zeolite L nanocluster suspensions can be used for the preparation of unsupported films by evaporation of the water. Supported films can also

100nm

8 hr

$39.4SiO_2:5.8TPA_2O:16800H_2O:40g$ seeds
160 °C

Figure 10. TEM micrographs from silicalite regrowth, as synthesized (left) and after 8 h regrowth (right).

be prepared by spin-coating or dip-coating with almost no limitations imposed on the choice of substrate, in contrast to *in situ* preparations. The pure zeolite films are microcrack free and translucent. The microcrack free here is used to describe the absence of submicron-sized cracks propagating along the surface or interior of the films. A thick (2-mm) unsupported film is shown in Fig. 11 (a); an FE-SEM top view is shown in Fig. 11 (b).

Nonsupported thick films prepared by these methods have a bimodal porous structure consisting of the zeolite micropores and pores in between the zeolite particles (intercrystalline porosity) as evident by FE-SEM in Fig. 11 (b). The films are fragile because the zeolite clusters are not bonded through solid bridges. The addition of a binder can enhance the mechanical strength and integrity of the films. Several binders have been employed including silica and alumina. For the preparation of zeolite L–Al_2O_3 films, a boehmite sol was added to a zeolite L suspension (33 g of zeolite/kg of H_2O) and then stirred. The boehmite suspensions were prepared by adding dropwise aluminum tributoxide to distilled water (1.51 drops/mol of alkoxide) heated to 80–85°C and stirred. After addition of the alkoxide, 0.070 mol of HNO_3/mol of butoxide was added to peptize the sols. The colloidal suspension was refluxed for 16 h to form a 1 M stable boehmite suspension.

Immediately upon mixing, we observed agglomeration caused by the opposite surface charges of zeolite and alumina particles because a typical zeta potential for an alumina suspension in water is ~40 mV and the zeta potential for the zeolite suspension used was recorded at −46.7.

The mixed dispersion was used for film casting. The resulting composite films were opaque and exhibited increased mechanical strength. We prepared films with compositions ranging from 80 to 97% in zeolite content. Nitrogen adsorption isotherms and pore size distributions reveal a decrease in interzeolitic porosity for composite zeolite–alumina films compared to pure zeolite films. Nitrogen adsorption isotherms and pore size distributions for a pure zeolite L film and a film with composition 93 wt % zeolite and 7 wt % alumina are shown in Fig. 12. It is evident that the addition of Al_2O_3 does not reduce the microporosity indicating no blockage of the zeolite pores. The pores in between the zeolite particles (interzeolitic porosity) have sizes on the order of the zeolite particle size. However, for molecular sieving applications the interzeolitic porosity should be eliminated.

13.3.4 Use of Secondary Growth for the Preparation of Asymmetric Films

The regrowth conditions identified in the preceding experiment were implemented to eliminate the interzeolitic porosity of the cast zeolite–alumina films. Film sections were held vertically in contact with 30 ml of a homoge-

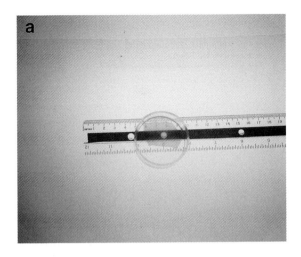

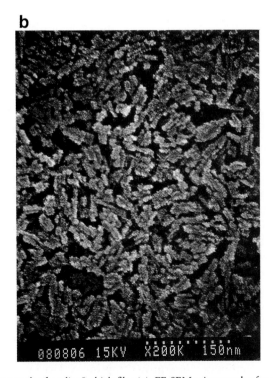

Figure 11. Photograph of zeolite L thick film (a); FE-SEM micrograph of zeolite L film (b).

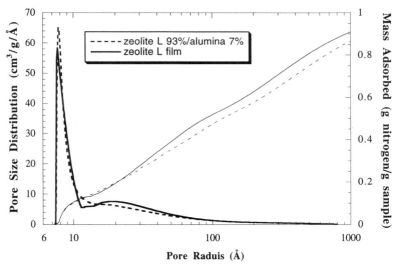

Figure 12. Nitrogen adsorption and pore size distribution: zeolite L and zeolite L–alumina films.

neous solution and heated in Teflon-lined stainless steel vessels. Films removed at various times were washed in distilled water, dried, and calcined up to 750°C. Field emission-scanning electron microscopy (FE-SEM) top views of a series of films obtained by this procedure, using a homogeneous solution with the composition of $10K_2O-1Al_2O_3-25SiO_2-1200H_2O$, are shown in Fig. 13.

Figure 13 (a) shows the initial calcined zeolite–alumina film. Figure 13 (b) is from the film after 6 h of secondary growth and reveals larger cylindrical crystals closely packed and intergrown. Small fragments of unconverted gel are observed on the surface emanating from the zeolite crystals. The surface has reduced interzeolitic porosity (no interzeolitic porosity to the resolution of the FE-SEM ~5 nm) and is chemically intergrown. The cylindrical morphology of these crystals is common to zeolite L, and TEM and microdiffraction confirm that the crystals present on the surface are zeolite L. Figure 14 (a) shows a TEM micrograph of the cross section of the film presented in Fig. 13. The cracks evident in Fig. 14 (a) are typical of samples prepared by microtomy, which is known to produce thickness variations (and consequently cracks) around every 1000 Å during sample sectioning [81]. The cross section shows a gradual decrease in crystal size across the thickness of the film. At higher magnifications the one-dimensional channels of zeolite L are apparent. The film is gradient in both porosity and grain size forming an asymmetric zeolite membrane with a thin (~200 nm) continuous intergrown top layer. The interplay of crystal

Figure 13. FE-SEM micrograph (top view) of zeolite L–alumina film (a); FE-SEM micrograph (top view) of regrown film, $10K_2O-1A1_2O_3-25SiO_2-1200H_2O$ (6 h heating @150°C) (b). Magnification 5,000×, insets at 50,000×.

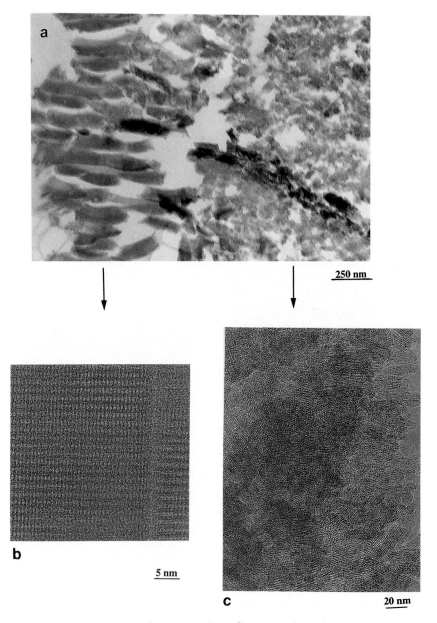

Figure 14. TEM cross section of regrown zeolite L film (regrowth conditions as in Fig. 9) (a); HREM of interior of film. (b); HREM of top layer (c).

growth, precursor gel formation, and transport of precursor species accounts for the asymmetric morphology of the regrown film. The asymmetric geometry is the desirable configuration for selective, but also highly permeable membranes. Therefore this method is promising for membrane formation. Evaluation of film performance in permeation experiments have yet to be performed.

Regrowth under different conditions results in different film microstructures. Figure 15 shows the top view FE-SEM of a regrown film with regrowth concentrations corresponding to those of Fig. 9a. These films have considerable interzeolitic porosity despite the regrowth, and

Figure 15. FE-SEM micrograph (top view) of regrown zeolite L film (regrowth composition $10K_2O-1Al_2O_3-20SiO_2-2000H_2O$).

therefore it is evident from the FE-SEM that they are not applicable for molecular sieving application, but may have uses as catalytic membranes.

The preceding results show that by the use of a zeolite nanocluster precursor and decoupling growth and film casting, a processing route is opened for efficient manipulation of film microstructure. Moreover, the ability to utilize suspensions of zeolite crystals for film preparation through sol–gel-type processing renders this technique a powerful tool for efficient film production.

Using a similar technique, the formation of thin silicalite films (~750 nm) from regrowth of nanocrystalline silicalite/alumina composite films has also been accomplished. The crystallites at the intergrown layer are oriented with their straight channels parallel to the substrate. The membranes show H_2/N_2 selectivities of about 60 at 150°C. This high selectivity is attributed to the orientation of the separating layer. Characterization and permeation results are reported elsewhere [82].

13.4 Conclusions and Directions for Further Development

There is an increased interest in the synthesis of zeolite nanoparticles. Driving forces for these developments are summarized as follows:

1. Studies of zeolite nanocrystal synthesis by SAXS, WAXS, DLS, and TEM provide an opportunity for a fundamental understanding of zeolite nucleation and growth.
2. Due to their short-channel lengths zeolite nanocrystals provide for fast diffusion times and increased activity. Small zeolite particles also provide high external to internal surface ratios resulting in increased resistance to deactivation and catalysis with molecules that are unable to enter the zeolite pores. The price to pay for catalysis from external surfaces is the loss of shape selectivity. A possible direction toward increasing selectivity lies in the control of interzeolitic porosity.
3. Zeolite nanoparticles appear ideal for thin film processing because they allow for film formation from suspensions as well as controlled regrowth.
4. Other applications of zeolite nanoparticles related to their ability to easily form stable suspensions are catalytic applications in liquid media and use in processing of zeolites as additives.

Although considerable progress has been achieved in preparing suspensions of zeolite particles with sizes ~50 nm, further reduction in size and economic production of monodisperse particles of 100 Å will require modifications of the currently available synthesis process through reactor design and further understanding of nucleation and growth mechanisms.

Acknowledgments

Acknowledgment is made to the donors of the Petroleum Research Fund, administered by the ACS, for support of this research, and to the National Center for Electron Microscopy at Lawrence Berkely Laboratory, for a NCEM/DOE fellowship.
This work was initiated at Caltech with funding provided by ARCO. We are grateful to Professors Tatsuya Okubo, Steven Suib, Dionisios Vlachos; and particularly Mark E. Davis for collaboration, suggestions, and assistance with various aspects of this work. We acknowledge the W. M. Keck Polymer Morphology Laboratory for use of its electron microscopy facilities.

References

1. Davis, M. E. (1991). *Ind. Eng. Chem. Res.,* 30, 1675.
2. Breck, D. W. (1973). *Zeolite Molecular Sieves,* Wiley, New York.
3. Schoeman, B. J., Sterte, J., and Otterstedt, J. E. (1994). *Zeolites,* 14, 208.
4. Schoeman, B. J., Sterte, J., and Otterstedt, J. E. (1994). *Zeolites,* 14, 110.
5. Persson, A. E., Schoeman, B. J., Sterte, J., and Otterstedt, J. E. (1994). *Zeolites,* 14, 557.
6. Schoeman, B. J., Sterte, J., and Otterstedt, J. E. (1995). *J. Colloid Interface Sci.,* 170, 449–456.
7. Persson, A. E., Schoeman, B. J., Sterte, J., and Otterstedt, J. E. (1995). *Zeolites,* 15, 611.
8. Meng, X., Zhang, Y., Meng, C., and Pang, W. (1993). *Proceedings of the 9th International Zeolite Conference* (R. von Ballmoos, *et al.,* eds.), p. 297 Butterworth-Heinemann.
9. Tsapatsis, M., Lovallo, M., Okubo, T., Davis, M. E., and Sadakata, M. (1995). *Chem. Mater.,* 7, 1734–1741.
10. Tsapatsis, M., Lovallo, M., and Davis, M. E. (1996). *Microporous Materials,* 5, 381–388.
11. Vaughan, D. E. W., and Strohmaier, K. G. (1994). Canadian Patent Application, 2 106 170.
12. Young, R. A. (ed.) (1993). *The Rietveld Method,* Oxford Science, New York.
13. Berne, B. J., and Pecora, R. (1976). *Dynamic Light Scattering,* Wiley, New York.
14. Dahneke, B. E. (ed.) (1983). *Measurement of Suspended Particles by Quasi-Elastic Light Scattering,* Wiley-Scientific, New York.
15. Yamamura, M., Chake, K., Wakatsuki, T., Okado H., and Fujimoto, K. (1994). *Zeolites,* 14, 643–649.
16. Ueda, S., Kageyama, N., and Koizumi, M. (1984). *Proceedings of the Sixth International Zeolite Conference* (D. Olson and A. Bisio eds.), p. 905, Butterworths.
17. Xu, W., Li, J., Li, W., Zhang, H., and Liang, B. (1989). *Zeolites,* 9, 468.
18. Zhandov, S. P. (1971). *Adv. Chem. Ser.* 101, 21.
19. Barrer, R. M. (1981). *Hydrothermal Chemistry of Zeolites,* Academic Press, London.
20. Ocelli, M. L., and Robson, H. E. (eds.) (1989). *Zeolite Synthesis,* ACS Symposium Series, 398, American Chemical Society, Washington, DC.
21. Kacirek, H., and Lechert, H. (1975). *J. Phys. Chem.* 79, 1589.
22. Bell, A. T. (1989). *Zeolite Synthesis* (M. L. Ocelli, and H. E. Robson eds.), p. 66, ACS Symposium Series 398, American Chemical Society, Washington, DC.
23. Vaughan, D. E. W. (1991). *Stud. Surf. Sci. Catal.,* 65, 275.
24. Davis, M. E., and Lobo, R. F. (1992). *Chem. Mater.,* 4, 756.
25. Burkett, S. L., and Davis, M. E. (1994). *J. Phys. Chem.,* 98, 4647.
26. Burkett, S. L., and Davis, M. E. (1995). *Chem. Mater.,* 7, 920.

342 / *Mark C. Lovallo and Michael Tsapatsis*

27. Subotic, B., and Graovac, A. (1985). *Zeolites, Synthesis, Structure, Technology and Application* (B. Drsaj, S. Hocevar, and S. Pejovnik eds.), p. 199, Elsevier Science, Amsterdam.
28. Gonthier, S., Gora, L, Guray, I., and Thompson, R. W. (1993). *Zeolites*, 13, 414.
29. Subotic, B., Skrtic, D., Smit, I., and Sekovanic, L. (1980). *J. Cryst. Growth*, 50, 498.
30. Subotic, B., Graovac, A., and Sekovanic, L. (1981). *5th International Conference on Zeolites* (R. Sersale, C. Colella, and R. Aille, eds.), p. 54, Naples, Italy.
31. Bronic, J., and Subotic, B. (1995). *Microporous Materials*, 4, 239.
32. Hirth, J. P., and Pound, G. M. (1993). *Condensation and Evaporation: Nucleation and Growth Kinetics*, Progress in Materials Science, Vol. 77 (B. Chalmers, ed.), Pergamon, New York.
33. Vlachos, D. G., unpublished.
34. Vlachos, D. G., Schmidt, L. D., and Aris, R. (1992). *J. Chem. Phys.*, 96, 6880.
35. Lechert, H., and Kacirek, H. (1993). *Zeolites*, 13, 192.
36. Schoeman, B. J., Sterte, J., and Otterstedt, J. E. (1994). *Zeolites*, 14, 568.
37. Twomey, T. A. M., MacKay, M., Kuipers, H. P. C. E., and Thompson, R. W. (1994). *Zeolites*, 14, 162.
38. Bogush, G. H., and Zukoski, C. F. (1988). *Ultrastructure Processing of Advanced Ceramics* (J. D. Mackenzie and D. R. Ulrich, eds.), p. 477, Wiley, New York.
39. LaMer, V. K., and Dinegar, R. H. (1950). *J. Am. Chem. Soc.*, 72(11), 4847.
40. Nielsen, A. E. (1964). *Kinetics of Precipitation*, Macmillan, New York.
41. Bogush, G. H., Dickstein, G. L., Lee, P., and Zukoski, C. F., IV. (1988). *Better Ceramics Through Chemistry III* (C. J. Brinker, D. E. Clard, and D. R. Ulrich, eds.), North-Holland, New York.
42. Towe, K. M., and Bradley, W. F. (1967). *J. Colloid Interface Sci.*, 24, 384.
43. Sugimoto, T., and Matijevic, E. (1980). *J. Colloid Interface Sci.*, 74, 227.
44. Murphy, P. J., Posner, A. M., and Quirk, J. P. (1976). *J. Colloid Interface Sci.*, 56, 284.
45. Dokter, W. H., van Garderen, H. F., Beelen, T. P. M., van Santen, R. A., and Bras, W. (1995). *Angew. Chem.*, 34, 73.
46. Dokter, W. H., Beelen, T. P. M., van Garderen, H. F., Rummens, C. P. J., van Santen, R. A., and Ramsay, J. D. F. (1994). *Colloids Surf. A*, 85, 89.
47. Treacy, M. M. J., and Newsam, J. M. (1987). *Ultramicroscopy*, 23, 411.
48. Terasaki, O., Thomas, J. M., and Ramdas, S. J. (1984). *J. Chem. Soc. Chem. Commun*, 4, 216.
49. de Gruyter, C. B., Verduijn, J. P., Koo, J. Y., Rice, S. B., and Treacy, M. M. J. (1990). *Ultramicroscopy*, 34, 102.
50. Fajula, F. (1989). *Guidelines for Mastering the Properties of Molecular Sieves*, (D. Barthoneuf, E. G. Derouane, and W. Hölderich, eds.), p. 53, NATO ASI Series, Plenum Press, New York.
51. Cundy, C. S., Henty, M. S., and Plaisted, R. J. (1995). *Zeolites*, 15, 353.
52. Cundy, C. S., Henty, M. S., and Plaisted, R. J. (1995) *Zeolites*, 15, 400.
53. Farcasiu, D., Hutchison, J., and Li, L. (1990). *J. Catal.*, 122, 34.
54. Barrer, R. M. (1978). *Zeolite and Clay Minerals*, p. 287, Academic Press, New York/London.
55. Inomata, M., Yamada, M., Okada, S., Niwa, M., and Murakami, Y. (1986). *J. Catal.*, 100, 2264.
56. Barrer, R. M. (1949). *Trans. Faraday Soc.*, 45, 358.
57. Suzuki, I., Namba, S., and Yashima, T. (1983). *J. Catal.*, 81, 485.
58. Farcasiu, M., and Degnan, T. F. (1988). *Ind. Eng. Chem. Res.*, 27, 45.

59. Fraenkel, D. (1990). *Ind. Eng. Chem. Res.*, 29, 1814.
60. Zecchina, A., Bordiga, S., Spoto, G., Scarano, D. *et al.* (1992). *J. Chem. Soc. Faraday Trans.*, 88(19), 2959.
61. Gilson, J. P., and Derouane, E. G. (1984). *J. Catal.*, 88, 538.
62. Shiralkar, V. P., Joshi, P. M., Eapen, M. J., and Rao, B. S. (1991). *Zeolites* 11, 511.
63. Verduijn, P. V., Mechilium, J., De Gruijter, C. B., Koetsier, W. T., and Van Oorschot, C. W. M. (1991). U. S. Patent, 5 064 630.
64. Tsapatsis, M., Okubo, T., Lovallo, M., and Davis, M. E. (1994). *Advances in Porous Materials*, (S. Komarnemi, D. M. Smith, and J. S. Beck, eds.), *MRS Symposium Proceedings*, Boston, Vol. 371 MRS, Pittsburgh.
65. Armor, J. N. (1989). *Appl. Catal.*, 49, 1–25.
66. Armor, J. N. (1994). *Chem. Mater.*, 6, 730.
67. Ozin, G. A., Kuperman, A., and Stein, A. (1989). *Angew. Chem. Int. Ed. Engl.*, 28, 359–376.
68. Bein, T., Brown, K., and Brinker, C. J. (1989). *Zeolites: Facts, Figures, Future*, Elsevier Science, Amsterdam.
69. Hennepe, H. J. C., Mulder, M. H. V., Smolders, C. A., Bargeman, D., and Schroeder, G. A. T. (1990). U. S. Patent, 4 925 562.
70. Jia, M., Peinmann, K. L., and Behling, R. D. (1991). *J. Membr. Sci.*, 57, 289–296.
71. Bein, T., and Brown, K. (1989). *J. Am. Chem. Soc.*, 111, 7641–7643.
72. Anderson, M. W., Pachis, K. S., Shi, J., and Carr, S. W. (1992). *J. Mater. Chem.*, 2, 255–256.
73. Sano, T., Kiyozumi, Y., Kawamura, M., Mizukami, F., Takaya, H., Mouri, T., Inaoka, W., Toida, Y., Watanabe, M., and Toyoda, K. (1991). *Zeolites*, 11, 842–845.
74. Yan, Y., Tsapatsis, M., Gavalas, G. R., and Davis, M. E. (1993). *J. Chem. Soc. Chem. Commun.*, 2, 227.
75. Geus, E. R., van Bekkum, H., Bakker, W. J. W., and Moulijn, J. A. (1993). *Microporous Mater.*, 1, 131–147.
76. Jia, M. D., Chem, B., Noble, R. D., and Falconer, J. L. (1994). *J. Membr. Sci.*, 90, 1–10.
77. Yamazaki, S., and Tsutsumi, K. (1995). *Microporous Mater.*, 4, 205–212.
78. Sano, T., Hasegawa, M., Kawakami, Y., Kiyozumi, Y., Yanagishita, H., Kitamoto, D., and Mizukami, F. (1994). *Stud. Surf. Sci. Catal.*, 85, 1175–1182.
79. Sano, T., Ejiri, S., Hasegawa, M., Kawakami, Y., Enomoto, N., Tamai, Y., and Yanagishita, H. (1995). *Chem. Lett.*, 2, 153–154.
80. Yan, Y., Davis, M. E., and Gavalas, G. R. (1995). *Ind. Eng. Chem. Res.* 34, 1652–1661.
81. Tsapatsis, M., and Gavalas, G. (1994). *J. Membr. Sci.*, 87, 281–296.
82. Lovallo, M. C., Boudreau, L., and Tsapatsis, M. *Preparation of Supported Zeolite Films and Layers: Processing of Zeolite Suspensions and In-Situ Growth from Homogeneous Solutions* (J. S. Beck, L. E. Iton, L. E. Corbin, R. F. Lobo, M. E. Davis, S. I. Zones, and S. L. Suib, eds.) *Microporous and Macroporous Materials* MRS Symposium Proceedings, Vol. 431, (accepted).

CHAPTER 14

Preparation of Pillared Clays and Their Catalytic Properties

Abraham Clearfield

Chemistry Department
Texas A&M University
College Station, Texas 77843

KEYWORDS: clay mineral, Keggin ion, montmorillonite, pillared clay, pillared interlayered clays, pillaring process, saponite, smectite

14.1 Introduction

Pillared clays represent a new class of microporous materials that have potential for use as catalysts. A great deal of progress has been made in our understanding of these materials, particularly as relates to their catalytic behavior. We will concern ourselves in this chapter to work relating to their preparation, characterization, and catalytic behavior. Earlier work has been summarized in a number of reviews and compilations [Pinnavaia, 1983; Clearfield, 1988; Burch, 1988; Michell, 1990; Clearfield and Kuchenmeister, 1992].

A brief introduction will be given for those new to the field. The impetus behind the original work was to create porous three-dimensional structures from two-dimensional clay minerals. It was felt that, because smectite clays have a low layer charge and if an inorganic polymeric cation were chosen that had a high positive charge, pores greater than those available in zeolites would be obtained. The initial choice of polymer [Vaughan *et al.*, 1979; 1980] was the aluminum Keggin ion

$[Al_{13}O_4(OH)_{24}(H_2O)_{12}]^{7+}$. A straightforward calculation [Clearfield and Kuchenmeister, 1992] based on the range of layer charges in smectite or swelling clays and a 7+ charge on the pillaring cations indicates that the distance between pillars should be 17–29 Å. Subtracting the width of the pillar, assumed to be about 7 Å after dehydration leaves a free distance of 10–22 Å. From the amount of aluminum incorporated the average charge lies between 4 and 5. Thus, the pores are smaller than the calculation indicates. However, many different pillaring agents have been used and the pore size depends on the choice of pillaring agent and several other factors. Therefore, a section will be devoted to this subject. We will begin with some information on the nature of the clay minerals used for pillaring.

14.2 Description of Smectites

The smectites or swelling clays have structures that consist of aluminum or magnesium–oxygen octahedra sandwiched between layers of silica tetrahedra (Fig. 1). In a unit cell formed from 20 oxygens and 4 hydroxyl groups there are eight tetrahedral sites and six octahedral sites. When all the octahedral sites are filled with Mg^{2+}, a magnesium talc of ideal formula $Mg_6Si_8O_{20}(OH)_4$ is the result. Such a talc has no imbalance of charge and no ion exchange properties. The smectites are distinguished by the type and location of their cation substitutions. For example, Li^+ may substitute for Mg^{2+} in the octahedral layer to produce a positive

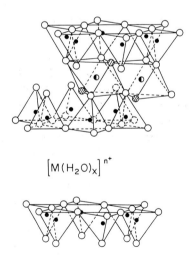

$$\left[M(H_2O)_x\right]^{n+}$$

Figure 1. Schematic drawing of smectite clay: O, octahedrally coordinated Mg^{2+} or Al^{3+}; O, tetrahedrally coordinated ion (Si); O, hydroxyl ion; and O, oxygen.

charge deficiency. This deficiency is compensated by cations, usually Na^+ or Ca^{2+}, residing between the layers. Substitutions may also take place in the tetrahedral sites (Al^{3+} for Si^{4+}). The principal smectite clay minerals and their ideal compositions are listed in Table 1. Typically the positive charge deficiency ranges from 0.4 to 1.2 e per Si_8O_{20} unit. These values translate to an ion exchange capacity range of $0.5-1.5$ meq/g. More usual values are $0.7-1.0$ meq/g. If the charge deficiency arises from octahedral substitution, then this excess negative charge is distributed over all the oxygens in the framework. These clays tend to be turbostratic, that is, the layers are randomly rotated about an axis perpendicular to the layers. However, should the smectite have been formed by substitution at tetrahedral sites, then the extra charge is more localized and the resultant clay tends to exhibit greater three-dimensional order [Mott, 1988]. Clay mineral compositions are often given in terms of 10 oxygens rather than 20. The formula in which the octahedral layers is a divalent ion such as Mg^{2+} would have three Mg^{2+} sites and is termed a trioctahedral clay. Similarly if a trivalent ion such as Al^{3+} dominates in the octahedral layer, two of the three available sites are occupied and the clay is termed dioctahedral. Smectites and mica layers are also termed 2:1 to indicate that there are two tetrahedral sheets to one octahedral sheet comprising the layer.

14.3   The Pillaring Process: Use of the Al_{13} Keggin Ion

14.3.1 Montmorillonite

The most frequently used pillaring agent is the Al_{13} Keggin ion (Fig. 2) together with montmorillonite as the host clay mineral. Commercially available clay minerals are impure. Quartz, dolomite, and iron oxide in varying amounts may be present. Carbonates can be removed by treatment

---

TABLE 1
Idealized Structural formulas for principal smectite clays[a]

Octahedrally substituted smectites

Montmorillonite: $M_{x/n}^{n+}[Al_{4-x}Mg_x](Si_8)O_{20}(OH)_4 \cdot yH_2O$

Hectorite: $M_{x/n}^{n+}[Mg_{6-x}Li_x](Si_8)O_{20}(OH)_4^a \cdot yH_2O$

Tetrahedrally substituted smectites

Montmorillonite: $M_{x/n}^{n+}[Al_4](Si_{8-x}Al_x)O_{20}(OH)_4 \cdot yH_2O$

Hectorite: $M_{x/n}^{n+}[Mg_6](Si_{8-x}Al_x)O_{20}(OH)_4 \cdot yH_2O$

[a] Some F^- may substituted for $(OH)^-$.

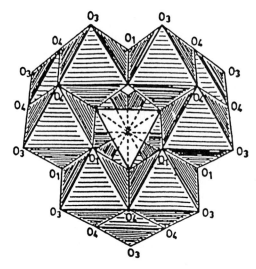

Figure 2. Schematic representation of the $[Al_{13}O_4(OH)_{24}(H_2O)_{12}]^{7+}$ ion. (Reprinted with permission from *Acta Chrmica Scandinavica*; 14, Johnson, G., 1960.)

with HCl and other impurities are separated by sedimentation [Newman, 1987]. Some care is required in the preparation of the Keggin ion. Two general methods have been used to obtain solutions containing the Keggin ion. Addition of base to a solution of $AlCl_3$ results in hydrolysis of the hexa-aquo aluminum ion $Al(H_2O)_6^{3+}$ to yield first a dimer and then the Al_{13} Keggin ion. The hydrolytic process may be followed by observing the ^{27}Al NMR spectrum of the solution. The Keggin ion exhibits a strong, sharp resonance at 62.8 ppm (Fig. 3A) arising from the single tetrahedrally coordinated aluminum in the center of the ion. The resonance due to the 12 octahedrally coordinated aluminum ions are quadrupolar relaxed, so that their resonance line is too broad to be observed at the sweep widths normally employed to obtain the spectra [Pinnavaia, 1984]. As seen in Fig. 3A at an OH:Al ratio of 2.42, the only species observed in the spectrum is that of the Keggin ion.

Many workers have used a commercially available aluminum chlorohydrate (ACH) solution prepared by the reaction of Al metal with a solution of $AlCl_3$. Figure 3B illustrates that the ACH solution is quite different than the base-hydrolyzed one. Even at an OH:Al ratio of 2.5 the hexa-aquo aluminum ion is still present and is probably in equilibrium with the dimer $Al_2(OH)_2(H_2O)_8^{4+}$ [Akitt and Farthing, 1981]. The very broad resonance reaching a maximum at 10.8 ppm must arise from polymeric species in which the aluminum is chiefly six-coordinate. The nature of these species is not known with certainty; however, an excellent treatment

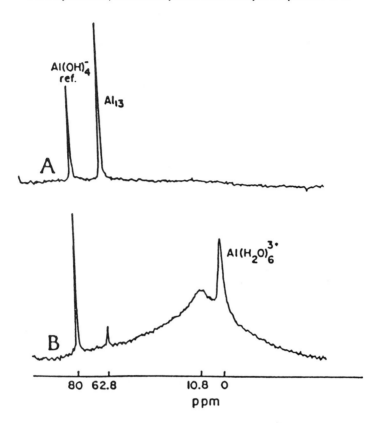

Figure 3. ^{27}Al spectra (46.9 MHz) of polyoxyaluminum-pillaring reagents. (A) Base-hydrolyzed AlCl$_3$, OH:Al = 2.42, (B) aluminum chlorohydrate, OH:Al = 2.50. The line near 80 ppm is the resonance of an external NaAlO$_2$ reference solution. (From *Heterogeneous Catalysis*, B. L. Shapiro (ed.), T. Pinnavaia, 1984, 145, with permission.)

of aluminum hydrolytic polymerization has been presented [Fu *et al.*, 1991]. A relatively complete summary of the literature to 1993 has been given by Wade [1993]. The Keggin ion is only a minor species in the ACH solution. Despite the large differences in the two solutions, the pillared products prepared from these solutions are fairly similar. The basal spacings range between 18.2 and 19 Å with surface areas of between 200 and 300 m^2/g.

A typical pillaring reaction is conducted as follows [Clearfield *et al.*, 1994]: The aluminum Keggin ion was prepared by dissolving 15 g of AlCl$_3$·6H$_2$O (0.062 mol) in 250 ml of distilled water. To this solution was added 350 ml of 0.43 M NaOH (0.15 mol) dropwise with vigorous stirring to achieve an OH/Al of 2.43. An ^{27}Al NMR spectrum showed a single resonance at 63.3 ppm. The Keggin ion solution was diluted to 0.0347 M and a 2% suspension of montmorillonite was added with stirring. The ratio of Keggin ion to clay was 5:1 but ratios as low as 1.5 gave similar results. Stirring was continued for 5 h. The collected solid was washed free of chloride ion, dried at 100°C, and calcined at 500°C for 3 h. The basal spacing was 18.5 Å, and the surface area was 253 m²/g with a total pore volume of 0.186 cc/g.

Many variations on this procedure have been conducted and the reader is referred to the literature for additional details [Pinnavaia et al., 1984; Bradley and Kydd, 1991; Sterte, 1988]. Of particular importance is the order of addition. It is claimed [Vaughan, 1988] that by adding the clay dispersion to the Keggin ion solution, concentrated clay dispersions may be used without gelation occurring. This point is important because any commercial manufacturing procedure would require inordinately large volumes of clay dispersion to achieve a reasonable production of catalyst. Other prerequisites required for synthesis of a real catalyst, stressed by Vaughan, are using clays with little or no purification and with no preexchange. Pillaring along these lines was conducted with a montmorillonite (westone-L) used as received [Molina et al., 1992]. Its cation exchange capacity was 73 meq/100 g and the Brunauer–Emmett–Teller (BET) (N₂) surface area 69 m²/g. The pillaring solution was prepared by addition of 0.5M NaOH to a 0.2 M aluminum nitrate solution to achieve a molar ratio OH:Al of 1.6. This solution was aged for 24 h before use. A 40% clay suspension held in dialysis bags was dipped into basic aluminum solutions adjusted to supply 25–100 meq Al^{3+} per gram of clay. The best results were obtained with a ratio of aluminum per clay of 75 meq/g. This ratio is quite high and unless the aluminum solution can be reused would present no real advantage over using more dilute clay suspensions. Surface areas of the pillared products were in the range of 229–277 m²/g with pore volumes of 0.22 cc/g. However, only one-third of the pore volume arose from micropores. What this study illustrates is that should a commercial use for a pillared clay catalyst arise, the preparative process is sufficiently flexible that a satisfactory catalyst could be manufactured at a reasonable cost.

Two additional factors are worthy of note. Molina et al. [1992] observed that if the aluminum-treated clay were not washed, pillaring would not occur. Similar observations have been made by Turney [1992] and the author [Tsai and Clearfield, 1995]. Apparently, the positively charged Keggin ions accumulate along the edges of the clay particles and only diffuse into the interlayer space during the washing procedure. This point requires further systematic investigation. The second point deals with the fact that both the ACH solutions and the base hydrolyzed solution yield essentially the same type, but not identical products [Pinnavaia, 1984], even though they differ greatly in their Keggin ion content. Apparently, once the process of pillaring begins the equilibrium in the solution shifts to create a continuing supply of the high charge species. This hypothesis needs to be reconciled with the lack of pillaring prior to the washing step.

14.3.2 Saponite

Montmorillonite is a clay mineral in which substitution of Al^{3+} by Mg^{2+} in the octahedral sites predominates. It is instructive to examine a tetrahedrally substituted clay because as we shall see the two types of clay minerals behave differently. Saponite is a magnesium (trioctahedral) clay mineral with an ion-exchange capacity close to 1 meq/g. One potential advantage is the fact [Herrero, 1985; 1986] that Al/Si substitutions have been shown to occur homogeneously. Thus, the arrangement of the Keggin ion pillars should be more regular than in montmorillonite. Very complete details on pillaring of saponite have been given by Chevalier *et al.* [1992]. The saponite was from Ballarat, U.S.A. (available from the Clay Minerals Repository, University of Missouri) and had a composition of $Na_{0.3}Ca_{0.014}K_{0.003}[Mg_{2.96}Fe^{2+}_{0.052}Ti_{0.003}Mn_{0.001}]$ $[Si_{3.63}Al_{0.369}]$ $O_{10}(OH)_2 \cdot nH_2O$. Its ion exchange capacity is 90 meq/100 g. The as-received clay was purified by sedimentation and the 2-μm fraction was chosen for the pillaring reactions. It was used as a colloidal suspension of 5 g/liter. The aluminum solution was the commercially available ACH (Reheis Chemical Company, Dublin) diluted to 0.1 M and aged for 2 h at 60°C with stirring. This ACH solution was added dropwise to the clay suspension, then kept at 80°C for 2 h with stirring, and left overnight at room temperature. The mixture was then dialyzed until free of Cl^-, centrifuged, and dried at 60°C while kept at the bottom of the centrifuge tube. Exchange of the Keggin ion was conducted with different ratios of Al to clay from 2.5 to 12.5 mmol/g at pH 4.8. The amount of Al fixed in the clay amounted to 2.2–4.1 mmol/g. An ion exchange capacity of 0.9 meq/g is equivalent to 1.67 mmol Al/g (1.8 mmol/g is given by Chevalier *et al.* [1992]). Thus, the amount of aluminum incorporated is always larger than calculated on the assumption that only the Keggin ion is exchanged and at a charge of 7+. Either the Keggin ion has hydrolyzed to a lower charge (5.3+ for 2.2 mmol/g; 2.86+ for 4.1 mmol/g) or species other than the Keggin ion have been intercalated. Evidence for severe hydrolysis of the Al_{13} is provided from experiments run at higher pH (i.e., 6 and 7). Corresponding increases in Al uptake were recorded, from 3.8–5.8 mmol/g at pH 6 and 2.8–10.3 at pH 7. These results suggest that higher polymers are also exchanged at the high pH values.

All of the samples gave basal spacings lying between 18.3 and 19.1 Å. Heating to 500°C was conducted at a rate of 36°C/h so as not to disorder the structure. Surface areas ranged from 230 to 330 m²/g. The best conditions were chosen as a clay concentration of 5 g/liter, an Al:clay ratio of < 5 mmol/g and a pH between 4.7 and 6.0.

14.3.3 Pillared Clays with Mixed Oxide Pillars

Pillared clays are often referred to as PILCs which, depending on the source, stands for pillared interlayered clays or pillared inorganic layered compounds. The latter definition is broader than the former and includes many other classes of layered compounds such as phosphates and sulfides. However, because we will only discuss clay-based materials, our use of PILC will refer only to pillared clays.

As we shall see when treating more details of alumina pillared clays, they have many shortcomings relative to their catalytic properties. We may wish to have more stable products, more acidic products, or larger pores. Whatever the reason, a host of mixed oxide pillars have been used to improve the properties of the aluminum Keggin ion pillared products. Among the mixed oxide clusters utilized are the gallium aluminum Keggin ion [Bradley and Kydd, 1991; Gonzalez *et al.*, 1992], silica–alumina [Zhao *et al.*, 1992; Occelli and Finseth, 1986], lanthanum–alumina [Sterte, 1991], iron–aluminum [Zhao *et al.*, 1993a; Bergaya *et al.*, 1993], and others that we will discuss under the question of stability of the PILCs. It is beyond the scope of this chapter to discuss all of these procedures. However, we shall illustrate the process with reference to hydroxysilicoaluminum (HSA) pillars [Zhao *et al.*, 1992; Occelli and Finseth, 1986]. The HSA solutions were prepared in two ways. In method A described by Wada [Zhao *et al.*, 1992] a silica sol was diluted with 2 liters and the pH adjusted to 4 to 5 by addition of 0.02 M NaOH over a 2-h period in an ultrasonic generator. To this solution was added a 0.117 $AlCl_3$ to achieve a fixed Si:Al ratio and the OH:Al ratio brought to 2 by addition of 0.1M NaOH. The resultant solution was aged for 2 weeks and then at 120°C for 4 h before use.

Method B involved a solution that contained hydrolyzed $AlCl_3$ with OH:Al = 2.4 that was aged for 14 days. Then calculated amounts of the diluted silica sol, prepared in A, were added over a period of 24 h and dispersed for 2 h in an ultrasonic generator. Solutions were prepared with Si:Al ratios 0, 0.2, 0.5, 1.0, and 2.0, and aged for 7 days and then at 120°C for 4 h before use.

The PILCs obtained by both methods gave very similar products. Two basal spacings at 26 Å and 18.6–19 Å were observed for each sample. The latter values are very close to those obtained by pillaring with the Keggin ion. In addition, the fraction with the larger d-spacing was only stable at 400°C. At 500°C only one value in the range of 17.5–18.3 Å remained. Surface areas of samples heated to ~200°C were on the order of 260 m²/g. It appears from these results that there was no advantage in using the mixed oxide pillar over that of a pure alumina pillar, but as we shall see the authors claimed a greater degree of Brønsted acidity for the aluminosilicate pillars.

14.3.4 Surfactant-Assisted Pillaring

Michot and Pinnavaia [Michot and Pinnavaia, 1992; Michot et al., 1993] have claimed that the use of a nonionic surfactant Tergitol 155-5 added to the Al_{13} solution produces a more ordered pillared clay. The surfactant is incorporated into the clay producing a nonporous product. Heating to 500°C removed the organic and yielded a product with a 15.3-Å basal spacing rather than the usual 17.5-Å spacing. However, the surface area was slightly higher (305 m²/g) relative to a nonsurfactant assisted PILC (279 m²/g). Also, the micropore size distribution was narrower in the surfactant preparation exhibiting a bimodel distribution with maxima of 6 and 7 Å. In addition, the mesopore volume was greater in the surfactant PILC, which the authors claim may prove beneficial in catalytic reactions by assisting diffusion of the reactant to the active site. The surfactant was found to improve the hydrolytic stability of the intercalated Al_{13} species preventing the formation of a hydrolytically induced chlorite phase if the PILC is not calcined. Suzuki *et al.* [1991] found that addition of polyvinyl alcohol to Al_{13}-pillaring solutions increased the basal spacing and the amount of aluminum taken up by a fluorhectorite. Although basal spacings as large as 28 Å were reported, no surface-area or pore-size data were provided.

14.4 Characterization and Properties of Al_{13}-Pillared Products

Before proceeding to describe pillared clays in which the pillars do not contain aluminum, it is well to consider some properties of the PILCs described to this point. This description may then be used as the basis for comparison of the many varied types of PILCs that have been synthesized.

14.4.1 Acidity

One of the most important properties to consider in relationship to catalysis is the acid character of the PILC. Exchange of the Keggin ion for the interlamellar cations results in a charge neutral product. On heating, the water and hydroxyl groups of the pillar must split out protons to satisfy the negative charge of the layers as the pillar approaches the neutral oxide state. In the process, considerable Brønsted acidity should be generated. Early studies of the acid character of aluminum PILCs were conducted by the pyridine sorption IR technique [Occelli and Tindwa, 1983; He *et al.*, 1988]. Dried Al_{13}-pillared montmorillonite exhibits an IR band at 1453 cm^{-1} (Lewis acid sites) and a band of much weaker intensity at 1550 cm^{-1} resulting from sorption of pyridine at Brønsted acid sites. Hammett acidity titrations with butylamine indicate that all of the sites lie between H_0 values if 3.3 and -3.0 and the ratio of Lewis to Brønsted sites is 4. Both the Lewis and Brønsted acid bands decrease in intensity with temperature, but a drastic decrease in the Brønsted acid band occurs at 350°C. Above that temperature the 1550 cm^{-1} band rapidly goes to zero. This decrease is somewhat correlated with the tunneling of H^+ into the octahedral clay layer [Tennakoon *et al.*, 1986; Jones, 1988]. The protons may be largely (more than 80%) recovered by exposure of the montmorillonite to NH_3 [Molinard *et al.*, 1994]. Ammonium ions form and nest in the hexagonal cavities of the clay layers. This gives rise to an IR band at 1440 cm^{-1}. On reheating the ammoniated PILC the deammonation process begins at 200°C and is complete at 500°C as shown by the change in intensity of the band at 1440 cm^{-1}. At the same time, a new peak forms at 3745 cm^{-1} and increases in intensity with increase in temperature. This band is attributed to silanol groups. As deamination occurs, some of the protons apparently interact with the Si—O bridges of the clay layer with formation of Si—OH groups.

In the case of saponite a very distinct band is present at 1549 cm^{-1} indicative of Brønsted acid sites as well as two strong bands for Lewis acid sites [Chevalier *et al.*, 1994a]. These bands diminish in intensity as the temperature of desorption is increased, but are still present to a small extent after heating to 500°C. These data were obtained on samples that were preheated to 500°C. Thus, the pillared saponite would appear to have stronger acid sites of both types than similarly pillared montmorillonite. Lewis acid sites originate on the pillars as a result of the dehydroxylation process and on the layers by defect formation. Brønsted acid sites result from loss of protons by the pillars to form Al—OH—Si type groups on the layers [Chevalier, 1994b; Plee *et al.*, 1985; 1987]. These strong acid sites are represented by two hydroxyl stretching bands at 3735 and 3695 cm^{-1}. From this and earlier work we may generalize that tetrahedrally substituted clays will exhibit stronger Brønsted acid sites than octahedrally substituted clays and that these Brønsted sites are more thermally stable in the former clay types.

14.4.2 Pore Size and Structure

The pore size of a PILC is determined by the interlayer distance and by the lateral distance or density of the pillars. A simplified representation of the pores generated by pillaring is shown in Fig. 4. The interlayer distance of a PILC depends on the size of the pillaring agent, while the lateral distance is regulated by the charge density of the layers and the effective charge on the pillars. A typical X-ray pattern of an aluminum-pillared saponite is shown in Fig. 5; and the N_2 sorption–desorption isotherm, after calcination at 550°C, is shown in Fig. 6 [Clearfield et al., 1994]. The surface area (SA) for this sample as determined by the BET N_2 sorption method at -177°C is 328 m²/g. Approximately 87% of the SA arises from micropores with a size range of 7–12 Å. However, the exact characteristics of the pore texture depend on thermal treatment of the raw pillared product and will be discussed in Sect. 14.4.3. We have already shown in the introduction that a simple calculation based on the amount of Al_{13} intercalated shows that the average charge on the aluminum Keggin ion lies between 4 and 5. A more quantitative analysis has now been given [Chevalier et al., 1994a and b]. The pillared saponite described in Sect. 14.3.2 (prepared at pH 4.8 and calcined at 500°C) was found to contain 2.8 mmol of Al_{13} per gram of saponite. The cation exchange capacity of the clay was 0.9 meq/g; and assuming that all the aluminum is present as the Keggin ion, the average charge per ion is 4.2+. One unit cell contains one $O_{20}(OH)_4$ unit (MW 777.1) equivalent to a charge of 0.662 e⁻. Thus, the number of unit cells per Al_{13} pillar is 4.2/0.662 = 6.5. The unit cell dimensions of the saponite within the layer are a = 5.29, b = 9.17 Å, and the area of 6.5 unit cells is 315.3 Å². This area represents a square of 17.75 Å on a

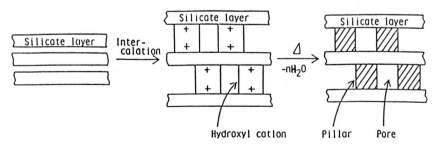

Figure 4. Schematic depiction of the pillaring process. (Reprinted from *Catal. Today*, 2, Kikuchi, E., and Matsuda, T., 297, 1988 with kind permission of Elsevier Science-NL, Sara Burgerhartstraat 25, 1055 KV Amsterdam, The Netherlands.)

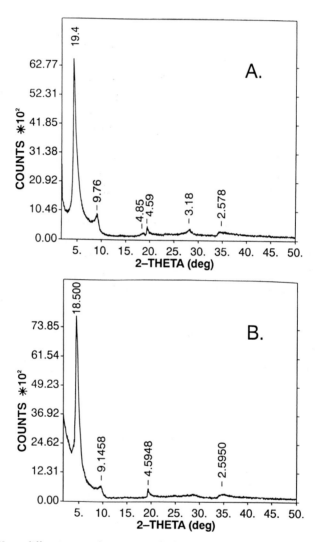

Figure 5. X-ray diffraction powder patterns of pillared saponite: (A) as prepared, (B) calcined at 500°C.

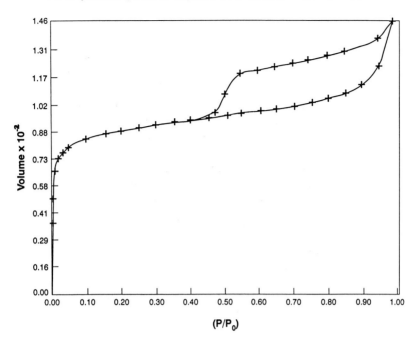

Figure 6. BET N_2 sorption–desorption isotherm for Al_{13}-pillared saponite preheated to 550°C (SA = 328 m²/g).

side. The Keggin ion can be considered to be a cylinder with a radius of 5.4 Å. Thus, if the pillars are placed at the corners of the square, the free space between them is ~ 7 Å. Dewatering the pillars at 500°C may add as much as 2 Å to this distance. Occelli *et al.* [1993] estimated the distance between pillars from atomic force microscopy (AFM) studies to be 6.6–8.7 Å.

The surface area of the top and bottom surfaces of the saponite layers is 659 m²/g. The layers are 9.4 Å thick and inclusion of this dimension raises the surface area to 751 m²/g. The area of the layers taken up by the Keggin ion, assumed to be a cylinder of 5.4 Å radius [Bottero *et al.*, 1982], is $(2.8 \times 10^{-3}/13 \times A) \cdot 2\Pi r^2 = 237$ m²/g. The increase in molecular weight on incorporation of the alumina (0.219 g/g of saponite) is 19% [Chevalier *et al.*, 1994b]. Thus, the expected surface area (751 - 237)/1.19 = 432 m²/g. The observed value by BET N_2 sorption was 425 m²/g. This calculation ignores the increased surface area arising from the length of the pillars. Furthermore, many Al_{13}-pillared clays have surface areas 100–200 m²/g lower. Thus, other factors must be operative.

Some interesting surface area data have been presented by Malla and Komarneni [1993] for montmorillonite and saponite (Table 2). The former clay mineral had a cation exchange capacity of 113 meq/100 g and the saponite value was 79 meq/100 g. However, reference to Table 2 shows a somewhat higher surface area and a higher pore volume for the pillared montmorillonite, whereas we would expect just the opposite. However, the two clays intercalated almost the same number of pillars, 0.08 mol for saponite and 0.083 mol for the montmorillonite. The average charge on the Keggin ion calculates to 5.66+ for montmorillonite and 4.38+ for saponite although both clays were pillared the same way.

Pore size estimates may be obtained from sorption of different sized molecules. Some early results obtained by Vaughan and Lussier [Vaughan and Lussier, 1980] for an Al_{13}-pillared montmorillonite is reproduced in Table 3. There is a falloff in absorption between molecules 7.6–8.0 Å in size. This cutoff agrees very well with pore size predictions. Sorption data presented by Occelli et al. [1985] for *n*-alkanes are shown in Table 4. The amount sorbed decreases as the length of the chain increases. Because one pore opens into another, the chains may still be accommodated in the interlamellar space even though they are longer than the free space in a single pore. However, the number of molecules sorbed is larger than the number of pores. Therefore, some of the molecules must be sorbed onto the outer faces and edges of the clay or more than one molecule enters the pores.

We have seen that the pore size is fairly uniform for a variety of clays pillared with the aluminum Keggin ion. To assess the effect of the layer charge on pore size one requires the same clay type with its layer charge varying from a low to a high value. This has been accomplished by ex-

TABLE 2

Nitrogen Adsorption Properties of Al_{13} and Mixed Silica–Titania (Si–Ti)-Pillared Montmorillonite (M) and Saponite (S)[a]

| | Surface area (m²/g) | | | | Surface area m²/g | | V_{micro} (cm³/g) | V_{total} (cm³/g) |
| | | | | | Micro | Meso | | |
	300°C	400°C	500°C	700°C	400°C	400°C	400°C	400°C
Al_{13} M	427	414	386	280	382	32	0.157	0.217
Al_{13} S	373	324	320	268	310	14	0.129	0.153
SiTi M	490	483	461	367	441	42	0.238	0.299
SiTi S	481	454	463	394	379	75	0.217	0.326

[a]After Malla and Komarneni, 1993, with permission.

TABLE 3
Pore Size of an Al_{13}-Pillared Montmorillonite as Determined by Sorption of Probe Molecules[a]

Probe molecule	Size (Å)	Sorption pressure (atm)	Amount sorbed (wt %)
n-Butane	4.6	0.79	8.0
Cyclohexane	6.1	0.079	8.4
Carbon tetrachloride	6.9	0.139	11.5
1,3,5-Trimethylbenzene	7.6	0.012	5.3
1,2,3,5-Tetramethylbenzene	8.0	0.009	0
Perfluorotributylamine	10.4	0.041	0

[a]Vaughan and Lussier, 1980.

changing NH_4^+ for all the exchangeable cations and then heating portions to different temperatures from 150 to 659°C. This treatment gradually reduced the cation exchange capacity from 68.5 to 3.6 meq/g. The samples were then pillared with ACH solutions to yield pillar densities of from 0.161 to 0.008 pillars per unit cell. The charge on the Keggin ions intercalated is the layer charge of the montmorillonite divided by the pillar density. The result was a constant value of $+3.15 \pm 0.10$. Unfortunately no surface area or pore size data were given.

TABLE 4
Comparison of the Sorptive Properties of H-ZSM-5 and ACH Bentonite[a]

Sorbate	Sorbate size (Å)[b]			Sorbate uptake (mmol g^{-1})	
	b	t	l	H-ZSM-5	ACH bentonite
n-C_6H_{14}	4.9	4.0	10.30	0.981	0.835
n-C_7H_{16}	4.9	4.0	11.51	0.774	0.558
n-C_8H_{20}	4.9	4.0	12.30	0.669	0.492
n-$C_{10}H_{22}$	4.9	4.0	15.24	0.650	0.401
Mesitylene	9.0	4.0	8.60	<0.10	0.542

[a]Occelli *et al.*, 1985.
[b]b = breadth; t = thickness; l = length.

The layer charge in montmorillonites can also be reduced by exchange of Li$^+$ followed by heating to transfer the Li$^+$ to the octahedral layer [Tennakoon et al., 1986; Jones, 1988]. It has also been found that Ni^{2+} can be fixed to the layers to reduce the layer charge [Suzuki et al., 1988]. However, in this case the surface area increased to a maximum of 310 m^2/g at a cation exchange capacity (CEC) of 1.0 meq/g and then decreased to ~ 200 m^2/g at a CEC of 1.4 meq/g.

Another factor affecting pore size is the drying process. Air-drying leads to a more crystalline product with relatively uniform pore sizes, whereas freeze-drying yields a more disordered product [Pinnavaia, 1984; Pinnavaia et al., 1984]. The latter process leads to more edge-to-edge and edge-to-face arrangements of the individual particles. Air-drying, in contrast, takes place more slowly and allows the particles to lie face-to-face or flat. As a result the pores are largely those resulting from the spacing of the pillars between the layers and only about 15% of the pores are mesopores, formed by other processes. Thus, shape selectivity in catalysis is more likely to occur with the air-dried and slowly heated PILCs. Of course, using a pillaring agent other than the Al$_{13}$ Keggin species produces different pore structures, but these cases will be discussed in Sect. 14.5.

Gil and Montes [1994] conducted a detailed study of the pore structure of several Al$_{13}$-pillared montmorillonites by obtaining nitrogen adsorption data at very low pressures, as well as adsorption–desorption isotherms up to P/P$_o$ = 1. They found that the microporosity could be characterized by Dubinin–Radushkevich plots and the Howarth–Kawazoe model for slitlike pores. Their analysis yielded a bimodel pore size distribution with maxima at a diameter of 8 Å (very narrow distribution) and 11 Å (relatively broad). However, there is a problem with this study. The aluminum solution pH was raised to a maximum of 4 with OH/Al = 2. This low pH leaves much of the aluminum in species other than the Keggin ion. In addition only about half the original sodium ion was exchanged leading to low levels of aluminum between the clay layers. As a result the pillaring process led to highly disordered PILCs containing a high level of mesopores. Although there is now great interest in mesoporous systems [Kresge, et al., 1992], no mesopore size distribution was presented.

14.4.3 Thermal and Hydrothermal Stability

The original intent in producing pillared clays was to provide a class of materials with pores in the 10–20-Å range. Such materials would allow the cracking of the gas oil and resid fractions of petroleum to supply the ever increasing gasoline market [Vaughan et al., 1979; 1980]. Early results were disappointing [Occelli and Tindwa, 1983; Vaughan and Lussier,

1980; Occelli *et al.*, 1985; Occelli, 1983; Occelli *et al.*, 1984]. Although pillared montmorillonites were found to be highly active in cracking heavy gas oils, they also exhibited high rates of coke formation. This proved to be a serious detriment to their use because on regeneration the layered structure usually collapsed [Occelli and Lester, 1984]. As a result much effort has been placed on increasing the thermal stability of the PILCs. An early treatment, as given by Occelli and Tindwa [1983], is reproduced as Fig. 7. The dashed line shows the weight losses for the unpillared bentonite (essentially montmorillonite). Interlamellar water is lost up to 150°C and dehydroxylation of the clay is in the range of 600–700°C. For the pillared bentonite pore water is lost first, but then the dewatering of the pillars occurs from ~ 150 to 675°C. There appears to be no separate weight loss due to the loss of the clay layer hydroxyls. The surface area of the PILC decreases slightly up to 525°C, but at higher temperature there is a rapid loss of surface area accompanied by about a 1-Å loss in interlayer spacing. At 700°C there is a total collapse of the layers.

Vaughan [1988] has given a relatively complete road map on obtaining more thermally stable products. We may summarize his results, which apply to Al_{13}-pillared montmorillonites, as follows (in order of increasing stability):

1. Heating the PILC in the presence of ammonia fixes the protons as ammonium ions followed by exchange of NH_4^+ by Ca^{2+}.
2. Treating the Al_{13} species with Mg^{2+} plus silicate anions produces a mixed aluminosilicate-pillaring polymer. The Mg^{2+} largely occupies the exchange sites after heat treatment.
3. Increasing the pH with ammonia further polymerizes the Keggin ion to a dimer, $[Al_{26}O_8(OH)_{52}(H_2O)_{20}]^{10+}$. This larger species creates stronger pillars less apt to collapse.
4. Add Mg metal and silicate ions to the Al_{26} species before pillaring.

The clays treated as previously described were always found to be more thermally stable in the ion-exchanged rather than the proton form. Not only thermal stability was improved, but also hydrothermal stability was improved. For example, at 625°C and 8-h treatment in 5-psi steam the Al_{13} PILC (SA = 270 m²/g) collapsed (SA = 20 m²/g) with the aluminum largely migrating to the clay surfaces. In contrast the PILCs prepared at pH 6 in the presence of magnesium and silicate exhibited higher surface areas (300–384 m²/g) and retained about two-thirds of their surface area after the steam treatment.

Another factor to consider is the destructive effect of the migrating protons on the montmorillonite layers. Apparently they combine with layer hydroxyls at lower temperatures to split out water. This is shown by a progressive loss of ion exchange capacity on heated and steamed PILCs. In contrast,

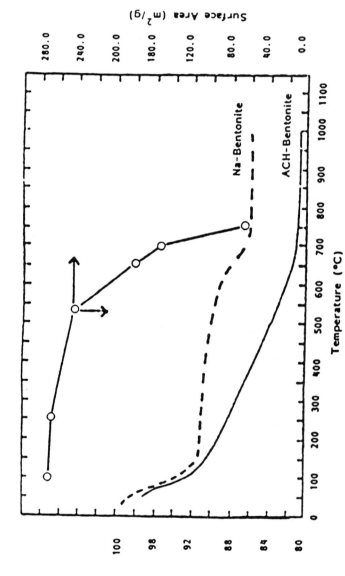

Figure 7. Thermogravimetric analysis of ACH bentonite showing the correlation between weight and surface area losses. (Reproduced from *Clays and Clay Minerals*, 31, Occelli and Tindwa, 22, 1983, with permission.)

saponite and beidellite are not subject to such self-destruction. By careful attention to the drying process and the thermal treatment it is possible to produce montmorillonites and saponites stable to 700°C without any special additions to the aluminum-pillaring solution. This is shown by the data in Table 2 and also by the data for saponite in Table 5 (Chevalier *et al.*, 1994b). It is evident that the pore structure has collapsed with the higher heating temperature. Our own work in progress [Tsai and Clearfield] confirms the results for montmorillonite: after heating at 700°C for 1.5 h the Al_{13} PILC maintained a basal spacing of 17.6 Å, but on heating to 750°C an XRD pattern of an amorphous solid was obtained. In contrast, the Al_{13} saponite maintained its integrity on heating to 750°C. The X-ray criterion is insufficient to indicate stability because the saponite sample PS 750°C in Table 5 has a basal spacing of 17.2 Å. Still the consensus is that the tetrahedrally substituted clays produce more stable PILCs. The increased stability is partly attributed to the stronger binding of the liberated protons to the outer tetrahedral oxygens where Al^{3+} substitutions for Si^{4+} have occurred. A second factor is thought to be the stronger binding or cross-linking of the alumina pillars to the layers.

Evidence for cross-linking of the Al_{13} pillars stems from NMR studies. Earlier work on beidellite and hectorite was conducted by Plee *et al.* [1985] and Fripiat [1988]. Briefly stated [Clearfield, 1988] chemical shifts occur in the ^{29}Si NMR spectra at about 400°C in the direction of increased concentrations of Si with one and two aluminum near neighbors. Pinnavaia *et al.* [1985] observed similar shifts for fluorhectorite. These shifts were interpreted as the onset of a cross-linking of the layers at silicate tetrahedral sites. The silica tetrahedra invert in the vicinity of an alumina pillar and form Si—O—Al linkages. A similar shift occurs to a smaller extent for the Si peak in hectorite and Gelwhite (octahedrally substituted clay minerals), but these shifts are the same as those that occur by tunneling of protons into the octahedral layer. Thus, cross-linking in the sense that new chemical bonds are formed, as for beidellite, does not appear to take place for octahedrally substituted clays. The mechanism of cross-linking involves protonation of the Si—O—Al link-

TABLE 5
Adsorption Data for the Initial Pillared Saponite (IS) and the PILC Heated to 500°C (PS 500) and 750°C (PS 750)[a]

Sample	Surface area (m^2/g)	V_{micro} (cm^3/g)	V_{total} (cm^3/g)	%V_{micro}
IS	425	0.158	0.234	67.5
PS500	403	0.149	0.235	63.4
PS750	145	0.044	0.090	48.9

[a]Reproduced from Chevalier *et al.*, 1994b, with permission.

age by protons generated in the dehydroxylation–dehydration of the pillars followed by inversion of the silicate group. Similar studies have now been conducted with saponite. Li *et al.* [1993] utilized a synthetic saponite with a high level of Al substitution ($Si_{6.53}Al_{1.47}$). The ^{29}Si NMR spectrum contained three resonances -84.5, -89.4, and -93.8 ppm representing Si—2Al, Si—1Al, and Si—0Al with intensities approximating 20:100:90. The pillared sample showed no interaction between the pillar and layer up to 400°C, but at 600°C significant changes occurred in the spectrum. The Si—1Al and Si—0Al at -89.4 and -93.8 ppm shifted to -90 and -95 ppm, respectively. The intensity of the Si—1Al peak decreased and the Si—0Al peak increased in intensity. The intensity of the Si—2Al peak at -84.8 ppm increased slightly and a small peak appeared at -79 ppm. This new resonance was assigned to Si(2 Als, Alp) where Als represents aluminum in the saponite layer and Alp aluminum in the pillar. Li *et al.* (1993) interpreted these shifts, which resemble those in beidellite, by the same cross-linking mechanism presented by Plee *et al.* [1987] and Fripiat [1988].

In another study by Lambert *et al.* [1994] the saponite had a much lower Al content (0.738 mol/O_{20}) and therefore the resonance at -84.5 ppm was missing. Nevertheless, chemical shifts similar to those found by Li were also observed for this natural saponite. These changes may also be correlated with the appearance of an IR band at $3594–3597$ cm^{-1} assigned to the protonation of Al—O—Si oxygens (Chevalier, 1994a). This protonation reaction is a necessary precursor to the inversion of the silicate tetrahedra followed by bonding to the pillars. In contrast to the clays with Al substitution in the tetrahedral sites, the octahedrally substituted clay minerals did not exhibit similar changes on heating. Thus, it is suggested that these pillars are not cross-linked with the layers, but instead undergo intense van der Waals interactions with layers because of the decreased basal spacing.

In utilizing PILCs for catalytic reactions, they may be exposed to acid or basic reagents and other aggressive environments. It is therefore of importance to inquire into the stability of the pillars to these conditions. In this connection we examined the behavior of an aluminum-pillared montmorillonite that had been calcined at 550°C. Portions of this PILC were treated with concentrations of HCl or NaCl varying from 0.005 to 0.100 M. These treatments resulted in solubilization of Al, increasing rapidly as the concentration of acid or base increased. Recalcination of the treated samples did not result in collapse of the pillars, but only in a decrease of 0.5 to 1 Å in basal spacing for the samples treated with 0.1 M reagent. The HCl-treated samples increased in surface area and pore volume, whereas the opposite effect was observed for the base-treated samples. The change in micropore size distribution with the different treatments is shown in Fig. 8 [Molinard *et al.*, 1994]. Both the smaller (less

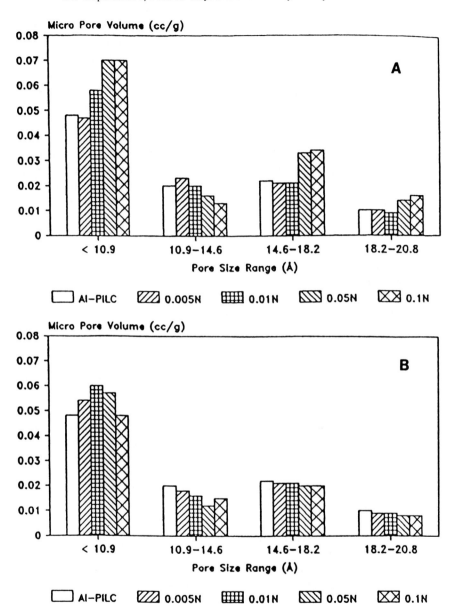

Figure 8. Discrete micropore size distribution of (A) acid-treated Al PILCs; and (B) base-treated Al PILC. (Reprinted from *Microporous Mater.*, 3, Molinard, A., Clearfield, A., Zhu, H. Y., and Vansant, E. F., Stability and Porosity of Alumina-Pillared Clay in Acid and Basic Solutions, 109–116, 1994 with kind permission of Elsevier Science-NL, Sara Burgerhartstraat 25, 1055 KV Amsterdam, The Netherlands.)

than 11 Å) and larger (14–20 Å) pores increased for the PILCs that were modified with HCl. The total micropore volume of the PILC samples modified in NaOH solutions first increased slightly and then decreased with increasing NaOH concentration.

In a separate experiment, an unpillared montmorillonite was treated with 0.1 M HCl. Only 0.065 mmol of Al per gram of swollen clay was extracted, compared to 1.43 mmol/g for the PILC. Thus, it is clear that the bulk of the aluminum extracted comes from the intercalated Al. We suppose that not only the Al_{13} Keggin ion but also smaller aluminum species exchanges with the clay. The effect of these fragments on pore size and blockage of access to the pore is shown in Fig. 9A. HCl treatment preferentially dissolves these fragments, creating more open pores as shown in Figure 9B. Although this model explains the results for the acid-treated Al PILC, it fails to provide an acceptable explanation for the base-modified samples. In the case of the NaOH treatment a much smaller amount of aluminum is solubilized and, instead of an increase, a decrease in surface area and micropore volume was observed. Moreover, X-ray diffraction (XRD) results show a larger decrease in the d_{001}-spacing. Thus, we may expect that a small breakdown in the structure occurs on NaOH treatment. Some pore blocking may occur due to formation of $Al(OH)_3$ solubilized and reprecipitated in the pores.

The ready solubilization of the aluminum pillars implies that calcination at 550°C does not convert the Keggin ions to a ceramic-like oxide. Instead the pillars remain reactive enough to readily interact with acids and bases. The extent of this reactivity must depend on the degree of rehydration or rehydroxylation that occurs after calcination. Evidence for the amount of rehydration was obtained for an Al_{13}-pillared saponite for which the X-ray pattern is shown in Fig. 5 [Clearfield et al., 1994]. Typical thermogravimetric weight loss curves for this saponite are shown in Fig. 10. The pillared, air-dried saponite lost 16.28% H_2O at 695°C. This weight loss includes water loosely held in the pores (10.9% at 200°C) and the water coordinated to the Al_{13} pillars together with the water formed by split out of pillar hydroxyl groups (200–695°C). A final weight loss occurs at temperatures above 695°C (2.46%) that arises from condensation of the layer hydroxyl groups. A sample of this pillared saponite was kept at 550°C for 3 h and then rehydrated by immersion in water for 24 h. Its thermal gravimetric analysis (TGA) curve is shown in Figure 10B. The weight loss to 695°C was 15.42% in this sample with 3.3% between the temperature of 200 and 695°C. Assuming that this 3.3% represents pillar hydroxyl condensation, it represents about a 62% recovery of the hydroxyl groups. Even when the sample was heated at 650°C, it rehydrated to about 41% of the original 5.34% weight loss. This rehydration is im-

Alumina Pillared Clay

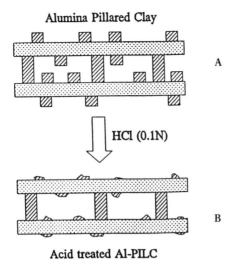

Acid treated Al-PILC

Figure 9. Schematic presentation of the opening of a "cage" due to the acid leaching of the alumina pillars. (From Molinard *et al.*, 1994, with permission.)

portant because it allows the pillars to participate in both cation and anion exchange reactions.

The reactivity of the pillars even after heating to high temperature explains why the Al_{13} PILCs are unstable to decoking processes. Pressurized steam would tend to favor rehydration of the pillars making them more susceptible to breakdown, especially if any residual acidity or basicity is present. Our hypothesis [Clearfield *et al.*, 1994] is that on calcination the pillars convert to a γ-alumina phase. This phase of alumina contains almost equal amounts of tetrahedral and octahedral Al. Evidence for its existence based on NMR spectroscopy is complicated by the presence of tetrahedral Al in the saponite layers. However, in the Al_{13}-pillared $NaCa_2Nb_3O_{10}$ perovskite the only aluminum present is that of the pillars. ^{13}Al NMR spectra taken at intervals of the heating process (Fig. 11) first resulted in an increase in tetrahedral Al until equal amounts of tetrahedral and octahedral aluminum were present at 350°C [Ram and Clearfield, in preparation]. At higher temperature the amount of octahedral aluminum increased markedly relative to the tetrahedral species. We interpret this result as indicating a slow transformation of the Keggin ion pillars into γ-Al_2O_3 followed by conversion to α-alumina. Some evidence for the transformation of the Al_{13} pillars into γ-Al_2O_3 is present in the NMR spectra provided by Chevalier *et al.* [1992]. Originally all the octahedral aluminum arises from the pillars, whereas the tetrahedral resonance stems mainly from the layers with a

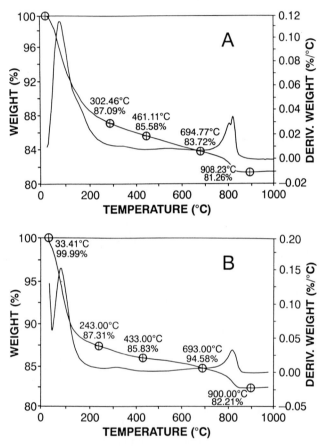

Figure 10. Thermogravimetric weight loss curves of (A) air-dried-pillared saponite and (B) the same sample heated at 550°C for 3 h and then rehydrated.

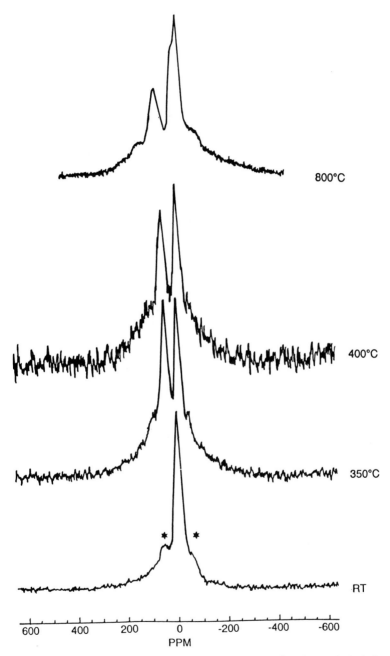

Figure 11. Solid-state MAS ^{27}Al NMR spectra of the alumina-pillared $[Ca_2Nb_3O_{10}]^-$-layered perovskite after heating to the indicated temperature.

small contribution from the pillars. On heating the PILCs the amount of octahedral Al decreases relative to the tetrahedral variety. This would be the case if the Al_{13} structure slowly transformed to γ-Al_2O_3. On rehydration the structure may not return to the Keggin ion structure, but to a hydroxylated γ-alumina such as γ-$Al_2O_x(OH)_{6-2x}$. Clearly, further studies are required to more fully elucidate the effect of thermal treatment on the pillared clays.

In Sect. 14.5 we shall describe PILCs in which the pillars are other than the Al_{13} species. We shall see that the choice of pillaring reagent and the method of pillaring have a profound influence on the pore structure.

14.5   PILCs Containing Non-Alumina Pillars

Many polyoxometal polymers have been used to pillar clays. Among the more important ones are those of zirconium, chromium, silicon, iron, and titanium.

14.5.1   Zirconium-Pillared Clays

Early work on the preparation of zirconium-pillared clays utilized solutions of zirconyl chloride, $ZrOCl_2 \cdot 8H_2O$, either at room temperature [Yamanaka and Brindley, 1979] or at reflux temperature [Bartley and Burch, 1985]. The latter technique leads to greater polymerization of the Zr, but yields more highly disordered PILCs. Much of this early work was summarized by Bartley [Bartley, 1988]. Interlayer spacings were on the order 16–22 Å with surface areas of 260–380 m^2/g. These surface areas were decreased by half on heating to 300°C, but collapse of the layers generally did not occur until 700°C.

In order to better understand the pillaring process we need to inquire as to the nature of the polymers formed by zirconyl chloride. The name, zirconyl chloride, is a misnomer and hails back to the time when it was thought that a zirconyl ion, ZrO^{2+}, exists. In actuality solid zirconyl chloride contains the tetrameric ion $[Zr(OH)_2 \cdot 4H_2O]_4^{8+}$. The metal ions form a square and are bonded to each other through double-ol bridges [Clearfield and Vaughan, 1956]. Each metal ion is also coordinated to four water molecules. The same ion exists in solution [Muha and Vaughan, 1960]. However, extensive hydrolysis occurs to relieve the high charge, resulting in highly acid solutions. Addition of base, aging, or heating the solution continues the hydrolytic process. Further polymerization can then occur by linking of tetramers, either in a plane [Clearfield, 1964] or by stacking [Fryer et al., 1970]. The exact nature of the polymers is unknown, but their size can be controlled to some extent.

Farfan-Torres *et al.* [1992] conducted the pillaring of montmorillonite by zirconyl chloride solutions in which they examined how variables such as temperature of pillaring solution, zirconium concentration, aging of the clay and the mixed solution, and washing procedure affected the final product. These workers confirmed the fact that pillaring with solutions kept at 100°C produced amorphous products. A temperature in the range of 40–60°C was appropriate. Too high a ratio of Zr to clay also produced broadened X-ray peaks indicative of disorder. Approximately 2 meq of Zr_4 per gram of clay was satisfactory. Aging the clay suspension or the pillaring mixture for up to 3 days did not change the outcome significantly. However, thorough washing of the reacted clay was absolutely necessary to achieve good pillaring. Otherwise no exchange of the Zr polymer with the interlamellar cations took place or partial exchange led to mixtures. Surface areas of ~ 200–225 m^2/g were obtained and these generally increased with heating to 300°C. Above this temperature the surface areas declined to about 100 m^2/g at 700°C. One interesting property is the dependence of CEC of the pillared clay on the amount of Zr intercalation. Values as high as 1.1 meq/g were obtained at a 40% ZrO_2 loading of the clay. This direct relationship of CEC to loading or density of pillars is a good indication that the ion exchange behavior stems from the pillars. Hydrous zirconia is a good anion exchanger [Ruvarac, 1982] as well as cation exchanger, but no attempt was made to determine anion exchange behavior.

Johnson *et al.* [1993] used a different approach. They pillared a tetrasilicic fluormica, $NaMg_{2.5}Si_4O_{10}F_2$, using a zirconium acetate of nomimal composition $ZrO(OH)_{0.5}(Ac)_{1.5}$. The reaction could be conducted at room temperature by mixing the reactants for 3 h. The best results were obtained with an excess of zirconium acetate (23 mmol/g of mica). Surface areas were generally 300 m^2/g and almost ideal type 1 isotherms were obtained. Furthermore, the PILCs so prepared showed excellent stability in high-temperature steam. A washed, calcined sample steamed at 700°C for 17 h had a surface area of 260 m^2/g (339 m^2/g original) and retained its 20-Å basal spacing. Prolonged steaming converted the pillars to tetragonal ZrO_2.

All the acetate-pillared products gave PILCs with a 22 Å basal spacing. Johnson *et al.* [1993] used a modeling approach to show that the zirconium tetramer would yield spacings of 16–17 Å. Therefore, they suggested that the more hydrolyzed $Zr_{18}O_4(OH)_{36}(SO_4)_{14}$ ion whose structure was recently reported [Squattrito *et al.*, 1987] serves as a model for the type of species present in basic zirconium acetate. By ionization of the acetate (sulfate) in solution a cation of large size becomes available for pillaring. The true nature of this species is unknown, but the acetate solution was at a relatively high pH (3) for zirconium solutions. Thus, it is not surprising that the major species would be a more highly polymerized basic cation of low charge.

Ohtsuka *et al.* [1993] used the same tetrasilicic fluormica as Johnson did, but pillared it with zirconyl chloride solutions. Reactions were run at room temperature (r.t.) both with unheated and refluxed (1-h) zirconyl solutions at a 5:1 zirconium to clay ratio. The zirconyl chloride solutions ranged from 0.1 M to 1.5 M. The mica pillared at r.t. in 0.1 M $ZrOCl_2 \cdot 8H_2O$ gave products with a 17.0-Å basal spacing, but they did not possess good temperature stability. However, when more concentrated solutions were used, a large shoulder emerged on the high angle side of the 001 reflection. This shoulder was attributed to the incorporation of smaller polymeric species than the tetramer, but the zirconium tetramer, which has dimensions of 8.98 Å in width and length and 5.82 Å in thickness, was the dominant pillaring species at high solution concentration. When the pillaring was conducted at reflux temperature with solutions 0.1–0.5 M in Zr, a PILC with a 23.3-Å basal spacing was obtained. Thus, a more polymerized Zr species must have been synthesized by the reflux procedure and this species produced the larger separation of the layers. This PILC attained its maximum surface area (BET N_2 sorption) of 300 m²/g at 600°C, a value that declined only slightly at 700°C (basal spacing of 19.5 Å). The solutions of higher concentration produced a PILC with a 21.3-Å basal spacing that reduced to 17.1 Å on heating to 500°C.

In one sense the results obtained by Ohtsuka *et al.* [1993] show that amorphous products or those of low basal spacing need not be obtained when using refluxed zirconyl chloride pillaring solutions. However, it may be that the fluorsilicate mica is more resistant to the highly acid solutions than montmorillonite is. Our own experience [Tsai and Clearfield, 1995] in preparing zirconia-pillared saponite is instructive in this regard. The results are shown in Table 6 and fall into two categories. PILCs prepared with very dilute solutions of zirconium and with a low ratio of Zr to clay produced products with a basal spacing somewhat larger than 19.5 Å. These PILCs had surface areas of ~ 300–325 m²/g, but were only stable to 650°C. The second type of PILC was prepared with more concentrated solutions and yielded products with much larger basal spacings (24.5–28.5 Å). When heated to 750°C, the basal spacings were reduced to 20.5–23.5 Å, but they retained high-surface areas. Elemental analysis showed that the PILCs with high basal spacings contained 1.5–2.6 mmol of Zr_4 per gram of clay whereas the PILCs that collapsed at 650°C contained between 0.4 and 0.5 mmol of Zr_4. Pillaring reactions were conducted either at ambient temperature or in the range of 60–80°C. Thus, it would appear that saponite in common with the fluorsilica mica is more resistant to attack by the acidity generated in the reaction.

TABLE 6
d-Spacings of Zr-Pillared Saponite with Different Synthetic Methods

Synthetic conditions	12A	12B	12H	12J	12M	12N
Conc of intercalant (meq/g)	2.5	4.0	20	20	30	20
pH of intercalant solution	1.50	1.30	0.88	0.88	0.70	0.88
Run temp (°C)	60	60	65	r. t.	70	75
Run time (h)	2.5	2	2.5	2.5	2.5	6
Aging time (h)						
Zr solution	45	45	90	90	83	70
meq Zr/g clay fixed	0.48	0.43	1.50	2.0	0.65	2.61
d-Spacing (Å)						
Dried at r. t.	19.67	19.85	26.16	25.82	24.87	28.5
Calc at 350°C	19.47	19.61	21.81	21.31	21.73	25.40
Calc at 550°C	18.54	18.61	20.69	21.02	20.30	23.54
Thermal stability	650°C	650°C	750°C	750°C	750°C	750°C
SA (m²/g) T (°C)	—	310 (450)	314 (450)	—	—	278 (350)
SA (m²/g) T (°C)	—	228 (650)	281 (650)	—	—	—

To explain the larger spacings they obtained, Ohtsuka *et al.* [1993] developed a theory of pillaring based on the hydrolytic polymerization ideas presented earlier by Clearfield (Clearfield, 1964; 1990). This mechanism is illustrated schematically in Figure 12A and B. A slow ordered joining of tetramer units takes place to form a layer. Two such layers join one on top of the other by an oxolation where hydroxyl groups in one layer combine with protons of another layer to split out water leaving oxo groups. Continuation of this process would eventually lead to tetragonal or cubic ZrO_2 having a fluorite structure. However, the process is terminated at an early stage by incorporation of the polymers between the clay layers. Intercalation of the tetrameric zirconyl ion would increase the basal spacing of the clay by 5.8 + 0.9 Å to allow for H-bonding of the polymer to the oxygens of the layer. This would result in a basal spacing of about 16.2 Å. Spacings of this size have been observed (Bartley, 1988). The removal of water by heating the ol bridges would determine the spacing, and this should amount to 5.2 Å or a basal spacing of 14.8 Å as observed by Ohtsuka *et al.* [1993]. Adding a second stack should increase the pillar size by about 3

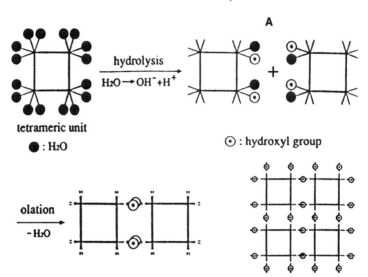

Figure 12. (A) Formation of the two-dimensional sheet polymeric species by oxolation between the tetrameric units. The solid-lined squares indicate the original tetrameric units and the dashed lines denote new ol bridges; (B) formation of (a) the three-dimensional polymeric species by oxolation between the tetrameric units, and (b) the fluorite structure of ZrO_2. (From Ohtsuka et al., 1993, with permission.)

Å ($Zr-Zr$ dis-tance in tetragonal ZrO_2). Thus, the basal spacing expected would be 19.3 Å in good agreement with our values at low loadings of Zr, but about 1 Å lower than the 20.4-Å PILC reported by Ohtsuka. A three-stack pillar should provide 25.3-Å spacings, which is close to our larger spacings. Additional modeling and determination of size reductions on heating are required in order to assess how well this theory fits or the direction in which it needs to be modified. Ohtsuka et al. [1993] found considerable amounts of Cl^- in their PILCs. This is not surprising because hydrous zirconia is an anion exchanger in acid solution (Ruvarac, 1982).

14.5.2 Titanium PILCs

The species formed in aqueous Ti(IV) solutions are less well characterized than the zirconium species are. No simple titanium ion has been isolated because the high charge to radius ratio would result in extensive hydrolysis. In fact, Ti(IV) solutions are more highly hydrolyzable than the corresponding Zr(IV) solutions. In crystalline $TiOSO_4 \cdot H_2O$ there are infinite zigzag $Ti-O-Ti-O$ chains. The prevalence of oxo rather than hydroxo groups is understandable because the high charge to radius ratio tends to displace protons from hydroxo groups.

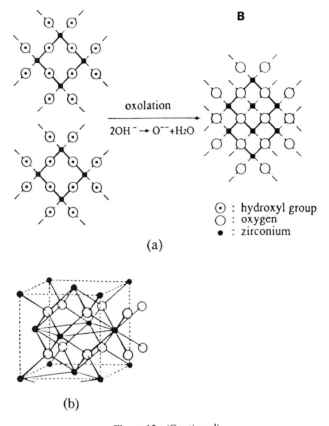

oxolation

$$2OH^- \rightarrow O^{--} + H_2O$$

⊙ : hydroxyl group
○ : oxygen
● : zirconium

(a)

(b)

Figure 12. *(Continued)*

Grange *et al.* [Bernier *et al.*, 1991; Castillo and Grange, 1993] prepared a titanium-pillared montmorillonite from a 4% aqueous suspension of the clay mixed with $TiCl_4$ solutions. The effect of pH, temperature, Ti(IV) concentration, and ratio of clay to Ti on the resultant PILC was assessed. Mixtures of unpillared montmorillonite ($d_{001} \cong 14$ Å) and a pillared clay ($d_{001} \cong 24.9$ Å) were almost always obtained. The best results were realized from a 0.82 M Ti(IV) solution that was 0.125 M in added HCl. The Ti to clay ratio was 10 meq/g and the temperature was 25°C. The maximum surface area was obtained at 400°C, 349 m^2/g, that reduced to 327 m^2/g at 600°C ($d_{001} \cong 21.6$ Å).

A different approach to pillaring with titanium was taken by Kijima *et al.* [1991]. They prepared a trinuclear titanium acetatochlorohydroxo complex and used it to pillar montmorillonite. The complex was prepared by adding 6.17 g of $TiCl_3$ to a mixture of 10 ml of ethanol and 5.8 ml of water to which was added 50 ml of acetic anhydride. This mix was stirred for 12 h and a black solid (2.9 g) of composition $[Ti(III)(CH_3COO)_{6.4}(OH)_{0.4}(Cl_{1.2}]Cl \cdot 11H_2O$ was recovered. An excess of this complex (10 g) was added to 100 ml of a 1% montmorillonite suspension and stirred for 24 h at room temperature. A two-phase product was obtained with basal spacings of 18.8 Å (Ti complex intercalated) and 12.2 Å (unpillared clay). When heated to 400°C, the latter spacing reduced to 10 Å as expected for an unpillared clay, but the intercalate expanded to 21.3–22.2 Å. This expansion was attributed to the fusion of two types of titanium complexes within the interlamellar space to form larger pillars. In the process oxidation to Ti(IV) took place. Surface areas of 105 and 128 m^2/g were obtained for two such samples. Additional information on titanium pillaring is given in Sect. 14.5.4.

14.5.3 Chromium PILCs

In early work [Brindley and Tamanaka, 1979] on pillaring of clays with chromium solutions, PILCs with small increases in basal spacing were obtained. These reactions were conducted at room temperature and it was pointed out that probably only polymers of a low degree of polymerization were present because of the slow rate of oligomerization at this temperature [Tzou and Pinnavaia, 1988]. Therefore, pillaring solutions were prepared by adding sodium carbonate to the chromium solutions and aging them at 95°C for up to 36 h. Solutions so aged and at pH values above 2 produced PILCs with a 27.6-Å basal spacing. The data in Table 7 are informative. The basal spacings were found to increase at any one temperature of preparation with the value of n, the ratio of carbonate to Cr(III) in the pillaring solution. In addition, the basal spacing increased with the increase in temperature at which the solutions were aged, and with the time of aging. It is clear from these data that the basal spacings track as the degree of polymerization. In order to maintain these large spacings on calcination, it is necessary to do so under vacuum or inert gas. Otherwise oxidation to Cr(VI) occurs with collapse of the layers.

Further insight into the pillaring process may be gained from the study by Drljaca *et al.* [1992]. Previous studies [Stunzi and Marty, 1983; Spiccia *et al.*, 1987; 1988] had shown that a dimer, a trimer, and tetramers (as shown in Figure 13) are present in hydrolyzed Cr(III) solutions. Solutions containing each of the species individually were prepared and allowed to react with a sodium montmorillonite (pH 2), first at ice temperatures and then at room temperature for 1 h. The integrity of the oligomer was established by reextraction and spectroscopic examination. The d_{001}-spacings

TABLE 7

Basal Spacings (nm) for Chromium Montmorillonites Prepared from Pillaring Reagent
Solutions Hydrolyzed under Various Conditions[a]

n^b	Hydrolysis temperature and time				
	25°C/3 weeks	60°C/2 h	95°C/1 h	95°C/6 h	95°C/36 h
0.0		1.45	1.338	1.60	1.63
0.5	1.45[c]	1.45	1.40	1.63	1.70
1.0	1.55	1.60	1.47	1.73	1.85
1.5	1.73	1.67	1.85	2.39	2.76
2.0	1.77	1.77	2.10	2.60	2.76
2.5	—	2.00	2.32	2.45	2.76

[a]From Tzou and Pinnavaia, 1988, with permission.
[b]meq CO_3^{2-} per mole of Cr(III) used in the hydrolysis reaction.
[c]All spacings are for air-dried samples prepared as oriented films on glass slides.

for the montmorillonite pillared with the dimer, trimer, and tetramer were 15.2, 14.6, and 15.7 Å, respectively. Either the dimer bonded to the layers with a split out of one water or it remained intact and assumed a tilted position relative to the layers. Similarly, the proposed structure of the trimer [Spiccia *et al.*, 1988] indicates a length of 6.5 Å and a depth of 5 Å. Because the d_{001}-spacing is 14.5 Å, the trimer must adopt a flat orientation with the long direction parallel to the layers. At pH 2 the tetramer solution contains roughly 70% of the open form (Fig. 13). Because an excess of chromium is used in pillaring, the higher charged, more abundant form would be expected to be intercalated. At pH 4.5 the tetramer is mainly in the closed form. The intercalation reactions conducted at pH 2 and 4.5 lead to basal spacings of 15.7 and 15.5 Å, respectively. The shortest dimension in either form of the tetramer is 6.5–7 Å and this fits the observed basal spacings. The addition of the fourth Cr octahedron increases the d_{001}-spacing by only 1 Å over that of the trimer-pillared PILC. When heated either in air or under nitrogen, the structures collapsed in the temperature range of 200–375°C. This experience is different from that reported by others [Tzou and Pinnavaia, 1988; Carrado *et al.*, 1986] where the pillared structures were maintained to 350°C in the absence of air. Also, dried chromium PILCs were able to swell when treated with glycol. Thus, Drljaca *et al.* [1992] believe that only one end of the chromium polymer bonds to the clay layers, allowing expansion in the presence of swelling agents. It is evident from these studies that the larger spacing

Chart I

1. DIMER

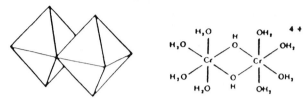

2. TRIMER

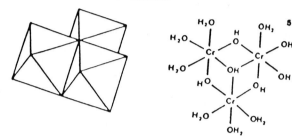

3a. OPEN TETRAMER

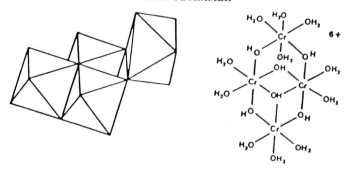

3b. CLOSED TETRAMER

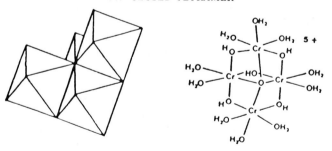

Figure 13. Representation of chromium hydroxy polymers by both octahedral and line drawings. (Reprinted with permission from *Inorg. Chem.*, 31, Drljaca *et al.*, 4894, Copyright 1992, American Chemical Society.)

products obtained by Tzou and Pinnavaia [1988] require much larger oligomers than the tetramer. Because the polymers tend to lie flat, at least an octamer would be required to achieve a spacing of 20–21 Å. Even larger interlayer spacings have been observed for chromium-pillared zirconium phosphates [Maireles-Torres *et al.*, 1991] and antimony phosphates [Wade, 1993]. Obviously, still larger polymers must be present in refluxed chromium acetate solutions. The use of such solutions in the pillaring of montmorillonite is described next [Jimenez-Lopez *et al.*, 1993].

The montmorillonite used for chromium pillaring had a CEC of 1.25 meq/g (calcined at 900°C), and the chromium acetate solutions varied in concentration from 9 to 366 mmol/liter. Each suspension was refluxed for 10 days before recovering the pillared clay. From 10 to 18.5 meq Cr^{3+} per gram was taken up with the high values being obtained with solutions in the range of 62.5–187 meq/g and solution concentrations of 0.091–0.274 mol/liter. Table 8 gives the compositions and basal spacings of the PILCs. Interestingly, the highest basal spacing was obtained at less than maximum chromium uptake. Given the amount of chromium incorporated into the clay samples, the low interlayer spacing is surprising. Remember that saponite intercalated 2.8 meq of Al as Al_{13} to yield a PILC with 18.5-Å spacing, whereas the montmorillonite took up six to seven times as much chromium. These amounts of chromium are much higher than the CEC and require a significant reduction in the charge per Cr in the pillaring species. The average charge per Cr can be calculated from the compositions listed in Table 8, on the assumption that all 1.25 meq of Na^+ per unit cell is exchanged.

The X-ray patterns of the chromium PILCs were of poor quality. This feature and the inability of the authors to derive empirical formulas for a single species led them to suggest the presence of more than one species in the interlayer space. The fact that the quantity of chromium intercalated does not correlate with the values of the basal spacing also argues in favor of this suggestion. Calcination either in air or under N_2 led to collapse of the pillars between 220 and 340°C. It was postulated that the release of protons on heating was responsible for the decomposition of the pillar. Calcination under NH_3 yielded microporous products stable to 625°C. The surface areas were on the same order of magnitude as those found by Tzou and Pinnavaia [1988] even though the basal spacings were smaller. Pore sizes ranged from 7.5 to 20 Å. They found that ther $Cr_{3.53}$ sample with a 21-Å basal spacing was able to sorb 0.74 mmol/g of perfluorotributylamine (kinetic diameter, 10.2 Å) and 1.56 mmol/g of neopentane (6.2-Å kinetic diameter). Because neopentane is a spherically shaped molecule, the pores must be at least 6.2 Å wide and more than 10.2 Å tall. Additional studies of this type are needed to obtain a more complete picture of the pore sizes.

TABLE 8
Preparative Conditions and Chemical Formulas of Chromium-Pillared Montmorillonite per unit of clay $[O_{20}(OH)_4]$ Using Chromium Acetate Pillaring Solutions[a]

Sample no.	Cr conc (mol/liter)	Meq Cr/g added	Meq Cr/g taken up	d(Å)	Formula
1	0.009	6.2	10.5	16.5	$Cr_{2.41}(OAc)_{0.45}(OH)_{8.98}(H_2O)_{6.25}$
2	0.027	18.7	15.1	18.9	$Cr_{3.46}(OAc)_{0.83}(OH)_{8.75}(H_2O)_{6.16}$
3	0.037	25.0	16.2	20.4	$Cr_{3.74}(OAc)_{1.04}(OH)_{9.38}(H_2O)_{4.76}$
4	0.091	62.5	18.5	17.7	$Cr_{4.36}(OAc)_{1.50}(OH)_{10.48}(H_2O)_{9.22}$
5	0.183	125.0	18.4	17.1	$Cr_{4.24}(OAc)_{1.68}(OH)_{10.24}(H_2O)_{7.42}$
6	0.274	187.5	17.1	16.4	$Cr_{3.94}(OAc)_{1.79}(OH)_{9.20}(H_2O)_{6.75}$
7	0.366	250.0	13.2	16.3	$Cr_{3.04}(OAc)_{2.18}(OH)_{6.14}(H_2O)_{3.42}$

[a]From Jimenez-Lopez et al., 1993, with permission.

14.5.4 Silica-Pillared Clays—DIMOS

The early work with pillared clays showed that the pores were in the general range of 6–12 Å. Although larger pores have been obtained by pillaring with zirconia and chromia, in 1988 these PILCs had not been well explored. Therefore, in an attempt to obtain larger pore spacings Moini and Pinnavaia [1988] attempted the direct intercalation of metal oxide sols (DIMOS). Silica sols diluted to 1% were allowed to react with aqueous suspensions of sodium montmorillonite. HCl was added to ensure that the sol particles were below the isoelectric point (pH $\cong 5$) and would be positively charged. The mixtures were kept at room temperature for 12–72 h. A second procedure was to dissolve tetraethylorthosilicate (TEOS), $Si(OC_2H_5)_4$, in alcohol and add it dropwise to an aqueous suspension of the clay held at pH 3. The evidence for pillaring is indirect because the X-ray patterns did not contain peaks at high d-values. However, the surface areas ranged from 259 to 450 m^2/g (silica sol) and 176 to 460 m^2/g for TEOS. By comparison of surface areas attainable with the pure silicas and those of the combined clay–silica composite, it was concluded that enlargement of the interlayer spacing must have occurred.

An extremely interesting pillaring procedure was developed by Yamanaka *et al.* [1992]. A small amount of a titania sol, prepared by hydrolysis of titanium tetraisopropoxide, $Ti(OC_3H_7)_4$, in 1 M HCl, was added to the silica sol. This SiO_2–TiO_2 mixed sol produced positively charged particles that readily exchanged with a sodium montmorillonite at 50°C. Basal spacings of 35–40 Å were readily obtained. These PILCs were stable to 500°C and gave surface areas of 250–400 m^2/g. If a pure titania sol were used in the pillaring reaction, the N_2 sorption–desorption isotherm would be of type IV. As the ratio of Si to Ti increased, the isotherm became more like a type I isotherm. In the titania-pillared montmorillonite [Yamanaka *et al.*, 1987] the basal spacings ranged from 24 to 27 Å and the isotherms indicated a high proportion of mesopores. In contrast, the isotherms for the high-silica-containing mixed sols, the pore sizes were in the range of 10–12 Å. This size range was established from sorption of molecules ranging in size from benzene to mesitylene. The adsorption isotherms for the large molecules fit the Langmuir linear plot, whereas those for smaller molecules (H_2O, CH_3OH) fit the BET plot. Yamanaka *et al.* [1992] proposed the model shown in Fig. 14 to explain the regular pore arrangement and size given the very large basal spacings.

We have now described a number of pillaring reactions in sufficient detail that the reader should have a good understanding of the processes involved. Many more variations have been tried, but our purpose is to be instructive instead of comprehensive. We have not discussed pillaring of

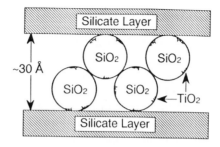

Figure 14. Schematic structural model of SiO_2–TiO_2-pillared clays. (From *Bulletin Chem. Soc. Japan*, 65, Yamanaka *et al.*, 2494, 1992, with permission.)

layered compounds other than clays, but significant progress has been made in that direction. Given the large number of layered materials available and the variety of pillars already utilized, a vast array of new materials is within the scope of our synthetic capabilities. Unexpected results are also a potential reward. For example, pillaring of zirconium phosphate with nickel acetate and mixtures of Ni—Co, Ni—Cu acetates produced PILCs that are ferromagnetic when calcined at 400°C [Shpeizer *et al.*, 1994]. However, because our concern here is catalysis, we will conclude with a treatment of that subject.

14.6   Catalytic Properties of PILCs

14.6.1 Catalytic Cracking Reactions

The original intent in creating PILCs was to provide a range of pore sizes such that heavy crude oils containing large refractory molecules could be processed by fluid catalytic cracking (FCC) methods. A constrained system such as that between the parallel plates of a clay and containing acidic sites similar to those in zeolites was felt necessary to create such suitable catalysts. Extensive examination of cracking reactions has been conducted by Occelli and Finseth [1986] and Occelli *et al.* [1986]. They showed that gas oil cracking by Al PILCs gives high selective yields of gasoline under moderate conditions. However, all of the catalysts exhibited high coke formation and instability to the decoking process. Occelli and Rennard [1988] also examined PILCs loaded with Ni and Mo both for cracking and as supports for these metals as hydrotreating catalysts. Some of their results are shown in Tables 9A and 9B. The pillared clays are much more reactive than the unpillared bentonite (largely a form of montmorillonite) and comparable to that of zeolite HY. However, the most active catalyst

TABLE 9A

Properties of Pillared Clays after Heating in Dry Air at 300°C for 10 h and Sizing to 100–235 Mesh Granules[a]

	ACH bentonite	ZACH bentonite	ALSI bentonite	Calcium bentonite	HY linde
Surface area (m²/g)	255.0	254.0	303.5	46.6	508
Pore volume (cm³/g)	0.16	0.17	0.21	0.08	0.32
Area % in pores with radius < 1 nm	88.0	90.0	87.7	25.2	94.9
1 < R < 10 nm	9.4	7.1	8.6	26.1	2.3
2 < R < 10 nm	2.4	2.4	3.2	29.1	1.4
> 10 nm	0.2	0.5	0.4	9.6	1.4

[a]After Occelli and Rennard, 1988.

was also the largest coke generator. In addition, as hydrocracking catalysts they were much less reactive than zeolite HY.

Cumene cracking can be used as a test reaction of the acidity and cracking ability of a particular catalyst. One group [He *et al.*, 1988] found that the activity of an Al_{13} montmorillonite PILC at 250°C was moderate, but increased from 9 to 11% conversion as the pretreatment temperature increased from 250 to 400°C. At pretreatment temperature greater than 400°C the activity declined to near zero. This effect was correlated with the disappearance of Brønsted acidity. We have already seen that Al PILCs heated to above 400°C essentially retain only their Lewis acid sites. In con-

TABLE 9B

Cracking Activity for Gas Oil Conversion before and after Pillaring[a]

	Calcium bentonite	ALSI bentonite[b]	ZACH bentonite[c]	ACH bentonite
Conversion (vol % FF[d])	28.4	67.9	73.8	82.1
Gasoline (vol % FF)	16.7	51.5	55.6	59.8
Furnace oil (vol % FF)	30.3	22.3	19.0	13.6
Slurry oil (vol % FF)	41.6	9.8	7.2	4.2
Hydrogen oil (wt % FF)	0.75	0.28	0.21	0.24
Carbon (wt % FF)	7.8	7.48	9.6	12.2

[a]Clays were heated at 400°C for 10 h in dry air and sized to 100–325 mesh before testing.
[b]ALSI stands for alumina–silica pillars.
[c]ZACH stands for an ACH solution containing Zr used as a pillaring solution.
[d]FF stands for fresh feed.

trast, reactions conducted with a group of silica–alumina mixed pillar PILCs [Zhao et al., 1992] exhibited conversions of 60 to 77% with high selectivity for benzene at 400°C. This level of conversion is still lower than to 90%+ achieved with zeolites.

However, the much greater activity of these Si—Al PILCs for cumene cracking is attributed to the greater number and stability of Brønsted acid sites. Yet Occelli and Rennard [1988] found that their Si—Al PILC was the least active for cracking reactions of those treated. This difference highlights the fact that careful attention to synthesis details is necessary. That is, it may be necessary to tailor a particular PILC as to its pore size, nature of its pillars, type and number of acid sites, and doping to achieve success for a particular reaction. Some efforts in this direction are discussed next.

14.6.2 Shape Selectivity

Shape-selective acid catalysis by PILCs was discussed by Kikuchi and Matsuda [1988]. Very few reproducible shape-selective reactions could be described. However, they were able to show some correlation between pore size of Al_{13} PILCs and production of 1,2,4,5-tetramethylbenzene by disproportionation of 1,2,4-trimethylbenzene. Differences in pore structure and shape selectivity were obtained by varying pillar heights or using different clays.

Shape selectivity or lack of it was demonstrated in a study of liquid-phase propene alkylation of biphenyl [Butruille and Pinnavaia, 1992]. The reactions were run at 250°C and 140 psi with an Al_{13} montmorillonite; an Al_{13} fluorhectorite; an aluminum-delaminated laponite; and an acid-leached montmorillonite and two zeolites, mordenite and ultrastable Y (USY). The most active catalyst was fluorhectorite (97% conversion), but it produced much higher levels of di-, tri-, and tetraalkyl products. The least active catalyst was the Al_{13}-pillared montmorillonite (A-PM). However, it was 60% selective for monoalkyl products. The product selectivity toward the three possible monoalkylated isomers was studied in the range of 50–80% conversion. The catalyst with few micropores (acid-washed clays) showed no selectivity toward the para- or meta-monosubstituted derivatives. In contrast, mordenite was selective toward the para-isomer (61%) with only 6% ortho-isomer produced. The A-PM catalyst exhibited results in between these extremes (17% ortho- and 43% para-isomer).

In a follow-up study [Butruille et al., 1993] it was shown that the Al_{13}-pillared hectorite exhibited very high acidity and activity for propene alkylation of biphenyl, but when preheated to 500°C this activity greatly declined. The loss of activity was attributed to removal of fluoride ion through interaction with water followed by further dehydroxylation between 350 and 500°C.

14.6.3 Alteration of PILC Properties

14.6.3.1 Stability

Almost all the initial cracking studies were done on Al PILCs or mixed oxide PILCs of which Al was an important component. Thus, efforts were directed toward overcoming the instability to high-temperature steam of these PILCs. We have already discussed methods of accomplishing more stable pillars provided by Vaughan [1988] in Sect. 14.4.3. Vaughan's methods appear to protect the pillars from extensive rehydration and structural changes associated with rehydration. We have also mentioned the very high stability of the zirconia-pillared fluorsilicic mica to high-temperature steam treatment (Sect. 14.5.1). We do not know whether this stability stems from the choice of the clay or the method of pillaring or both. It would be worthwhile to investigate these points further and also to conduct cracking reactions with this PILC. Finally, it should be remembered that beidellite and saponite are more thermally stable than montmorillonite or hectorite PILCs.

One of the concerns relative to stability was the behavior of the clay layers. This instability we have stressed arises from protons tunneling into the octahedral layer of octahedrally substituted smectites and the consequent irreversible condensation of the layer hydroxyl groups to water. Rectorite is a mixed-layer clay consisting of high-charge mica layers alternating with montmorillonite layers. The interlamellar ions of the mica layer are not exchangeable, so effectively the thickness or basal spacing is 18 Å. Pillaring of this clay mineral did indeed produce a more stable PILC [Johnson and Brody, 1993]. Surface areas of 150–175 m^2/g have been obtained with thermal and hydrothermal stability similar to zeolites. After steaming at 760°C for 5 h (100% steam, 1 atm) the pillared rectorites retain their pillared structure and activity.

Guan and Pinnavaia [1994] pillared a rectorite clay in the presence of polyvinyl alcohol using ACH solutions. They achieved basal spacings in the range of 42–52 Å when dried in air. Treatment at 800°C in 100% steam reduced the basal spacings by 10 Å, but they retained high surface areas and high activity for cracking of gas oil (62–64% conversions). This stability and activity after 800°C steaming is interesting in light of what we have already said about the transformation of alumina pillars under different treatment conditions. Therefore, further examination of the pillars producing these super galleries is in order.

14.6.3.2 Acidity

One potential method of increasing the acidity of the PILCs is by introduction of acidic species to the pillars. In this connection Farfan-Torres *et al.* [1992] introduced sulfate ions to zirconia-pillared montmorillonite. It is well known that sulfated zirconia behaves as a superacid [Arata and

Hino, 1990]. Although the source of the superacidity is still controversial, there is much evidence to show enhancement of Brønsted acidity in sulfated zirconia [Clearfield et al., 1994]. In fact, Farfen-Torres et al. [1992] found that after calcination of the sulfated Zr PILC at 400°C it had a Lewis/Brønsted (L/B) acid ratio of one as compared to two for the unsulfated PILC. This further compares to Al-pillared montmorillonite with L/B 4–8 to 1. Similarly, attempts to increase the acidity by adding phosphorus species to Al montmorillonites were also conducted [Shen et al., 1990]. Both H_3PO_3 and H_3PO_4 were added to the Al_{13}-pillaring solution to produce PILCs containing about 6% P_2O_5. Both the total acidity and Brønsted acidity were reduced in the PILC prepared from the phosphoric acid-treated solution. This reduction was attributed to its much lower surface area. In fact, this PILC was thermally unstable. However, on the basis of acid sites per unit area it was the most acidic of the three PILCs. Part of the problem lies in the lower pH of the solution containing H_3PO_4. The phosphorous acid sample had a significantly larger Brønsted acidity, but a decreased Lewis acidity. The acidity of the three PILCs for n-butanol dehydration correlated well with the Lewis acidity of the catalysts: the greater the Lewis acidity, the higher the percentage of conversion. This dependence on Lewis acidity may mean that the reaction is limited by the nucleophilic attack on the carbonation and not by the initial protonation step.

In our own studies we impregnated an Al_{13}-pillared montmorillonite, precalcined to 550°C, with ammonium phosphate [Cahill and Clearfield, 1995]. The pillars sorbed about 4.5% phosphate ion and in the process was reduced in surface area by about 80 m^2/g. Therefore, we treated a second sample first with sulfuric acid to increase the surface area as described in Sect. 14.4.3 and then impregnated the acid-treated sample with phosphate as before. The two changes in surface area offset each other, so that the final phosphated PILC had a surface area close to 250 m^2/g. Pyridine sorption studies indeed showed a higher L/B acid ratio than the untreated PILC. However, its activity in an isopropanol dehydration reaction was lower than that of the unphosphated PILC.

Zirconia-pillared saponite readily sorbed phosphate ion on impregnation [Tsai and Clearfield, 1995] with ammonium phosphate. ^{31}P NMR spectra indicated that ions of the type $H_2PO_4^{2-}$ were present on the pillar surface, but heating the PILCs shifted the peaks upfield, indicating water split out with formation of HPO_4^{2-}. Further characterization of these changes and of the effect on catalytic performance is in progress.

A different approach was taken by Mokaya and Jones [1995]. They acid-washed montmorillonite samples with different ratios of H_2SO_4 to clay and then pillared the acid-treated samples with ACH. This treatment removed from 14 to 48% of the octahedral ions for acid:clay ratios of 0.25:0.60. Very large increases in Brønsted acidity were also observed and attributed to enhanced acidity of the layers. The role of the pillars is to make these acid layer sites more accessible. Surface areas and pore values are shown in Table 10. The pillared but not acid-washed clay exhibited a much lower activity (19.5%) than the acid-washed and pillared samples (30–40%) for butanol dehydration (Table 11). All of the samples increased in activity with temperature reaching 82% for the Al PILC and 90–96% for the acid-washed catalysts. Cumene-cracking activity was also greatly improved, reaching 33% for the Al PILC and 35–46% for the acid-washed products at 400°C. Furthermore, the acid-washed PILCs were more highly selective for benzene.

14.6.3.3 Cation Modification of PILCs

An extremely useful way of modifying the behavior of PILCs is through ion exchange of a catalytically active cation. This step may or may not be utilized to reduce the cation to produce metal clusters. We illustrate with a few examples. Hydrocracking is a process that requires milder operating conditions than cracking, and coke formation can be prevented by use of sufficient hydrogen. To assess the utility of PILCs for this reaction Doblin *et al.* [1991] used an ammonium ion exchanged Al_{13} montmorillonite treated with a 0.16% solution of $Pt(NH_3)Cl_2$. The catalyst was then kept under oxygen while heating to 150°C and then to 350°C, holding this temperature for 3 h. The platinum was then reduced under H_2 at 300°C. This catalyst was examined for hydrocracking of *n*-octane and 2,2,4-trimethylpentane (224TMP). Conversions of 80–95% were achieved at 270–300°C. For *n*-octane the major reaction at 44% conversion was isomerization with cracking amounting to ~5% of the product. At higher conversions more cracking to C_4 and C_5 products was obtained. The opposite was true for 224TMP, the cracking reaction being dominant at all conversion levels. The catalyst showed very little deactivation over a period of 215 h. However, the product mix changed continuously over the first 30 h for the *n*-octane reaction, and then remained stable.

Hydrocracking of gas oils from synthetic crude was examined for a Ni—Mo catalyst supported on Al_{13} montmorillonite [Monnier *et al.*, 1993]. This catalyst was deemed superior to a similar one supported on zeolite–Y alumina because it utilized H_2 more efficiently and produced more liquids in the middle distillate and naphtha ranges.

Molina *et al.* [1994] examined four pillared clays for their activity in

388 / *Abraham Clearfield*

TABLE 10
Surface Area and Pore Volume of Pillared Materials[a]

Pillared clay	Surface area (m²/g)	Micropore area (m²/g)	Pore volume (cm³/g)	Micropore volume (cm³/g)
PPE	341	295 (87)[b]	0.20	0.12 (60)[c]
PPE25[d]	387	207 (54)	0.33	0.08 (24)
PPE30	374	160 (43)	0.34	0.06 (18)
PPE35	364	135 (37)	0.38	0.053 (14)
PPE40	353	122 (35)	0.39	0.048 (12)
PPE45	350	110 (31)	0.41	0.043 (10)
PPE60	315	44(14)	0.48	0.02 (4)

[a]From Mokaya and Jones, 1995.
[b]Figures in parentheses are precentage of micropore area.
[c]Figures in parentheses are precentage of micropore volume.
[d]PPE25 means pillared Peruvian montmorillonite, washed with 25% H_2SO_4.

the hydroisomerization and hydrocracking of decane. The catalysts were Al_{13}- and Ga_{13}-pillared montmorillonite, Al_{13} beidellite, and two transition metal-rich clays in which substitution occurred in both the octahedral and tetrahedral sites. The Brønsted acidity, as measured by the intensity of the IR band at 1440 cm^{-1} for the ammonia sorbed PILCs, was on the order:

TABLE 11
Butanol Dehydration Conversion over Various Clay Catalysts[a]

Clay	Conversion (%)		
	250°C	350°C	450°C
PPE	19.5	57.5	82.0
PPE25	37.5	78.1	95.3
PPE30	40.1	80.2	96.4
PPE35	32.4	76.5	92.7
PPE40	31.7	77.2	92.3
PPE45	31.0	75.4	91.7
PPE60	29.7	75.0	89.0

[a]From Mokaya and Jones, 1995, with permission.

$$Al\ PM \cong Ga\ PM < Al\ PN_4 < Al\ PN_7 < Al\ PB,$$

where PM is pillared montmorillonite, PB is pillared beidellite, and PN_4 and PN_7 represent the transition metal-substituted clays. These pillared clays were loaded to the 1% level with Pt. The catalytic activity was found to parallel the Brønsted acidity. For example, at the maximum decane isomerization yield temperature for Al PB (259°C), the conversion level was 79.2% with 77.9% selectivity for isomerization. The comparable values for Al PM were 60% conversion and 35.2% selectivity at 349°C. The catalytic activity of the Al_{13}-pillared beidellite was greater than ultrastable zeolite Y (USY) with Si/Al 15 and slightly less than USY with Si/Al 30. Zhao *et al.* [1993b] found that lanthanum-exchanged Al_{13} montmorillonite was more thermally stable than the unexchanged PILC. Although the acidity was decreased by lanthanum doping, it increased the cracking activity for cumene (400°C) and the selectivity for disproportionation of 1,2,4-trimethylbenzene. The addition of Pd and use of H_2 as a carrier gas were found to largely prevent deactivation of an Al_{13}-montmorillonite during disproportionation and alkylation of 1,2,4 trimethylbenzene [Kikuchi and Matsuda, 1988].

From the preceding brief discussion it is obvious that addition of cations to PILCs can profoundly influence the course of catalytic reactions. Several studies [Carrado *et al.*, 1986; Tokarz and Shabtai, 1985; Comets and Kevan, 1993; Bergaoui *et al.*, 1995] have been conducted on the siting and structure of cation-exchanged PILCs, but in many cases they have not been connected to catalytic behavior. More work along these lines needs to be done. For example, it was found [Occelli *et al.*, 1993] that during cracking of a vanadium containing crude the V was initially taken up in the micropores of a pillared rectorite (Al_{13} and Zr pillars). On steaming, the V caused decomposition of the pillars with collapse of the microporous structure. Thus, if the PILCs are to be used as FCC catalysts, the pillars must be protected against the destructive effects of metal incorporation.

14.6.3.4 Additional Catalytic Studies

A large number of catalytic reactions have now been examined with a variety of PILCs. It is not our purpose to discuss them all, but for the interested reader a partial list has been given by Molina *et al.* [1994]. We conclude by noting some unusual reactions catalyzed by PILCs. An Al_{13} montmorillonite impregnated with Ni^{2+} was presulfided to provide a catalyst that displayed equal or higher activity for the hydrodesulfurization of thiophene than for alumina- or carbon-supported NiS [Kloprogge *et al.*, 1993]. A titanium-pillared montmorillonite catalyzed the asymmetric oxidation of sulfides to sulfoxides in CH_2Cl_2 by tertiary butyl hydroperoxide at $-20°C$. Yields of 60–90% were obtained for a variety of sulfides.

14.7   Conclusions

Commercial applications for PILCs have been slow in coming. However, they represent a diversified range of materials with a very large number of possible variations: choice of layer, pillar, cation substitution, polymer–pillar combination, pre- and postpillaring treatments, etc. It is hard to imagine that useful applications will not arise from such a diversified class of materials. We need to learn how to better control the synthetic process, to elucidate the structures in more detail, and to better stabilize the composites. Additionally, uses outside the field of catalysis need to be explored (e.g., in the field of organic synthesis, sorption, and separation of gases, and as sorbents of organic toxic compounds). Some ideas along these lines have been explored in the book edited by Mitchell [1990]. Yamanaka and Makita [1995] were able to lithiate the titania pillars of a Ti–montmorillonite PILC to produce a rechargeable Li^+ battery with the PILC as cathode.

An even larger range of PILCs to be explored includes those that can be derived from non-clay-layered compounds. This literature is now expanding very rapidly and is sure to lead to materials with very interesting properties. However, that is a subject for a separate review.

Acknowledgment

The author wishes to thank the National Science Foundation, grants number DMR 9107715 and 9407899, for financial support of PILC-related research.

References

Akitt, J. W., and Farthing, A. (1981). *J. Chem. Soc. Dalton Trans*, 1617;1624.
Arata, K., and Hino, M. (1990). *Mater. Chem. Phys.*, 26, 213.
Bailey, S. W. (1980). *Crystal Structures of Clay Minerals and Their X-ray Identification* (G. W. Brindley, and G. Brown, eds.), Mineral Society, London.
Bartley, G. J. J. (1988). *Catal. Today*, 2, 233.
Bartley, G. J. J., and Burch, R. (1985). *Appl. Catal.*, 19, 175.

Bergaoui, L., Lambert, J. -F., Suquet, H., and Che, M. (1995). *J. Phys. Chem.*, **99**, 2155.
Bergaya, F., Hassoun, N., Barrault, J., and Gatineau, L. (1993). *Clay Miner.*, **28**, 109.
Bernier, A., Admaiai, L. F., and Grange, P. (1991). *Appl. Catal.*, **77**, 269.
Beson, G. A., Mifsud, A., Tchoubar, C., and Mering, J. (1974). *Clays Clay Miner.*, **22**, 379.
Bottero, J. Y., Tchoubar, D., Cases, J. M., and Fiessinger, F., (1982). *J. Phys. Chem.*, **86**, 3667.
Bradley, S. M., and Kydd, R. A. (1991). *Catal. Lett.*, **8**, 185.
Brindley, G. W., and Yamanaka, S. (1979). *Am. Miner.*, **64**, 830.
Burch, R. (ed.) (1988). *Catalysis Today*, Special Issue, Vol. 2, Elsevier, Amsterdam.
Butruille, J. -R., and Pinnavaia, T. J. (1992). *Catal. Lett.*, **12**, 187.
Butruille, J. -R., Michot, L. J., Barres, O., and Pinnavaia, T. J. (1993). *J. Catal.* **139**, 664.
Cahill, R. A., and Clearfield, A. (1995). ACS National Meeting, Anaheim, CA.
Carr, M. (1985). *Clays Clay Miner.*, **33**, 357.
Carrado, K. A., Suib, S. L., Skoularikis, N. D., and Coughlin, R. W. (1986). *Inorg. Chem.*, **25**, 4217.
Carrado, K. A., Kostapapas, A., and Suib, S. L. (1986). *Solid State Ionics*, **22**, 117.
Castillo Del, H. L., and Grange, P. (1993). *Appl. Catal. A.*, **103**, 23.
Chevalier, S., Suquet, H., Franck, R., Marcilly, C., and Barthomeuf, D. (1992). *Expanded Clays and Other Microporous Solids* (M. L. Occelli, and H. Robson, eds.), p. 32, Van Nostrand-Reinhold, New York.
Chevalier, S., Franck, R., Suquet, H., Lambert, J. -F., and Barthomeuf, D. (1994a). *J. Chem. Soc. Faraday Trans.*, **90**, 667.
Chevalier, S., Franck, R., Lambert, J. -F., Barthomeuf, D., and Suquet, H. (1994b). *Appl. Catal.*, **110**, 153.
Clearfield, A., and Vaughan, P. A. (1956). *Acta Crystallogr.*, **9**, 555.
Clearfield, A. (1964). *Rev. Pure Appl. Chem.*, **14**, 91.
Clearfield, A. (1988). *Surface Organometallic Chemistry: Molecular Approaches to Surface Catalysis* (J. M. Basset *et al.*, eds.), p. 271, Kluwer Academic, Dordrect, The Netherlands.
Clearfield, A. (1990). *J. Mater. Res.*, **5**, 161.
Clearfield, A., and Kuchenmeister M. (1992). *Supramolecular Architecture* (T. Bein, ed.), ACS Symposium Series 499, American Chemical Society, Washington, D. C.
Clearfield, A., Kuchenmeister, M. E., Wade, K., Cahill, R., and Sylvester, P. (1992). *Synthesis of Microporous Materials*, Vol. 2 (M. L. Occelli, and H. E. Robson, eds.), Van Nostrand-Reinhold, New York.
Clearfield, A., Aly, H. M., Cahill, R. A., Serrette, G. P. D., Shea, W. -L. and Tsai, T. -Y. (1994). *Zeolites and Microporous Crystals* (T. Hattori, and T. Yashima, eds.), p. 433. Kodansha/Elsevier, Tokyo.
Clearfield, A., Serrette, G. P. D., and Khazi-Syed, A. H. (1994). *Catal. Today*, **20**, 295.
Comets, J. -M., and Kevan, L. (1993). *J. Phys. Chem.* **97**, 12004.
Doblin, C., Mathews, J. F., and Turney, T. W. (1991). *Appl. Catal.*, **70**, 197.
Drljaca, A., Anderson, J. R., Spiccia, L., and Turney, T. W. (1992). *Inorg. Chem.*, **31**, 4894.
Farfan-Torres, E. M., Sham, E., and Grange, P. (1992). *Catal. Today*, **15**, 515.
Fripiat, J. J. (1988). *Catal. Today*, **2**, 281.
Fryer, J. R., Hutchison, J. L., and Paterson, R. (1970). *J. Colloid Interface Sci.*, **34**, 238.
Fu, G., Nazar, L. F., and Bain, A. D. (1991). *Chem. Mater.*, **3**, 602.
Gil, A., and Montes, M. (1994). *Langmuir*, **10**, 291.
Gonzalez, C., Pesquera, C., Blanco, C., Benito, I., and Mendioroz, S. (1992). *Inorg. Chem.*, **31**, 727.

Guan, J., and Pinnavaia, T. J. (1994). *Mater. Sci. Forum,* **152–153,** 109–113.

He, M. -Y., Liu, Z., and Min, E. (1988). *Catal. Today,* **2,** 321.

Herrero, C. P., Sanz, J., and Serratose, J. M. (1985). *J. Phys. C,* **18,** 13.

Herrero, C. P., Sanz, J., and Serratose, J. M. (1986). *J. Phys. C,* **19,** 4169.

Jimenez-Lopez, A., Maza-Rodriguez, J., Olivera-Pastor, P., Maireles-Torres, P., and Rodriguez-Castellon, E. (1993). *Clays Clay Miner.,* **41,** 328.

Johansson, G. (1960). *Acta Chem. Scand.,* **14,** 769.

Johnson, J. W., and Brody, J. F. (1993). unpublished research, Exxon Research and Engineering Company.

Johnson, J. W., Brody, J. F., Alexander, R. M., Yacullo, L. N., and Klein, C. F. (1993). *Chem. Mater.,* **5,** 36.

Jones, W. (1988). *Catal. Today,* **2,** 357.

Jones, J. R., and Purnell, J. H. (1993). *Catal. Lett.,* **18,** 137.

Kijima, T., Nakazawa, H., and Takenouchi, S. (1991). *Bull. Chem. Soc. Jpn.,* **64,** 1395.

Kikuchi, E., and Matsuda, T. (1986). *New Aspects of Spillover Effects in Catalysis* (T. Inui *et al.,* eds.), Elsevier Sci., Amsterdam.

Kikuchi, E., adn Matsuda, T., (1988). *Catal. Today,* **2,** 297.

Kloprogge, J. T., Welters, W. J. J., Booy, E., de Beer, V. H. J., van Santen, R. A., Geus, J. W., and Jansen, J. B. H. (1993). *Appl. Catal. A,* **97,** 77.

Kresge, C. T., Leonowicz, M. E., Roth, W. J., Vartuli, J. C., and Beck, J. S. (1992). *Nature (London),* **359,** 710.

Lahav, N., and Shabtai, J. (1978). *Clays Clay Miner.,* **26,** 107.

Lambert, J. -F., Chevalier, S., Franck, R., Suquet, H., and Barthomeuf, D. (1994). *J. Chem. Soc. Farad. Trans.,* **90,** 675.

Li, L., Liu, X., Ge, Y., Xu, R., Rocha, J., and Klinowski, J. (1993). *J. Phys. Chem.,* **97,** 10389.

Maireles-Torres, P., Olivera-Pastor, P., Rodriguez-Castellon, E., Jiminez-Lopez, A., and Tomlinson, A. A. G. (1991). *J. Mater. Chem.,* **1,** 739.

Malla, P. B., and Komarneni, S. (1993). *Clays Clay Miner.,* **41,** 472.

Michell, I. V. (ed.) (1990). *Pillared Layered Structures,* Elsevier Appl. Sci., New York.

Michot, L. J., and Pinnavaia, T. J. (1992). *Chem. Mater.,* **4,** 1433.

Michot, L. J., Barres, O., Hegg, E. L., and Pinnavaia, T. J. (1993). *Langmuir,* **9,** 1794.

Moini, A., and Pinnavaia, T. J. (1988). *Solid State Ionics,* **26,** 119.

Mokaya, R., and Jones, W. (1995). *J. Catal.,* **153,** 76.

Molina, R., Vieira-Coelho, A., and Poncelet, G. (1992). *Clays Clay Miner.,* **50,** 480.

Molina, R., Moreno, S., Vieira-Coelho, A., Martens, J. A., Jacobs, P. A., and Poncelet, G. (1994). *J. Catal.,* **148,** 304.

Molinard, A., Clearfield, A., Zhu, H. Y., and Vansant, E. R. (1994). *Microporous Mater.,* **3,** 109.

Monnier, J., Charland, J. P., Brown, J. R., and Wilson, M. F. (1993). *New Frontiers in Catalysis, Proceedings 10th International Congress Catalysis,* Budapest, Hungary.

Mott, C. J. B. (1988). *Catal. Today,* **2,** 199.

Muha, G. M., and Vaughan, P. A. (1960). *J. Chem. Phys.,* **33,** 194.

Newman, A. C. D. (1987). *Chemistry of Clays and Clay Minerals,* Monograph 6, Mineralogical Society, Longman, London.

Occelli, M. L. (1983). *Ind. Eng. Chem. Prod. Res. Dev.,* **22,** 553.

Occelli, M. L., and Tindwa, R. M. (1983). *Clays Clay Miner.,* **31,** 22.

Occelli, M. L., and Lester, J. E. (1984). *Ind. Eng. Chem. Prod. Res. Dev.,* **24,** 27.

Occelli, M. L., Parulekar, V., and Hightower, J. (1984). *Proceedings 8th International Congress Catalysis,* Berlin.

Occelli, M. L., Innes, R. A., Hure, F. W. W., and Hightower, J. W. (1985). *Appl. Catal.,* **14,** 69.

Occelli, M. L., and Finseth, D. H. (1986). *J. Catal.,* **99,** 316.

Occelli, M. L., Hsu, J. T., and Gayla, L. G. (1986). *J. Mol. Catal.,* **35,** 377.

Occelli, M. L., and Rennard, R. J. (1988). *Catal. Today.,* **2,** 309.

Occelli, M. L., Dominguez, J. M., and Ekert, H. (1993). *J. Catal.,* **141,** 510.

Occelli, M. L., Drake, B., and Gould, S. A. C. (1993). *J. Catal.,* **142,** 337.

Ohtsuka, K., Hayashi, Y., and Suda, M. (1993). *Chem. Mater.,* **5,** 1823.

Pinnavaia, T. J. (1983). *Science,* **220,** 365.

Pinnavaia, T. J. (1984). *Heterogeneous Catalysis* (B. L. Shapiro, ed.) Proceedings Symposium Industry-University Cooperative Chemistry Program (IUCCP), Texas A&M University Press, College Station, TX.

Pinnavaia, T. J., Tzou, M. -S., Landau, S. D., and Raythatha, R. H. (1984). *J. Mol. Catal.,* **27,** 195.

Pinnavaia, T. J., Landau, S. D., Tzou, M. -S., Johnson, I. D., and Lipsicas, M. (1985). *J. Am. Chem. Soc.,* **107,** 7222.

Plee, D., Borg, F., Gatineau, L., and Fripiat, J. J. (1985). *J. Am. Chem. Soc.,* **107,** 2362.

Plee, D., Schutz, A., Poncelet, G., and Fripiat, J. J. (1985). *Catalysis by Acids and Bases* (B. Imeliks *et al.,* eds.), p. 343, Elsevier, Amsterdam.

Plee, D., Schutz, A., Poncelet, G., and Fripiat, J. J. (1987). *Clays Clay Miner.,* **35,** 251.

Ram, R. A. M., and Clearfield, A., manuscript in preparation.

Ruvarac, A. (1982). *Inorganic Ion Exchange Materials* (A. Clearfield, ed.), CRC Press, Boca Raton, FL.

Shen, Y. -F., Ko, A. -N., and Grange, P. (1990). *Appl. Catal.,* **67,** 93.

Shpeizer, B., Poojary, D. M., Ahn, K. Runyan, C. E., Jr., and Clearfield, A. (1994). *Science,* **266,** 1357.

Spiccia, L., Stoeckli-Evans, H., Marty, W., and Giovanoli, R. (1987). *Inorg. Chem.* **26,** 474.

Spiccia, L., Marty, W., and Giovanoli, R. G. (1988). *Inorg. Chem.,* **27,** 2660.

Squattrito, P. J., Rudolf, P. R., and Clearfield, A. (1987). *Inorg. Chem.,* **26,** 4240.

Sterte, J. (1988). *Catal. Today,* **2,** 219 (Part of Ref. 3).

Sterte, J. (1991). *Clays Clay Miner.,* **39,** 167.

Stunzi, H., and Marty, W. (1983). *Inorg. Chem.,* **22,** 2145.

Suzuki, K., Horio, M., and Mori, T. (1988). *Mater. Res. Bull.,* **23,** 1711.

Suzuki, K., Masakazee, H., Masuda, H., and Mori, T. (1991). *J. Chem. Soc. Chem. Commun.,* 873.

Tennakoon, D. T. B., Jones, W., and Thomas, J. M. (1986). *J. Chem. Soc. Faraday Trans. 1,* **82,** 3081.

Tennakoon, D. T. B., Jones, W., Carpenter, T. A., Ramdas, S., and Thomas, J. M. (1986). *J. Chem. Soc. Faraday Trans. 1,* **82,** 545.

Tokarz, M., and Shabtai, J. (1985). *Clays Clay Miner.,* **33,** 89.

Tsai, T. -Y., and Clearfield, A. (1995), work in progress.

Turney, T. (1992). CSIRO, Melbourne, private communication.

Tzou, M. S., and Pinnavaia, T. J. (1988). *Catal. Today,* **2,** 243.

Vaughan, D. E. W., Lussier, R. J., and Magee, J. S. (1979). U.S. Pat. 4 176 090; (1980). 7th Canadian Symposium on Catalysis Preprints, *Chem. Inst. Canada,* 80.

Vaughan, D. E. W., and Lussier, R. J. (1980). *5th International Conference Zeolites,* Naples, Italy, Heyden, London.

Vaughan, D. E. W. (1988). *Catal. Today,* **2,** 187.

Wade, K. L. (1993). Ph.D. dissertation, Texas A&M University, College Station, TX.

Yamanaka, S., and Brindley, G. W. (1979). *Clays Clay Miner.,* **27,** 119.

Yamanaka, s., Nishihara, T., Hattori, M., and Suzuki, Y. (1987). *Mater. Chem. Phys.,* **17,** 87.

Yamanaka, S., Inoue, Y., and Hattori, M. (1992). *Bull. Chem. Soc. Jpn.,* **65,** 2494.

Yamanaka, S., and Makita, K. (1995). *J. Porous Mater.,* **1,** 29.

Zhao, D., Yang, Y., and Guo, X. (1992). *Inorg. Chem.,* **31,** 4727.

Zhao, D., Wang, G., Yang, Y., Guo, X., Wang, Q., and Ren, J. (1993a). *Clays Clay Miner.,* **41,** 317.

Zhao, D., Yang, Y., and Guo, X. -X., (1993b). *Mater. Res. Bull.,* **28,** 939.

CHAPTER 15

Microporous Metal–Oxygen Cluster Compounds (Heteropoly Oxometalates)

Synthetic Variables, Nature and Source of the Porosity, Catalytic Applications, and Shape Selectivity

J. L. Bonardet*, K. Carr, J. Fraissard*, G. B. McGarvey,
J. B. McMonagle, M. Seay, and J. B. Moffat

Department of Chemistry and the Guelph–Waterloo Centre for Graduate Work in Chemistry,
University of Waterloo, Waterloo, Ontario, Canada, N2L 3G1
*Laboratoire de Chimie des Surfaces, Université Pierre et Marie Curie, Paris, France

KEYWORDS: Heteropoly oxometalates and their derivatives, structure, synthesis, stability, acidity, microporosity, surface and catalytic properties, diffusivity, ion exchange, shape selectivity

15.1 Introduction

Heteropoly oxometalates (also known as metal–oxygen cluster compounds, MOCC) have been known for more than 150 years [1,2]. Although another form of oxygen polyanion exists, namely the isopolyanions $[M_mO_y]^{P-}$, the heteropoly ions $[X_xM_mO_y]^{q-}$ $(x \leq m)$ are those that contain two different elements in addition to oxygen. The present work is concerned with the latter species.

The stoichiometry of the heteropolyanions may be represented as $[X_xM_mO_y]^{q-}$ $(x \leq m)$, where M is usually molybdenum or tungsten or sometimes vanadium, niobium, or tantalum [1]. Pope has noted that the peripheral metal elements (M) in the heteropolyanions require rather special ionic radii and charges and the capability of forming $d\pi-p\pi$ M—O bonds, while in contrast a large number of elements have been employed as the central atom (X) [1].

15.2 Historical Background

What is now known as ammonium 12-molybdophosphate, $(NH_4)_3PMo_{12}O_{40}$, was first produced in 1826 by Berzelius as a yellow precipitate from the addition of ammonium molybdate to phosphoric acid [3]. This heteropoly oxometalate was employed by Svanberg and Struve in 1848 for use in analytical chemistry [4]. However, the analytical compositions of the heteropoly acids were not determined until the discovery of the tungstosilicic acids and their salts by Marignac in 1862 [5]. It is interesting to note that no tungstosilicate species not previously described by Marignac were found until 1974.

The early explanations of the composition of heteropolyanions were based on Werner's coordination theory advanced by Miolati and Pizzighelli [6] and developed by Rosenheim [7]. In 1929 Pauling challenged the early work of Rosenheim and proposed a structure for the 12:1 complexes based on an arrangement of 12 MO_6 tetrahedra surrounding a central XO_4 tetrahedron [8]. In 1933 Keggin showed, by X-ray diffraction, that the octahedra of the heteropoly anion in $H_3PW_{12}O_{40}\cdot5H_2O$ were linked by shared edges as well as corners [9].

Bradley and Illingworth [10] confirmed the structure in 1936. The next new polyanion structure, due to Evans, did not appear until 1948 [11]. It was not until the 1970s that the interest in the heteropoly oxometalates intensified. By 1980 a number of research groups were active in this area.

15.3 Structure

While many possible structures are known for the heteropoly anions, that elucidated by Keggin has been more commonly studied for its catalytic and surface properties. A number of isomers of the Keggin structure are known; of these the α form appears to be the most stable and as a consequence catalytic investigations have generally focused on this isomer. In what follows reference will be made to the α-isomer unless otherwise stated and the α notation will be omitted. For simplicity the anions with Keggin structures will be referred to as Keggin Units (KU).

The Keggin structure has a central atom (X) bonded to four oxygen atoms arranged tetrahedrally. Most of the surface and catalytic studies have employed P or Si as the central atom although, as noted earlier, many possibilities exist. The selection of P or Si as the central atom undoubtedly relates to factors of thermal stability, ease of preparation, and cost. The central XO_4 tetrahedron is surrounded by 12 octahedra with a peripheral metal atom (M) at each of their approximate centers and oxygen atoms at their vertices. For catalytic purposes the more common peripheral metal atoms are W, Mo, and V. The octahedra share oxygen atoms with each other and with the central atom. The 12 octahedra are arranged in four groups of three edge-shared octahedra, M_3O_{13}, which are linked by shared corners to each other and to the central XO_4 tetrahedron (Fig. 1). There are two types of bridging oxygen atoms, those connecting the central atom to the peripheral metal atoms and those joining each of the peripheral metal atoms. It should be noted that the peripheral metal atom is not found at the exact centers of the octahedron containing it but rather is displaced outward toward the exterior of the anion structure. The third type of oxygen atom in the Keggin structure is probably best described as termi-

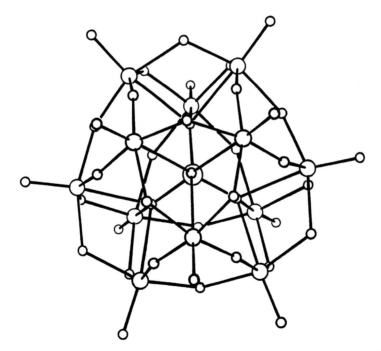

Figure 1. The Keggin Structure for the α-$[XM_{12}O_{40}]^{n-}$ anion. Bigger circles, central and peripheral atoms; smaller circles, oxygen atoms.

nal in the sense that it is bound to only one atom, the peripheral metal atom. The bond lengths for the latter fall in the range from 1.65 to 1.75 Å, while those bridging the peripheral metal atoms are approximately 2.40 ± 0.05 Å in length. Those for the X—O bonds are 1.60 ± 0.05 Å.

It is important to note that the anion, in spite of its cagelike appearance, is a packed structure. Workers in catalysis familiar with cages in zeolites may speculate concerning the possibility of small gaseous species entering the Keggin anion. In this context it should be noted that there is no evidence for such a possibility nor would such seem likely in view of the crystallographic data available for these anions, which demonstrates that the bond lengths of the interior pairs of atoms are similar to those normally found and consequently void space is not present within the anion.

Unfortunately, relatively few single-crystal crystallographic data are available for the acidic forms of the MOCC. However, an X-ray and neutron diffraction study of 12-tungstophosphoric acid has provided considerable valuable information, not only confirming Keggin's original analysis, but also locating the protons and water molecules in the solid [12]. Brown and co-workers investigated the six hydrate, that is, the solid in which two water molecules are associated with each proton (Fig. 2). Although a twofold thermal disorder of the water molecules evidently exists, each of these molecules is hydrogen bonded to the proton in a quasi-planar arrangement. Although not reported in the aforementioned publication, three or four water molecules may occupy hydrogen-bonded positions in the structure [13]. The water molecules are, in turn, hydrogen bonded through their hydrogen atoms to the terminal oxygen atoms of the anion. For 12-tungstophosphoric acid the protons and anions are arranged in a cubic (*Pn3m*) structure.

A recent single-crystal XRD investigation of 12-molybdophosphoric acid [14] has shown the structure of the $PMo_{12}O_{40}^{3-}$ anion to be consistent with that determined previously [15,16]. Both the cubic arrangement of the anions and the interstitial void spaces between the anions were clearly evident (Fig. 3), although the hydrate of 12-molybdophosphoric acid that was examined was found to belong to a face-centered, not a primitive, space group. It should be noted that a portion of the water molecules expected to be present could not be located.

Although the crystallographic structures show interstitial voids and thus appear to be open, the BET N_2 (78 K) surface areas for 12-tungstophosphoric, 12-molybdophosphoric, and 12-tungstosilicic acids ($H_3PW_{12}O_{40}$, $H_3PMo_{12}O_{40}$, and $H_4SiW_{12}O_{40}$, abbreviated as HPW, HPMo, and HSiW) are 8, 4, and 4 m²/g, respectively. Although water molecules within the interstitial voids would be expected to reduce the surface areas, anhydrous or quasi-anhydrous forms of the aforementioned acids do not show significantly higher surface areas.

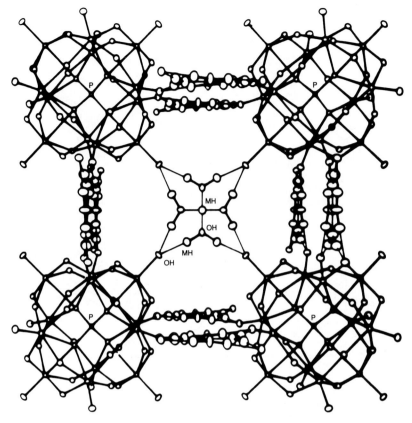

Figure 2. Crystallographic arrangement of protons, anions, and water molecules in hydrated 12-tungstophosphoric acid.[12]

15.4 Synthesis

The methods of synthesis for the common heteropoly oxometalates are generally simple in principle but, as so often happens in such cases, less simple in practice. The most common procedure employs the acidification of aqueous solutions of compounds containing the desired elements. The stoichiometric equation representing the preparative method for 12-tungstophosphoric acid exemplifies the apparent simplicity of the method:

$$12WO_4^{2-} + HPO_4^{2-} + 23H^+ \rightarrow (PW_{12}O_{40})^{3-} + 12H_2O.$$

Although, as has been noted, the phosphorus and silicon-containing Keggin structures are generally easily prepared, variables such as tempera-

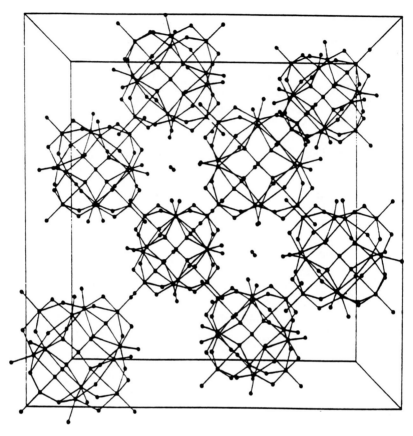

Figure 3. View down the 100 axis of $Ba^{2+}/PMo_{12}O_{40}{}^{3-}$. Reprinted from *J. Molecular Catal.*, 80, 59, G. B. McGarvey, N. Taylor, and J. B. Moffat, 1993 with kind permission of Elsevier Science-NL, Sara Burgerhartstraat 25, 1055 KV Amsterdam, The Netherlands.

ture, pH, excess of one reagent or the other, and the order of adding reagents may be important in preparing pure material.

While some heteropoly oxometalates appear to be unable to be isolated from solution, those of P or Si with Mo or W do not suffer from this problem. Indeed, their free acids can be crystallized from aqueous solution by the use of diethyl ether. Addition of diethyl ether to the acidified reaction mixture yields three layers: an upper ether layer, an aqueous layer, and a heavy oily etherate. The lowest layer is drawn off, shaken with excess ether to remove the entrained aqueous solution, and separated again. The etherate is decomposed by the addition of water followed by the removal of ether, and the aqueous solution of the heteropoly acid is allowed to evaporate until crystallization occurs.

Salts of the heteropoly acids are also in principle easily obtained from the parent acid and the appropriate cation-containing salt. However, as will be emphasized later, the salts prepared in this manner almost inevitably are nonstoichiometric in the sense of containing two cations, the one desired as well as residual protons.

15.5 Bulk and Surface Properties

15.5.1 Effects of pH

The importance of control of pH in the preparation and subsequent treatment of heteropoly anions may be illustrated from NMR and IR spectroscopic measurements in solution [17]. The ^{31}P NMR chemical shift is a convenient parameter for the assessment of the phosphorus environment in these anions (Fig. 4). For an aqueous solution of $WO_4{}^{2-}$ and $HPO_4{}^{2-}$, the ^{31}P chemical shift displays three plateaus as the pH is decreased from 10 to 0, indicating that changes in the environment of the central atom are occurring as the acidity increases. The lacunary or defect polyanion $PW_{11}O_{39}{}^{-7}$ forms from free $WO_4{}^{2-}$ and $HPO_4{}^{2-}$ between pH 8 and 7.5 and exists in solution as the pH is decreased to approximately 2.5. Between pH 2.5 and 1.0 several NMR peaks are observed in addition to those associated with $PW_{11}O_{39}{}^{-7}$ and $PW_{12}O_{40}{}^{-3}$ (Fig. 5). The peak at approximately 12.8 ppm is attributed to the $P_2W_{21}O_{71}{}^{-6}$ anion, known to be stable in acidic solution [1]. The peak associated with this species vanishes at pH 1. It is evident that the defect structure can be prepared from either the Keggin structure or the inorganic reagents employed in the preparation of the Keggin structure. When the pH was increased from 0 to 10, a set of NMR peaks similar to the aforementioned was observed at the same values of pH, thus demonstrating the reversibility of the process. Although not discussed here, the effects of pH on 12-molybdophosphoric acid and its constituent ions were also studied with ^{31}P NMR [17]. Results of ^{75}As NMR studies on the arsenic-containing heteropoly anion were reported (1990) [18]. The ^{75}As chemical shifts for 12-tungstoarsenic acid and 12-molybdoarsenic acid show a dependence on the nature of the peripheral metal atom that is similar to the dependence shown by the ^{31}P chemical shifts for 12-tungstophosphoric and 12-molybdophosphoric acids. The relatively narrow linewidths demonstrate the stabilizing effect that the surrounding M_3O_{13} trimetalate units exert on the tetrahedral arsenate species at the center of the Keggin anion.

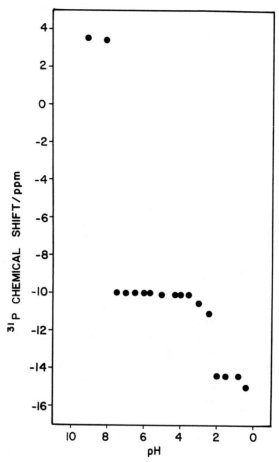

Figure 4. ^{31}P NMR chemical shifts of an aqueous solution of 12-tungstophosphoric acid for various values of pH. Reprinted from *J. Molecular Catal.*, 69, 137, G. B. McGarvey and J. B. Moffat, 1991 with kind permission of Elsevier Science-NL, Sara Burgerhartstraat 25, 1055 KV Amsterdam, The Netherlands.

15.5.2 Thermal Stability

The thermal stability of the heteropoly oxometalates is of considerable interest and importance, particularly in applications of these solids as heterogeneous catalysts. In view of the cagelike structures of the anions, it may be anticipated that degradation would occur at relatively low temperatures.

The heteropoly oxometalates of Keggin structure fortunately have characteristic spectra in the infrared region (Fig. 6) [19]. Unfortunately, however, as noted earlier the solid acids have low BET surface areas, of the

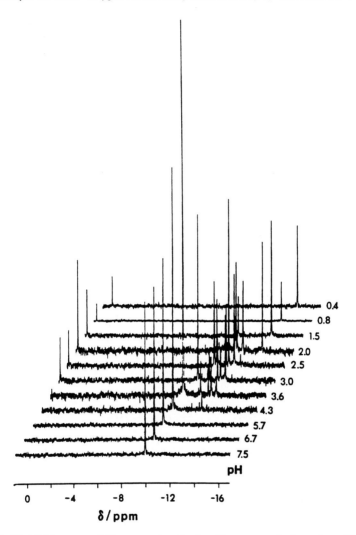

Figure 5. ^{31}P NMR spectra of 12-tungstophosphoric acid aqueous solutions after addition of various quantities of NaOH. Reprinted from *J. Molecular Catal.*, 69, 137, G. B. McGarvey and J. B. Moffat, 1991 with kind permission of Elsevier Science-NL, Sara Burgerhartstraat 25, 1055 KV Amsterdam, The Netherlands.

order of 10 m²/g. Consequently the use of pressed disks for spectroscopic studies is inappropriate. KBr and nujol mulls may be used but these may introduce unwanted interactions and should be avoided.

Photoacoustic spectroscopy (PAS) relies on a calorimetric mode of detection and therefore is particularly advantageous for application to strongly scattering and/or opaque materials [20,21]. It is surprising how

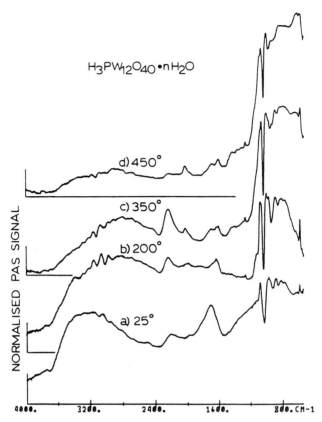

$H_3PW_{12}O_{40} \cdot nH_2O$

d) 450°

c) 350°

b) 200°

a) 25°

NORMALISED PAS SIGNAL

4000.   3200.   2400.   1600.   800. CM-1

Figure 6. PAS FTIR spectra of $H_3PW_{12}O_{40} \cdot nH_2O$ showing characteristic bands in the 1100–800 cm^{-1} region and the effect of heating *in vacuo* (°C).[19]

little this method has been used in catalytic studies. The PAS FTIR spectra of 12-tungstophosphoric acid ($H_3PW_{12}O_{40} \cdot nH_2O$) are shown in Fig. 6 after heating *in vacuo* to various temperatures [19]. A group of five or six bands appearing below 1100 cm^{-1} can be attributed to the heteropoly anion. Bands at 1080 and 980 cm^{-1} can be attributed to the triply degenerate asymmetric stretching vibration of the central PO_4 tetrahedron and the stretching vibration of the tungsten terminal oxygen bond, respectively [22]. These bands remain even at 450°C, although there is some slight diminution in intensities. Consequently it may be concluded that the heteropoly anion structure in 12-tungstophosphoric acid is largely intact at 450°C.

The thermal stability of the heteropoly oxometalates is dependent both on the composition of the anion and on the nature of the cation [23].

The tungsten-based heteropoly oxometalates are more stable than those containing molybdenum, although the anions containing only one peripheral metal element are generally more stable than those containing two of these elements. The salts are found to be more stable than the parent acids [23]. In particular, the ammonium salts appear to be the most stable for a given anion.

Unfortunately there has been little work reported on the products formed from the decomposition of the MOCC. However, there is evidence that the products consist primarily of the oxides of Mo(VI) and W(VI) [24].

15.5.3 Heteropoly Oxometalates as Three-Dimensional Intercalates

The heteropoly acids display a property similar to that which would be expected in a three-dimensional intercalate. Direct evidence for this has been obtained from photoacoustic FTIR spectroscopy [19,25,26]. When 12-tungstophosphoric acid, preevacuated at 200°C, is exposed stepwise to gaseous ammonia at 150°C, the PAS spectrum of HPW displays, as expected, the characteristic five to six bands in the $800–1100$ cm^{-1} range (Fig. 7). The bands at ≈ 3200 and 1420 cm^{-1} characteristic of the ammonium ion begin to develop as well as the sharpening of the bands produced by the anions. Ultimately a maximum of three molecules of NH$_3$ per anion is reached. The PAS evidence for the formation of the ammonium ion, together with the $1:1$ correspondence between the total number of protons present in the solid and the number in the NH$_3$ taken up, demonstrates that the ammonia molecules are penetrating into the bulk structure of the solid acid.

Similar PAS FTIR results have been obtained with pyridine, although unlike the observations with NH$_3$, evidence is found for the formation of protonated dipyridine [25].

On exposure of HPW to gaseous methanol at room temperature, as many as eight molecules of methanol per anion are sorbed, but on evacuation the number decreases to three (Fig. 8). The PAS FTIR spectra show that the methanol has been protonated and, as with ammonia and pyridine, has apparently entered the bulk structure of the solid to interact with the protons contained therein [26]. Further PAS evidence demonstrated that the protonation of methanol that occurred at temperatures as low as 25°C is the first stage in the production of higher hydrocarbons [27]. When the sample of HPW containing the sorbed and protonated methanol is heated, scission of the C—O bond occurs with the production of water and CH$_3{}^+$, the latter of which methylates the anion by bonding with its terminal oxygen atoms [27].

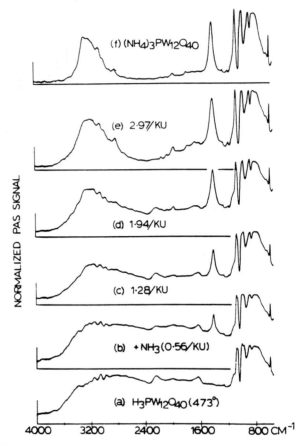

Figure 7. Effect of dosing NH_3 (molecules sorbed/KU) stepwise, (b)–(e), at 423 K on $H_3PW_{12}O_{40}$ preevacuated at 473 K (a); NH_3 salt for comparison (f).[19]

15.5.4 Acidity

Extended Hückel calculations [28–34] have shown that the magnitude of the negative charge on the terminal oxygen atoms of the anions is higher where the peripheral metal atoms are molybdenum than with tungsten (Fig. 9). In the absence of water it appears reasonable to assume that the interaction between the proton and the terminal oxygen atoms is primarily Coulombic in nature. Consequently the aforementioned calculated results predict that the proton in molybdenum-based heteropoly acid is more tightly bound than in those acids containing tungsten. Thus the proton in the latter case is more mobile and hence more acidic. Although direct and quantitative acidity data are difficult to obtain [35] (particularly for the colored molybdenum-containing heteropoly oxometalates), indirect

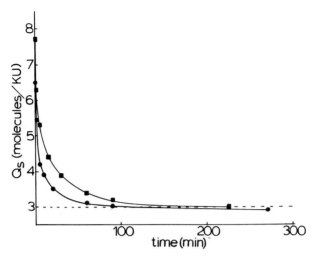

Figure 8. Desorption of methanol from 12-tungstophosphoric acid (preevacuated 330°C) dosed in excess at 25°C and evacuated at 25°C.[26]

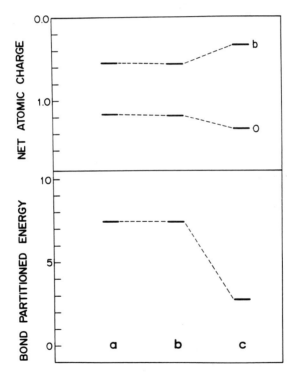

Figure 9. Results of extended Hückel calculations on heteropoly anions. *Top:* Magnitude of negative charges on bridging (b) and outer (o) oxygen atoms for (a) PW, (b) SiW, (c) PMo anions. *Bottom:* Binding energies between terminal oxygen atoms and peripheral metal atoms. Reprinted from *J. Molecular Catal.,* 26, 385, J. B. Moffat, 1984 with kind permission of Elsevier Science-NL, Sara Burgerhartstraat 25, 1055 KV Amsterdam, The Netherlands.

408 / J. L. Bonardet et al.

supporting evidence for the theoretical predictions comes from the studies of methanol conversion for which 12-tungstophosphoric acid is effective in conversions to hydrocarbons, while 12-molybdophosphoric acid yields largely oxidation products [36].

As has been reiterated [34,35], while a number of methods for the assessment of the acidity of solids are available, the correlation of these methods is complex and little success in such endeavors has been achieved. Thus infrared spectroscopy, nuclear magnetic resonance, titrations employing Hammett indicators, and calorimetry are among such techniques currently available, but the relationship between the results from these and the acidity parameters usually employed with, for example, homogeneous systems is tenuous. Indeed, many workers currently believe that, at least where heterogeneous catalysis is of principal concern, the use of catalytic reactions for the evaluation of the acidic properties of solid surfaces may be preferable [34,35].

The first direct evidence for the superacid properties of 12-tungstophosphoric and 12-tungstosilicic acids was obtained from microcalorimetric measurements of the differential heat of the sorption of ammonia (Fig. 10) [37]. The differential heat, after evacuation at 423 K, for less than 0.2 molecules of NH_3/KU sorbed, is approximately 180–190 kJ mol^{-1}. For larger amounts of NH_3 taken up, the differential heat remains approximately constant up to 1.25 NH_3/KU, at which the value drops to approxi-

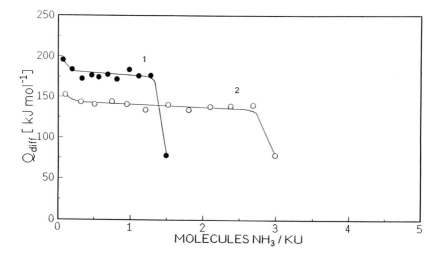

Figure 10. Differential heats of ammonia sorption on HPW at 323 K. (1) Activation: 423 K/2h/UHC; (2) activation: 523 K/1h/UHV. Reprinted from Microporous Materials, 1, 313, L. C. Jozefowiez, H. G. Karge, E. Vasilyeva, and J. B. Moffat, 1993 with kind permission of Elsevier Science-NL, Sara Burgerhartstraat 25, 1005 KV Amsterdam, The Netherlands.

mately 75 kJ mol^{-1}. In contrast, after pretreatment at 523 K the initial values for the differential heat are approximately 150 kJ mol^{-1}. With increases in the quantity of NH$_3$ sorbed, the differential heat decreases slightly but then remains virtually constant up to approximately 3 NH$_3$/KU at which point Q_{diff} decreases abruptly to approximately 75 kJ mol^{-1}.

It is interesting to note that other workers have obtained a relationship between the acid strength (H_0) and the differential heat of adsorption of ammonia on silica–alumina [38,39]. They reported that differential heats of 137.0 and 76.1 kJ mol^{-1} correspond to H_0 values of -14.5 and -5.6, respectively. Since a superacid is conventionally defined as any acid stronger than 100% H$_2$SO$_4$ ($H_0 < -12$) [40], it appears that HPW can be, at least on the basis of the microcalorimetric results, classed as superacid.

The microcalorimetric results suggest that the presence of water, presumably in the form of H$^+$(H$_2$O)$_n$($n \leq 4$) effectively guards the proton from interaction with sorbing species and therefore accounts for the precipitous drop in Q_{diff} at a quantity of NH$_3$ sorbed less than stoichiometric with the HPW pretreated at 423 K. The precipitous drop in Q_{diff} at 3 NH$_3$/KU with the HPW pretreated at 523 K supports the contention, derived from the PAS studies [19], that NH$_3$ is capable of penetrating into the bulk structure to interact with both surface and bulk protons. The higher adsorption energies found with the HPW pretreated at 423 K can be attributed to an inductive effect of the residual water on the acidic strengths of the accessible protons and/or quantitative alteration in the crystallographic structure, evidence of which is found from the decrease in the lattice parameter as the ammonium salt forms [19,37].

Semiquantitatively similar results have been obtained in measurements of the differential heat of ammonia adsorption of 12-tungstosilicic acid (HSiW) and 12-molybdophosphoric acid (HPMo), although the value of Q_{diff} observed at the plateau for HPMo is significantly smaller than that observed for either HPW or HSiW, consistent with the expectations from both theoretical [28–34] and experimental [36] studies.

Although somewhat tangential to the present discussion, it may be useful to note that surfaces of heterogeneous catalysts are expected to possess distributions of acidic strengths. These are particularly difficult to measure and the methods for so doing are small in number. Perhaps the most frequently utilized method for this purpose involves the titration of the acidic sites with a nonaqueous solution of a weak organic base in the presence of a set of surface-adsorbed indicators. As has been frequently noted, this method is not without its deficiencies [34,35].

15.5.5 Catalytic Properties in the Conversion of Methanol to Hydrocarbons

Evidence for the distinctive catalytic behavior of the salts of the heteropoly acids was first reported in 1983 from studies of the conversion of methanol to higher hydrocarbons [26,27,36,41–43]. Although the heteropoly acids were found to be active and selective in this process, the selectivity is strongly dependent on the peripheral metal element. While HPW yields significant quantities of linear and branched hydrocarbons containing up to six carbons, with HPMo large quantities of CO are produced, although the catalyst is quite active [36].

Surprisingly, at least at the time, the ammonium salt of HPW was found to be a more effective catalyst for methanol conversion to hydrocarbons than its parent acid. In comparison with the parent acid, HPW, the ammonium salt (abbreviated as NH_4PW) produced considerably higher yields of hydrocarbons [41,43] (Fig. 11). The total yield of C_1–C_5 hydrocarbons was 76.9% for NH_4PW at 350°C, while that for HPW was 32.4%, the more strongly acidic free acid showing evidence of larger quantities of irreversibly chemisorbed materials, primarily in the form of carbonaceous substances. In contrast to the results with HPW with which olefins were observed, the hydrocarbons produced on NH_4PW were largely paraffinic except for a small amount of ethylene (yield 2.0–2.8%). The major product was butane, in which the iso content was 86.5, 80.7, and 75.9% at 325, 350, and 400°C, respectively.

The surface area of NH_4PW begins to decrease between 300 and 400°C, while the amount of chemisorbed ammonia reaches a maximum at 400°C [41,43]. Acidic sites with pKa as low as -8.2 have been found on both the free acid and the ammonium salt. The surface concentration of acidic sites on NH_4PW of approximately 2×10^4 meq/cm^2 compares closely with data reported in the literature on other acidic catalysts.

15.6   Microporosity

The rather surprising results obtained with NH_4PW in the methanol-to-hydrocarbons process spawned a continuing project of preparing and studying the salts of the heteropoly acids. Because NH_4PW was found to have a significantly higher surface area, in contrast to that of the parent acid, studies of the conversion of methanol were carried out with a sufficiently high mass of HPW to mimic that of NH_4PW as used in the reactor [41–43]. The results show that, although the higher surface area of NH_4PW is important in rationalizing the aforementioned differences in the

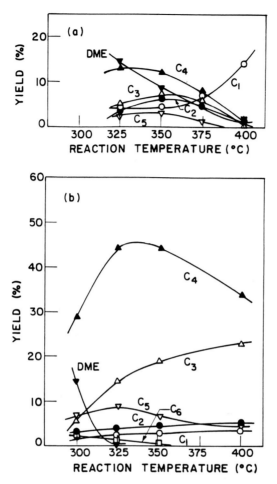

Figure 11. Product distribution from methanol conversion over (a) $H_3PW_{12}O_{40}$ and (b) $(NH_4)_3PW_{12}O_{40}$.[43]

results obtained with HPW and NH₄PW, it is evidently not the principal factor in explaining these observations.

15.6.1 Monovalent Cations from Group IA

The study of a number of salts prepared with monovalent cations has demonstrated that high surface areas are not restricted to NH₄PW [31,33,44–50]. Typical N_2 (78 K) adsorption–desorption isotherms are shown in Fig. 12 for several of the salts of HPW prepared with monova-

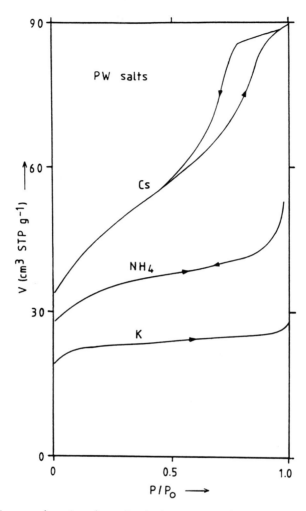

Figure 12. Nitrogen adsorption–desorption isotherms (78 K) for the high-surface-area cesium, ammonium, and potassium salts of HPW.[44]

lent cations. These display large uptakes of gas at very low relative pressures, typical of the behavior expected where micropores are present. The surface areas are found to be dependent primarily on the nature of the cation and secondarily on the composition of the anion. Of the anions studied the surface areas of the NH_4^+, Rb^+, and Cs^+ salts are all relatively high, as compared with those of the parent acids (Table 1). In addition, the surface areas of the potassium salts of the PMo, PW, and AsW anions are also relatively high.

TABLE 1
Surface Areas[a] of Heteropoly Oxometalates

Cation	Anion			
	$SiW_{12}O_{40}{}^{-4}$	$PMo_{12}O_{40}{}^{-3}$	$PW_{12}O_{40}{}^{-3}$	$AsW_{12}O_{40}{}^{-3}$
H	3.2	8	8	6.5
Na	1.3	3.5	3.7	—
Ag	—	—	3.0	—
K	3.3	39.9	90.0	46.0
NH_4	116.9	193.4	128.2	82.1
Rb	116.3	—	—	101.4
Cs	153.4	145.5	162.9	65.4
$MeNH_3$	3.0	1.3	3.0	2.1
Me_4N	—	—	4.5	—

[a]BET N_2 78 K; m^2g^{-1}.

It is interesting to note that high surface areas are not found with all salts prepared from monovalent cations. Neither the sodium nor the methylammonium salts possess high surface areas, the former being representative of smaller cations, the latter of larger cations.

The IUPAC-approved method for obtaining the surface areas of solids makes use of physisorption isotherms. These may be obtained with either a gravimetric or volumetric apparatus, although the latter is probably more common. Brunauer, Emmett, and Teller developed a multilayer adsorption theory and an equation that may be placed in a linear form,

$$\frac{P}{N(P_o - P)} = \frac{1}{CN_m} + \left(\frac{C-1}{CN_m}\right)\frac{P}{P_o}, \qquad (1)$$

where P is the pressure at adsorption equilibrium; P_0 is the vapor pressure of the adsorbate as a liquid at the adsorption temperature; N is the number of moles of gas adsorbed at the pressure P; N_m is the number of moles required to cover the solid with a single layer of adsorbed molecules; and C, a constant at the adsorption temperature, is proportional to the difference between the energy of adsorption in the first layer and the heat of liquefaction of the adsorbing gas.

Because the experimental measurables are N as a function of P, the left-hand side of Eq. (1) may be plotted versus P/P_0, the relative pressure. Although the data in this form are usually linear, the range of relative pressures over which the equation fits the data can vary considerably, depending on the nature of the solids studied.

The values of the BET C parameter are also found to be generally high for those salts having high surface areas (Table 2). This is not unexpected, because according to the BET multilayer theory, the C parameter is, for a given adsorbate and adsorption temperature, a function of E_1, the energy of adsorption of the first layer of adsorbate. However, since calculation of the C parameter requires a value for the intercept as obtained from the linear form of the BET equation, values of this parameter are subject to considerable error. The inaccuracies in the values of the intercept are further compounded with these materials because of the reduced linear ranges of the data when fitted to the linear form of the BET equation.

Micropore size distributions were calculated from the N_2 adsorption–desorption isotherms by application of a method due to Mikhail, Brunauer, and Bodor, usually referred to as the MP method [51]. Values for the thickness of the adsorbed layer as a function of the relative pressure for the corresponding nonporous solid are required for the calculation. To simulate these data, nitrogen adsorption–desorption isotherms were measured for the nonporous parent acids of the salts. Reference isotherms were also obtained by a method that related these to the value of the BET C parameter [52]. The resulting micropore size distributions calculated by either of the methods mentioned previously were found to be quite similar, and as a consequence, the subsequent calculations employed reference isotherms obtained by the latter method.

The pore size distributions are typically unimodal with a sharp but relatively broad peak (e.g., Fig. 13). The average pore radii were calculated from

TABLE 2
BET C Parameter for Heteropoly Salts

Cation	Anion			
	$SiW_{12}O_{40}^{-4}$	$PMo_{12}O_{40}^{-3}$	$PW_{12}O_{40}^{-3}$	$AsW_{12}O_{40}^{-3}$
H	100	—	—	58
Na	473	142	18	—
Ag	—	—	101	—
K	127	1070	2430	22
NH_4	1919	877	760	1292
Rb	489	—	—	149
Cs	—	568	1720	1359
$MeNH_3$	—	—	261	292
Me_4N	—	—	113	—

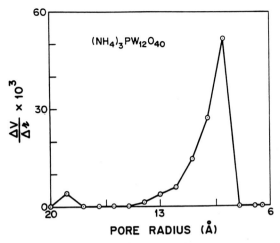

Figure 13. Micropore size distributions for $(NH_4)_3PW_{12}O_{40}$.[44]

$$\bar{r} = \frac{\sum (\Delta V / \Delta r) r}{\sum (\Delta V / \Delta r)}. \tag{2}$$

Further information on the microporosity may be obtained by fitting the adsorption data to the finite layer BET equation,

$$N = \frac{N_m C_{BET} x [1 - (n + 1)x^n + nx^{n+1}]}{(1 - x)[1 + (C_{BET} - 1)x - C_{BET} x^{n+1}]}, \tag{3}$$

where x is the relative pressure P/P_0 and n is the number of adsorbed layers. The values of n obtained in this way fall in the range from 1.5 to 5.0 for the high area salts (Table 3), which corresponds to pores of approximately 5–17 Å, again providing corroborative evidence for the presence of micropores.

15.6.2 Ion Exchange Properties

The ion exchange properties of microporous solids are of considerable interest. It is well known that the zeolites show such capabilities. Zeolite A, for example, in its sodium form has pores of approximately 4 Å in diameter, that are enlarged to 5 Å on exchange of the sodium by calcium ions. The ion exchange properties of the microporous salts of the heteropoly acids are of similar interest. Two questions are pertinent. Is ion exchange of the cations in these solids possible and to what extent? Is the structure—crystallographic and morphological—retained if such ion exchange occurs? Both questions may be answered in the affirmative. Exchange of a monovalent cation is possible, although com-

TABLE 3
Optimized n Values from Finite Layer BET Equation

Cation	Anion			
	$SiW_{12}O_{40}^{-4}$	$PMo_{12}O_{40}^{-3}$	$PW_{12}O_{40}^{-3}$	$AsW_{12}O_{40}^{-3}$
K	—	1.8	1.5	2.1
NH_4	1.6	2.6	2.0	2.3
Rb	1.7	—	—	1.6
Cs	2.1	5.0	3.0	2.0

plete exchange appears to be difficult, if not impossible, with retention of both the microporous structures and high surface areas. Microporosity and surface area are modified, consistent with the size of the exchanged-in cation [34,53–57]. For example, as the potassium ion is exchanged into NH_4PW, both the surface areas and lattice parameters that were applicable to NH_4PW decrease as they approach those for KPW (Figs. 14 and 15). Although it is clear that ion exchange with structure retention is possible, it should be noted that the extent of this exchange appears to be dependent upon the relative sizes of the entering and leaving cations. The maximum exchange capacity decreases as the radius of cation in the solid phase increases so that, for example, the exchange of the cesium ion from the solid to the liquid phase by the ammonium or potassium ion is relatively difficult, whereas the potassium and ammonium ions can be exchanged from the solid by cesium ions with relative ease. The ammonium and potassium ions, differing in radii by only

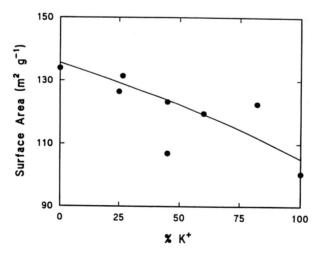

Figure 14. BET surface area as a function of the cation composition for the ion-exchanged $K^+/NH_4^+/PW_{12}O_{40}^{-3}$ system.[55]

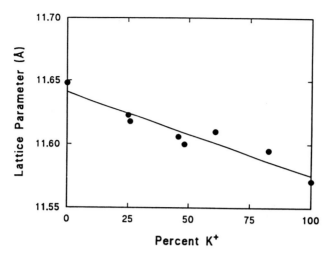

Figure 15. Lattice parameter as a function of cation composition for the $K^+/NH_4^+/PW_{12}O_{40}^{-3}$ ion-exchange system.[55]

0.1 Å, can be interchanged equally readily but to an extent of approximately 50%.

15.6.3 Evaluation of the Microporosity from ^{129}Xe NMR

Confirming evidence of the existence of a microporous structure in the salts of the heteropoly acids prepared from monovalent cations has been obtained from ^{129}Xe NMR [58,59]. Xenon isotherms were obtained with a classical volumetric apparatus at 300 K and *in situ* at 223 and 273 K for the ammonium and cesium salts of 12-tungstophosphoric, 12-molybdophosphoric, and 12-tungstosilicic acids as well as the potassium salt of the first acid. Although small amounts of xenon were adsorbed, the NMR signal is readily detected and for all adsorption temperatures and equilibrium pressures. The NMR spectra show only one remarkably narrow signal (Fig. 16). The plots of chemical shift versus the quantity of xenon adsorbed were linear for all samples and the slopes are inversely proportional to the void volume. The strength and narrowness of the signals and the linearity of the $\delta = f(n_{Xe})$ plots strongly suggest that the microporosity is "organized," homogeneous, and as closed as that found in zeolites. The three ammonium salts appear to have similar pore openings of 9 Å, regardless of the composition of the anion, whereas those for the cesium salts are found to be dependent on the nature of the anion. Because the lattice parameters for the ammonium salts fall in the range from 11 to 13 Å, and conditional on the validity of the source of the microporosity discussed later, the pore sizes obtained from Xe^{129} NMR appear to be more reasonable than those

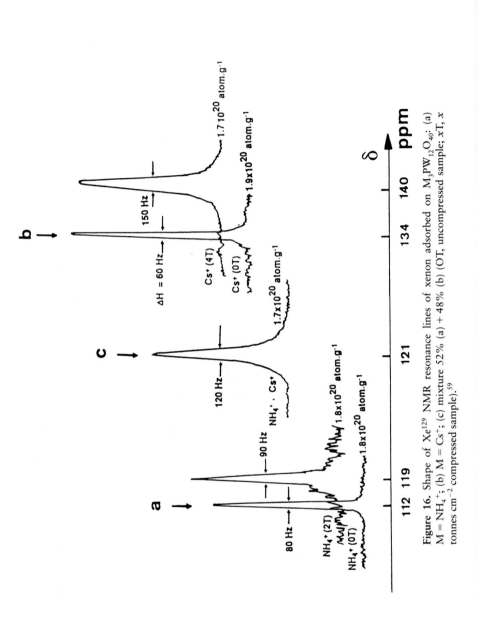

Figure 16. Shape of Xe^{129} NMR resonance lines of xenon adsorbed on $M_3PW_{12}O_{40}$: (a) $M = NH_4^+$; (b) $M = Cs^+$; (c) mixture 52% (a) + 48% (b) (OT, uncompressed sample; xT, x tonnes cm^{-2} compressed sample).[59]

calculated from the N_2 adsorption–desorption isotherms. However, the slopes of the chemical shift versus xenon-adsorbed plots for the three ammonium salts correlate well with the volumes of the micropores obtained previously from analysis of the N_2 adsorption–desorption isotherms (Fig. 17).

15.6.4 Diffusion and Diffusivities in the Parent Acids and the Microporous Salts

Comparison of the results of measurement of the sorption capacities for, and diffusivities of, various organic molecules has provided further information on both the parent acids and their microporous derivatives [60–63]. The diffusivities were calculated from data obtained gravimetrically in a controlled atmosphere, controlled pressure system. The diffusion equation of Fick can be truncated and differentiated to show that

$$\lim_{t \to 0} \frac{d(n/n_\infty)}{d\sqrt{t}} = \frac{2}{\sqrt{\pi}} \left(\frac{D}{L^2}\right)^{1/2}, \tag{4}$$

where n/n_∞ is the amount adsorbed at time t relative to that at equilibrium, D is the diffusivity, and r is the radius of the sorbent crystal. L is the characteristic length (V/a), where a is the external surface area of the crystal and V is the volume of the crystal.

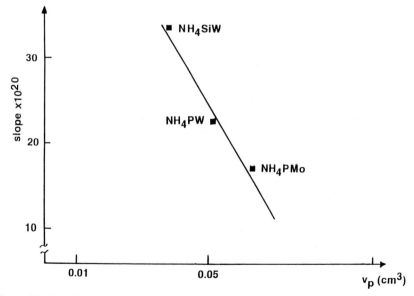

Figure 17. Correlation between the slope of the $\delta = f(n_{Xe})$ curve and the microporous volume of NH_4^+ MOCC samples.[59]

Diffusivities ($\times 10^{-11}$ cm^2 sec^{-1}) of hexane, 3-methylpentane, cyclohexane, n-heptane, n-octane, and isooctane fall in the range from 2 to 7 at temperatures from 20 to 50°C on HPW, HSiW, and NH$_4$PW [60]. In contrast, values from 10 to 30 and from 45 to 160 are observed with NH$_4$PMo and NH$_4$SiW, respectively. Sorption capacities of these organic compounds vary only slightly with the size of the molecule and differ little between heteropoly acids (Fig. 18a). In contrast, with the microporous ammonium salts the sorption capacities are 10–25 times larger (Fig. 18b), providing further evidence of the inability of the alkanes to penetrate into the bulk structures of the heteropoly oxometalates, as well as reflecting both the higher surface areas and the pore structure of the ammonium salts. Recall that the surface areas of NH$_4$PW, NH$_4$PMo, and NH$_4$SiW are 128.2, 193.4, and 116.9 m^2/g, respectively. The isosteric heats of sorption of the saturated hydrocarbons on all the sorbents are 9.7 ± 0.6 kcal/mol.

With the alkenes, 1-hexene, 2,3-dimethylbut-1-ene, 1-heptene, 1-octene, cyclohexene, and 4-methylcyclohex-1-ene, the diffusivities are 1–5, 3–22, and 14–96 for NH$_4$PW, NH$_4$PMo, and NH$_4$SiW, respectively [61] at the same temperatures. Sorption capacities for the three heteropoly acids are approximately five times those observed with the saturated hydrocarbons and the adsorption is now irreversible, which would be expected for sorbate molecules penetrating into the bulk solid. The sorption capacities are similar with all the alkenes for HPW and HSiW but approximately 50% less for HPMo. The sorption capacities of the ammonium salts for the alkenes fall within 0.3–0.6 mmol/g with little difference seen for NH$_4$PW and NH$_4$PMo, but larger values found for NH$_4$SiW. Note that the last-mentioned solid has the largest average micropore radius.

With aromatic hydrocarbons—benzene, toluene, xylene, mesitylene, and m-diethylbenzene—the diffusivities on the heteropoly acids and their ammonium salts decrease with increasing size of the sorbate molecule [62]. The diffusivities on NH$_4$PMo are 24 times higher than those on HPW, HSiW, HPMo, and NH$_4$PW but about 20% of those for NH$_4$SiW. The sorption capacities of the aromatics on the salts are approximately 20–30 times those on the acids and again are highest for NH$_4$SiW.

The diffusivities of alcohols on the ammonium salts fall in the order NH$_4$SiW > NH$_4$PMo > NH$_4$PW [63]. The sorption capacities are 2–5 times those observed with the alkenes, again with NH$_4$SiW showing the largest values.

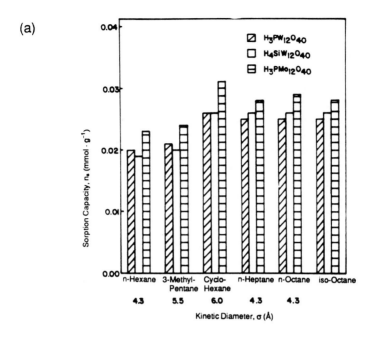

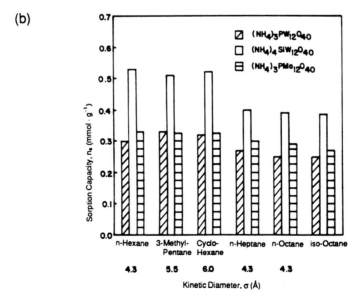

Figure 18. Sorption capacities of *(a)* heteropoly acids and *(b)* ammonium salts of heteropoly acids for aliphatic saturated hydrocarbons at 293 K.[60]

15.6.5 Divalent Cations

To investigate the possibility that high-surface-area, microporous derivatives of the heteropoly acids could be prepared with divalent cations, the aforementioned preparative technique was utilized with magnesium, calcium, strontium, and barium hydroxide and 12-tungstophosphoric and 12-molybdophosphoric acids [14,64]. The N_2 adsorption–desorption isotherms of these preparations showed significant differences from those obtained for the high-surface-area monovalent salts of these acids. At both low and high relative pressures relatively small quantities of N_2 were adsorbed in contrast with those adsorbed with the aforementioned monovalent salts. These results are reflected in the BET surface areas and values of the C_{BET} parameter, both of which are small (Table 4).

Infrared spectroscopic studies of the preparations with divalent cations confirmed the presence of Keggin anions and provided no evidence of species that would be expected to result from the decomposition of the anions. Powder and single-crystal X-ray diffraction studies suggest that the alkaline earth cations are not incorporated into the lattice and thus do not form salts with the trivalent 12-heteropoly anions. In particular, the single-crystal study on the preparation with the barium ion showed no location or even the existence of the barium cation. Further, the measured lattice parameters for the various preparations with the divalent cations are nearly identical to those of the parent acids, providing additional evidence that the cations have not been incorporated into the heteropoly lattice. Differential thermal analysis measurements suggest that the parent acid ex-

TABLE 4
Nitrogen Adsorption Data for Alkaline
Earth/12-Heteropoly Anions

Alkaline earth/heteropoly anion	$S_{BET}(m^2g^{-1})$	C_{BET}
$Mg^{2+}/PW_{12}O_{40}{}^{3-}$	18.5	6
$Ca^{2+}/PW_{12}O_{40}{}^{3-}$	17.9	5
$Sr^{2+}/PW_{12}O_{40}{}^{3-}$	20.5	5
$Ba^{2+}/PW_{12}O_{40}{}^{3-}$	1.8	90
$Mg^{2+}/PMo_{12}O_{40}{}^{3-}$	11.5	11
$Ca^{2+}/PMo_{12}O_{40}{}^{3-}$	19.4	5
$Sr^{2+}/PMo_{12}O_{40}{}^{3-}$	19.3	4
$Ba^{2+}/PMo_{12}O_{40}{}^{3-}$	3.1	66

ists in a two-phase system with the cation present as another salt. In view of the crystallographic structures of the parent acids, the introduction of divalent cations would be difficult, if not impossible, on both electrical neutrality and geometrical grounds.

15.6.6 Cations of the IB and IIIA Groups

The aforementioned results on microporous salts are concerned primarily with cations of the alkali metals or group IA elements for which only one oxidation state is possible. It seemed both worthwhile and interesting to investigate the possibility of forming high-surface-area microporous solids from other elements of the periodic table as cations. For this purpose silver and thallium were chosen, as representatives of the IB and IIIA groups, respectively [65]. The BET surface areas and C_{BET} values for the silver and thallium salts of 12-tungstophosphoric, 12-molybdophosphoric, and 12-tungstosilicic acids are shown in Table 5. It is evident that with all the salts, except that of silver 12-molybdophosphate, both the surface areas and C values are relatively high as compared with the parent acids. The surface areas for the silver and thallium salts together with those for the aforementioned salts of the monovalent cations are shown plotted versus the cation diameters (Fig. 19). Although the surface areas are clearly dependent on the cation diameter, a second-order dependence on the composition of the anion is also evident.

The mean micropore radii for the various salts are plotted versus the values of n, the number of adsorbed layers (Fig. 20) as obtained from the finite layer BET equation (Eq. 3). The points generally lie on or near a

TABLE 5
Surface Area and C_{BET} Parameters for
Stoichiometric Salts of Ag and Tl

Sample	$S_{BET}(m^2/g)$	C_{BET}
$Ag_3PW_{12}O_{40}$	100.9	7,400
$Ag_3PMo_{12}O_{40}$	1.5	14,000
$Ag_4SiW_{12}O_{40}$	106.0	35,600
$Tl_3PW_{12}O_{40}$	131.6	2,700
$Tl_3PMo_{12}O_{40}$	157.0	900
$Tl_4SiW_{12}O_4$	97.5	5,000

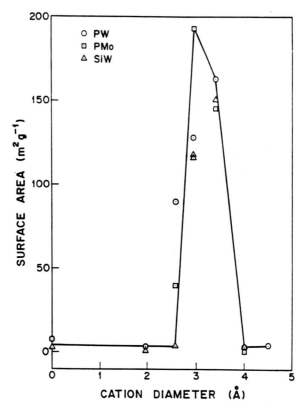

Figure 19. Surface area versus cation diameter for microporous salts of HPW, HPMo, and HSiW. Reprinted from *Polyhedron*, 5, J. B. Moffat, 261, Copyright (1986), with kind permission from Elsevier Science Ltd., The Boulevard, Langford Lane, Kidlington 0X5 1GB, UK.

straight line passing through the origin and with a slope of approximately 3.5 Å, which would be expected for nitrogen as the adsorbate. Because values of $\bar{r}$ and n are obtained independently, these observations provide evidence for the internal self-consistency of both the experimental results and the methods employed to calculate the pore size distribution.

In view of the results of earlier work that demonstrated that the relative amounts of the preparative reagents in the syntheses of the salts have a strong influence on the morphology of the resulting solid and the concentrations of residual protons, similar studies with the silver and thallium salts were undertaken [19,25,65–67]. The effect of the stoichiometry of the preparative reagents on the yields and surface areas is illustrated for the thallium salts in Figs. 21 and 22, respectively. The yields increase with increasing quantity of the thallium cation relative to that of the Keggin anion up to a 1:1 ratio of the latter two quantities and, at least approxi-

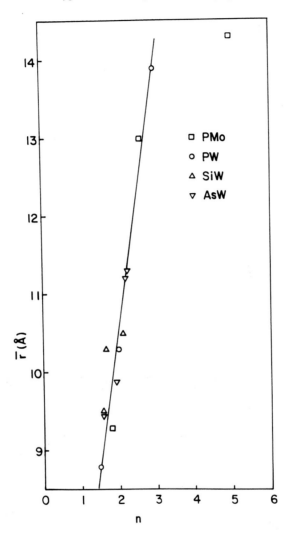

Figure 20. Average radius ($\bar{r}$) versus number of adsorbed layers (n) for microporous salts.

mately, remain relatively constant for further increases in the stoichiometry. The surface area increases up to preparations with a 15% deficiency in the quantity of the cation, remains approximately constant up to a 15% excess of the cation, and then decreases.

The effect of the preparative stoichiometry on the pore size distribution is illustrated in Fig. 23 for the deficit, stoichiometric, and excess salts of thallium 12-tungstophosphate (TlPW). The peak in the distribution re-

Effect of Tl(I):Acid ratio on Yields
(Based on initial amount of Acid)

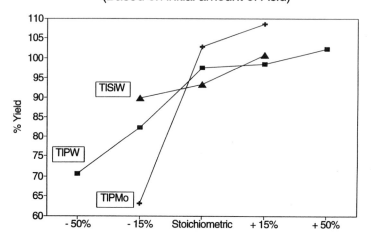

Figure 21. Yield of $Tl_3PW_{12}O_{40}$ for three reactant stoichiometries—based on initial amount of acid.

Tl(I):Acid Ratio Effect on Surface Area

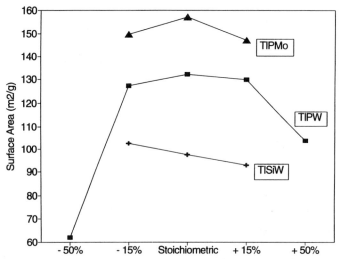

Figure 22. Surface areas (S_{BET}) for the thallium salts of the acids shown, as a function of the cation-to-acid ratio.[65]

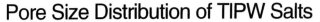

Pore Size Distribution of TlPW Salts

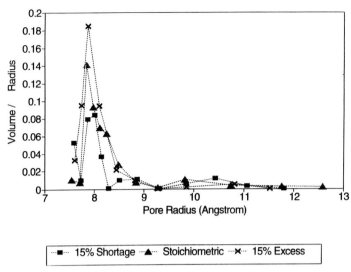

Figure 23. Pore size distribution for $Tl_3PW_{12}O_{40}$, prepared from three reactant stoichiometries.[65]

mains at approximately 8 Å regardless of the stoichiometry, but the magnitude of the peak increases with the relative quantities of the cation used in the preparation.

The effect of the preparative stoichiometry is also evident from ^1H MAS NMR on the TlPW samples (Fig. 24). Both the chemical shift and the intensities of the peaks decrease with increase in the relative quantities of the thallium cation, indicating that the concentration of residual protons is decreasing with the stoichiometry of the preparative mixture.

It is well known that unambiguous information on the microporous structures of solids is difficult to obtain. Since no *ab initio* method is as yet available, various semiempirical approaches are employed. Frequently these make use of the intuitively reasonable assumption that micropores fill by adsorption layer thickening whereas others attempt to construct potential functions for the interaction between adsorbate and adsorbent. Not surprisingly each method has its advantages and disadvantages and ideally should be applied concomitantly to provide confidence in the data generated.

Work in this laboratory has compared a number of methods for the analysis of adsorption–desorption isotherms of microporous solids by application of these techniques to thallium 12-tungstosilicate and 12-molybdophosphate [67]. The methods examined include the BET, MP, and Du-

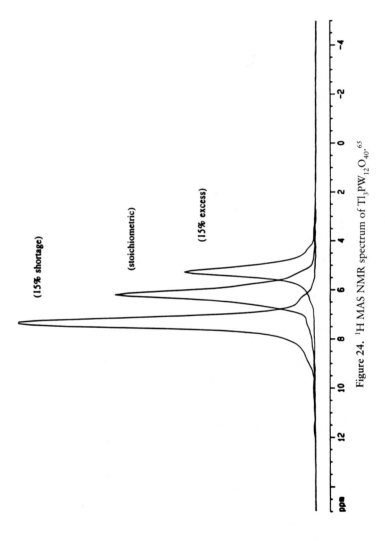

Figure 24. ^1H MAS NMR spectrum of $Tl_3PW_{12}O_{40}$.[65]

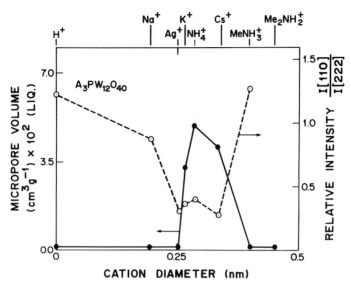

Figure 25. Correlation of estimated micropore volume and XRD intensity ratio [110]:[222] with monovalent cation diameter for 12-tungstophosphates.

binin–Radushkevich (DR) and that due to Horvath, Kawazre, Saito, and Foley. In view of the limited space here, the interested reader may wish to consult the original publication for further information.

15.6.7 Nature and Source of the Microporosity

The nature and source of the microporosity are of considerable interest. Although differences in the lattice structures do not apparently result from the formation of the microporous materials, variations in the intensities of the X-ray diffraction patterns are found (Fig. 25). The intensity of the [110] reflection relative to that of the [222] plane, the latter being the most intense reflection, is relatively high for the small-surface-area, nonmicroporous solids but low where a microporous structure is present, in an approximately reciprocal relation to the micropore volumes. With increasing cation size, the micropore volumes and the XRD I[110]:I[222] ratios pass through a maximum and minimum, respectively, in the cation diameter range of approximately 2.5–4.0 Å.

In view of the cubic crystallographic structure possessed by the parent acids, interstitial voids that appear to be separated from one another by the terminal oxygen atoms of the Keggin anions are present in these solids. Because the cubic lattice parameter of the salts increases with increasing cation size, it is expected that separation of the anions would result in widening of the interstitial voids, changes in the anionic bonding pattern, and reorientation of the anions within the secondary structure. Hence the positions of the anionic terminal oxygen atoms will no longer separate the interstitial voids, the XRD intensity of the [110] plane will be decreased, and channels both parallel with and normal to this plane will be formed.

For completeness it is important to note that the existence of MOCC with microporous structures is not universally accepted. For example, the high surface area found with some MOCC has been attributed to the small size of the particles of the solid and not to the presence of intrinsic micropores in the crystal structure [68].

15.6.8 Catalytic Applications of the Microporous Salts and Evidence of Shape Selectivity

The aforementioned studies of the conversion of methanol to hydrocarbons on HPW, HSiW, HPMo, and the microporous derivatives of these solid acids showed that, e.g., the microporous ammonium salts of HPW produced higher conversions of methanol than those obtained with the parent acids, a result that could not be attributed solely to the higher surface area of the former as compared with the latter [26,27,36,41–43]. The additional observation that the products obtained with NH_4PW were largely saturated as opposed to the alkenes obtained with HPW suggested the occurrence of hydride transfer on the microporous salt, which has been attributed in zeolites to the enhanced concentration of reactant species in the micropores and the second-order kinetics of the transfer process. Three factors—the surface areas, the microporosity, and the distribution of Brønsted acid strengths—must be considered in rationalizing the differences observed in methanol conversion to hydrocarbons on HPW and NH_4PW.

The catalysis of methanol and ethanol conversions on alkylammonium and ammonium derivatives of HPW and HPMo has also been investigated [23]. The salts of HPMo produced selectivities to acetaldehyde from ethanol ranging from 30 to 50% at 623–693 K. Ethane, not normally a feature of the dehydration/dehydrogenation of ethanol, was observed with selectivities as high as 33%, again reminiscent of the saturated hydrocarbons found with methanol conversion.

The double bond and cis-trans isomerizations of C_6–C_8 alkenes are strongly catalyzed by HPW, NH_4PW, and HSiW but weakly by HPMo, NH_4SiW, and NH_4PMo in the liquid phase at or near ambient temperatures [69]. No skeletal isomerization of any of the olefins was observed at temperatures up to 343 K. These results appear to be primarily a reflection

TABLE 6

Product Compositions from the Alkylation of Toluene with Methanol on NH$_4$PW, NH$_4$SiW, Theta-1,[a] and ZSM-5[a] Catalyst

	NH$_4$PW	NH$_4$SiW	NH$_4$PW	NH$_4$PW[b]	Theta-1	ZSM-5
Conditions						
Temperature (°C)	200	250	200	160	550	600
TOl/MeOH molar	1/1	1/1	2/1	1/1	2/1	2/1
W/F [mg cat.min/ml He]	1.43	1.43	1.43	1.43	1(h^{-1})	4.8(h^{-1})
Conversion (%)						
Toluene	17.3	17.0	20.5	0.9	23.9	31.8
Liquid product (wt %)						
Benzene	0.0	0.0	0.0	0.0	0.7	3.4
Toluene	79.0	79.7	79.5	98.9	70.8	67.0
p-xy	3.0	4.6	4.3	0.40	10.2	11.1
m-xy	2.8	3.7	6.1	0.25	6.5	11.1
o-xy	2.5	1.9	3.9	0.15	6.0	4.6
1.3.5-TriMe	2.1	1.1	1.1	0.05	0.4	0.2
1.2.4-TriMe	4.4	4.0	3.3	0.09	4.3	2.4
1.2.3-TriMe	0.9	0.8	0.9	0.01	N/D[c]	N/D
Xylene composition (%)						
p-xy	36.2	45.1	30.1	49.9	44.9	41.4
m-xy	33.7	36.0	42.6	31.3	28.6	41.4
o-xy	30.1	18.9	27.3	18.8	26.5	19.2

[a] Reference (71).
[b] 200 μl NH$_3$/50 mg cat. added to catalyst at 160°C prior to reaction.
[c] Not detected.

of the relatively high acidic strengths rather than pore structure, since studies of the liquid phase adsorption of such alkenes on other acidic solids showed no evidence for isomerization.

Studies of the alkylation of toluene provide evidence for the presence of shape-selective effects with the microporous derivatives of the heteropoly acids [70]. The results of Ashton *et al.* for the alkylation of toluene on several zeolites [71] provide relevant comparative data (Table 6). Ashton *et al.* note that H-*theta*-1 is a highly selective catalyst for the production of *p*-xylene (column 5). It is interesting to note that NH$_4$PW and NH$_4$SiW (columns 2 and 4) are capable of achieving *p*-xylene selectivities in the xylene isomers as high as those obtained with H-*theta*-1, suggesting the existence of a shape selective factor with these microporous heteropoly salts.

The conversions of methylethylbenzenes on NH$_4$PW and NH$_4$SiW at temperatures up to 400°C show significant differences in the results obtained for the two catalysts, which can be attributed to differences in the distribution of the acidic sites as well as those of the pore structures [72].

The oxidative dehydrogenation of isobutyric acid to methacrylic acid has been examined on ion-exchange-modified microporous derivatives of the heteropoly acids [73]. Effects due both to the nature of the cation and to the consequent differences in microporous structure are seen, and somewhat surprisingly, with ammonium salts of the heteropoly acids nitriles are found in the product streams [74].

The cracking of C$_6$–C$_8$ alkenes on NH$_4$PW [75] and of butylbenzenes [76] have been studied. With the *t*-, *s*-, and *n*-butylbenzenes the rate constants per surface area are higher for NH$_4$PW than HPW by factors of approximately 6.0, 3.0, and 1.5, respectively, demonstrating the influence of differences in both the acidities and porosities.

Acknowledgment

The financial support of the Natural Sciences and Engineering Research Council of Canada is greatly acknowledged.

References

1. Pope, M. T. (1983). *Heteropoly Oxometalates*, Springer-Verlag, Berlin.
2. Pope, M. T., and Müller, A. (1991). *Angew. Chem. Int. Ed. Engl.*, **30**, 34.
3. Berzelius, J. (1826). *Pogg. Ann.*, **6**, 369, 380.
4. Svanberg, L., and Struve, H. (1848). *J. Prakt. Chem.*, **44**, 257, 291.
5. Marignac, C. (1862). *R. Acad. Sci.*, **55**, 888, (1862). *Ann. Chim.*, **25**, 362.
6. Miolati, A., and Pizzighelli, R. (1908). *J. Prakt. Chem.*, **77**, 417.

7. Rosenheim, A., and Jaenicke, H. (1917). *Z. Anorg. Allg. Chem.*, **100**, 304.
8. Pauling, L. C. (1929). *J. Am. Chem. Soc.*, **51**, 2868.
9. Keggin, J. F. (1933). *Nature*, **131**, 908, (1934). *Proc. R. Soc. London*, **A144**, 75.
10. Bradley, A. J., and Illingworth, J. W. (1936). *Proc. R. Soc.*, **A157**, 113.
11. Evans, H. T., Jr. (1948). *J. Am. Chem. Soc.*, **70**, 1291.
12. Brown, G. M., Noe-Spirlet, M.-R., Busing, W. R., and Levy, H. A. (1977). *Acta Crystallogr.*, **B33**, 1038.
13. Brown, G. M., private communication.
14. McGarvey, G. B., Taylor, N. J., and Moffat, J. B. (1993). *J. Mol. Catal.*, **80**, 59.
15. Strandberg, R. (1975). *Acta. Chem. Scand.*, **A29**, 359.
16. Allman, R. (1976). *Acta Chem. Scand.*, **A30**, 152.
17. McGarvey, G. B., and Moffat, J. B. (1991). *J. Mol. Catal.*, **69**, 137.
18. McGarvey, G. B., and Moffat, J. B. (1990). *J. Magn. Reson.*, **88**, 305.
19. Highfield, J. G., and Moffat, J. B. (1984). *J. Catal.*, **88**, 177.
20. Rosencwaig, A. (1975). *Anal. Chem.*, **47**, 592A.
21. Kirkbright, G. F., and Castleden, S. L. (1980). *Chem. Brit.*, **661**.
22. Rocchiccioli, C., Thouvenot, R., and Franck, R. (1976). *Spectrochim. Acta.*, **32A**, 587.
23. McMonagle, J. B., and Moffat, J. B. (1985). *J. Catal.*, **91**, 132.
24. West, S. F., and Audrieth, L. F. (1955). *J. Phys. Chem.*, **59**, 1069.
25. Highfield, J. G., and Moffat, J. B. (1984). *J. Catal.*, **89**, 185.
26. Highfield, J. G., and Moffat, J. B. (1985). *J. Catal.*, **95**, 108.
27. Highfield, J. G., and Moffat, J. B. (1986). *J. Catal.*, **98**, 245.
28. Moffat, J. B. (1984). *Catalysis on the Energy Scene, Studies in Surface Science and Catalysis* (S. Kaliguine, ed.), Vol. 19, p. 77, Plenum, New York.
29. Moffat, J. B. (1984). *J. Mol. Catal.*, **26**, 385.
30. Moffat, J. B. (1984). *Proceedings, 8th Iberoamerican Symposium. Catalysis*, p. 349, Lisbon.
31. Highfield, J. G., Hodnett, B. K., McMonagle, J. B., and Moffat, J. B. (1984). *Proceedings, 8th International Congress Catalysis*, pp. 5, 611, Berlin.
32. Moffat, J. B. (1985). *Catalysis by Acids and Bases, Studies in Surface Science and Catalysis* (B. Imelik, C. Nacache, G. Courdurier, Y. Ben Taarit, and J. C. Vedrine, eds.), Vol. 20, p. 157, Elsevier, Amsterdam.
33. Moffat, J. B. (1987). *Preparation of Catalysts IV, Studies in Surface Science and Catalysis* (B. Delmon, P. Grange, P. A. Jacobs, and G. Poncelet, eds.), Vol. 31, p. 241, Elsevier, Amsterdam.
34. Moffat, J. B. (1994). *Acidity and Basicity of Solids: Theory, Assessment and Utility* (Kluwer, Dordrecht, Holland, eds.), p. 213.
35. Moffat, J. B. (1994). *Acidity and Basicity of Solids: Theory, Assessment and Utility* (Kluwer, Dordrecht, Holland, eds.), p. 237.
36. Hayashi, H., and Moffat, J. B. (1982). *J. Catal.*, **77**, 473.
37. Jozefowicz, L. C., Karge, H. G., Vasilyeva, E., and Moffat, J. B. (1993). *Microporous Materials*, **1**, 313.
38. Taniguchi, H., Masuda, T., Tsutsumi, K., and Takahashi, B. (1978). *Bull. Chem. Soc. Jpn.*, **51**, 1970.
39. Taniguchi, H., Masuda, T., Tsutsumi, K., and Takahashi, B. (1980). *Bull. Chem. Soc. Jpn.*, **53**, 2463.
40. Gillespie, R. J., and Peel, T. E. (1971). *Adv. Phys. Org. Chem.*, **9**, 1; Gillespie, R. J., and Peel, T. E. (1973). *J. Am. Chem. Soc.*, **95**, 5173; Gillespie, R. J., Peel, T. E., and Robinson, E. (1971). *J. Am. Chem. Soc.*, **93**, 5083.

41. Hayashi, H., and Moffat, J. B. (1983). *J. Catal.*, **83**, 192.
42. Hayashi, H., and Moffat, J. B. (1983). *J. Catal.*, **81**, 61.
43. Moffat, J. B. (1984). *Catalytic Conversions of Synthesis Gas and Alcohols to Chemicals* (R. G. Herman, ed.), p. 395, Plenum, New York.
44. McMonagle, J. B., and Moffat, J. B. (1984). *J. Colloid Interface Sci.*, **101**, 479.
45. Taylor, D. B., McMonagle, J. B., and Moffat, J. B. (1985). *J. Colloid Interface Sci.*, **108**, 278.
46. Moffat, J. B. (1986). *Polyhedron*, **5**, 261.
47. Moffat, J. B. (1988). *Solid State Ionics*, **26**, 101.
48. Moffat, J. B. (1989). *J. Mol. Catal.*, **52**, 169.
49. McGarvey, G. B., McMonagle, J. B., Nayak, V. S., Taylor, D., and Moffat, J. B. (1988). *Proceedings, 9th International Congress on Catalysis*, 1804, Calgary.
50. McGarvey, G. B., and Moffat, J. B. (1988). *J. Colloid Interface Sci.*, **135**, 51.
51. Mikhail, R. Sh., Brunauer, S., and Bodor, E. E. (1968). *J. Colloid Interface Sci.*, **26**, 45.
52. Lecloux, A., Pirard, J. P. (1979). *J. Colloid Interface Sci.*, **70**, 265.
53. McGarvey, G. B., and Moffat, J. B. (1991). *J. Catal.*, **128**, 69.
54. Moffat, J. B., McGarvey, G. B., McMonagle, J. B., Nayak, V., and Nishi, H. (1990). *Guidelines for Mastering the Properties of Molecular Sciences* (D. Barthomeuf, E. G. Derouane, and W. Hölderich, eds.), p. 193, Plenum Press, New York.
55. McGarvey, G. B., and Moffat, J. B. (1991). *J. Catal.*, **130**, 483.
56. Lapham, D., McGarvey, G. B., and Moffat, J. B. (1992). *Stud. Surf. Sci. Catal.*, **73**, 261.
57. McGarvey, G. B., and Moffat, J. B. (1993). *Multifunctional Mesoporous Inorganic Solids* (C. A. C. Sequeira, and M. J. Hudson, eds.), p. 451, Kluwer, Dordrecht, Holland.
58. Bonardet, J. L., McGarvey, G. B., Moffat, J. B., and Fraissard, J. (1993). *Colloids and Surfaces*, **A72**, 191.
59. Bonardet, J. L., Fraissard, J., McGarvey, G. B., and Moffat, J. B. (1995). *J. Catal.*, **151**, 147.
60. Nayak, V. S., and Moffat, J. B. (1988). *J. Colloid Interface Sci.*, **122**, 475.
61. Nayak, V. S., and Moffat, J. B. (1988). *J. Phys. Chem.*, **92**, 2256.
62. Nayak, V. S., and Moffat, J. B. (1987). *J. Colloid Interface Sci.*, **120**, 301.
63. Nayak, V. S., and Moffat, J. B. (1988). *J. Phys. Chem.*, **92**, 7097.
64. McGarvey, G. B., and Moffat, J. B. (1992). *Catal. Lett.*, **16**, 173.
65. Parent, M., and Moffat, J. B., (in press).
66. Lapham, D., and Moffat, J. B. (1991). *Langmuir*, **7**, 2273.
67. Parent, M., and Moffat, J. B. *Langmuir*, **11**, 4474.
68. Misono, M. (1987). *Catal. Rev.—Sci. Eng.*, **29**, 269.
69. Nayak, V. S., and Moffat, J. B. (1988). *Appl. Catal.*, **36**, 127.
70. Nishi, H., Nowinska, K., and Moffat, J. B. (1989). *J. Catal.*, **116**, 480.
71. Ashton, A. G., Barri, S. A. I., Cartlidge, S., and Dwyer, J. (1986). *Chemical Reactions in Organic and Inorganic Constrained Systems* (R. Setton, ed.), Reidel, Holland.
72. Nishi, H., and Moffat, J. B. (1989). *J. Mol. Catal.*, **51**, 193.
73. McGarvey, G. B., and Moffat, J. B. (1991). *J. Catal.*, **132**, 100.
74. McGarvey, G. B., and Moffat, J. B. (1991). *Catal. Lett.*, **10**, 41.
75. Nayak, V. S., and Moffat, J. B. (1991). *Appl. Catal.*, **77**, 251.
76. Donsig, H. A., and Moffat, J. B., to be published.

CHAPTER 16

Preparation of Bulk and Supported Heteropolyacid Salts

S. Soled,* S. Miseo,* G. B. McVicker,* J. E. Baumgartner,*
W. E. Gates,* A. Gutierrez,† and J. Paes†

*Exxon Research and Engineering Company
Annandale, New Jersey 08801
†Exxon Chemical
Linden, New Jersey 07036

KEYWORDS: acid salt catalyst, heteropolyacid salt, solid acid catalyst, supported catalyst

16.1 Introduction

In the last several years, a worldwide renaissance in solid acid catalysis has developed, driven by advances in materials science and growing environmental challenges associated with conventional liquid or halide-containing acids [Misono and Okuhara, 1993; Thomas, 1992]. For example, environmentally compatible oxide-based solid acids would make a dramatic impact on catalytic processes based on hydrogen fluoride, sulfuric acid, aluminum trichloride, or boron trifluoride, all of which present either disposal, transport, or toxicity problems.

Heteropoly compounds have attracted interest as solid acid catalysts [Ono et al., 1981; Ai, 1981; Hayashi and Moffat, 1982; Misono, 1983]. Heteropoly compounds consist of polyoxoanions separated by ion-exchangeable cations; in the parent acids, the cations consist of (hydrated) protons. A variety of polyanion structures exist, each with a different molecular arrangement. The relative acid strength of heteropolyacids depends on anionic charge. Generally, as the charge of a conjugate base (here, the heteropolyanion) decreases, its stability increases, which increases the strength of the "conjugate" acid. Thus, in the Keggin family of heteropoly-acids, $H_3PW_{12}O_{40}$ forms the strongest acid because the $PW_{12}O_{40}^{3-}$ anion has the lowest negative charge. Several reviews have appeared detailing the structural and catalytic features of 12-tungstophosphoric (HPW) or 12-molybdophosphoric acids (HPMo), which contain a central phosphorus-oxygen tetrahedron surrounded by 12 tungsten (or molybdenum) oxygen octahedra [Misono, 1982; Misono, 1987a; Misono, 1987b; Moffat, 1985].

Partial proton exchange produces acid salts. Such salts retain acidic properties, even in the presence of alkali metal or ammonium cations, as long as some of the protons remain. The surprising retention of strong acidity in alkali-exchanged bulk heteropolyacid salts is opposite the trend observed with alkali-exchanged zeolites and macroporous oxides, and re-sults from the close association of exchanged cations with the anions. The high activity of Cs acid salts in acid-catalyzed aromatic alkylations has been pointed out [Nishimura et al., 1991].

Particle physical morphology also changes with differences in intersti-tial cation size. For example, the parent protonic acids as well as salts or acid salts with small cations, such as Fe, Co, Ni, and Na, are water soluble and produce large particles with low-surface areas (1–5 m^2/g) when crys-tallized by solution evaporation. In contrast, acid salts with large cations, such as Cs, Rb, K, and NH_4, form as submicron ($\sim$ 100-Å) water- (or po-lar organic-) insoluble particles with surface areas exceeding 100 m^2/g [Misono, 1987a and b].

Supported heteropolyacids are easily prepared by aqueous impregna-tion techniques [Nowinska et al., 1991; Schwegler et al., 1992; Baba and Ono, 1986]. Such catalysts have been studied for benzene alkylation, acetic acid esterification, 2-propanol dehydration, cumene cracking, and toluene disproportionation [Izumi et al., 1983; Henke and Sebulsky, 1967; Nomiya et al., 1990; Sebulsky and Henke, 1971]. The insolubility of the large cation acid salts, on the other hand, makes conventional impregna-tion procedures impractical; and for applications in packed-bed reactors, it would be necessary to support the submicron acid salts on larger carriers. Here we describe synthetic techniques for accomplishing this task.

16.2   Experimental

X-ray diffraction spectra, collected on a Rigaku D-Max diffractometer using CuK_α radiation, identified crystalline phases. The 12-tungstophosphoric acid either was synthesized by acid condensation from sodium tungstate and sodium phosphate [Bailar, 1978] or was obtained from J. T. Baker. Bulk ammonium and cesium salts and acid salts of 12-tungstophosphoric acid were prepared by dropwise addition of ammonium and cesium carbonate solutions into an HPW solution, air-drying the resulting precipitate, and then calcining at 300°C.

Silica-supported ammonium salts were prepared by three techniques: gas-phase reaction with ammonia, hydrothermal precursor synthesis, and urea coimpregnation. In the first route, a 40% HPW/SiO$_2$ sample (Davison 62 SiO$_2$ powder, surface area 270 m^2/g), prepared by incipient wetness, was heated in a 5% NH$_3$/N$_2$ mixture at 150°C with a space velocity and for a time sufficient to allow a threefold excess of NH$_3$ to titrate all the HPW protons. The sample was purged in nitrogen and heated to 300°C. In the hydrothermal route, a water soluble precursor was prepared by reacting ammonium metatungstate [(NH$_4$)$_6$H$_2$W$_{12}$O$_{40}$·xH$_2$O] with ammonium phosphate dibasic [NH$_4$H$_2$PO$_4$] in a 12:1 W:P molar ratio. Both starting materials were dissolved in water and heated in an autoclave at 150°C overnight. A solution was produced that on drying at 120°C formed an amorphous solid product. Calcination to 300°C yielded a crystalline salt isomorphous with the conventionally prepared ion-exchanged ammonium salt. This soluble precursor, when impregnated onto silica powder and calcined at 300°C, produced a supported ammonium acid salt catalyst. In the third technique, urea (H$_2$NCONH$_2$) was incorporated into an HPW solution in a mole ratio of urea to HPW of 0.5:2/1 and impregnated onto silica. The resulting samples were dried and calcined at 300°C.

The silica-supported Cs acid salt was prepared by sequential impregnation and *in situ* reaction on the support. First, cesium carbonate was impregnated by aqueous incipient wetness onto silica powder (Davison 62) or extrudates (1/16-in. diameter, 260 m^2/g), dried at 110°C, and calcined at 300°C. Following this, the 12-tungstophosphoric acid was impregnated by a similar aqueous impregnation route, dried at 110°C, and calcined at 300°C.

Ammonia temperature-programmed desorption experiments (TPD) were conducted in a unit containing a quartz reaction tube, a programmable furnace, and a Balzer's quadrupole mass spectrometer. A gas containing 4010 ppm of NH$_3$ in N$_2$ served as a calibration standard. Approximately 0.5 g of catalyst (40% HPW/SiO$_2$) was weighed and loaded into the reaction cell, heated in helium to 300°C, and then cooled to 110°C. A flow of 50 cm^3/min of 5% NH$_3$ in N$_2$ was passed through the catalyst for 30 min to ensure sufficient NH$_3$ to titrate all acid sites. A 100% N$_2$ flow (80 cm^3/min) at 100°C for 150 min stripped off most physisorbed NH$_3$. The furnace was heated to 700°C at 10°C/min and the gaseous effluent was analyzed quantitatively.

All catalysts were evaluated using the isomerization of 2-methylpent-2-ene (2MP2) as a probe reaction, as described previously [Kramer and McVicker, 1986]. The formation rates and rate ratios of the product hexene isomers formed in this test reaction reflect the relative acid site concentration and strength, respectively. The hexene isomers formed include both cis- and trans-4-methylpent-2-ene (t-4MP2), trans-3-methylpent-2-ene (t-3MP2), and 2,3 dimethylbut-2-ene (2,3DMB2). Only a double-bond shift, a reaction occurring on weak acid sites, is required by 4MP2. A methyl group shift (i.e., stronger acidity requirement than double-bond shift) is required by 3MP2, whereas the double-branched 2,3DMB2 product requires even stronger acidity. For a homologous series of solid acids, differences in t-3MP2 rates, normalized with respect to surface area, reflect changes in the density of acid sites possessing strengths sufficient to catalyze the skeletal isomerization. Because skeletal isomerization rates generally increase with increasing acid strength, the ratio of methyl group migration rate to double-bond shift rate should increase with increasing acid strength. The use of rate ratios, in lieu of individual conversion rates, is preferable because differences in acid site populations are normalized. Figure 1 outlines the reaction pathways for this model compound test.

Some catalysts were also evaluated for the liquid-phase aromatic alkylation of 1,3,5-trimethylbenzene (mesitylene) with cyclohexene, as described earlier [Nishimura et al., 1991]. Bulk catalyst (0.5 g) or an equivalent loading of heteropolyacid on supported catalyst (freshly calcined at

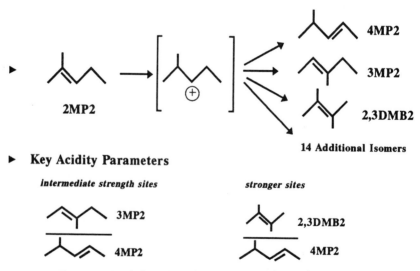

Figure 1. 2-Methylpent-2-ene isomerization acidity probe reaction.

300°C) was loaded with 160 cm^3 of a premixed 5% cyclohexene in mesitylene solution into a 300 cm^3 Parr autoclave, heated to 100°C under N$_2$ at 15 psi, and mixed at 600 rpm. Two hours after the temperature reached 100°C, the heat and stirrer were turned off and the sample was withdrawn and analyzed by GC.

16.3   Results and Discussion

16.3.1 Formation of Bulk Heteropolyacid Salts of Alkali and Ammonium Cations

Misono [1987a and b] has pointed out that alkali or ammonium heteropolyacid salts are conveniently formed by the addition of aqueous alkali or ammonium solutions to the acid. When large alkali cations are used (K, Rb, Cs), a fine, insoluble precipitate that can be isolated by solvent evaporation forms. The as-formed precipitate is nonuniform in Cs composition, but becomes more homogeneous after thermal treatment [Misono, 1987]. We have monitored pH and conductometric titrations for the addition of cesium or ammonium carbonate to a solution of HPW. As Cs$_2$CO$_3$ is added to HPW, a precipitate forms as Cs ions exchange for protons in the acid. The pH remains low until the Cs:PW$_{12}$O$_{40}^{3-}$ molar ratio approaches 3 and then rises rapidly (Fig. 2). At this point the capacity of the acid to accept Cs ions is exceeded. As the number of Cs atoms approaches three per Keggin anion, one would expect that the Brønsted acidity of the phase would disappear because no protons remain. The corresponding conductometric curve also passes through a minimum at three Cs atoms per each Keggin ion (Fig. 3).

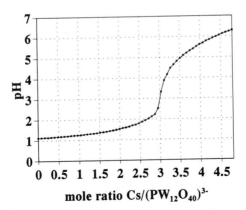

Figure 2. pH Titration of HPW with Cs$_2$CO$_3$.

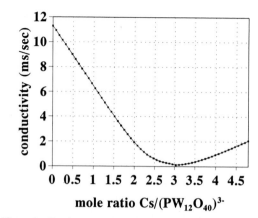

Figure 3. Conductometric titration of HPW with Cs_3CO_3.

The behavior with ammonium exchange differs. As Figs. 4 and 5 show, titration of HPW with ammonium carbonate goes to completion only when the nominal $NH_4^+/PW_{12}O_{40}^{3-}$ mole ratio exceeds three. The ammonium carbonate solution has been calibrated with a standardized 0.1 N HCl. We find that formation of $(NH_4)_3PW_{12}O_{40}$ requires in excess of three ammonium ions in solution per Keggin ion. Consequently, a preparation of $(NH_4)_3PW_{12}O_{40}$ formed by a nominally stoichiometric ion exchange produces a solid with an actual composition close to $(NH_4)_{2.2}H_{0.8}PW_{12}O_{40}$.

This ammonia content clearly shows up in the conversion test of 2MP2 as illustrated in Fig. 6 where the formation rate of 2,3DMB2 increases for $(NH_4)_xH_{3-x}PW_{12}O_{40}$ as x varies from a nominal value of 0 to

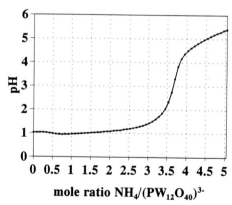

Figure 4. pH Titration of HPW with $(NH_4)_2CO_3$.

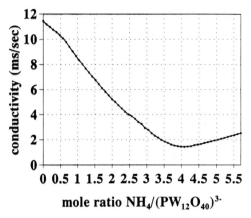

Figure 5. Conductometric titration of HPW with $(NH_4)_2CO_3$.

3. Because the rate of isomerization should decrease as full proton exchange is approached ($x = 3$ if stoichiometric exchange occurred), this behavior is consistent with the actual NH_4 content being less than the nominal value. The increasing ratio appears to parallel the behavior of the Cs system in the region of $2 < x < 2.5$ described as follows.

Figure 7 shows the behavior for the Cs acid salts of 12-tungstophosphoric acid, where the 2,3DMB:4MP2 rate ratio passes through a clear maximum at $x = 2.5$. Studies suggest that this maximum represents a compromise between increasing pore size (accessibility) of reactants as x

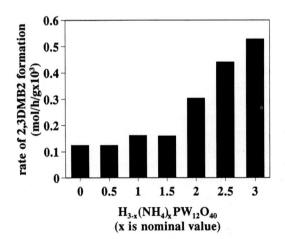

Figure 6. Rate of formation of 2,3DMB2 over ammonium acid salts of 12-tungstophosphoric acid at 250°C, 1 h reaction time.

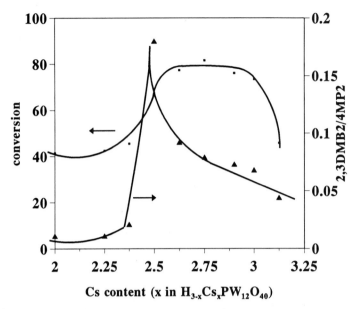

Figure 7. Conversion and 2,3DMB2:4MP2 rate ratio for Cs acid salts of 12-tungstophosphoric acid at 200°C, 5 min reaction time.

increases from 2.0 to 2.5, and decreasing proton concentration as x increases from 2.5 to 3 [Okuhara, 1995]. Figure 7 also suggests that between $x = 2.5$ and 3.0, the sites strong enough to catalyze the more difficult acid reactions are replaced by ones capable of catalyzing facile reactions (like double-bond migration).

16.3.2 Supported Acid Salt Catalysts

As described in the preceding section, the partial ion exchange of large cations such as ammonium, cesium, potassium or rubidium for protons in 12-tungstophosphoric acid produces water-insoluble salts. Precipitation by base addition to an aqueous solution of the acid has been the traditional method for preparing these salts. Reaction tests with 2MP2 confirm literature reports that show high acid catalyzed isomerization activity is retained with these acid salts (Fig. 8). However, because these salts are water insoluble, they cannot conveniently be placed on supports by traditional, aqueous precursor–impregnation, followed by drying and calcination. These interesting acid salts also form extremely small (~ 100-Å) particles, which makes it necessary to find a way of supporting them on large particles to avoid unacceptable pressure drops in fixed-bed reactors or filtration difficulties in slurry reactors and also to allow easy particle separation from the mother liquor during synthesis.

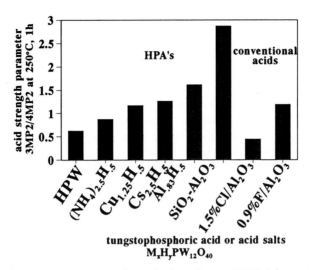

Figure 8. 3MP2:4MP2 rate ratio for acid salt catalysts at 250°C, 1 h reaction time.

In choosing a support for the acid salts, we examined the behavior of 12-tungstophosphoric acid on silica and alumina. At the high loadings we employed, HPW forms small crystalline particles on silica, as observed in the X-ray diffraction pattern shown in Fig. 9. On alumina, an amorphous pattern appears, similar to what is observed with a WO_3 precursor (ammonium metatungstate). This suggests that HPW either forms small crys-

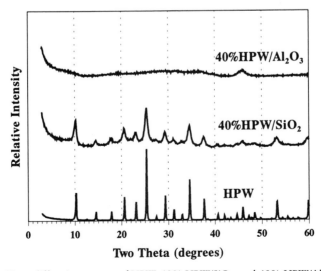

Figure 9. X-ray diffraction spectra of HPW, 40% HPW/SiO₂, and 40% HPW/Al₂O₃.

tallites (<40 Å) on Al$_2$O$_3$ or substantially decomposes. Acidity probe reactions and NH$_3$ TPD support the latter possibility and agree with prior studies that attributed a higher thermal stability to silica-supported heteropolyanions than to alumina-supported ones [Kasztelan, 1990].

For example, as shown in Fig. 10, NH$_3$ desorbs from the unsupported HPW sample (with a molar ratio of NH$_3$ to total protons near unity) at temperature near 575°C. The desorption pattern of HPW on silica resembles that of bulk HPW and the number of ammonia molecules corresponds closely to the number of protons (Table 1).

The supported WO$_3$ catalysts adsorb ~ 1 NH$_3$ molecule per two tungsten oxide groups. The desorption spectrum of HPW supported on alumina behaves similarly to WO$_3$ supported on alumina both in the temperature of desorption and in the number of ammonia molecules adsorbed per gram (672 versus 790 μmol). These results together with the X-ray diffractogram suggest that HPW has substantially decomposed to WO$_3$ on the alumina support. This instability may relate to the surface acidity of the support in aqueous suspension. Silica is more acidic in aqueous suspension than alumina is (isoelectric point of 2 versus 8), so that in the presence of the basic sites on alumina, HPW decomposes predominantly to WO$_3$. Consequently, we chose silica as the support for heteropolyacid salts.

In the experimental section, we described three techniques for preparing supported ammonium salt catalysts, namely (1) gaseous NH$_3$ reaction on supported HPW, (2) hydrothermal synthesis of soluble precursors, and (3) coimpregnation of an HPW solution with urea.

Figure 11 shows the clearly distinguishable X-ray spectrum of bulk (NH$_4$)$_{3-x}$H$_x$PW$_{12}$O$_{40}$ made by ion exchange. Note that the unit cell of the ammonium acid salt has contracted compared to the hexahydrated starting

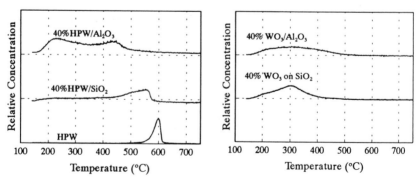

Figure 10. TPD spectra of HPW, 40% HPW/SiO$_2$, 40% HPW/Al$_2$O$_3$, 40% WO$_3$/SiO$_2$ and 40% WO$_3$/Al$_2$O$_3$.

TABLE 1
NH₃ Adsorption Titers on HPW, and Silica- and
Alumina-Supported HPW and WO₃

Catalyst	μmol NH₃/g	μmol H⁺/g	μmol WO₃/g
Bulk HPW	860	1042	
40% HPW/Al₂O₃	672	436	(1744)
40% HPW/SiO₂	441	436	(1744)
40% WO₃/Al₂O₃	790		1725
40% WO₃/SiO₂	791		1725

acid, because of the smaller size of the NH_4^+ versus $(H_5O_2)^+$ cation. The figure also shows that 24% HPW/SiO₂ converts easily to the supported ammonium salt by treatment in NH₃.

Figure 12 shows the X-ray spectrum of the bulk and supported phase formed by hydrothermal preparation. The dry reaction of ammonium metatungstate and ammonium phosphate dibasic does not form this phase nor does a simple solution reaction at 80°C in water. Hydrothermal synthesis at 150°C is necessary to produce the acid salt. The broadening of the diffraction peaks at low loadings (here at 8.1 wt %) has been attributed previously in the case of HPW/SiO₂ to the formation of small acid particles [Izumi *et al.*, 1983].

Figure 13 shows the synthesis of the supported ammonium salt on silica using the urea addition technique. This procedure is the simplest be-

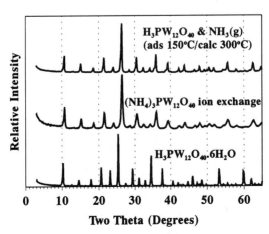

Figure 11. Preparation of supported $(NH_4)_{3-x}H_xPW_{12}O_{40}$ by reaction of 40% HPW/SiO₂ with NH₃.

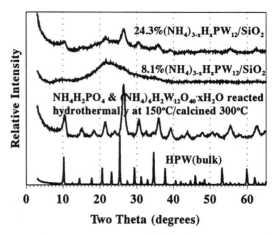

Figure 12. Preparation of $(NH_4)_{3-x}H_xPW_{12}O_{40}$ bulk and SiO_2 supported by hydrothermal synthesis.

cause it uses an HPW solution with addition of inexpensive urea. During the thermal decomposition of the urea, the NH_3 generated reacts with HPW to form the acid salt.

The bulk Cs acid salt suffers from the same limitations as the ammonium salt: it forms submicron particles that are difficult to separate by filtering and difficult to use in catalytic reactors without creating large pressure drops. Because the salt is not water soluble, the traditional aqueous

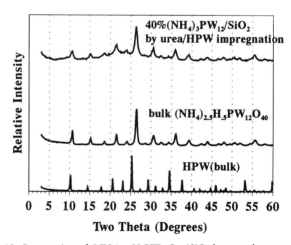

Figure 13. Preparation of $(NH_4)_{3-x}H_xPW_{12}O_{40}/SiO_2$ by urea decomposition.

impregnation route of depositing a soluble precursor onto a suitable support cannot be employed. However, we have found that a supported Cs acid salt can be prepared by an *in situ* reactive deposition technique. A soluble cesium salt is first impregnated onto silica, dried, and calcined; and the support is then impregnated with the soluble (HPW) acid, dried, and calcined. The Cs acid salt apparently forms by a reaction between cesium carbonate and acid on the support. The X-ray diffraction spectrum of the Cs acid salt supported on silica clearly shows the presence of the Cs acid salt (Fig. 14). The pattern resembles that of the bulk Cs acid salt and the reflections are clearly shifted to higher two-theta values relative to the hexahydrate acid form, indicating that the unit cell has contracted.

The unusual feature of this simple preparation method is the internal thin rings (eggwhite distribution) that are formed when the impregnations are conducted on a silica extrudate (Fig. 15). In this figure, the extrudate shown has a diameter of 1/16 in. with a surface area of 260 m^2/g, a pore volume of 1.11 cm^3/g, and a median pore diameter of 180 Å. Of the Cs acid salt, 80–90% is present in a ring about 10 μm thick (appearing white in Fig. 15) located about half way into the 1/16-in. diameter extrudates. Electron backscattering and energy dispensive spectroscopy (EDS) measurements indicate that Cs and W are present in the same location. Subsequent impregnation with heteropolyacid produces the ring.

During the initial impregnation, at high pH, Cs ions apparently exchange with the protons of the surface hydroxyls creating an even distribution of surface Si—O—Cs groups. EDS confirms an even Cs distribution

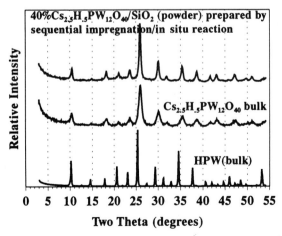

Figure 14. Preparation of 40% $Cs_{2.5}H_{0.5}PW_{12}O_{40}/SiO_2$ by sequential impregnation and *in situ* reaction.

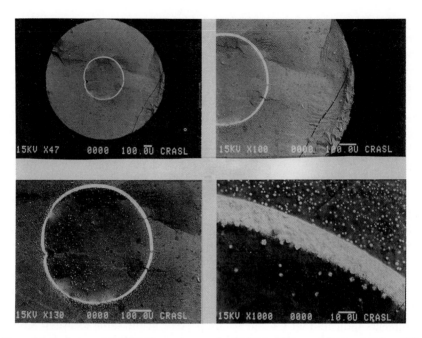

Figure 15. Backscattering SEM photomicrograph of 1/16-in. SiO_2 extrudate containing 40% $Cs_{2.5}H_{0.5}PW_{12}O_{40}$ located in narrow eggwhite distribution.

following impregnation and calcination, but before acid addition. During the subsequent impregnation with 12-tungstophosphoric acid, protons from the acid exchange with and release Cs ions attached to the surface, so that a homogeneous rapidly moving front of composition $H_{3-x}Cs_xPW_{12}O_{40}$ passes into the extrudate. As the front progresses further into the extrudate, x increases. When a critical content of Cs is reached, the acid salt precipitates, creating a well-defined eggwhite pattern within the extrudate. The proposed mechanism is illustrated in Fig. 16.

Not surprisingly, transport limitations affect the observed reaction rates with this distribution of heteropolyacid salt. Table 2 compares 40% $Cs_{2.5}H_{0.5}PW_{12}O_{40}$ supported on SiO_2 powder (0.17-mm diameter) with 40% $Cs_{2.5}H_{0.5}PW_{12}O_{40}$ supported on silica extrudates (1.4-mm diameter) for the liquid-phase aromatic alkylation of 1,3,5 trimethylbenzene (mesitylene) with cyclohexene at 100°C after 2 h of reaction. The diffusional effects of having the active phase inside the extrudate are substantial for this

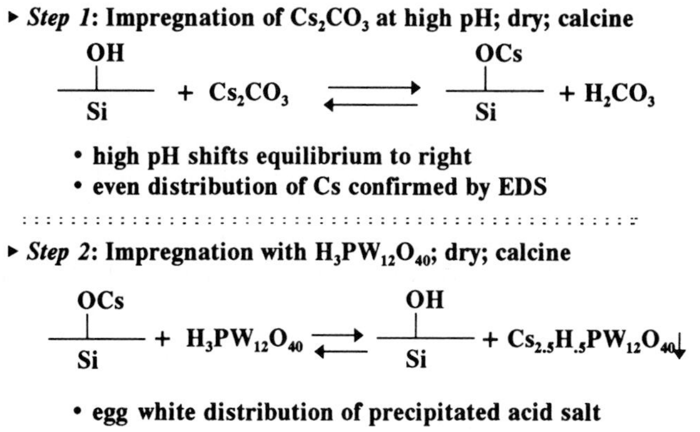

▶ *Step 1*: **Impregnation of Cs₂CO₃ at high pH; dry; calcine**

$$\underset{\text{Si}}{\text{OH}} \;+\; \text{Cs}_2\text{CO}_3 \; \underset{\longleftarrow}{\longrightarrow} \; \underset{\text{Si}}{\text{OCs}} \;+\; \text{H}_2\text{CO}_3$$

• **high pH shifts equilibrium to right**
• **even distribution of Cs confirmed by EDS**

▶ *Step 2*: **Impregnation with H₃PW₁₂O₄₀; dry; calcine**

$$\underset{\text{Si}}{\text{OCs}} \;+\; \text{H}_3\text{PW}_{12}\text{O}_{40} \; \underset{\longleftarrow}{\longrightarrow} \; \underset{\text{Si}}{\text{OH}} \;+\; \text{Cs}_{2.5}\text{H}_{.5}\text{PW}_{12}\text{O}_{40}\!\!\downarrow$$

• **egg white distribution of precipitated acid salt**

Figure 16. Proposed mechanism for formation of eggshell distribution of $Cs_{2.5}H_{0.5}PW_{12}O_{40}$ in SiO_2 extrudate.

liquid-phase reaction. For the gas-phase reaction of 2MP2 over these catalysts, the differences were smaller. We measured the reaction over 10–20 mesh pills of the powder, which are nominally the same size as the 1/16-in. extrudates. The activity of the the eggwhite extrudate was only about 10% lower than that of the catalysts prepared on silica powder. A detailed description of the activity of these catalysts in aromatic alkylation and the accompanying issues regarding deactivation is described elsewhere [Soled, 1996].

TABLE 2
Reaction of Bulk and Supported Cs Acid Salts

Catalyst	% Cyclohexene conversion to alkylate[a] (catalyst amount)	2MP2 conversion[b]
Bulk $Cs_{2.5}H_{0.5}PW_{12}O_{40}$	90 (0.5 g)	67.4
40% $Cs_{2.5}H_{0.5}PW_{12}O_{40}/SiO_2$ powder	91 (1.25 g)	45.7
40% $Cs_{2.5}H_{0.5}PW_{12}O_{40}/SiO_2$ extrudate	38 (1.25 g)	41.2

[a]@100°C, 2 h.
[b]200°C, 5 m, 1 g cat, powder catalysts run at 10–20 mesh.

16.4   Conclusions

Supported catalysts containing the water insoluble ammonium and cesium heteropolyacid salts of 12-tungstophosphoric acid have been prepared by gas-phase reaction, hydrothermal synthesis, precursor decomposition, and *in situ* reaction and precipitation. These supported versions of bulk insoluble acid salts provide a set of new strong acid catalysts that can be evaluated in large-scale fixed-bed or slurry reactors without the pressure drop or separation problems associated with submicron-sized bulk acid salts.

Acknowledgments

We are thankful to John Ziemiak, Angela Klaus, Greg DeMartin, and Ken Riley for contributions to this study.

References

Ai, M. (1981). *J. Catal.*, **71**, 88–94.

Baba, T., and Ono, Y. (1986). *Appl. Catal.*, **22**, 321–324.

Bailar, J. C. (1978). *Inorganic Syntheses* (H. S. Booth, ed.), Vol. 1, pp. 132–133, Robert E. Krieger Publishing, Huntington, New York.

Bartoli, M. J., Monceaux, L., Bordes, E., Hecquet, G., and Courtine, P. (1992). *New Developments in Selective Oxidation by Heterogeneous Catalysis* (P. Ruiz and B. Delmon, eds.), pp. 81–90, Elsevier, New York.

Hayashi, H., and Moffat, J. B. (1992). *J. Catal.*, **77**, 473–479.

Henke, A. M., and Sebulsky, R. T. (1967). U. S. patent 3 346 657.

Izumi, Y., Hasebe, R., and Urabe, K. (1983). *J. Catal.*, **84**, 402–409.

Kasztelan, S., Payen, E., and Moffat, J. B. (1990). *J. Catal.*, **125**, 45–54.

Kramer, G. M., and McVicker, G. B. (1986). *Acc. Chem. Res.*, **19**, 78–87.

Misono, M. (1982). *Proceedings 4th International Conference on Chemistry of Mo* (H. F. Barry and P. C. H. Mitchell, eds.), pp. 289–295, Climax Molybdenum, Ann Arbor, MI.

Misono, M. (1983). *J. Catal.*, **83**, 121–129.

Misono, M. (1987a). *Catal. Rev. Sci. Eng.*, **29** (2–3), 269–321.

Misono, M. (1987b). *Mater. Chem. Phys.*, **17** (1–2), 103–120.

Misono, M., and Okuhara, T. (1993). *Chemtech*, 23–29.

Moffat, J. B. (1985). *Stud. Surf. Sci. Catal.*, **20**, 157–166.

Nishimura, T., Okuhara, T., and Misono, M. (1991). *Appl. Catal.*, **73**, L7–L11.

Nomiya, K., Sugaya, Y., Sasa, S., and Miwa, M. (1990). *Bull. Chem. Soc.*, **53**, 2089–2093.

Nowinska, K., Fiedorow, R., and Adamiec, J. (1991). *J. Chem. Soc. Faraday Trans.*, **87** (5), 749–753.

Okuhara, T., Nishimura, T., and Misono, M. (1995). *Chem. Lett.*, (2), 155.

Ono, Y., Mori, T., and Keii, T. (1981). *Proceedings 7th International Congress Catalysis* (T. Seiyama and K. Tanabe, eds.), pp. 1414–1418, Elsevier, Amsterdam.

Schwegler, M. A., Vinke, P., van der Eijk, M., and van Bekkum, H. (1992). *Appl. Catal. A,* **80**, 41–57.
Sebulsky, R. T., and Henke, A. M. (1971). *Ind. Eng. Chem. Process Dev.* **2**, 272–276.
Soled, S., Miseo, S., McVicker, G., Gutierrez, A., and Paes, J. (1996). *J. Chem. Eng.,* in press.
Thomas, J. M. (1992). *Sci. Am.,* **266** (4), 112–120.

CHAPTER 17

Preparation of
Iron/Molybdenum/Molecular Sieves
by Chemical Vapor Deposition

Jin S. Yoo, J. A. Donohue, and C. Choi-Feng
Amoco Research Center
Naperville, Illinois 60422

KEYWORDS: Catalyst preparation, CVD, impregnation, Fe/Mo/DBH, gas-phase oxidation, silica coated, terephthaldehyde, xylenes

17.1   Introduction

The laser-induced chemical vapor deposition (CVD) and photochemical vapor depositions of a variety of metals, Ga, In, Sb, Au, W, Al, and Te; and metalloids such as As, P, and Si have been extensively applied to the semiconductor and solar cell industry. In particular, the chemical vapor deposition of silica film using tetraalkoxysilane becomes a key process in producing semiconductor devices [1–5].

Extensive efforts are also being made to develop H_2, O_2, CO, N_2, H_2S, and CO_2 permselective microporous inorganic membranes by applying the same CVD technique. In general, silane derivatives are used for a thin layer of silica deposition, and the sol–gel method is also used to coat a thin silica film on the top of the substrate materials.

The ultrathin layer of silica was prepared by the CVD of $Si(OMe)_4$ on γ-Al_2O_3 [6,7], titania, and zirconia [8]. The monolayer consisting of the 1:1 bonds of M—O—Si—OH (M = Al, Ti, or Zr) in the network of siloxanes (Si—O—Si) generated Brønsted acidity [6–10]. The structure and acidic properties of SiO$_2$ layers have been studied by other authors [11–14]. The resulting silica monolayers on these metal oxides were applied for the double-bond isomerization [7], and the Beckmann rearrangement of cyclohexanone oxime [15].

The mechanism of the CVD Si(OMe)$_4$ on γ-Al$_2$O$_3$ was further elucidated in detail by conducting the CVD at 323°C and low temperature (<200°C) [16]. The activity for isomerization of butene was observed over the sample prepared at 323°C, while the sample obtained at 200°C exhibited no activity. It was proposed that the Al—O—Si—OH in the network of Si—O—Si was the site for the Brønsted acidity, whereas the dispersed silica formed at lower temperature did not show acidity. Additional works on the CVD silica monolayer preparation [12–15], the characterization [17], and the application [18] were reported.

The similar ultrathin layer of germanium oxide with monoatomic order thickness was prepared by CVD of germanium alkoxide on γ-Al$_2$O$_3$, and the structure of the oxide monolayer and the origin of acidity were analyzed by the extended X-ray absorption fine structure (EXAFS) [19]. The same group extended the CVD of germanium alkoxide on zeolite in an effort to control the opening size of the micropore [19,20].

Also, the CVD technique has attracted unusual attention from the catalyst manufacturing industry. The technique has been employed for depositing highly and uniformly dispersed fine particles of nanosize of the catalytic materials such as metal and metal oxide particles on a supporting matrix. These CVD catalysts tend to exhibit better catalytic stability, and novel catalytic performance different from the counterparts prepared by the conventional techniques [9,10]. The technology was also applied to modify the catalyst surface structure with the silica layer [23,24], and the V$_2$O$_5$ thin film [25].

The chemical vapor deposition of silicon alkoxide has been employed for controlling pore-opening size of zeolites such as mordenite [26]. A platinum-loaded H-mordenite has been modified by the same CVD of silica to improve its shape–selectivity for the hydrocracking of paraffins. The finely controlled modification of pore size by the silica deposition via the CVD would have a wide application to shape selective catalytic reactions over metal-loaded zeolites. Other applications include cracking, isomerization, alkylation, hydrocracking, disproportionation, and oxidation.

Alkylation of toluene with methanol and toluene disproportionation to make p-xylene have been conducted over HZSM-5 modified with CVD silicon alkoxide (Si/HZSM-5) [27,28]. As the silica deposition amount increased, the selectivity to p-xylene in the xylene product increased to more than 98%. Also, methoxytripropylsilane was found to be effective to make the CVD Si/HZSM-5 for controlling pore size on the exterior surface [28]. Iwasawa [29] published an excellent review article in which the structure and catalysis of CVD-prepared catalysts were thoroughly discussed.

We previously reported some novel oxidation reactions observed with the CVD Fe/Mo/partially deboronated borosilicate molecular sieve (DBH) catalyst. These include selective synthesis of terephthaldehyde from p-xylene [30], gas-phase O_2 oxidations of alkylaromatics in the presence of CO_2 [31], preferential oxidation of p-xylene from a xylene mixture containing ethylbenzene [32], one-step hydroxylation of benzene to phenol [33], effect of supporting matrices on p-xylene oxidation [34], *para*-selective oxidation of polymethylbenzenes [35], and characterization of the CVD Fe/Mo/DBH catalyst [36].

In this chapter, we focus on the procedure of the preparation of the CVD Fe/Mo/DBH catalyst, its silica-coated system, and the impregnated counterpart. The catalyst performances of these three systems are compared based on the results of the p-xylene oxidation in the gas phase under comparable conditions.

Details of the catalyst preparation, the surface properties, and *para*-selective oxidation property of xylenes over the CVD Fe/Mo/DBH catalyst are compared with those of other CVD mixed metal oxide catalysts supported on various zeolite matrices. The CVD Fe/Mo/DBH, impregnated catalyst prepared by the conventional incipient wetness method, and the CVD Fe/Mo/DBH catalyst modified with tetramethylorthosilicate (TMS) were characterized by Raman spectroscopy and electron micrography (EM) [36]. The active catalytic species, $Fe_2(MoO_4)_3$, is uniformly dispersed as fine particles (2–40 nm) on the surface of the CVD Fe/Mo/DBH catalyst, and this is reflected on the superior catalyst performance and stability over its impregnation counterpart for the *para*-selective gas-phase O_2 oxidation of p-xylene to aldehydes.

17.2   Experimental

17.2.1 Preparation of CVD Fe/Mo/DBH

The use of silanols on the borosilicate molecular sieve as anchoring sites for metallic elements or compounds is an area of continuing interest. The borosilicate molecular sieve, HAMS-1B-3 (containing 1.21 wt % boron, and 440 ppm Al, 256 ppm Fe, and 157 ppm Pb as impurities), has been used as a main anchoring matrix for the catalyst preparation. The HAMS-1B-3 was partially deboronated to produce the deboronated borosilicate (DBH) containing two levels of the boron content, 0.26 and

0.013 wt %. The resulting deboronated samples have been determined to have 19 and 20 wt % silanol groups, respectively, by ^{29}Si MAS NMR [37]. The level of silanol on the borosilicate matrix depends on the level of boron in the lattice framework.

However, the deboronation from the borosilicate molecular sieve, HAMS-1B-3, occurs mainly *in situ* at the initial steps of the CVD operations, in particular, the FeCl$_3$ vapor deposition step. In short, the HAMS-1B-3 sample was used as a starting supporting matrix for the preparation of most of the CVD catalysts discussed in this work. The CVD Fe/Mo/DBH catalyst was prepared from FeCl$_3$ and MoO$_2$Cl$_2$, MoOCl$_4$, or MoCl5 with HAMS-1B-3 by the CVD technique.

As shown in Scheme 1, there are three main steps involved in the preparation of catalyst. The first step involves the exchange reaction with ammonium acetate (NH$_4$OAc) to adjust the silanol content on the sieve surface by removing some boron. Ammonium acetate (153.2 g) was dissolved in 3500 ml of deionized water in a 4-liter pyrex beaker using a mechanical stirrer. The sample (350 g) of borosilicate molecular sieve containing 1.70% boron was added and the white slurry was stirred at room temperature for one-half hour. The resulting slurry was filtered, air-dried on a filter, and then oven-dried at 110–120°C overnight. The dried product contained 1.26% boron and it was calcined in air at 660°C for 16 h to get rid of the excess acetate ions and to adjust the silanol content.

In the second step, the calcined product was used for FeCl$_3$ vapor deposition. The calcined sample was loaded into the horizontally mounted quartz tube with quartz wool plugs securing the sample in place. About 6.0 g of FeCl$_3$ was placed with additional quartz wool plugs at the end of the reactor in a dry bag that was constantly being purged with N$_2$. The reactor was placed back into the tube furnace and a N$_2$ line was hooked up to the end, close to the FeCl$_3$ at a flow of about 60 ml/min. The reactor temperature was raised to 460°C in 1 h, and was kept at this temperature for 1 h to allow FeCl$_3$ vapor to pass through the molecular sieve bed. The flow of nitrogen was reversed from the other cool end of the reactor, where some of unreacted FeCl$_3$ vapor was condensed. The nitrogen flow was reversed to reuse the condensed FeCl$_3$ and to ensure the uniform deposition of FeCl$_3$ throughout the molecular sieve bed. The resulting bed became uniformly yellow in color. After the resulting yellow iron-deposited solid was unloaded from the reactor, it was added to 800 ml of deionized water to form a slurry, washed three times, and oven-dried at 120°C. The dried iron-deposited Fe/DBH weighed 42.9 g and contained 41% Si, 0.157% boron, and 1.42% Fe.

Commercial borosilicate molecular sieve, HAMS-1B-3

Treatment	Purpose
NH₄OAC Exchange	Remove excess boron that is blocking pores.
↓	
Calcination	Reduce silanol concentration to limit iron uptake.
↓	

Treated molecular sieve

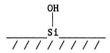

Heat reactor containing FeCl₃	Vapor deposit iron
↓	
Cool and wash	Remove chloride and boron to provide Si-OH for Mo attachment
↓	

Fe/DBH

Heat reactor containing MoO₂Cl₂	↓	Vapor deposit molybdenum
Calcine	↓	Stabilize Mo for washing
Wash or steam	↓	Remove chloride
Prolonged calcination	↓	Activation

Fe/Mo/DBH

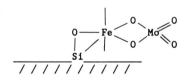

Scheme 1

In the third step, a significant amount of boron was removed from 1.25 to 0.157% (following completion of the second step). The resulting Fe/DBH sample (30.9 g) was loaded into the reactor and calcined at 400°C under a N_2 flow, and became 26.1 g. The same vapor deposition procedure as in the second step was repeated with 6.5 g MoO_2Cl_2 for 4 h at 200°C and one-half hour at 200–300°C for the reversed operation. After 6.5 h at

300°C, N_2 was replaced with air and kept the reactor at 300°C for additional 8 h. The reactor temperature was then raised to 650°C and kept at that temperature for the additional 8 h before it was cooled down to an ambient temperature. The resulting Fe/Mo/DBH was analyzed; it contained 4.6 wt % Mo and 1.41 wt % Fe, and the Mo:Fe ratio was 1.9:1, which became ready for evaluation of the p-xylene oxidation. If the Mo:Fe ratio is 2.2:1 or higher, it is preferred to subject the sample to further calcination at 650–690°C for a prolonged period until the ratio reaches 1.6 ~ 2.0:1 for the final activation.

Many batches of CVD Fe/Mo/DBH catalysts having different compositions and various metal loading levels were prepared according to the preceding CVD technique. Catalyst G was coated with tetramethylorthosilicate (TMS, 1.4 wt % gain as silica) *in situ* in the microreactor according to the procedure described elsewhere [32]. Other catalysts were also prepared by varying the metal components as well as the supporting matrix by the CVD method. Mo/Sb/DBH, W/Fe/DBH, Mo/Sb/dealuminated Y-zeolite (DAY), Mo/Cr/DBH, Mo/Ti/DBH, V/Fe/CBH, Fe/Mo/ZSM-5, Fe/Mo/silicalite, Fe/Mo/β-zeolite, and Fe/Mo/DAB (dealuminated β-zeolite) are among them. These catalysts were evaluated for the gas-phase O_2 oxidation of p-xylene, and the representative results are summarized in Table 1.

The results showed that the DBH-based catalyst, namely, CVD Fe/Mo/DBH, exhibited the most promising results for the aldehyde syntheses among various supporting matrices. We decided to focus mainly on the oxidation reaction of xylenes for the evaluation of the catalyst performance in this work. Physical properties of five CVD Fe/Mo/DBH catalysts, A, B, C, D, and E, which have different levels of metal loading; catalyst F (the silica-coated catalyst); and catalyst G prepared by the incipient wetness impregnation method are compared in Table 2.

17.2.2 Characterization of Catalysts

17.2.2.1 Raman Spectroscopy

All Raman scattering measurements were conducted using the 514.5-nm line from an Ar ion laser, and a Jobin-Yvon Ramalor 1000 Raman spectrometer with a 2400 grooves per millimeter grating or a Dillor Omars 89 Raman spectrometer equipped with a multichannel detector. The laser beam intensity was typically 100–150 mW, and a sample spinner was used during analysis at a speed of ca. 3000 rpm. An Ar ion laser operated at a wavelength of 5145 Å and 25 mW was used for analysis. The sampling area was approximately 5–10 μm^2. All peak width and line positions were determined by a least-squares fit to a Lorentzian line shape.

TABLE 1
Effect of Supporting Matrix on p-Xylene Oxidation Catalyst: CVD Fe/Mo/Supporting Matrix[a]

Supporting matrix	ZSM-5			Silicalite		β-Zeolite (DAB)[b]	HAMS-1B-3 (DBH)	US-Y (DAY)[c]
Temp (°C)	350	350	550	350	400	350	350[d]	350
pX conv. (mol %)	56	1.6	65.5	13.1	43.8	74.7	21.4	>60
Product selectivity (mol %)								
TPAL	0	6.6	33.4	6.4	8.8	39.4	18.7	
TOAL	0	0	27.0	34.5	23.6	24.9	46.4	
Toluene	66			3.8	5.2		2.2	
PSCUME	3.8			0.8	0.2			
MA				3.9	4.9	9.8	0.2	
BZAL						2.5	0.8	
CO				10.5	15.2	5.3	5.9	
CO$_2$				38.5	39.8	14.0	26.1	
CO+CO$_2$	21.0	61.5	39.8					>95

[a]TPAL: terephthaldehyde, TOAL: p-tolualdehyde, PSCUME: pseudocumene, MA: maleic anhydride, BZAL: benzaldehyde, DBH: deboronated borosilicate, DAB: dealuminated β-zeolite.
[b]Deboronated β-zeolite.
[c]Dealuminated US-Y.
[d]Impregnated catalyst.

TABLE 2
Physical property of CVD Fe/Mo/DBH and Impregnated Fe/Mo/DBH

Catalyst	Metal wt %		Surface area (m²/g)	Radius (Å)	Micropore	
	Mo	Fe			Area (m²/g)	Vol (cc/g)
HAMS-1B-3[a]						
Support	—	—	330	43	—	0.136
CVD Fe/Mo/DBH[b]						
A[c]	9.9	1.7	247	38	207	0.099
B	8.7	2.2	271	27	212	0.096
C	6.2	1.9	270	27	196	0.090
D	5.8	1.4	316	27	229	0.104
E	9.4	1.6	—	—	—	—
SiO₂-coated CVD Fe/Mo/DBH						
F[d]	9.5	1.6	299	—	—	—
Impregnated Fe/Mo/DBH						
G[e]	5.4	0.99	303	34	262	—

[a]Borosilicate molecular sieve matrix.
[b]Deboronated borosilicate molecular sieve (DBH).
[c]Catalyst calcined at 400°C (not fully activated).
[d]Catalyst E was coated with SiO₂.
[e]Fe and Mo were impregnated on DBH by incipient wetness technique.

17.2.2.2 Electron Micrography

To study the pathway for deposition and the distributions of Fe and Mo in the final CVD Fe/Mo/DBH catalyst, three samples, the fresh catalyst, the activated catalyst calcined in air for a prolonged period, and the spent catalyst resulting from the catalyst stability runs, were characterized. These samples were ground, embedded in LR-White resin, and cured for 60 min at 90°C. Transmission electron microscopy (TEM) thin sections were then obtained by ultramicrotomy, which were coated lightly with carbon prior to the EM examination. The thin sections were examined with a JEOL JEM 2000EX high-resolution electron microscope (HREM) and analyzed with a Philips 420 analytical electron microscope (AEM) that is equipped with a Noran 5502 energy dispersive X-ray emission spectrometry (EDXS).

17.2.2.3 Ammonia TPD

An ammonia temperature-programmed desorption (TPD) study was conducted on a series of borosilicate molecular sieves (HAMS-1B-3), iron-deposited Fe/DBH, and final Fe/Mo/DBH to monitor the surface change at each step in the CVD procedure. The sample was pretreated at 550°C for an hour, cooled to 70°C, and dosed with 5% ammonia. The excess ammonia was purged from the system for 1 h to remove weakly bound ammonia. The temperature was then raised at 10°C/min to a maximum of 500°C. The results are shown in Fig. 1.

The HAMS-1B-3 (the supporting matrix) sample showed a single peak at relatively low temperature (172°C), which was interpreted as weakly acid material. During the step that iron was deposited on the HAMS-1B-3 and washed in the subsequent treatment, the partial deboronation occurred and Fe/DBH was prepared. A second peak developed at 319°C on the resulting Fe/DBH. This strong peak was generated as a result of interaction between ferric chloride and the surface silanol group on the matrix. When molybdenum was deposited to Fe/DBH at the final step to prepare Fe/Mo/DBH, the strong peak at 319°C was lost and only the weak peak at 195°C was generated. In other words, the strong peak at 319°C on the Fe/DBH disappeared, presumably via the chemical interaction between oxides of iron and Mo to form ferric molybdate and molybdenum trioxide phase.

17.2.3 Evaluation of Catalysts

These catalysts prepared according to the preceding procedure were evaluated for the gas-phase O_2 oxidation of *p*-xylene in a microreactor with 0.5 g of catalyst and regular reactor with 5–10 g of catalyst under the standard conditions. The procedures for catalyst evaluation, reactor operation, and product analyses with an on-line gas chromatography (GC) are fully described elsewhere [30,31].

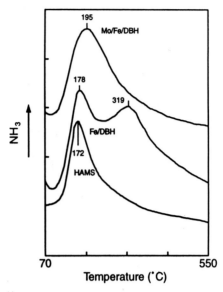

Figure 1. Ammonia TPD, ammonia desorption rate as a function of temperature. Fe/DBH vs Fe/Mo/DBH (deboronated borosilicate molecular sieve). Instrument: multifunctional TPD/TPR pulse chemisorption unit.

17.3   Results and Discussions

17.3.1 Chemical Vapor Deposition

In the chemical vapor deposition procedure, the order of metal deposition was very important. The deposition of $FeCl_3$ that preceded the deposition of MoO_2Cl_2 on the HAMS-1B-3 matrix produced an active catalyst. However, when the order of metal deposition was reversed, namely, the MoO_2Cl_2 deposition was followed by the deposition of $FeCl_3$, the resulting catalyst became unstable and caused an extensive burning in the gas-phase O_2 oxidation of p-xylene. The strong interaction of the Si—OH group on the supporting matrix with $FeCl_3$ provided the sites for the MoO_2Cl_2 vapor to interact in the subsequent step. This interaction led to formation of the ferric molybdate species, which was identified as an active phase. On the other hand, the molybdenum deposition in the initial step failed to provide the sites on which iron could interact to form the active system. The dramatically different catalyst behaviors observed with these two catalysts can be explained by Raman spectra shown in Fig. 2. The normal order of metal deposition formed two phases: ferric molybdate, $Fe_2(MoO_4)_3$, and MoO_3, which were active sites for the aldehydes formation; and the reverse order generated the α-Fe_2O_3 phase; which caused an extensive combustion.

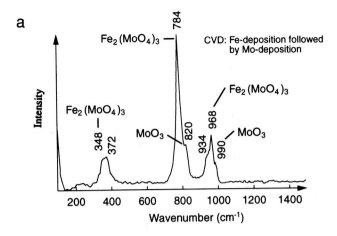

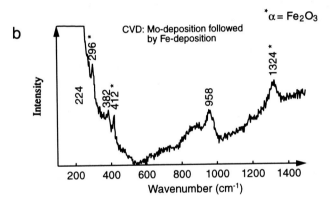

Figure 2. Effect of the order of metal vapor deposition (a) Fe deposition followed by Mo deposition; (b) Mo deposition followed by Fe deposition.

17.3.2 Activation of CVD Fe/Mo/DBH by Air Calcination

The catalyst precursor was prepared from $FeCl_3$, MoO_2Cl_2, and borosilicate molecular sieve (HAMS-1B-3) according to Scheme 1. To remove the residual chloride and to stabilize it by inducing the chemical interaction between Fe and Mo moieties, the catalyst precursor was subjected to the steam treatment at approximately 150°C for several hours. The steamed and dried catalyst precursor contains 14.5 wt % Mo, 2.55 wt % Fe, 0.05 wt % boron, atomic ratio of Mo:Fe = 3.3:1, 0.45 wt % Cl, and surface area of 283 m^2/g. The resulting catalyst precursor was activated in the subsequent step by calcining in air at elevated temperature for a prolonged period. The precursor (1 ml [0.521 g]) was loaded in a quartz microreactor and again calcined *in situ* in the reactor at 400°C for 2 h before the gas-phase O_2 oxidation of *p*-xylene was studied. The catalyst resulting from this step is referred to as catalyst A. Catalyst A was calcined *in situ* in the same reactor at 680–695°C by prolonging the calcination time consecutively in a series. The atomic ratio by Mo:Fe was changed from the initial value of 3.3:1 to 1.8:1 via calcination at 680°C for 1 day. This change was attributed to the loss of excess MoO_3, which sublimed off the matrix in the form of $MoO_3.H_2O$ during the calcination step. The ratio was maintained even after it was further calcined at 695°C for another 5 days. These catalysts resulting from each step were evaluated for the gas-phase O_2 oxidation of *p*-xylene at 375°C in the same reactor before the subsequent calcination started. The results summarized in Table 3 suggest that the calcination and reaction temperature have a profound effect on the catalytic activity and selectivity to terephthaldehyde and burning. When the catalyst was calcined *in situ* in the reactor at 680°C for 1 day, the catalyst precursor was practically fully activated.

Both the catalyst precursor and the activated catalyst resulting from the calcination step at 680°C were characterized by Raman spectroscopy, and the results are shown in Fig. 3. It appears that both polymeric and monomeric surface molybdate species in the precursor were transformed mainly to MoO_3, the majority of which was sublimed off the matrix as $MoO_3.H_2O$, and the remaining portion interacted with the "bare" iron sites to form the $Fe_2(MoO_4)_3$ phase.

In another study, a series of calcinations were successively conducted with catalyst A in the flow of air at a designated temperature for 12–16 h in each step. The temperature was raised from 400 to 800°C. The changes in the physical properies that occurred during these calcination steps were monitored and summarized in Table 4.

The phase changes occurring at each calcination step were also tracked by the Raman spectroscopy as shown in Fig. 4. Variation of the ratios of two phases, $MoO_3:Fe_2(MoO_4)_3$, were determined from the Raman spectra, and plotted as a function of the calcination temperature in Fig. 5. These results confirmed the early finding that the active catalyst was prepared by subliming off excess molybdenum from its precursor during the prolonged calcination period. The sublimation of MoO_3 continued during the subsequent calcination steps at elevated temperature until a limited portion of MoO_3 remained on the matrix. The final activated catalyst resulting from this study has an atomic ratio of Mo:Fe = 1.8:1, approaching the stoichiometry of $Fe_2(MoO_4)_3$ (Mo:Fe = 1.5:1).

TABLE 3
A Series of Consecutive Oxidations of p-Xylene over Calcined Batches of Catalysts in a Microreactor[a]

Run no.	Calcination[b]		Oxidation			Product selectivity (%)[c]				
	T (°C)	Time	T (°C)	O$_2$/pX	Conv	TPAL	TOAL	CO$_2$	BYPROD	
1	400	1 h	375	10	20	39	18	37	6	
2	680	1 day	375	10	68	47	21	19	13	
3	695	1.8 days	375	38	32	54	28	14	4	
4	695	4 days	375	38	17	58	27	13	2	
5			425	38	43	60	18	18	4	
6			500	38	74	64	9	21	6	
7			564	38	83	60	6	28	6	
8			510	38	79	67	9	19	5	

[a] Catalyst A (Mo : Fe = 3.3 : 1) and Activated Catalyst (Mo : Fe = 1.8 : 1).
[b] The catalyst in the reactor was treated in situ and consecutively through runs 1 to 8.
[c] TPAL: terephthaldehyde, TOAL: p-tolualdehyde, BYPROD: byproducts.

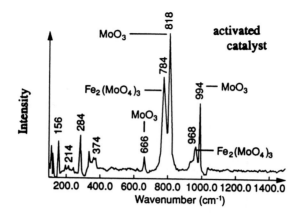

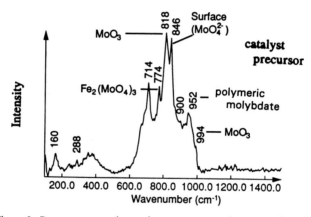

Figure 3. Raman spectra, the catalyst precursor vs the activated catalyst.

17.3.3 Silica-Coated Catalyst

Catalyst E was coated with tetramethylorthosilicate (TMS) *in situ* in the reactor, and the resulting catalyst (referred to as catalyst F) contained approximately 1.4 wt % silica. Catalysts E and F were subjected to the gas-phase O_2 oxidation of the mixed xylene feed containing ethylbenzene in the microreactor [32]. The results with catalysts E and F were plotted in Fig. 6 and 7, respectively. These plots showed that *p*-xylene was preferentially oxidized over its isomer and ethylbenzene with uncoated catalyst E; however, the conversions of ethylbenzene, *o*-xylene, and *m*-xylene remained low at lower temperature (350°C), but increased to a substantial level at higher temperature (above 400°C). With the silica-coated catalyst F, the preferen-

TABLE 4
Effect of Calcination on Catalyst A, CVD Fe/Mo/DBH

Catalyst	Calcination temp (°C)	Metal			BET SA m²/g	Micropore		Av pore radius (Å)	
		Mo[a]	Fe[a]	Mo:Fe[b]		Area (m²/g)	Vol (cc/g)	Adsorp	Desorp
Precursor		14.5	2.55		283				
1	400	9.9	1.71	3.3	247	207	0.099	38	43
2	500	9.0	1.75	3.0	274	190	0.092	24	25
3	600	7.9	1.77	2.6	268	183	0.088	17	17
4	650	7.0	1.84	2.2	242	181	0.087	17	15
5	700	6.9	1.91	2.1	195	143	0.068	18	17
6	750	6.2	1.85	2.0	158	116	0.056	19	19
7	800	5.6	1.85	1.8	146	109	0.052	21	18

[a] wt %.
[b] Atomic ratio.

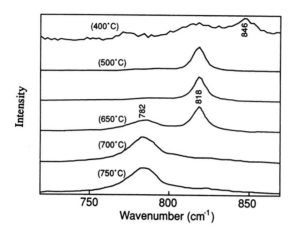

Figure 4. Raman spectra of catalyst A, the phase transformation with temperatures.

tial oxidation of *p*-xylene was noticeably improved by keeping the conversion of *o*-, and *m*-xylene and ethylbenzene at minimal levels at a temperature below 400°C. One drawback of silica-coated catalyst F is that it seems to burn the substrate more than the uncoated counterpart does.

17.3.4 Impregnated Fe/Mo/DBH

Catalyst G was prepared by impregnating $Fe(NO_3)_3$ and ammonium *p*-molybdate onto HAMS-1B-3 by the incipient wetness technique. It was activated by calcination in air at 650°C overnight. The Raman spectrum of catalyst G was compared with that of the CVD catalyst in Fig. 8. Two phases, $Fe_2(MoO_4)_3$ and MoO_3, were identified in catalyst G, as in the CVD counterpart by the Raman spectroscopy.

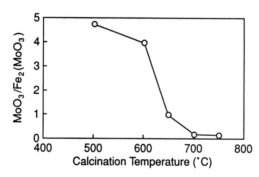

Figure 5. Variation of the ratio of $MoO_3 : Fe_2(MoO_4)_3$ with calcination temperatures.

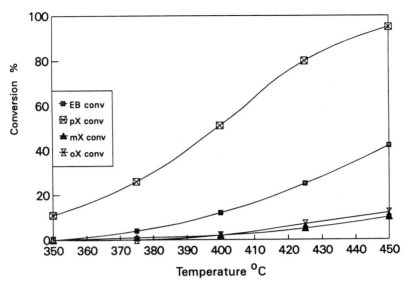

Figure 6. The oxidation of xylene mixture with ethylbenzene over catalyst E, CVD Fe/Mo/DBH.

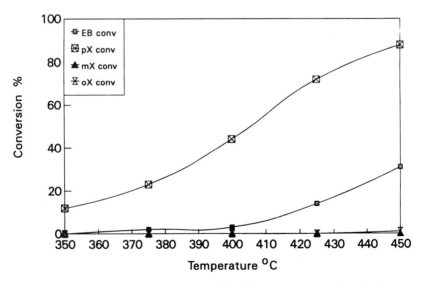

Figure 7. The oxidation of xylene mixture with ethylbenzene over catalyst F, silica-coated catalyst E.

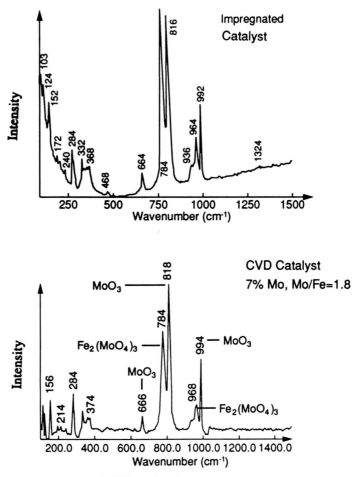

Figure 8. Raman spectra of CVD Fe/Mo/DBH and impregnated counterpart, catalyst G.

The resulting calcined catalyst contained 4.2 wt % Mo, 0.99 wt % Fe, and 0.82 wt % boron, had a ratio of Mo:Fe = 2.5:1, and was evaluated under the standard oxidation conditions. Effects of steam on the catalytic activity and product selectivity were also examined. The performance of catalyst G in two reactors, namely, microreactor over 0.5 g of catalyst with 0.21-s contact time in the feed containing 0.10 vol % p-xylene, and regular reactor over 5.0 g of catalyst with a shorter contact time (0.11 s) in another feed containing 0.21 vol % p-xylene, was summarized in Table 5. Quite different oxidation results were obtained with the same catalyst

TABLE 5
Comparison of Catalyst Perfoemances in Two Reactors[a]

Reactor	I[b]			II[c]		
Catalyst load (g)	0.51	0.51	0.50	5.0	5.0	5.0
Temp (°C)	350	375	400	400	425	450
pX % in feed	0.10	0.10	0.10	0.21	0.21	0.21
Contact time (s)	0.21	0.21	0.21	0.11	0.11	0.11
WHSV (h^{-1})	0.23	0.23	0.23	0.95	0.95	0.95
pX conversion (mol %)	21.4	38.5	63.1	24.9	51.1	86.0
Product selectivity (%)						
Benzaldehyde	0.8	1.1	1.0	0.7	1.1	2.0
p-Tolualdehyde	46.4	46.7	44.7	24.9	20.5	12.9
p-Toluic acid	0.0	0.4	0.4	—	—	—
Terephthaldehyde	18.7	16.8	18.1	44.6	51.7	59.1
Toluene	1.9	2.0	1.9	2.4	2.5	2.2
CO	5.9	6.6	7.6	5.7	5.6	6.1
CO_2	26.1	26.3	25.4	20.6	16.9	14.9

[a]Calcined catalyst G: 4.2 wt % Mo, 0.99 wt % Fe, Mo:Fe = 2.5:1.
[b]Reactor I: microreactor loaded with 0.5 g (1.4 ml); gas feed: 0.1% pX, 1.0% O_2, 1.0% N_2 in He.
[c]Reactor II: regular quartz reactor loaded with 5.0 g; gas feed: 0.21% pX, 4.4% O_2, 15.4% N_2 in He.

in reactors I and II. The concentration of *p*-xylene in the feed—0.1% in reactor I and 0.21% in reactor II—and the contact time—0.21 versus 0.11 s—are believed to be key factors along with the reaction temperature that contribute to the different results in the *p*-xylene conversion and the product selectivity. In general, it is believed that experimental results from the reactor loaded with a larger amount of the catalyst are more readily reproducible.

These results are also compared with those of the CVD catalyst in Table 6. It shows that the CVD catalyst performs much better than the impregnated catalyst counterpart. In order to explain why the CVD catalyst performs better than its impregnation counterpart, the electron micrography study was undertaken as part of the catalyst characterization.

Electron micrographs of catalysts G and the CVD counterpart are shown in Fig. 9. The high-resolution electron micrographs for the fresh CVD Fe/Mo/DBH catalyst reveal that the metal components are uniformly dispersed as fine particles (2–40 nm) onto the DBH matrix, primarily along micropore channel. Practically all iron and molybdenum reside in-

TABLE 6
Evaluation of Catalyst Performances, Impregnation versus CVD Technique

Temp (°C)	325	350		375		400	
Catalyst prep	CVD	Impr	CVD	Impr		CVD	Impr
pX conv (%)	(43.3)	21.4	(74.7)	32.6 [69.2][b]		(96.7)	61.6
Product selectivity (%)							
TPAL	(48.3)	18.7	(39.4)	3.5	[4.1]	(23.3)	18.1
TOAL	(33.7)	46.4	(24.9)	42.6	[48.6]	(24.9)	44.7
Benzal	(2.0)	0.8	(2.5)	1.0	[0.9]	(25)	1.0
MA	(5.7)	0.2	(9.8)	0.0	[0.0]	(17.0)	0.9
CO	(2.7)	5.9	(5.3)	8.1	[6.9]	(11.1)	7.6
CO_2	(6.6)	26.1	(14.0)	32.3	[33.5]	(28.2)	25.4

[a]Run conditions: contact time, 0.21 s; WHSV, 0.23/h; premixed gas: 0.10% p-xylene, 1.0% O_2, and 1.0% N_2 in He.
[b][ ]: With ˜2% steam.

side channel structure of the DBH sieve. However, the iron and Mo moieties migrated to the exterior surface of the DBH matrix during the activation step by the prolonged calcination. Contrary to the CVD catalyst, the aggregated species were deposited nonuniformly on the exterior surface of the impregnated counterpart (Fig. 9b).

Catalyst G was subjected to the 8-h run, and then calcined *in situ* in the reactor at 650°C for 4 h in 1 day; this procedure was repeated for 4 additional straight days. The catalyst performance became better as the hours onstream increased. The resulting activated catalyst G was tested for the catalyst stability for 60 h in an aging run, and the results are shown in Fig. 10.

The data indicate that the catalytic activity and selectivity are still maintained during 60 h onstream. In comparison to the activated CVD counterpart, catalyst G activated by the preceding manner required higher temperature to attain the equivalent level of p-xylene conversion, but selectivity toward aldehydes was as good as the CVD counterpart. However, the spent catalyst resulting from this run contained 2.7 wt % Mo, 0.82 wt % Fe, 0.82 wt % boron, and Mo:Fe = 1.9:1, as shown in Table 7.

Although the performance of catalyst G was maintained during the aging run, the catalyst composition was significantly altered by the loss of molybdenum. Raman spectra of both fresh (catalyst G) and spent catalyst (catalyst G_s) were shown in Fig. 11. The α-Fe_2O_3 phase has already appeared in the Raman spectra of the spent catalyst. The impregnated catalyst also required higher temperature than the CVD counterparts for the equal level of p-xylene conversion, and this should become another cause

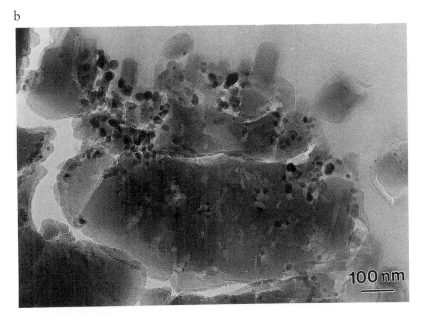

Figure 9. TEM (a) CVD Fe/Mo/DBH versus (b) impregnated Fe/Mo/DBH.

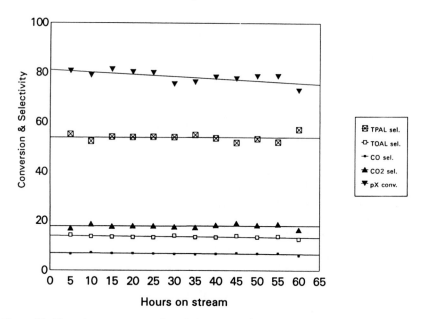

Figure 10. The aging run over catalyst G, impregnated Fe/Mo/DBH, for the *p*-xylene oxidation. TPAL: terephthaldehyde. TOAL: *p*-tolualdehyde.

for the accelerated deactivation. Thus, one could expect to see a sign of abrupt deactivation if the aging run with catalyst G were continued for a prolonged period, longer than 60 h.

Catalysts B, C, and D, which had already proved to be equally active and selective for aldehyde formation from the *p*-xylene oxidation runs, were also subjected to the aging stability test for the *p*-xylene oxidation under similar conditions. Both fresh and spent catalysts resulting from these aging runs were analyzed, and these results are combined together in Table 7. These data suggest that the rate of Mo leaching appears to be much faster with catalyst G, an impregnated catalyst, than with the CVD counterparts.

17.4 Conclusions

A series of the binary metal oxides catalysts were prepared by utilizing the surface silanols on various zeolite matrices by the chemical vapor deposition (CVD) technique. Among the various matrices tried, such as borosilicate, ZSM-5, β-zeolite, silicalite, and Y-zeolite, the partially deboronated borosilicate (DBH) played a unique role for the *para*-selective

TABLE 7
Comparison of Fresh and Spent Catalysts

Catalyst	Metal				BET SA (m²/g)	Cumulative pore vol (cc/g)	Average pore radius (Å)	Micropore	
	Mo[a]	Fe[a]	Mo:Fe[b]	B (ppm)				Area (m²/g)	Vol (cc/g)
B	8.7	2.2	2.3	920	271	—	27	212	0.096
Bs[c]	8.7	2.4	2.1	940	—	—	—	—	—
C	6.2	1.9	1.9	1070	270	0.11	27	196	0.090
Cs[d]	5.7	1.7	2.0		305	0.10	27	144	0.067
D	5.8	1.4	2.4	1080	316	0.15	27	224	0.104
Ds[e]	3.7	1.3	1.7		158	0.078	29	117	0.054
G	5.4	0.92	3.5	8200	303	0.07	34	262	0.12
Gs[f]	2.7	0.82	1.7		302	0.05	26	254	0.116

[a] wt %.
[b] Atomic ratio.
[c] Spent B.
[d] Spent C: ran about 200 h on stream.
[e] Spent D: ran 130 h in stream.
[f] Spent G.

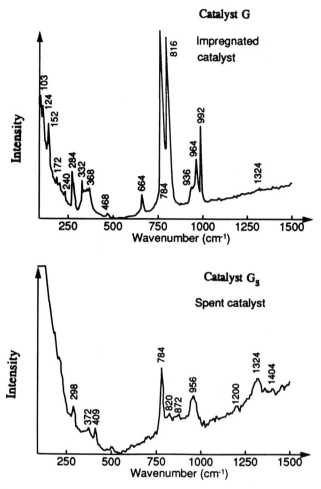

Figure 11. Raman spectra, impregnated catalyst (G) and its spent catalyst (G$_s$) resulting from a prolonged aging run.

oxidation of *p*-xylene. The catalyst, CVD Fe/Mo/DBH, was activated by subliming off the excess molybdenum over the ferric molybdate stoichiometry during the prolonged calcination in air at 650–695°C. The activated catalyst maintained good activity and excellent stability for the gas-phase O$_2$ oxidation of *p*-xylene to produce terephthaldehyde and p-tolualdehyde, despite the unusually high-surface area of 250–316 m^2/g.

The order of metal vapor deposition was an important factor for producing the active catalyst in the CVD procedure. The reversed order, the deposition of MoO_2Cl_2 followed by $FeCl_3$ deposition, resulted in forming the α-Fe_2O_3 phase, which caused extensive combustion.

The silica-coated catalyst preferentially oxidized *p*-xylene to aldehydes over its isomers and ethylbenzene. The catalytic species dispersed nonuniformly in a more aggregated state on the exterior surface of the impregnated catalyst, while finer particles of 5-10 nm were uniformly dispersed on the surface of the CVD catalyst. This difference observed between the CVD catalyst and its impregnated counterpart is reflected on catalyst stability and performance for the gas-phase O_2 oxidation.

References

1. Desu, S. B. (1989). *J. Am. Chem. Soc.*, 79, 1615.
2. Maeda, K., and Fischer, S. M. (1993). *Solid State Technol.*, 83.
3. Fujino, K., Nishimoto, Y., Tokumasu, N., and Maeda, K. (1989). *J. Electrochem. Soc.*, 139, 79.
4. Ikeda, K., Nakayama, S., and Maeda, M. (1994). *Semicond. World*, 60.
5. Nakajima, T., Matsuoka, M., Mishima, S., and Matsuzaki, I. (1994). *Nippon Kagaku Zashi*, 12, 1134.
6. Niwa, M., Hibino, T., Murata, H., Katada, N., and Murakami, Y. (1989). *J. Chem. Soc. Chem. Commun.*, 289.
7. Niwa, M., Katada, N., and Murakami, Y. (1990). *J. Phys. Chem.*, 94, 6441.
8. Niwa, M., Katada, N., and Murakami, Y. (1992). *J. Catal.*, 134, 340.
9. Hattori, T., Matsuda, M., Suzuki, K., Miyamoto, A., and Murakami, Y. (1988). *Stud. Surf. Sci. Catal.*, 4, 1640.
10. Inumaru, K., Okuhara, T., and Misono, M. (1990). *Chem. Lett.*, 1207.
11. Jin, T., Okuhara, T., and White, J. M. (1987). *J. Chem. Soc. Chem. Commun.*, 1248.
12. Imizu, Y., Tada, Y., Tanaka, K., and Toyoshima, I. (1988). *Shokubai*, 30, 525.
13. Sato, S., Toita, M., Sodesawa, T., and Nozaki, F. (1988). *Appl. Catal.*, 62, 73.
14. Niwa, M., Morimoto, S., Kato, M., Hattori, T., and Murakami, Y. (1984). *Proc. 8th Int. Congr. Catal.*, IV, 701.
15. Jin, T., and White, J. M. (1988). *Surf. Interface Anal.*, 11, 517.
16. Katata, N., Toyama, T., and Niwa, M. (1994). *J. Phys. Chem.*, 98, 7647.
17. Sheng, T. C., and Gay, I. D. (1994). *J. Catal.*, 145, 10.
18. Xu, B. Q., Yamaguchi, T., and Tanabe, K. (1989). *Chem. Lett.*, 149.
19. Katada, N., Tsubouchi, T., Niwa, M., and Murakami, Y. (1995). *Appl. Catal.*, 124, 1.
20. Katada, N., and Niwa, M. (1995). *Catal. Lett.*, 32, 131.
21. Hibino, T., Niwa, M., Murakami, Y., and Sano, M. (1989). *J. Chem. Soc. Faraday Trans.*, 1, 85, 2327.
22. Hibino, T., Niwa, M., Murakami, Y., Komai, S., and Hanaichi, T. (1989). *J. Phys. Chem.*, 93, 7847.
23. Sato, S., Sodesawa, T., and Nozaki, F. (1990). *Shokubai*, 32(6), 342.
24. Katada, N., Niwa, M., and Murakami, Y. (1990). *Shokubai*, 32(2), 59.
25. Inumaru, K., Okuhara, T., and Misono, M. (1990). *Shokubai*, 32(2), 51.
26. Niwa, M. Kato, S., Hattori, T., and Murakami, Y., (1985). *J. Chem. Soc. Faraday Trans.*, 1, 84, 2757.

27. Niwa, M., Kato, M., Hattori, T., and Murakami, Y. (1986). *J. Phys. Chem.,* **90**, 6233.
28. Hibino, T., Niwa, M., and Murakami, Y. (1991). *J. Catal.,* **128**, 551.
29. Iwasawa, T. (1995). *Shokubai,* **37**(4), 292.
30. Yoo, J. S., Donohue, J. A., Kleefisch, M. S., Lin, P. S., and Elfline, S. D. (1993). *Appl. Catal. A,* **105**, 85.
31. Yoo, J. S., Lin, P. S., and Elfline, S. D. (1993). *Appl. Catal. A,* **106**, 259.
32. Yoo, J. S., Donohue, J. A., and Kleefisch, M. S. (1994). *Appl. Catal. A,* **110**, 75.
33. Yoo, J. S., Sohail, R. A., Grimmer, S. S., and Choi-Feng, C. (1994). *Catal. Lett.,* **29**, 299.
34. Yoo, J. S., Choi-Feng, C., and Donohue, J. A. (1994). *Appl. Catal. A,* **118**, 87.
35. Yoo, J. S., Lin, P. S., and Elfline, S. D. (1995). *Appl. Catal. A,* **124**, 139.
36. Zajac, G. W., Choi-Feng, C., Faber, J., Yoo, J. S., Patel, R., and Hochst, H. (1995). *J. Catal.,* **151**, 338.
37. Donohue J. A., and Ray, G. J., unpublished work.

CHAPTER 18

Chemical Vapor-Phase Deposition and Decomposition

Preparation of Metal and Metal Oxide Catalysts in a
Fluidized-Bed Reactor

M. Baerns,* S. Termath, and M. Reiche
Lehrstuhl für Technische Chemie
Ruhr-Universität Bochum
D-44780 Bochum, Germany

KEYWORDS: CVD, metal acetylacetonate catalyst precursor, preparation of catalysts, supported Cr_2O_3 catalysts, supported Pt catalysts, supported V_2O_5 catalysts

18.1   State of the Art

Chemical vapor phase deposition (CVD) has become an often applied method for producing small metal and metal oxide particles or thin films on various kinds of support materials. In some cases it also serves the purpose of preparing finely divided particles that are then used for subsequent processes. Usually the deposition of a precursor is followed by its decomposition. Besides the preparation of catalysts, applications are mainly found in the electronics industry (e.g., doping of semiconductive material) and in the machinery industry (depositing of anticorrosive or hardening films on, e.g., pipes and tools). Catalysis is emphasized in this contribution.

*Present address: Institut für Angewandte Chemie Berlin-Adlershof e.V., D-12484 Berlin, Germany.

*Advanced Catalysts
and Nanostructured Materials*

479

In this state-of-the-art review that starts with general aspects of the CVD technique, some mention of other areas related to catalysis, particularly with respect to preparation methods for materials with specific electronic properties, is made in Sects. 18.1.4 and 18.1.5. Following this introductory section, Sect. 18.2 also deals with the CVD technique in catalysis as illustrated by some specific examples for the preparation of supported metal and metal oxide catalysts that were obtained in the authors' laboratory.

18.1.1 CVD Techniques

In principle, all chemical compounds volatile at the required temperature of deposition may be used. Generally, it is advantageous when decomposition does not already occur during volatilization to avoid losses of the usually expensive material. According to the partial pressure of the precursor compound that usually corresponds to the vapor pressure at the temperature of volatilization or other experimental boundary conditions, the pressure of CVD ranges from high vacuum to ambient conditions.

The decomposition of metals from pyrolytic decomposition of a molecular precursor in the gas phase dates back to the Mond process for preparing high-purity nickel via its carbonyl [1]. Later on, the deposition of gallium arsenide from trimethylgallium and arsine for microelectronic purposes was another important application of CVD [2]. Nowadays, a large variety of precursor compounds is being used in CVD; among those are metal acetylacetonates, β-diketonates, alkoxides, homoleptic allyl and alkyl complexes, or even more complex compounds. As examples, reference is made to some of those. There exist also different approaches in obtaining the desired compound on the support material. The precursor molecule may be decomposed to the final species at the temperature of exposure or only after additional heating up; these processes may be performed under inert, reductive, or oxidative gas atmosphere.

In a review article Kirss [3] discusses extensively the use of homoleptic allyl ligands (i.e., C_3H_5) containing metal precursors for CVD. These compounds are easily available through one-pot syntheses using the respective metal halides and allyl Grignard reagents or by alkylation of alkali metal salts. The relatively low molecular mass of a C_3H_5 ligand contributes to high-vapor pressures, while the stability of the allyl radical should reduce the decomposition temperature. From the work described it is concluded that films prepared from compounds containing selenium, tellurium, and possibly palladium and platinum may prove useful for electronic devices; films from the respective compounds of rhodium, iridium, tungsten, and molybdenum lead to unacceptable contamination of the film by carbon. In the latter case the presence of gas-phase hydrogen or of its radicals is required to obtain high-quality films. Kirss points out that the relationships between reagent structure, pyrolysis pathways, and film purity are likely to remain a fruitful area for CVD research.

Yamazaki *et al.* [4] introduced a new class of Pb, Zr, and Ti oxide precursors for forming ferroelectric thin films on (100) silicium and (100) magnesium surfaces. Such films of lead zirconate titanate (PZT) have a potential for application to nonvolatile random access memory. Although CVD has the advantage of good step coverage over a patterned substrate, there are still some problems originating from the precursor materials. $Pb(C_2H_5)_2$, which is presently used is a high-risk poison; $Zr(i\text{-}OC_3H_7)_4$; and $Ti(i\text{-}OC_3H_7)_4$ easily hydrolyze in air. The new material consists of Pb-bis-dipivaloylmethane $Pb(DPM)_2$, $Zr(DPM)_4$, and $Ti(DPM)_2(i\text{-}OC_3H_7)_2$. These DPM sources are not poisonous and are less reactive in the vapor phase prior to deposition in the presence of oxygen, which is, however, required for forming the oxide film. There are some feeding problems due to the low-vapor pressures of the materials that might, however, be overcome by flash evaporation of the DPM source material according to the authors. When carrying out CVD on (100) MgO substrates with the DPM materials, PZT films grown at 773 K were amorphous; at 823 K a mixed phase of a-axis- and c-axis-oriented perovskite, and at 873 K a single-phase c-axis-oriented perovskite were obtained. These results illustrate the effect of reaction variables on the CVD process.

The application of fluidized-bed reactors for CVD processes was reported in 1994. Moene *et al.* [5] and the present authors [6] described the feasibility of different CVD processes; the latter authors studied particularly the influence of the preparation conditions on dispersion of the deposited catalytic compounds (Pt, Cr_2O_3, V_2O_5 on SiO_2 and Al_2O_3).

18.1.2 Mechanisms of CVD Processes

As for other heterogeneous reactions various processes may take place, that is, mass transfer of the precursor to the support surface, its adsorption and decomposition, and release or desorption of decomposition products. Finally, physical processes of the deposited target compound such as spreading, crystallization, or sintering by surface transport may take place on the support surface. Besides, often gas-phase processes may interfere with the gas–surface reactions of the metal-organic (or organometallic) species. At best, these phenomena may be studied separately; very often an accurate discrimination between these processes is, however, not possible.

Various techniques are applied to study the surface processes occurring in CVD when preparing thin films or small particles on a support material.

From *in situ* IR spectroscopy and temperature-programmed reaction combined with analysis of the evolving gaseous products mechanistic insights can be gained. For illustration, one study follows here and also in Sect. 8.2, although related work is included in many of the studies cited in Sect. 18.1.3.

Nickl *et al.* [7] comprehensively studied CVD of vanadyl alkoxide onto TiO_2, SiO_2, Al_2O_3, and ZrO_2 as supports. The chemical and structural properties of the obtained catalysts and their respective precursors were characterized by using vibrational spectroscopy, temperature-programmed reduction by hydrogen, and temperature-programmed reaction and desorption of ammonia. Results derived from the different catalytic performances of the various catalysts for the reduction of NO by ammonia were interpreted.

The chemical processes occurring in the reactions related to CVD include a combination of homogeneous gas-phase and heterogeneous gas-surface reactions as outlined previously. These complex phenomena are difficult to model and to understand; a novel metal-organic MOCVD technique was developed by Aitchison *et al.* [8] that addresses these problems and enables a more thorough mechanistic understanding of the heterogeneous decomposition pathways of metal-organic compounds. At low pressure the gaseous precursor is pulsed over the support avoiding gas-phase reactions; mainly surface-mediated reactions dominate. Both single metal oxides (TiO_2) and binary oxide systems (TiO_2/SiO_2) have been investigated on a variety of substrates to elucidate the effects of precursor chemistry, type of substrate, temperature, and pressure on film, composition, and morphology.

Attempts have been made to model CVD processes including mass transfer, adsorption, and desorption as well as chemical transformations for identifying the determining factors by simulation on the basis of such models.

Durst *et al.* [9]developed a mathematical model for epitaxial growth in MOCVD that is based on the conservation equations for mass, momentum, heat, and chemical species including thermodiffusion and chemical reactions. In the specific case of preparing epitaxial layers of GaAs from mixtures of $Ga(CH_3)_3$ and AsH_3 in the presence of H_2, it was assumed that the growth rate does not depend on the arsine concentration but that it is limited by the mass transfer of the gallium compound.

A generalized model for low-pressure MOCVD (LPMOCVD) of silica from silane on structured wafer surfaces was introduced by Wille *et al.* [10] and may easily be extended to other LPMOCVD systems. The model includes options to alter the features of the near-surface transport mechanisms of the precursor deposition according to different process–gas chemicals and deposition parameters. Process–gas molecules impinging on the surface of a structured wafer can be reflected or scattered to a certain, material-dependent degree. They also may be adsorbed and may diffuse a short distance on the surface. Furthermore, they can be remitted from the surface several times with a characteristic angular distribution. If the precursors do not escape into the process–gas, the deposition precursors are finally incorporated in the growing film. By comparing the simulations on the basis of the various model modifications, a deeper understanding of the transport and growth mechanisms in CVD processes may be obtained.

18.1.3 Catalysis and Catalysts

In catalysis CVD is used for preparing or modifying supported and unsupported catalysts. Various CVD precursors have been applied. In some instances surface processes of depositing and decomposing the catalyst precursors have been identified. An illustrative overview is presented on catalytic materials prepared by CVD and catalytic reactions conducted on these materials.

18.1.3.1 Supported Catalysts

Various studies have been reported on the preparation of supported nickel catalysts and their application. Nickel or its oxide was deposited on preexisting supports or it was codeposited with the support. Omata *et al.* [11] prepared Ni on active carbon by nickel nitrate followed by activation with hydrogen. This catalyst was applied for methanol carbonylation. Ooi *et al.* [12] produced nickel oxide/silica particles by using all gaseous components, for example, $NiCl_2$, $SiCl_4$, and O_2. NiO was highly dispersed in a silica matrix; the particle diameter of NiO was as low as 3 to 8 nm. Similarly, Yano *et al.* [13] obtained NiO (1–3 nm) in silica (20-nm) matrix; the material was claimed to be superior to conventionally produced catalysts for CO methanation and methanol decomposition.

From a methodical point of view (see also the preceding paragraph) the work of Moene *et al.* [14] on depositing Ni on a porous support is of interest. The authors developed a model for the mathematical description of the deposition of the catalytic material from the gas phase in porous structures. Gaseous $NiCl_2$ is infiltrated into a catalyst pellet or a monolith that is then decomposed to Ni by elevating the temperature. The simulations indicate that the preparation of catalysts with homogeneous and inhomogeneous active sites distributions is possible. Associative adsorption of the active nickel compound gives rise to a catalyst with active site distributions that vary between a degenerated eggshell and well-formed eggshell. Dissociative adsorption results in eggshell, eggwhite, and nearly eggyolk activity profiles. The same considerations are valid for CVD of metals in monolithic reactors. Application of the concept of a generalized Thiele modulus shows a correlation between the modulus and the location of maximum deposition when dissociative adsorption is assumed. These general results should always be taken into account if intraparticle transport processes are expected to occur in CVD. Related to this work, Mirua et al. [15] observed in CVD that Ru was deposited in the outer shell of a Pd/Al$_2$O$_3$ catalyst where Pd was distributed in the same region. It may be assumed that Pd catalyzed the decomposition of the Ru precursor and hence its deposition.

Vanadia catalysts for selective catalytic reduction of NO by ammonia were prepared by vapor deposition of vanadyl alkoxide onto various supports as already outlined [7]. Similarly, Inumaru et al. [16] prepared vanadium oxide overlayers on high-surface-area silica from VO(OC$_2$H$_5$)$_3$; in their work they put emphasis on identifying the elementary surface reactions in CVD preparation. Vanadia/alumina catalysts were also prepared by CVD of VOCl$_3$ as reported by Hattori et al. [17]; these catalysts were then compared with a conventional impregnation catalysts by applying them to NO reduction by NH$_3$, oxidation of H$_2$, oxidation of NH$_3$, and oxidation of benezene. The catalytic activities as described by the turnover frequency of the two types of catalyst were different; the authors could, however, not present an unambiguous explanation for their observation.

Hattori et al. [18] prepared an SnO$_2$-on-silica catalyst by reaction of SnCl$_4$ vapor with silica. This catalyst was used for the oxidative dehydrogenation of ethylbenzene to styrene. The CVD catalyst was superior to its conventional counterpart made by impregnation; selectivity to styrene was higher and no coke deposition leading to catalyst deactivation was observed.

For catalytic synthesis of unsaturated nitriles from NO-alkane and NO-alkene Inoue et al. [19] used a Sn–Pt/SiO$_2$ catalyst obtained from reaction of Sn(CH$_3$)$_4$ with Pt particles on SiO$_2$. The selectivities of the bimetallic CVD catalyst to acrylonitrile from (NO + propene) and of methacrylonitrile from (NO + isobutene) and NO + isobutane) amounted to 70 to 93% in contrast to negligible activity and selectivity (being about one order of magnitude lower) with the monometallic Pt/SiO$_2$ catalyst.

In another application Sato et al. [20,21] prepared a boria-on-silica catalyst by CVD of B(OC$_2$H$_5$)$_3$ on silica for converting cyclohexanone oxime to ε-caprolactam via a Beckmann rearrangement reaction. This catalyst was more efficient than silica–boria and alumina–boria that had been obtained by the usual impregnation method.

When depositing a silica overlayer on alumina by utilizing surface silanization with CH$_3$Si(OCH$_3$)$_3$ and subsequent decomposition and oxygen treatment, Imizu and Tada [22] observed an increase in the catalytic activity compared to a commercial silica–alumina catalyst.

A rather interesting application of CVD was introduced by Nariman *et al.* [23] who deposited by hot-filament-enhanced CVD a diamond layer on inorganic oxide catalysts to improve thermal conductivity on their surfaces and hence within the bed of catalysts. Simulations were conducted for determining the thermal conductivity effect on the phthalic anhydride selectivity in o-xylene oxidation, which is a severely exothermic and temperature-sensitive reaction. Although it may presently not be economically feasible to coat catalyst particles with porous diamond, with the advances that are being made in synthetic diamond deposition technologies and CVD, in general, it is conceivable that new hybrid materials consisting of a highly conductive, nonreactive, abrasion-resistant porous coating on the outside of a catalyst particle eventually may provide a means to improve the performance of packed beds beyond what is possible today with conventional materials.

18.1.4 Modification of Zeolites

Modifications of zeolites by CVD aim at different targets that may be summarized as follows: (1) fine control of pore-opening size (or alternatively called pore–size engineering, see e.g., [24], (2) modification of the zeolite properties, and (3) preparation of bifunctional catalysts by taking advantage of zeolite catalysis combined with an additional catalytic function obtained by depositing a catalytic compound on the external or within the internal zeolite structure.

18.1.4.1 Control of Pore-Opening Size

A reduction in outer pore sizes of a mordenite in its H form was achieved by Niwa *et al.* [25] when depositing $Si(OCH_3)_4$ on the external surface of the zeolite and decomposing it to SiO_2; because the molecular size of silicon methoxide is larger than the pore-opening size of the mordenite, the precursor cannot enter the zeolite pores. The outer pore–size opening was reduced by ca. 0.1 and 0.2 nm, depending on the number of silica monolayers formed (1 to 2 and 3 layers, respectively). The same group later showed that a surface activation by water is required for depositing SiO_2 and hence for achieving pore–size tuning of Na mordenite [26]; the pore size was derived from the product distribution obtained in the cracking of different hydrocarbons. Chun *et al.* [27] also confirmed that for the H form of a β-zeolite no internal deposition of silica occurred within the zeolite crystal when using $Si(OCH_3)_4$. Similar to pore–size engineering of zeolites, the permeability of membranes may be modified by CVD [28].

18.1.4.2 Modification of Zeolite Properties

Acidity of zeolites is affected by depositing various materials on the external or internal surfaces. Simon et al. [29] deposited sodium clusters in NaX zeolite by CVD of sodium metal at 498 K and 10^{-4} torr. The cluster size amounted to $Na_{65}{}^+$. In this complex the sodium ions are arranged in octahedral geometries in the sodalite cages as determined by electron spin resonance (ESR), magic angle spinning-nuclear magnetic resonance (MAS-NMR), Fourier transform infrared (FTIR), luminescence spectroscopy, and X-ray diffraction (XRD). These ionic clusters have an electron trapped in the center of the octahedra, giving rise to a colored center. From IR studies of pyridine adsorption it was derived that Brønsted acidity of the untreated NaX zeolite was low and that it was completely lost after treatment with sodium vapor. The catalytic mechanism for isomerization of cyclopropane to propylene on this material is discussed in terms of participation of defect sites that may initiate radical formation.

Ito et al. [30] studied the nature of active titanium species for epoxidation of olefins with H_2O_2 and tert-butyl hydroperoxide for titania deposited on a silica support by CVD and compared the results with titanium silicalite (TS-1) and titania–silica prepared by a sol–gel method. IR and X-ray adsorption near-edge spectroscopy (XANES) analyses suggested that Ti in titania–silica and in the CVD catalyst has a tetrahedral configuration bonded to SiO_2 and that Ti in TS-1 has a configuration composed of >Ti-O or related to it. These configurations are closely related to the catalytic activities in olefin epoxidation; the former works with tert-butyl hydroperoxide and the latter, with H_2O_2.

Chamoumi et al. [31] silanated an offretite with octamethylcyclotetrasiloxane (OMCTS) and tetramethylsiloxane (TMS) using a CVD method. The parent and silanated offretites were characterized by scanning electron microscopy (SEM), XRD, internal reflection spectroscopy (IRS), microporous volume determination, and stepwise thermal desorption of ammonia. OMCTS and TMS precursors transform the more accessible silanol groups into I and II grafted groups, respectively. OMCTS was found to be a suitable agent for the selective silanation of external acidic sites, leading to unilayer grafting of the surface without reducing the internal site numbers and the intracrystalline voids within the zeolite.

Hölderich and Paczkowski [32] could show that the catalytic activity of various zeolites (HZSM-5, USY, pentasil) was enhanced by modification of their external surface by deposition of Si alkoxides using CVD. For illustration they used the rearrangement of m-dioxanes to neo-alkyl aldehydes at 523–673 K.

18.1.4.3 Bifunctional Zeolite Catalysts

Yoo *et al.* [33,34] prepared Fe/Mo oxide catalysts on a borosilicate molecular sieve matrix by CVD using $FeCl_3$ and MoO_2Cl_2 or $MoCl_5$ as precursors. The catalyst obtained was found to be active for the gas-phase oxidation of alkylaromatics to produce aldehydes in high yields. When using a mixture of isomeric xylenes the *para*-compound was preferentially oxidized to toluylaldehyde by Fe/Mo oxide on a partially deboronated borosilicate zeolite [35]. If the catalytic compounds were deposited on ZSM-5, disproportionation of *p*-xylene became predominant while deposition on silicalite resulted in only low activity and poor aldehyde selectivity [36]. Similarly, an Fe/Mo oxide on deboronated borosilicate molecular sieve catalyst prepared by CVD was used for the selective formation of phenol by gas-phase N_2O oxidation of benzene; this catalyst was more active than the counterpart prepared by liquid-phase impregnation [37]. A thorough characterization of the Fe/Mo oxide on deboronated borosilicate catalysts using Raman spectroscopy, electron microscopy, X-ray powder diffraction, X-ray photoelectron spectroscopy (XPS), and *in situ* total electron yield near edge spectroscopy was reported [38].

Dossi *et al.* [39,40] introduced selectively platinum hexafluoroacetylacetonate inside the channels of KL zeolite (Linde type L) via CVD in a flow of Ar at 343 K. Small nonacidic Pt clusters of nucleophilic nature were formed. The catalyst showed remarkably high activity and selectivity in the conversion of methylcyclopentane to benzene at 773 K; only little deactivation took place due to reduced coke formation and very slow sintering.

CVD preparation of Ga/HZSM-5 using $GaCl_3$ and HZSM-5 leads to replacement of protons by $(GaO)^+$ as reported by Kwak and Sachtler [41]. As more protons are being substituted by $(GaO)^+$, the conversion of propene to aromatics decreases, but that of propane passes through a maximum, as does the aromatics selectivity. This is assumed to be consistent with a bifunctional mechanism in which Brønsted acid sites catalyze oligomerization and ring closure, but Ga, in concert with protons, acts as a dehydrogenation site.

A fundamental study was conducted on manganese-promoted rhodium/NaY zeolite catalysts prepared by ion exchange and CVD $(Rh(CO)_2(acac)$ and $Mn_2(CO)_{10})$, followed by thermal decomposition. The catalysts that were used for CO hydrogenation were presented by Beutel *et al.* [42]. One essential finding of the authors was the fact that catalysts prepared by CVD do not contain any Mn^{2+} ions opposite to the ion-exchanged one because the manganese carbonyl complex is decomposed on previously deposited Rh^0 particles, thus forming bimetallic particles the surface of which is presumably enriched by Mn. The number of Rh_3 ensembles is therefore low, and bridging CO ligands are not found in the presence of CO. The possible relevance of these results are discussed for CO hydrogenation.

18.1.5 Thin Films

CVD is also frequently used for the growth of thin films for different purposes. Some selected illustrations of this work are given in the following sections because there is relevance to catalysis.

Cobalt–iron–chromium oxide films with high uniformity and abrasion resistance as well as broad chemical stability have been deposited on float glass windows for their good solar attenuation. In this context, Greenberg [43] reported on an interesting modification of the usually applied CVD procedure by using a liquid carrier (i.e., chlorinated hydrocarbons) for the acetylacetonates of Co, Fe, and Cr. A pneumatic spray of this solution traverses the moving hot float glass surface consisting of pristine soda–lime–silica. At 773 K a film of several 10 nm is obtained on the hot surface. The solvent most likely acts, to a great extent, only as a carrier. During the flight of the droplet toward the surface, the solvent is evaporized and the metal-organic material starts decomposing before it hits the surface as some kind of a hollow or porous particle where it is then further transformed to the film.

Thin films of stoichiometric early transition metal nitrides exhibit extreme hardness, high-melting points, excellent chemical durability, and interesting optical properties as described by Toth [44]. Also the respective oxides are of similar importance [45]. A comprehensive review of preparing thin films of transition metal and main group nitrides from their homoleptic amido complexes, $M(NR_2)_n$, at about 473–723 K and atmospheric pressure has been given by Hoffman [46]. Amido complexes were chosen as precursors on the basis of their reactivities in solution. Only little carbon contamination was observed.

Thin Ta_2O_5 films have been deposited on chemically cleaned n-type (100) silicon substrates in a hot-wall-type vertical furnace reactor by LPMOCVD between 650 and 770 K and then annealed in oxygen at temperature up to 1270 K. The total pressure amounted to 30 Pa and the precursor partial pressure to 3×10^{-2} Pa. Tantalum ethylate, $Ta(OC_2H_5)_5$, was used as the metal-organic precursor by Rausch and Burte [47]. This process that was studied with respect to its application as a capacitor dielectric material in low-power, high-density dynamic random access memories resulted in a tantala layer thickness of about 80 nm on the silica waver. The layer showed a very good step coverage that is required for application as a dielectric material in three-dimensional stacked capacitors; after the annealing step the films were polycrystalline and showed orthorhombic structure.

Also for high-temperature superconducting materials thin films on a support are applied. Sugimoto [48,49] reviewed the progress that has been made in various types of CVD processes such as MOCVD, plasma-enhanced MOCVD, laser-enhanced MOCVD, etc. for superconducting materials (e.g., Y–Ba–Cu–O, Bi–Sr–Ca–Cu–O, Tl–Ba–Ca–Cu–O). Zhang and Erbil [50,51] point out that the advantages of MOCVD manifest themselves in the quality of the thin films produced and in the economy of the process, but they also mention some difficulties in developing a large-scale manufacturing technology for superconducting materials based on MOCVD.

A further application of CVD in the area of thin films on, for example, AlN and SiC is the deposition of diamond substrates for small discrete devices such as laser diodes as reported by Pickrell *et al.* [52,53]. Such substrates offer extremely high thermal conductivities coupled with good dielectric properties. High thermal conductivity might be also of interest for highly exothermic catalytic reactions (see preceding text).

Finally, the deposition of high-quality cubic boron nitride films on a nickel substrate applying a hot-filament-assisted radio frequency plasma has been mentioned [54]. This CVD process is of special interest to catalysis because nickel probably had a catalytic effect on the nucleation and growth of the c-BN phase and inhibited the formation of a hexagonal BN phase.

18.2   Preparation of Catalysts on Support Materials in a Fluidized-Bed Reactor

18.2.1 Introduction

Catalysts preparation by means of chemical vapor-phase decomposition (CVD) is conducted by vaporizing a suitable precursor and adsorbing the gaseous compound on the support material. Subsequently, the adsorbate is transformed to the catalytic compound [7,17,21,55], as has already been outlined in Sect. 18.1. In this section the preparation of supported Pt, Cr_2O_3, and V_2O_5 catalysts by vapor-phase adsorption of metal acetylacetonates — $Pt(acac)_2$, $Cr(acac)_3$ and $V(acac)_3$ — on SiO_2 and γ-Al_2O_3 and the subsequent decomposition of the adsorbates to the catalytic compound, is described [6,55]. The process was conducted in a fluidized-bed reactor by this technique. All particles were evenly exposed to the vapor ascertaining homogeneous loading of the catalytic compound on the support material; this is considered to be an essential improvement compared to fixed-bed or similar technologies. To study the adsorption of metal acetylacetonates on the supports and the subsequent transformation, *in situ* DRIFT spectroscopy supplemented by mass-spectrometric analysis of volatile decomposition products was applied. To elucidate the influence of preparation conditions on dispersion of the catalytic compound, several silica- and alumina–supported Pt, Cr_2O_3, and V_2O_5 catalysts were prepared by varying the amount of adsorbed metal acetylacetonate. The Pt(acac)$_2$ adsorbates were decomposed in N_2 as well as in air, whereas the adsorbates of Cr(acac)$_3$ and V(acac)$_3$ were decomposed in air only. Also supported Pd catalysts were prepared by decomposition of adsorbed Pd(acac)$_2$ on the support material. However, the application of the method was less successful because Pd(acac)$_2$ tends to decompose in the gas phase before adsorption is accomplished [55].

18.2.2 Experimental

The quartz-made fluidized-bed reactor used for catalyst preparation was electrically heated; a cyclone was incorporated into the freeboard of the reactor to prevent elutriation of fines. By using fluidized beds it was ascertained that the particles of the support materials were ($75 < d_p/\mu m \leq 100$), on the average, exposed to the same precursor concentration. A sketch of the apparatus used for CVD catalyst preparation is shown in Fig. 1. The support, which was fluidized by N_2, was first thermally pretreated to remove physisorbed water. Then the vaporized metal acetylacetonate was adsorbed on the support at 400 K and at constant partial pressure in a flow of N_2 for

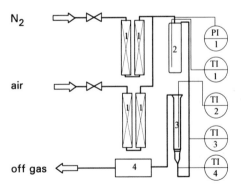

Figure 1. Apparatus for catalyst preparation by CVD. (1) Molecular-sieve adsorber for gas purification; (2) constant-temperature saturator, containing a mixture of solid metal acetylacetonate and quartz; (3) fluidized-bed reactor; (4) IR analysis for CO and CH_4 in effluent.

a given period of time. The subsequent decomposition of the adsorbate was conducted by increasing the fluidized-bed temperature with a rate of 4 K/min. Pt was deposited in either N_2 or air at 573 K; Cr_2O_3 and V_2O_5 were deposited in air at 673 K. Pt catalysts that were decomposed in N_2 were additionally treated in air at 573 K (Pt/SiO$_2$ catalysts) and at 773 K (Pt/Al$_2$O$_3$ catalysts). More detailed preparation conditions as well as conditions for elemental and XRD analysis have been described elsewhere [6,55].

Dispersion of the deposited material was derived from the amounts of adsorbed CO on Pt catalysts, and O_2 on Cr_2O_3 and V_2O_5 catalysts, respectively, using samples pretreated with H_2 (see Table 1) [6].

DRIFT spectra (Perkin-Elmer 1710 spectrometer; Spectra Tech 003-102 diffuse reflectance cell) of adsorbed probe molecules were recorded at 298 K (before adsorbing CO, the solid samples were pretreated in H_2 at 473 K). The *in situ* decomposition experiments were conducted in the temperature range from 295 to 773 K. Volatile products were detected by means of a quadrupole mass spectrometer (Micromass PC, VG Quadrupoles) as reported elsewhere [6]. Temperature-programmed desorption (TPD) of NH_3 adsorbed at 298 K was performed after pretreatment of the samples at 673 K in air in the temperature range between 300 and 1000 K.

18.2.3 Results

18.2.3.1 Surface Processes in Metal Acetylacetonate Adsorption and Decomposition

Surface processes are strongly influenced by the nature of the support material. When adsorbing Pt(acac)$_2$, Cr(acac)$_3$, and V(acac)$_3$ on SiO$_2$, the IR spectra are to a large extent in agreement with those of the unsupported

TABLE 1
Experimental Conditions for Gas Adsorption on
Pt, Cr_2O_3, and V_2O_5 Catalysts

	Pt catalysts	Cr_2O_3 catalysts	V_2O_5 catalysts
Probe molecule	CO	O_2	O_2
Assumed stochiometry	$Pt:CO = 1:1$	$Cr:O_2 = 2:1$	$V:O_2 = 2:1$
Catalyst pretreatment	H_2, 473 K, 2 h	H_2, 623 K, 2 h	H_2, 623 K, 6 h
Adsorption temperature	298 K	298 K	623 K
Method	GC pulse	Microbalance	GC pulse

Reprinted from (Poncelet, G. *et al.*, eds.), *Preparation of Catalysts VI Scientific Bases for the Preparation of Heterogeneous Catalysts*, 1995, pp. 1009–1016, with kind permission from Elsevier Science, Sara Burgerhartstraat 25, NL-1055 KV Amsterdam, The Netherlands.

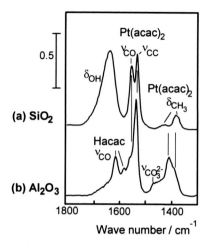

Figure 2. DRIFT spectra of Pt(acac)$_2$ adsorbed (a) on SiO$_2$ (4.6 wt % Pt) and (b) on Al$_2$O$_3$ (1.6 wt % Pt) at 298 K. Reprinted from (Poncelet, G. et al., eds.), *Preparation of Catalysts VI Scientific Bases for the Preparation of Heterogeneous Catalysts*, 1995, pp. 1009–1016, with kind permission from Elsevier Science, Sara Burgerhartstraat 25, NL-1055 KV Amsterdam, The Netherlands.

precursor. The detailed analysis of the various adsorbate spectra showed that Pt(acac)$_2$ and Cr(acac)$_3$ adsorbed without decomposition on SiO$_2$ (see Fig. 2 (a)). Because fourfold coordinated Pt(II) as well as sixfold coordinated Cr(II) cannot form an additional coordination to the surface, it is proposed that H-bridges between surface silanol groups and acetylacetonate ligands are formed (see Fig. 3). In addition, on SiO$_2$-supported V(acac)$_3$ (see Fig. 4 (a)), weakly adsorbed surface compounds like acetylacetone, acetone, and acetic acid were detected along with solid VO(acac)$_2$. It was concluded that (1) one acetylacetonate ligand was split

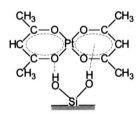

Figure 3. Proposed adsorbate structure of Pt(acac)$_2$ on SiO$_2$: H-Bridges between surface silanoles and the delocalized π-electrons and oxygen of the acac ligands. Reprinted from (Ponce, G. et al., eds.), *Preparation of Catalysts VI Scientific Bases for the Preparation of Heterogeneous Catalysts*, 1995, pp. 1009–1016, with kind permission from Elsevier Science, Sara Burgerhartstraat 25, NL-1055 KV Amsterdam, The Netherlands.

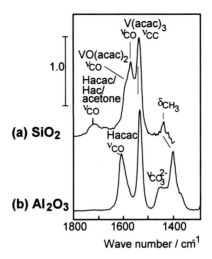

Figure 4. DRIFT spectra of V(acac)$_3$ adsorbed (a) on SiO$_2$ (0.3 wt % V) and (b) on Al$_2$O$_3$ (0.3 wt % V) at 298 K. Reprinted from (Poncelet, G. *et al.*, eds.), *Preparation of Catalysts VI Scientific Bases for the Preparation of Heterogeneous Catalysts*, 1995, pp. 1009–1016, with kind permission from Elsevier Science, Sara Burgerhartstraat 25, NL-1055 KV Amsterdam, The Netherlands.

off when adsorbing V(acac)$_3$ on SiO$_2$, and (2) surface-oxygen coordinated V(acac)$_2$ and acetylacetone are formed according to:

$$\text{Si-OH}_{surf} + \text{V(acac)}_{3gas} \rightarrow \text{Si-O-V(acac)}_{2surf} + \text{Hacac}_{ads}$$

Acetylacetone may be further decomposed to acetone and acetic acid. The surface reactions are caused by the lesser stability of this metal acetylacetonate compared to Pt(acac)$_2$ and Cr(acac)$_3$.

In contrast to the adsorbates on SiO$_2$, the DRIFT spectra of Al$_2$O$_3$-supported Pt(acac)$_2$ and Cr(acac)$_3$ showed a drastic change in relative band intensities of the acetylacetonate ligands and, moreover, additional vibrations that are due to acetylacetone adsorbed in its enolate form and to surface carbonate on Al$_2$O$_3$. Although in the spectra of adsorbed Pt(acac)$_2$ and Cr(acac)$_3$ bands that are characteristic for the precursor were still observed (see Fig. 2 (b) for illustration), the spectrum of adsorbed V(acac)$_3$ (see Fig. 4 (b)) corresponded completely to the spectrum of acetylacetone adsorbed in its enolate form. It was reasoned that Pt(acac)$_2$ and Cr(acac)$_3$ were partly and V(acac)$_3$ was completely decomposed on Al$_2$O$_3$ by catalytic action of support surface sites. The surface reaction involves the split of acteylacetonate ligands that may adsorb on Al^{3+} sites in their enolate form and possibly the further conversion to acetone and acetic acid strongly adsorbed on Al^{3+} sites (see Fig. 5).

$$H_3C \quad \overset{H}{\underset{C}{C}} \quad CH_3$$

Figure 5. Acetylacetone adsorbed on surface Al^{3+} sites as an enolate. Reprinted from (Poncelet, G. *et al.*, eds.), *Preparation of Catalysts VI Scientific Bases for the Preparation of Heterogeneous Catalysts*, 1995, pp. 1009–1016, with kind permission from Elsevier Science, Sara Burgerhartstraat 25, NL-1055 KV Amsterdam, The Netherlands.

On heating, the adsorbates decomposed. In N_2 atmosphere adsorbed Pt(acac)$_2$ decomposed between 420 and 570 K on SiO$_2$ as well as on Al$_2$O$_3$ (see Fig. 6); CO was the main gaseous decomposition product besides acetylacetone (only on SiO$_2$), acetone, and CO$_2$[55]. CO adsorbs at least partly on Pt particles already formed, which is reflected by the DRIFT spectra (see Fig. 7). A detailed analysis of the respective Pt–CO vibrations is described elsewhere [6] and leads to the conclusion that Pt is first deposited in its oxidized state, but then is reduced by the reductive gas atmosphere formed during decomposition. Adsorbed Cr(acac)$_3$ was completely transformed in air to chromium oxide at 670 K, and adsorbed V(acac)$_3$ was completely transformed to vanadium oxide in air at 570 K; no organic residue was detected by DRIFTS (see Fig. 8). V(acac) decomposition on SiO$_2$ was accompanied by the evolution of acetylacetone, acetone, acetic acid, and traces of CO and CO$_2$. On Al$_2$O$_3$, diacetyl, acetic acid, and small amounts of acetone were formed. In the metal acetylaceto-

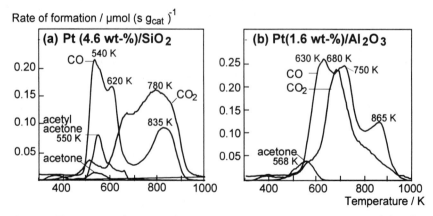

Figure 6. TPD spectra of Pt(acac)$_2$ decomposition on (a) SiO$_2$ (4.6 wt % Pt) and (b) Al$_2$O$_3$ (1.6 wt % Pt).

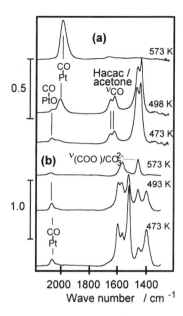

Figure 7. *In situ* DRIFT spectra of Pt(acac)$_2$ decomposition on (a) SiO$_2$ (4.6 wt % Pt) and (b) Al$_2$O$_3$ (1.6 wt % Pt) when increasing temperature. Reprinted from (Poncelet, G. *et al.*, eds.), *Preparation of Catalysts VI Scientific Bases for the Preparation of Heterogeneous Catalysts*, 1995, pp. 1009–1016, with kind permission from Elsevier Science, Sara Burgerhartstraat 25, NL-1055 KV Amsterdam, The Netherlands.

nate transformation on Al$_2$O$_3$, carbonaceous products were formed; during further heating carbonates and carboxylates became visible. As shown by the characteristic bands in the *in situ* DRIFT spectra (see Fig. 8), the surface compounds were stable after final decomposition, but were removable by subsequent air treatment of the samples at 773 K. Also because detectable amounts of elementary carbon were deposited and CO adsorbed on Pt particles when the decomposition was performed in N2, additional air treatment as applied for catalyst preparation (see preceding text) was required.

18.2.3.2 Dispersion and Particle Sizes of Pt, Cr$_2$O$_3$, and V$_2$O$_5$ on SiO$_2$ and Al$_2$O$_3$

Dispersions of differently loaded Pt, Cr$_2$O$_3$, and V$_2$O$_5$ catalysts that were prepared by the fluidized-bed technique are listed in Table 2. Dispersions of Pt and Cr$_2$O$_3$ on SiO$_2$ and on Al$_2$O$_3$ were affected to different extents by the amount of the catalytic compound on the support. On SiO$_2$ the dispersion of Pt and Cr$_2$O$_3$ decreased steadily with increased loading. By increasing the amount of the catalytic compound on the support, ag-

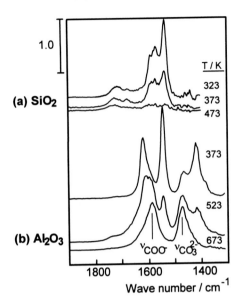

Figure 8. *In situ* DRIFT spectra of V(acac)₃ decomposition on (a) SiO₂ (0.3 wt % V) and (b) Al₂O₃ (0.3 wt % V) when increasing temperature. Reprinted from (Poncelet, G. *et al.*, eds.), *Preparation of Catalysts VI Scientific Bases for the Preparation of Heterogeneous Catalysts*, 1995, pp. 1009–1016, with kind permission from Elsevier Science, Sara Burgerhartstraat 25, NL-1055 KV Amsterdam, The Netherlands.

glomeration of the primary, small Pt or Cr_2O_3 clusters was promoted on the rather inert silica surface; it may be also assumed that the primary clusters act as seeds for further crystallite growth. Thus, with increasing metal or metal oxide coverage of the surface, larger crystallites are formed. The dispersion of Pt depended significantly on the atmosphere in which the adsorbate was decomposed. At, for example, 0.6 wt % Pt on SiO_2, the dispersion amounted to 21% when decomposing the adsorbate in air but to 36% when using N_2 instead. This effect can be explained by a higher mobility of small PtO clusters during $Pt(acac)_2$ decomposition in air that does not exist in an insert atmosphere. Histograms of the number of particles as function of their size are given for Pt on silica and alumina in Fig. 9.

As shown by the transmission electron microscopy (TEM) photograph of a Pt/SiO_2 catalyst (see Fig. 10) plain, irregular as well as hemispherical particles exist on SiO_2 with sizes between 1 and 25 nm. From a statistical interpretation of the micrograph it can be determined that of 37 particles 20% exhibit a size of 2.5 nm, 46% are smaller than 9 nm, and 13% have sizes larger than 20 nm. The mean particle size amounts to 9 nm. A slight nonuniformity of Pt particles, particularly when decomposed in air and at low loading (0.6 wt % Pt), is indicated by the splitting of IR bands of CO.

TABLE 2
Pt and Cr_2O_3 Dispersion on Silica and Alumina Support

	On SiO_2			On Al_2O_3		
	Loading[a] (wt %)	Adsorbed gas[b] (μmol g^{-1})	D[a] (%)	Loading[a] (wt %)	Adsorbed gas[b] (μmol g^{-1})	D[a] (%)
Pt	0.6	11	36	0.3	4	26
deposited in N_2	2.6	35	26	0.7	11	30
	4.8	65	26	1.2	21	34
	6.0	61	20	1.6	33	40
	8.5	62	14	2.0	33	32
Pt	0.1	2	39	0.2	4	39
deposited in air	0.2	3	29	0.6	8	26
	0.4	5	27	1.2	14	23
	0.6	7	21	1.4	21	29
	0.9	9	19	2.0	34	33
Cr_2O_3	1.2	28	53	0.9	30	77
deposited in air	3.8	38	22	3.5	77	49
	9.1	24	6	6.8	96	32
	16.9	30	4	11.8	118	22

[a]Referred to Pt in the case of Pt catalysts and referred to Cr_2O_3 in the cast of Cr_2O_3 catalysts.
[b]CO for Pt catalysts and O_2 for Cr_2O_3 catalysts.

Reprinted from (Poncelet, G. *et al.*, eds.), *Preparation of Catalysts VI Scientific Bases for the Preparation of Heterogeneous Catalysts*, 1995, pp. 1009–1016, with kind permission from Elsevier Science, Sara Burgerhartstraat 25, NL-1055 KV Amsterdam, The Netherlands.

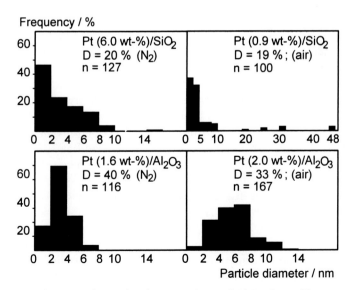

Figure 9. Pt particle size distribution on SiO_2 and Al_2O_3 obtained by CVD.

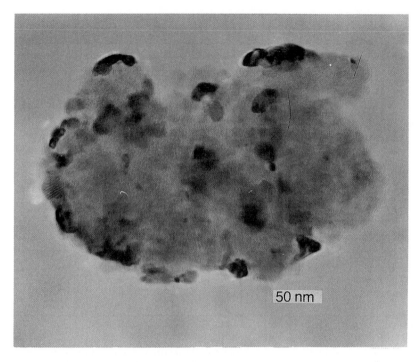

Figure 10. TEM result of Pt (0.6 wt %) on SiO$_2$ particles.

Although the Pt loading varied on Al$_2$O$_3$, the Pt dispersion changed randomly between 23 and 40% (i.e., no dependency on loading was observed [see Table 2]). On N$_2$-decomposed as well as air-decomposed samples, the same average dispersions were determined. For Pt particles on Al$_2$O$_3$ it is proposed that Pt exists preferentially in a disperse state that is stabilized by Al$_2$O$_3$. This was first of all confirmed by DRIFT spectra of CO adsorbed on Pt/Al$_2$O$_3$ catalysts (see Figs. 11 and 12). A CO band at 2064 cm^{-1} (decomposition in nitrogen) was attributed to adsorption on very small particles exhibiting a high Pt dispersion and a further band at 2083 cm^{-1} was attributed to adsorption on larger particles [56]; this phenomenon was less marked when decomposition took place in air. On the basis of the intensity relation between the two IR bands, the amount of finely dispersed particles was estimated to be at least 75% of the total amount of particles. The stabilization of dispersed Pt on Al$_2$O$_3$ may be tentatively ascribed to the Lewis acidity of the support by which the mobility of surface metal species during decomposition and crystallite growth may be limited. Because donor–acceptor interactions between deposited Pt and Al^{3+} sites may exist and result in stabilization of Pt by the support, all the other preparation

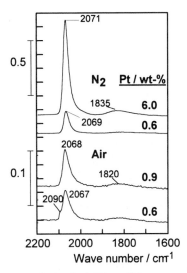

Figure 11. *In situ* DRIFT spectra of adsorbed CO on SiO$_2$-supported Pt prepared by decomposition in N$_2$ and air.

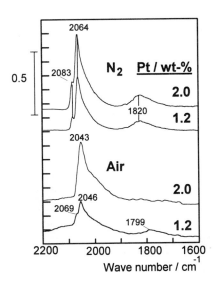

Figure 12. *In situ* DRIFT spectra of adsorbed CO on Al$_2$O$_3$-supported Pt prepared by decomposition in N$_2$ and air.

conditions are obviously overruled. Although the Cr_2O_3 dispersion on alumina decreased with increasing Cr_2O_3 loading, the effect is less marked in comparison with SiO_2 (see Table 2), which is first of all ascribed to the absence of crystalline particles as confirmed by XRD. Amorphous Cr_2O_3 on Al_2O_3 exhibits higher catalytic surfaces than the compact corundum-type crystallites that were observed by XRD on SiO_2.

For supported V_2O_5 catalysts the dispersion was neither affected by the support material nor by the V_2O_5 loading although the latter was varied over a broad range (see Table 2). On the other hand, TPR results suggested that the V_2O_5 particle size is affected by the loading [57], which was inconsistent with the V_2O_5 dispersions determined. Furthermore, XRD analysis confirmed that the same phenomena concerning the crystallinity of the catalytic compound as observed for Cr_2O_3 occur on SiO_2; crystallites are formed while on Al_2O_3, the compounds are amorphously dispersed.

18.2.3.3 *Acid Surface Sites of Cr_2O_3 and V_2O_5 on SiO_2 and Al_2O_3*

The supported metal oxides as well as the pure supports exhibited Lewis acid sites after oxidative pretreatment; no Brønsted acid sites were present as confirmed by IR spectra of adsorbed pyridine. Catalysts consisting of 1.2 wt % Cr_2O_3 and 0.9 wt % V_2O_5 on SiO_2, respectively, showed an increase in the number of Lewis acid sites (6.9×10^{17} per m^2 and 5.1×10^{17} per m^2, respectively), compared to SiO_2 (2.2×10^{17} sites per m^2). New Lewis acid surface sites generated by the deposition of metal oxide on SiO_2 as expected may be derived from TPD profiles (Fig. 13); however, on Al_2O_3 the total number of Lewis acid sites was not affected by the deposi-

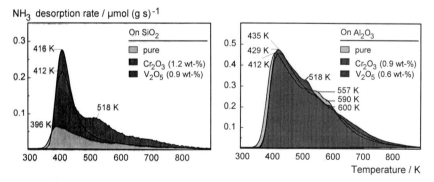

Figure 13. TPD spectra of NH_3 desorption from SiO_2- and Al_2O_3-supported Cr_2O_3 and V_2O_5.

TABLE 3
NH$_3$ Adsorption on SiO$_2$- and Al$_2$O$_3$-Supported Cr$_2$O$_3$ and V$_2$O$_5$ Catalysts

	Adsorbed NH$_3$ (μmol g^{-1})	Number of NH$_3$ adsorption sites (1×10^{17}m^{-2})	TPD (T/K)
On SiO$_2$			
Pure	31	2.2	396
1.2 wt % Cr$_2$O$_3$	94	6.9	412, 518
0.9 wt % V$_2$O$_5$	72	5.1	416
On Al$_2$O$_3$			
Pure	273	14.0	412, 557, 590
1.2 wt % Cr$_2$O$_3$	264	15.3	435, 600
0.9 wt % V$_2$O$_5$	284	14.4	429, 518

tion of Cr$_2$O$_3$ (0.9 wt %) and V$_2$O$_5$ (0.6 wt %), respectively, because the samples exhibited on the average $(14.6 \pm 0.6) \times 10^{17}$ sites per square meter. The data on ammonia desorption are summarized quantitatively in Table 3. It was concluded from the NH$_3$ desorption profiles (see Fig. 13) that Al^{3+} sites of stronger Lewis acidity (T$_{Des}$ = 530 K) were preferentially covered by Cr$_2$O$_3$ and V$_2$O$_5$, respectively, which resulted in an increased number of weakly Lewis acid sites (T$_{Des}$ = 420–430 K). The latter were also detected for SiO$_2$ catalysts.

18.3 Conclusions

Surface properties of the support material influence the adsorption states of Pt(acac)$_2$, Cr(acac)$_3$ and V(acac)$_3$, the decomposition pathways of the adsorbates, and finally the dispersion of the catalytic compound. The deposited particles are more mobile on SiO$_2$ and hence are more capable of agglomerating Al$_2$O$_3$-supported Pt is presumably stabilized by coordinatively unsaturated Al^{3+} surface ions; similar arguments probably apply for the stabilization of amorphous Cr$_2$O$_3$ and V$_2$O$_5$ of alumina. Because the metal acetylacetonate decomposition is accompanied by deposition of carbonaceous compounds, an additional air treatment of the samples is required to accomplish a carbon-free surface.

Acknowledgment

This work was partly supported by Fonds der Chemischen Industrie and Deutsche Forschungsgemeinschaft. T; this has been greatly appreciated.

References

1. Mond, H. J. (1895). *Soc. Chem. Ind.*, 14, 945.
2. Manasavit, H. M. (1968). *Appl. Phys. Lett.*, 12, 136.
3. Kirss, R. U. (1992). *Appl. Organomet. Chem.*, 6, 609.
4. Yamazaki, H., Tsuyama, T., Kobayashi, I., and Sugimori, Y. (1992). *Jpn. J. Appl. Phys.*, 31 (Pt. 1, no. 9b), 2995.
5. Moene, R., Kramer, L. F., Schoonmann, J., Makkee, M., and Moulijn, J. A., (1995). *Preparation of Catalysts VI, Scientific Bases for the Preparation of Heterogenous Catalysts* (G. Poncelet *et al.*, eds.), p. 379, Elsevier, Amsterdam.
6. Köhler, S., Reiche, M., Frobel, C., and Baerns, M., (1995). *Preparation of Catalysts VI, Scientific Bases for the Preparation of Heterogenous Catalysts* (G. Poncelet *et al.*, eds.), p. 1009, Elsevier, Amsterdam.
7. Nickl, J., Dutoit, D., Baiker, A., Scharf, U., and Wokaum, A. (1993). *Ber. Bunsenges. Phys. Chem.*, 97, 217.
8. Aitchison, K. A., Barrie, J. D., and Ciofalo, J. (1992). Better Ceramics through Chemistry V, *Mater. Res. Soc. Symp. Proc.*, 271, 957.
9. Durst, F., Kadinskii, L., Peric, M. and Schäfer, M. (1992). *J. Crystal Growth*, 125, 612.
10. Wille, H., Burte, E., and Ryssel, H. (1992). *J. Appl. Phys.*, 71, 3532.
11. Omata, K., Mazaki, H., Yagita, H., and Fujimoto, K. (1990). *Catal. Lett.*, 4, 123.
12. Ooi, H., Otsuki, A., Yano, M., and Harano, Y. (1990). *Kagaku Kogaku Ronbunshu*, 16, 579.
13. Yano, M., Nishikawa, T., Harano, Y., and Ohi, H. (1990). *Chem. Lett.*, 12207.
14. Moene, R., Makkee, M., and Moulijn, J. A. (1993). *Chem. Eng. J. (Lausanne)*, 53, 13.
15. Mirua, H., Oki, K., Ochiai, H., and Kimura, J. (1989). *Shokubai*, 31, 417.
16. Inumaru, K., Okuhara, T., and Misono, M. (1990). *Chem. Lett.* 1207.
17. Hattori, T., Matsuda, M., Suzuki, K., Miyamoto, A., and Murakami, Y. (1988). *Proceedings. 9th International Congress in Catalysis*, Vol. 4, p. 1640, The Chemical Institute of Canada, Ottawa.
18. Hattori, T., Itoh, S., Tagawa, T., and Murakami, Y. (1987). *Preparation of Catalysts IV* (B. Delmon, P. Grange, P. A. Jacobs, and G. Poncelet, eds.), p. 113, Elsevier, Amsterdam.
19. Inoue, T., Tomishige, K., and Iwasawa, Y. (1995). *J. Chem. Soc. Chem. Commun.*, no. 3, 329.
20. Sato, S., Sakurai, H., Urabe, K., and Izumi, Y. (1985). *Chem. Lett.*, 277.
21. Sato, S., Sakurai, H., Urage, K., and Izumi, Y. (1986). *J. Catal.*, 102, 99.
22. Imizu, Y., and Tada, A. (1989). *Chem. Lett.*, 1793.
23. Nariman, K. E., Lerou, J. J., Bischoff, K. B., and Foley, H. C. (1993). *Ind. Eng. Chem. Res.*, 32, 263.
24. Vansant, E. F. (1990). *Pore Size Engineering in Zeolites*, Wiley, New York.
25. Niwa, M., Kato, S., Hattori, T., and Murakami, Y. (1984). *J. Chem. Soc. Faraday Trans.*, 80, 3135.
26. Hibino, T., Niwa, M., Hattori, A., and Murakami, Y. (1988). *Appl. Catal.*, 44, 95.
27. Chun, Y., Chen, X., Yan, A.-Z., and Xu, Q.-H. (1994). *Stud. Surf. Sci. Catal.*, 84, 1035.
28. Megiris, C. E. and Glezer, J. H. E. (1992). *Ind. Eng. Chem. Res.*, 31, 1293.
29. Simon, M. W., Edwards, J. C. and Suib, S. L. (1995). *J. Phys. Chem.*, 99, 4698.
30. Ito, T., Kanai, H., Nakai, T., and Imamura, S. (1994). *React. Kinet. Catal. Lett.*, 52, 421.
31. Chamoumi, M., Brunel, D., Fajula, F., Geneste, P., Moreau, P., and Solofo, J. (1994). *Zelites*, 4, 282.
32. Hölderich, W. F., Paczkowski, M. E. (1994). *Stud. Surf. Sci. Catal.*, 83, 399–406.

33. Yoo, J. S., Donohue, A., Kleefisch, M. S., Lin, P. S. and Elfline, S. D. (1993). *Appl. Catal. A*, **105**, 83.
34. Yoo, J. S., Lin P. S., and Elfline, S. D. (1993). *Appl. Catal. A*, **106**, 256.
35. Yoo, J. S., Donohue, J. A., and Kleefisch, M. S. (1994). *Appl. Catal. A*, **110**, 75.
36. Yoo, J. S., Chin, C. F., and Donohue, J. A. (1994). *Appl. Catal. A*, **118**, 87.
37. Yoo, J. S., Sohail, A. R., Grimmer, S. S., and Choi-Feng, C. (1994). *Catal. Lett.*, **29**, 299.
38. Zajac, G. W., Choi-Feng, C., Faber, J., Yoo, J. S., Patel, R., and Hochst, H. (1995). *J. Catal.*, **151**, 338.
39. Dossi, C., Psaro, R., Bartsch, A., Fusi, A., Sordelli, L., Ugo, R., Bellatreccia, M., Zanoni, R., and Vlaic, G. (1994). *J. Catal.*, **145**, 377.
40. Dossi, C., Psaro, R., Bartsch, A., Brivo, E., Galasco, A., and Losi, P., (1993). *Catal. Today*, **17**, 527.
41. Kwak, B. S., and Sachtler, W. M. H. (1994). *J. Catal.*, **145**, 456.
42. Beutel, T., Knözinger, H., Trevino, H., Zhang, Z. C., Sachtler, W. M. H., Dossi, C., Psaro, R., and Ugo, R. (1994). *J. Chem. Soc. Faraday Trans*, **90**, 1335.
43. Greenberg, C. B. (1985). *J. Electrochem. Soc. Solid State Sci. Technol.*, **132**, 1394.
44. Toth, L. E. (1971). *Transition Metal Carbides and Nitrides*, Vol. 7, (J. L. Margrave, ed.), Academic Press, New York.
45. Eversberg, K. R. (1994). *Galvanotechnik*, **85**, 76.
46. Hoffman, D. M. (1994). *Polyhedron*, **13**, 1169.
47. Rausch, N., and Burte, E. P. (1993). *Microelectron. J.*, **24**, 421.
48. Sugimoto, T. (1993). *Mater. Sci. Forum*, 137–139.
49. Sugimoto, T. *Advances in High-Temperature Superconductors*, pp. 395–453.
50. Zhang, K., and Erbil, A. (1993). *Mater. Sci. Forum*, 130–132.
51. Zhang, K., and Erbil, A. *Synthesis and Characterization of High-Temperature Superconductors*, pp. 255–268.
52. Pickerell, D. J., Santini, P. J., Kimock, F. M. (1993). *Proceedings SPIE-International Society Optics Engineers*, 2105.
53. Pickrell, D. J., Santini, P. J., and Kimock, F. M. (1993). *International Symposium on Microelectronics*, 405–406.
54. Zhang, F., Guo, Y., Song, Z., and Chen, G. (1994). *Appl. Phys. Lett.*, **65**, 971.
55. Termath, S. (1994). Doctoral dissertation, Ruhr-University Bochum, Germany.
56. Rothschild, W. G., Yao, H. C., and Plummer, H. K., Jr. (1986). *Langmuir*, **2**, 588.
57. Hattori, T., Matsuda, M., Suzuki, K., Miyamoto, A., and Murakami, Y. (1988). *Stud. Surf. Sci. Catal.*, **4**, 1640.

C H A P T E R   1 9

Synthesis of Catalytic Materials by Spray Pyrolysis

Aaron Wold, Y-M. Gao, Daniel Miller, Robert Kershaw,
and Kirby Dwight

Brown University
Providence, Rhode Island 02912

KEYWORDS: catalyst synthesis, polycrystalline powder preparation, spray pyrolysis, thin-film preparation

19.1   Introduction

For a number of years there has been a need to develop new low-temperature (below 500°C) syntheses for the preparation of potentially important oxides and sulfides. Catalyst preparation was regarded as an art and most of the details were available only in the patent literature until the first international conference devoted to catalyst synthesis was held 20 years ago [1]. Since that time new techniques have been developed for catalyst preparation and the effort is continuing as exemplified in this chapter. Increased insight into the nature of catalyst preparation has led to the synthesis of a number of new catalysts with improved activity, selectivity, and stability properties.

Despite progress, there still is a need to develop new low-temperature syntheses of metal oxides and sulfides. Research at Brown University has been directed to the study of the pyrolysis of volatile transition metal precursors. Chemical vapor deposition (CVD) at atmospheric pressure has been used to prepare thin films and polycrystalline powders of a number of oxides including TiO_2 and modified TiO_2 photocatalysts. This has resulted in the design of a simple reactor in which a volatile precursor is introduced as a spray and then pyrolyzed in a hot zone of the reactor. In these studies, which will be discussed in detail, only solutions containing volatile precursors have been utilized. Ceramic powder synthesis by spray pyrolysis has recently received considerable attention and Messing et al. [2] have reviewed the status of this technique.

In this chapter spray pyrolysis will be discussed only in terms of pyrolyzing a suitable vapor of precursors that when decomposed give oxides with interesting catalytic properties. Whereas the principal emphasis will be directed toward the preparation of TiO_2 and modified TiO_2, Sect. 19.3 is devoted to the preparation of other oxides and sulfides.

The initial studies conducted at Brown emphasized the preparation of thin films. The application directed toward the removal of harmful organic compounds from water necessitated a process in which the water flowed through glass pipes that were coated with TiO_2 or modified TiO_2 and exposed to sunlight. Undoubtedly, there will be a growing need for the use of catalysts that are deposited as thin films and this will therefore be an area of increased research effort. The modification of the process to yield polycrystalline powders rather than thin films is trivial and indeed it is far simpler to prepare such powders than to deposit uniform high-quality thin films.

Spray pyrolysis was initially studied by Chamberlain and Skarman [3] and has been used for the preparation of oxide and sulfide thin films, particularly in the area of photovoltaic devices. Spray pyrolysis is a relatively simple and an inexpensive preparative method [4,5].

For the preparation of a thin film, spray pyrolysis involves spraying an atomized solution containing an appropriate precursor onto a heated substrate. The composition and properties of the resulting deposited film depends on the atmosphere, temperature, and constituents of the sprayed solution. For the production of a polycrystalline powder, the solution is sprayed into a silica tube that is maintained at a desired temperature. The desired product deposits on the wall of the tube and can be readily removed on completion of the process.

Solution atomization is a major consideration in spray pyrolysis. This determines the droplet size and size distribution that play important roles in determining the nature and composition of the product formed. Blandenet et al. [6] have shown that ultrasonic nebulization is an effective method for achieving a narrow size distribution of droplets. A commercial ultrasonic humidifier was used in this study and provided efficient and inexpensive nebulization of various precursor solutions. A major limitation of ultrasonic nebulization has to do with the nature of the solution being sprayed. Because the atomization process depends on setting a liquid film into motion, the more viscous the liquid, the more difficult it becomes to create vibratory motion sufficient for atomization. To some extent this situation can be offset by increasing input energy, making the atomizing surface vibrate with greater amplitude. With more sophisticated spray nozzles [7], solutions with viscosities up to 50 centipoise (cP) can be atomized. However, such systems can cost up to $6000 compared to approximately $50 for an inexpensive humidifier unit.

Viguie and Spitz [8] have shown that the conversion of the liquid droplets to vapor before pyrolysis of the precursor occurs, offers an advantage over spraying droplets onto the surface followed by condensed-phase reaction. In spray pyrolysis a solution is nebulized ultrasonically, and sprayed continuously or in pulses. Viguie and Spitz [8] suggested four possible growth mechanisms for the spray pyrolysis process as a function of substrate temperature. These are shown in Fig. 1 and can be related to the growth of oxide and sulfide films reported in this study. In the first process the spray droplet impinges directly on the substrate, followed by evaporation of the solvent and decomposition of the precursor to the oxide. In the second process the solvent is evaporated just prior to contacting the sub-

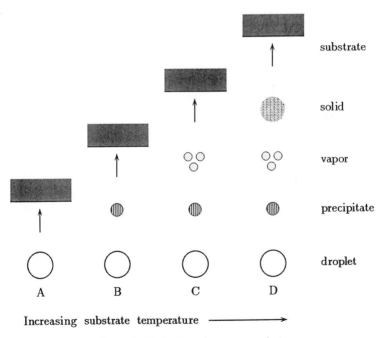

Figure 1. Mechanisms for spray pyrolysis

strate, followed by decomposition to the oxide. The third mechanism involves volatilization of the dried metallic precursor salt, diffusion of the vapor to the substrate, and followed by decomposition to the oxide. This process is referred to as low-temperature chemical vapor deposition (LTCVD). The fourth process is a homogeneous nucleation of the vapor phase forming the oxide particle, which then deposits on the substrate.

The formation of the oxide films reported in this study probably proceeds by the third mechanism because the deposited films are smooth, are homogeneous, and have a mirror-like appearance. Albin and Risbud [9] have suggested that the deposited films obtained by the last process would have a powdery appearance. Viguie and Spitz [8] have shown that films grown by the first two processes have a rougher microstructure than films grown by the third process.

This chapter will focus on the preparation of thin films of a number of transition metal oxides. The application of the method to the preparation of polycrystalline powders will be discussed using TiO_2 as an example. Finally, the method will be applied to the synthesis of zinc sulfide thin films.

19.2   Experimental

19.2.1 Preparation of TiO_2 Films

Titanium oxide films have been prepared by an ultrasonic nebulization and pyrolysis technique developed in this laboratory [10]. Titanium acetylacetonate was synthesized following Yamamoto's method [11]. A 0.005 M solution of titanium acetylacetonate in ethanol was nebulized by a commercial ultrasonic humidifier (Holmes Air) and was carried into a horizontal reactor by argon (Fig. 2). The substrate was held perpendicular to the gas flow in the furnace (two-zone mirror furnace, Trans-Temp, Chelsea, Massachusetts) by a silica holder that was rotated by a low-speed motor to achieve best uniformity. Both the efficiency of the deposition and the uniformity of the films were affected by the deposition parameters used. Typical reaction parameters were furnace temperature (380°C), argon flow rate (3.5 liter/min), and distance between nozzle and substrate (6.4 cm). The nebulized solution was delivered to the substrate in 15-s pulses with 5-s intervals between pulses. After deposition, the film was annealed in a flowing oxygen atmosphere at the same temperature for 30 min. Both (100) silicon and silica wafers were used as substrates. Cleaning of the silicon substrate was conducted just prior to the deposition according to the procedure described by Fournier *et al.* [12] Silica substrates were cleaned with hydrochloric acid, distilled water, and semiconductor grade acetone prior to use.

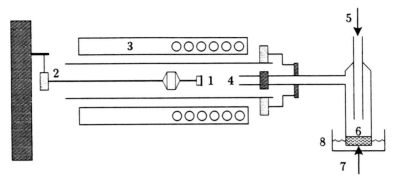

1 Substrate
2 Substrate rotation motor
3 Furnace
4 Spray nozzle

5 Carrier gas
6 Solution
7 Membrane
8 Ultrasonic Humidifier

Figure 2. Spray reactor with rotating substrate for films.

19.2.2 Film Characterization

X-ray diffraction patterns of the films were obtained using a Philips diffractometer with monochromated high-intensity CuKα_1 radiation (λ = 1.5405 Å). Diffraction patterns were taken with a scan rate of 1° 2θ/min over 10° <2θ <60°.

The thickness of the films on silicon substrates was determined by ellipsometry using a Rudolph Research Auto E1-II Ellipsometer [13], whereas those on silica substrates were measured by interference fringes in their optical spectra.

The surface morphology of the films was examined by means of a JEOL-840F scanning electron microscope (SEM) operating at 10 kV. Before measurement, each sample was coated with a thin film of platinum in order to reduce charge accumulation. Optical measurements of the films on silica substrates were performed using a Cary model 17 dual-beam ratio-recording spectrophotometer in the range of 300–1000 nm. Measurements were made in the transmission mode. The optical bandgap was deduced from the transmittance near the absorption edge.

19.2.3 Preparation of Polycrystalline TiO$_2$

A schematic diagram of the experimental apparatus for the preparation of polycrystalline samples is shown in Fig. 3. A solution containing a titanium precursor was nebulized by a commercial ultrasonic humidifier (Holmes Air). The generated mists were carried by the carrier gas and fed into a horizontal silica tube with an inside diameter of 1 in. and an alu-

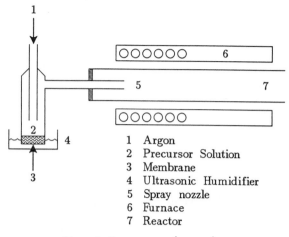

1	Argon
2	Precursor Solution
3	Membrane
4	Ultrasonic Humidifier
5	Spray nozzle
6	Furnace
7	Reactor

Figure 3. Spray reactor for powder.

mina liner with an inside diameter of 5/8 in. The length of the liner was 6 in. The precursor $TiCl_4$ with oxygen was fed into the center of the reactor via an alumina nozzle with an inside diameter of 3/8 in. The nozzle was sheathed by an alumina liner to protect the silica tube from attack by such precursors as $TiCl_4$ at elevated temperature. Wet oxygen was generated by bubbling the oxygen through water at room temperature and was introduced into the reactor. The flow rate of water-saturated oxygen was 0.13 liter/min. The reaction between the $TiCl_4$ mist and the wet oxygen was conducted at the center of the hot zone, which was kept at 830°C. The product was collected from the cool zone of the silica tube. If an alkoxide of titanium were used (e.g., pure titanium(IV) isopropoxide) as the precursor, the nebulized mist would be carried by a stream of argon gas with a flow rate of 1.6 liter/min. The temperature within the reaction tube was maintained at 900°C. Oxygen was also introduced into the reactor by a separate tube and this maintains an oxidizing atmosphere during pyrolysis. The oxygen flow rate was maintained at 0.032 liter/min. The collected powder was postannealed at 600°C for 30 min in an oxygen atmosphere with an oxygen flow rate of 1 liter/min to oxidize any carbon residue in the sample.

When the alkoxide is solid or does not readily nebulize, it is necessary to dissolve it in alcohol. The solution containing the alkoxide can then be readily nebulized. However, the nebulization must be conducted with argon (1.6 liter/min) as the carrier gas, as well as the gas fed into the pyrolysis reactor. The collected powder is black and must be postannealed at 600°C in an oxygen atmosphere (flow rate 1 liter/min) in order to eliminate carbon from the sample.

19.2.4 Characterization of TiO$_2$ Powders

19.2.4.1 Measurement of Surface Area

The surface area was measured by the single-point BET method using the Flow Sorb(II) 2300 (Micromeritics Instrument Corp., Norcross, Georgia). The samples were degassed at 170°C for 2 h and the adsorbate gas consisted of a mixture of 30% N$_2$ and 70% He.

19.2.4.2 Measurement of Photocatalytic Activity

The photocatalytic activities of various titanium(IV) oxide samples were evaluated by degradation of *p*-dichlorobenzene (DCB). A saturated DCB solution was prepared by adding an excess of DCB to distilled water and stirring the solution for 8 h while heating at 60°C. The solution was capped and preserved in darkness. A sample (0.0015 wt % of TiO$_2$) was ultrasonically dispersed in distilled water and 0.2 ml of a saturated aqueous solution of DCB was added to 3 ml of this TiO$_2$ suspension in a quartz cuvette. The reference cuvette, to be used as a blank for the absorption measurement, contained 0.2 ml of distilled water to which 3 ml of the TiO$_2$ suspension was added. The samples were then irradiated with a 150-W xenon short-arc lamp whose output was passed through a Pyrex filter. The differential absorbance at 224 nm (maximum absorption peak of DCB) was measured using a Perkin-Elmer 552A spectrophotometer.

19.2.4.3 Measurement of Dispersion

An indication of the amount of catalyst interacting with the DCB was measured by the absorption of the reference cuvette containing 0.2 ml of distilled water and 3 ml of the TiO$_2$ solution at 240 nm using a Perkin-Elmer 552A spectrophotometer. The dispersion measurement was taken prior to initiating the activity measurement.

19.2.4.4 Measurement of Surface Acidity

The surface acidity of titanium(IV) oxide samples was determined by titration with *n*-butylamine following a procedure established by Tamele [14]. Approximately 0.4 g of catalyst powder was dispersed in 50 ml of benzene using a glass stirring bar. Seven drops of a 0.05 N methyl red indicator solution in benzene were added to the flask. Titrated against the powder was 0.008 N *n*-butylamine, and the amount of titer necessary to effect the color change on the surface of the powder was recorded. The acidity was then calculated in terms of millimole per gram of catalyst. Although the titration technique has intrinsic limitations, precautions were taken to handle the samples in a consistent manner to ensure that the results were reproducible and could be confidently compared on a relative basis.

19.3   Results and Discussion

Samples of TiO_2 were prepared by spray pyrolysis of titanium iso-propoxide. The properties of the TiO_2 obtained were compared with those of P25 (Degussa TiO_2) and TiO_2 prepared by evaporation of titanium iso-propoxide followed by pyrolysis. Table 1 lists the properties of three TiO_2 products, and it can be seen that they were found to be nearly identical in surface area, acidity, dispersion, and photoactivity. Table 1 also lists two other precursors, that is, titanium(IV) chloride and diisopropoxy(tita-nium(IV)-bis-acetylacetonate. However, the catalytic properties of TiO_2 obtained from these precursors were lower than those formed from the isopropoxide. Ultrasonic nebulization pyrolysis compares favorably with thermal evaporation pyrolysis. However, the latter process cannot always be conducted.

It has been shown that by means of spray nebulization–pyrolysis members of the systems MoO_3/TiO_2 and WO_3/TiO_2 can be prepared using precursors of molybdenum and tungsten as well as titanium [15,16]. These results are given in Table 2. It can be seen that the activities of the modi-fied TiO_2 products formed were considerably higher than those obtained for Degussa TiO_2 (P25).

Films, unlike powders, can be used repeatedly for the measurement of photocatalytic activity, and unlike highly dispersed powders, films present no problem in terms of separating the catalyst from the solution after degradation of the organic impurities. TiO_2 on Pyrex glass substrate were prepared and evaluated as photocatalysts [17,18]. Nebulization of diiso-propoxy(titanium)-bis-acetylacetonate resulted in the production of films with good adherence to the substrate and uniform thickness. The films produced were evaluated by measuring the degradation of salicylic acid. These were shown to be capable of photodecomposing organic impurities

TABLE 1
Characterization of Titanium(IV) Oxide Samples

Precursor	Method	Photocatalyst activity	Surface acidity (mmol/g)	Surface area (m$_2$/g)
P25 (Degussa TiO_2)	—	39	0.010	46
Titanium isopropoxide	UNP[a]	34	0.006	38
Titanium isopropoxide	TEP[b]	36	0.005	42
Titanium(IV) chloride	—	33	—	—
Diisoproproxy(titanium)(IV)-bisacetylacetonate	—	30	—	—

[a]UNP = Ultrasonic nebulization pyrolysis.
[b]TEP = Thermal evaporation pyrolysis.

TABLE 2
Photocatalytic Activity, Surface Area for Modified TiO$_2$ Samples

Precursors	System	Activity	Surface area
—	P25	40	49
0.5 M Ethanol solution of molybdenyl acetylacetonate diisopropoxy(titanium)-bis-acetylacetonate)	(3/97) MoO$_3$/TiO$_2$	59	33
0.5 M isopropanol solution of tungsten(V) pentaethoxide and titanium(IV) tetraisopropoxide	(3/97) WO$_3$/TiO$_2$	60	30

in water. Hence, it has been shown that ultrasonic nebulization followed by pyrolysis is a useful technique for the preparation of thin oxide films of high quality as well as powdered titanium(IV) oxide photocatalysts. This procedure has also been used to prepare thin films and powdered samples of other oxides (e.g., CuO, ZnO, Cr$_2$O$_3$, Fe$_2$O$_3$) [19–22].

It has also been shown that sulfides can be prepared by the conversion of zinc oxide films by treatment with H$_2$S [23]. Zinc oxide films were first prepared by ultrasonic nebulization of a zinc acetate solution followed by decomposition of the zinc acetate to zinc oxide. Uniform films of zinc sulfide were obtained by annealing the oxide film in a mixture of hydrogen sulfide and argon (H$_2$S:Ar = 1:1) in a horizontal tube furnace. The flow rate of the gas mixture was 50 cc/min. The temperature of the furnace was raised gradually from room temperature to 500°C in 4 h. The furnace was maintained at 500°C for another 3 h and then cooled down slowly in the H$_2$S–Ar atmosphere. Completion of the conversion was verified by X-ray diffraction analysis and infrared spectroscopy of the films.

Thin films of zinc sulfide can also be prepared by ultrasonically spraying a toluene solution of bis(diethyldithiocarbamato) zinc(II) onto silicon, sapphire, or gallium arsenide substrates at 460–520°C [7]. The films prepared on silicon or sapphire were found to have a highly oriented hexagonal structure, while those deposited onto cubic (100) gallium arsenide showed a highly oriented cubic structure. There appears to be at least three important parameters that influence the zinc sulfide polymorph formed in film fabrication processes: temperature, sulfur fugacity, and nature or orientation of the matched substrate. On amorphous or poorly lattice-matched substrates, the factors that dominate are the temperature and the sulfur fugacity, while on closely lattice-matched substrates, the cubic phase can be formed even under conditions (i.e., high temperature) that normally favor the hexagonal phase.

The examples utilizing spray pyrolysis have shown the broad versatility of the technique in producing oxides and sulfides that are of commercial interest. The equipment needed is relatively inexpensive and the products are as pure as those that can be obtained by more elaborate procedures. Mass production of large quantities of simple or complex oxides or sulfides, suitable as catalysts, can readily be made by this process.

Acknowledgment

This research was partially supported by the National Science Foundation Grant DMR 9401562.

References

1. International Symposium Scientific Basis Catalysis Preparation, Brussels, (1975). October 14–17.
2. Messing, G. L., Zhang, S.-C., and Jayanthi, G. V. (1993). *J. Am. Ceram. Soc.*, 2707.
3. Chamberlain, R. R., and Skarman, J. S. (1966). *J. Electrochem. Soc.*, 113(1), 86.
4. Mooney, J., and Radding, S. (1982). *Annu. Rev. Mater. Sci.*, 12, 81–101.
5. Albin, D., and Risbud, S. H. (1987). *Adv. Ceram. Mater.*, 2(3A), 243.
6. Blandenet, G., Court, M., and Lagrade, Y., (1981). *Thin Solid Films*, 77, 81.
7. Sonno-Tek Ultrasonic Atomizing Nozzle Systems, Sonno-Tek Corp., Poughkeepsie, NY.
8. Viguie, J. C., and Spitz, J. (1975). *J. Electrochem. Soc.*, 122, 585.
9. Albin, D., and Risbud, S. H. (1987). *Adv. Ceram. Mater.*, 2(3A), 243.
10. Wu, P., Gao, Y.-M., Kershaw, R., Dwight, K., and Wold, A. (1990). *Mater. Res. Bull.*, 25(3), 357.
11. Yamamoto, A., and Kambara, S. (1957). *J. Am. Chem. Soc.*, 79, 4344.
12. Fournier, J., DeSisto, W., Brusasco, R., Sosnowski, M., Kershaw, R., Baglio, J., Dwight, K., and Wold, A. (1988). *Mater. Res. Bull.*, 23, 131.
13. Brusaco, R., Kershaw, R., Baglio, J., Dwight, K., and Wold, A. (1986). *Mater. Res. Bull.*, 21, 301.
14. Tamele, M. W. (1950). *Discuss. Faraday Soc.*, 8, 270.
15. Lee, W., Do, Y. R., Dwight, K., and Wold, A. (1993). *Mater. Res. Bull.*, 28, 1127.
16. Do, Y. R., Lee, W., Dwight, K., and Wold, A. (1994). *J. Solid State Chem.*, 108, 198.
17. Lee, W. L., Gao, Y.-M., Dwight, K., and Wold, A. (1992). *Mater. Res. Bull.*, 27, 685.
18. Gao, Y.-M., Shen, H.-S., Dwight, K., and Wold, A. (1992). *Mater. Res. Bull.*, 27, 1023.
19. DeSisto, W., Sosnowski, M., Smith, F., DeLuca, J., Kershaw, R., Dwight, K., and Wold, A. (1989). *Mater. Res. Bull.*, 24, 753.
20. Gao, Y.-M., Wu, P., Baglio, J., Dwight, K., and Wold, A. (1989). *Mater. Res. Bull.*, 24, 1215.
21. Qian, Y.-T., Kershaw, R., Dwight, K., and Wold, A. (1990). *Mater. Res. Bull.*, 25, 1243.
22. Qian, Y.-T., Niu, C.-M., Hannigan, C., Yang, S., Dwight, K., and Wold, A. (1991). *J. Solid State Chem.*, 92, 208.
23. Gao, Y.-M., Wu, P., Baglio, J., Dwight, K., and Wold, A. (1989). *Mater. Res. Bull.*, 24, 1215.

CHAPTER 20

Formation of Nanostructured V_2O_5-Based Catalysts in Flames

Philippe F. Miquel and Joseph L. Katz
Department of Chemical Engineering
The Johns Hopkins University
Baltimore, Maryland 21218

KEYWORDS: counterflow diffusion flame burners, formation of catalysts in flames, vanadium oxide based catalysts

20.1   Introduction

Vanadium oxide based catalysts are widely used for the selective oxidation of hydrocarbons [1]. For example, V_2O_5–TiO_2 is used for the selective oxidation of *o*-xylene to phthalic anhydride, and vanadium–phosphorus oxides (VPO) are used for the selective oxidation of butene and *n*-butane to maleic anhydride. High selectivity and activity are achieved when the vanadia forms a bidimensional layer, called "monolayer," on the surface of the support [2]. For VPO catalysts, the selectivity toward maleic anhydride is related to the presence of specific and known crystalline VPO phases, and, in particular, to the presence of $(VO)_2P_2O_7$ [3,4]. To obtain these specific crystalline phases, "traditional" methods of preparation typically require a very high number of steps, such as pretreatment, mixing, chemical reaction, filtration, purification, drying, and calcination. These methods thus can be expensive and can cause waste treatment difficulties.

Over the past decade we have developed the counterflow diffusion flame burner into a device very well suited to produce and study the formation of nanostructured oxides and mixed oxides [5–8], and have used it to produce vanadium oxide based powders [9,10]. This novel technique offers significant advantages over traditional methods. One can form powders of desired particle size, morphology and crystalline structure by selecting the appropriate flame temperature, precursor concentration ratio, and precursor. Flame synthesis also reduces the powder production process to a single-step operation, thus drastically reducing the processing time and the number of environmentally detrimental side streams. Finally, it allows one to obtain powders with a high degree of purity and high-surface area.

20.2 Literature Survey

Nanostructured materials are of growing interest because they can have properties quite different from those of conventional materials [11]. Materials are called "nanostructured" when their average grain or structural domain sizes are below 100 nm. The physical and chemical properties of these new structural materials have drawn considerable attention from researchers in the field of catalysts, semiconductors, and ceramics [12]. Nanostructured catalysts not only have higher surface areas, but also are expected to show improved catalytic activity and selectivity [13]. Such catalysts are of particular interest for structure-sensitive reactions, such as mild oxidation reactions, which require particles of specific size and structure. Nanostructured semiconductors also are expected to have electronic properties that differ significantly from their conventional counterparts if they consist of crystallites that are small in comparison to bulk electron delocalization lengths (10–100Å) [14]. The discovery of carbon nanotubes by Iijima [15] has triggered intensive research in this area. These nanotubes are reported to be metallic or semiconducting depending on their diameters. Glasses doped with nanosized oxide crystallites have shown high speed optical responses and large third-order optical susceptibilities [16]. Nanostructured ceramics have grain sizes that are much smaller than those of conventional ceramics. This results in an increase in hardness, yield, and fracture strength; a decrease in ductile–brittle transition temperature; and thus results in materials requiring lower sintering temperatures and having finer microstructures [11].

The number of laboratory investigations on the physics and performances of nanostructured materials demonstrates the growing interest in these materials. However, as a panel report from the U.S. Department of Energy pointed out, a major challenge in the field of nanostructured materials is the design of processes appropriate for industrial-scale production [12] that are able to produce materials with specific particle size and morphology or crystalline structure. For example, TiO_2 produced by flame synthesis is used for a variety of applications (e.g., in paint opacifiers, catalysts, or ceramics) that depend on the powder characteristics. For use as paint opacifier, TiO_2–rutile particles (about 150 nm in diameter) are coated with a 10- to 20-nm layer of SiO_2. This coating layer deactivates the catalytic ability of TiO_2 to oxidize the organic polymer that binds paint films. TiO_2–anatase is a well-known support for catalysts. For such application a high-surface-area powder is necessary. TiO_2 is also a component of ceramics such as aluminum titanate. In ceramic processing, the particle size, the particle size distribution, and the presence or lack of agglomerates are critical to the rate of sintering and microstructure development.

The synthesis of nanostructured materials has been conducted most frequently by aerosol processes [17]. These aerosol processes include sputtering [18], plasma reactors [19], laser ablation [18], gas condensation method [20], spray pyrolysis, and flame and furnace reactors [17]. Spray pyrolysis is a process whereby a precursor solution is atomized into small droplets that are directed toward a heated region by a carrier gas. Inside the furnace the solvent evaporates and intraparticle reaction occurs to form the product material. This process allows one to produce multicomponent powders with uniform particle composition, a critical advantage for the formation of complex multicomponent oxides such as superconductors [21]. It also allows the formation of non-oxide powders such as nitride particles [22], fullerenes [23], or metal powders [24]. Flame and furnace processes also allow the formation of a wide range of nanostructured materials. Combustion processes, which historically have been viewed almost exclusively as energy generators, are now seen to offer a unique environment for chemical synthesis. Combustion synthesis offers significant advantages over traditional methods: the powders are produced in a single-step operation, in a very short processing time, with a high degree of purity, and without most of the environmentally detrimental side streams. Combustion synthesis does not involve the tedious steps and large volumes of liquid byproducts of conventional wet chemistry (sol–gel operations). It also provides a highly controlled environment; the energy necessary for vaporization, decomposition and reaction; and the oxygen source necessary for the formation of oxides.

Cuer *et al.* [25,26] were one of the first to point out the advantages of using a flame burner to produce structural materials. They produced Al_2O_3 particles in an open flame burner and analyzed the particle size and surface area as a function of flame temperature and precursor loading. They also mentioned the possibility of using the flame burner to produce mixed oxides of SiO_2 and Al_2O_3 [27]. However, extensive studies on mixed oxides did not begin until fairly recently [6–8]. These studies were made using the rectangular version of the counterflow diffusion flame burner [5]. Hung *et al.* [7,8] and Miquel *et al.* [9,10,28] investigated oxide formation in flames, with particular emphasis on mixed oxides. The oxides Al_2O_3, GeO_2, P_2O_5, SiO_2, TiO_2, V_2O_5, and their mixed oxides were produced, and their formation mechanisms were correlated to flame temperature, residence time, and precursor concentration. These investigations resulted in such a sufficiently good understanding of the processes occurring in the counterflow diffusion flame burner, that the production of materials, which are of great importance as catalysts, was possible in a single- or at most, a two-step operation process.

Other investigators also have recognized the advantages of using a flame process to produce valuable structural materials. Calcote *et al.* [29] used premixed burner stabilized flames of $SiH_4/NH_3/N_2$ to produce nano-structured Si_3N_4 for advanced ceramics. Ravankar *et al.* [30] have designed a hydrogen–chlorine flame reactor to produce ultrafine ceramic powders of tungsten and tungsten carbides. Akhtar and Pratsinis [31] also used a flame reactor to produce nanostructured Al-doped titanium oxide powders for use as pigments and paint opacifiers.

Over the past four decades, the oxide powder industry has recognized the advantage of using combustion synthesis. Large quantities of oxide powders currently are produced by combustion of their chloride precursors [32]. Tens of thousands of tons of SiO_2, TiO_2, and Fe_2O_3 are produced annually using combustion processes. Flame-generated SiO_2–GeO_2 particles are used in the production of optical fibers. These particles are deposited as a preform, treated to remove water and other undesirable substances, melted, and drawn [33]. They also are used as the starting materials for the manufacturing of high-tech ceramics. These ceramics are used in extreme environments such as internal combustion, gas turbine, jet, and rocket engines. Their special electrical properties also make them valuable materials for use as capacitors, piezoelectric devices, thermistors, and solar cells [34].

20.3   Experimental Setup

20.3.1 The Counterflow Diffusion Flame Burner

We are using the rectangular version of the counterflow diffusion flame burner [5]; it is shown in Fig. 1. This burner consists of two vertically opposed tubes of rectangular cross section separated by a distance of 15 mm. The fuel (hydrogen diluted with argon) flows upward from the lower tube, and the oxidizer (oxygen diluted with argon) flows downward

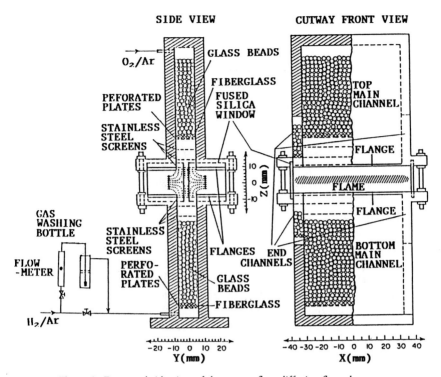

SIDE VIEW   CUTWAY FRONT VIEW

Figure 1. Front and side view of the counterflow diffusion flame burner.

from the upper tube. A visible flame is generated in the gap between the two tubes, in the region where the two opposed gas streams impinge. Each tube consists of three channels, a central main channel, and two side channels. Two pairs of flat, fused silica plates connect the outside of the two side channels of the opposed tubes, thus forcing the combustion gases to flow out only through the front and back. A 4-mm gap between each pair of plates allows one to send light beams across the burner, without significantly affecting the flow. To compensate for heat losses from the silica plates, the oxidizer and fuel streams in the side channels are enriched in oxygen and hydrogen, respectively, to precisely the extent required to obtain a temperature distribution that is uniform in the direction normal to the two fused silica plates. Flanges fitted to both top and bottom of the burner minimize entrainment of surrounding air and keep the gas outflow parallel to the burner surfaces.

A key advantage of this geometry is that the flame is very flat and uniform in the horizontal plane (the X–Y plane). Thus, both temperature and concentration distributions are also uniform in the horizontal plane, and the gas flow along the stagnation streamline $(Y = 0)$ can be considered essentially one dimensional. This flatness also makes it easy to use nonintrusive measurement techniques. Thus temperature and some vapor concentrations can be accurately measured using absorption spectroscopy. Valuable information also can be obtained using emission spectroscopy and laser light-scattering. Measurements (temperature, concentration, and light-scattering intensity) typically are made along the stagnation streamline $(Y = 0)$, characterized by the Z axis $(Z = -7.5$ mm is the fuel side burner mouth, $Z = 0.0$ mm is the center of the burner, and $Z = +7.5$ mm is the oxidizer side burner mouth). To scan the vertical direction, the burner is mounted onto a motorized vertical positioner (Oriel Encoder Mike) that provides a resolution of 0.1 μm. Because of the flatness of the flame, all properties are very uniform in the $X - Y$ plane; they vary significantly only in the Z (i.e., the vertical) direction. (We will use the word "profile" to refer to the variation of any property of interest along the Z axis.)

The fuel stream is hydrogen diluted in argon and the oxidizer stream is oxygen diluted in argon. Hydrogen is used as the fuel because its combustion does not produce condensing species and because its combustion is a very fast process. Thus, the assumption of local chemical equilibrium is reasonably valid. Hydrocarbons also could be used as the primary fuel, but their combustion often leads to the formation of soot and can be quite slow.

The temperature profile is controlled by the relative flow rates of hydrogen to oxygen and by the relative flow rates of argon in each stream. The profiles resulting with two such sets of gas flow rates are shown in Fig. 2. They will be referred to as Flame 1 or the high-temperature flame, and Flame 2 or the low-temperature flame. These temperature profiles are measured both by thermocouples (open symbols) and by a spectroscopic technique (filled symbols) discussed as follows. This figure shows that one can vary the temperature of a flame in the counterflow burner by changing the ratio of fuel or oxidizer in the respective streams and still be able to keep its variation with position constant.

A gas stagnation plane (GSP) is a characteristic of counterflow diffusion flames. It is the horizontal plane where the two opposed gas streams impinge and flow out of the burner. At the location of the gas stagnation plane, the velocity component in the Z direction is essentially zero. The location of this plane can be adjusted by adjusting the momentum ratio (the ratio of the momenta of the fuel to oxidizer streams), which is accomplished by adjusting the relative flow rates of the argon in the fuel and oxidizer streams.

One typically starts with a liquid precursor of the oxide involved. Typical precursors are $Al(CH_3)_3$, $GeCl_4$, PCl_3, $SiCl_4$, $TiCl_4$, $VOCl_3$, and VCl_4. These were chosen because they are liquids that have a relatively high va-

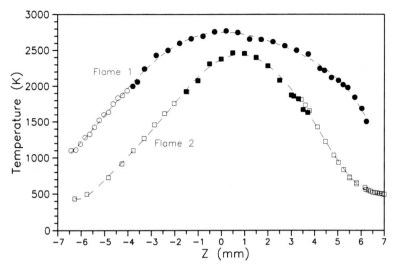

Figure 2. Temperature as a function of elevation for two sets of gas flow rates (flame 1: fuel stream, $H_2 = 2263$ cc/min and $Ar = 1200$ cc/min; oxidizer stream, $O_2 = 1818$ cc/min and $Ar = 167$ cc/min. Flame 2: fuel stream, $H_2 = 1239$ cc/min and $Ar = 2730$ cc/min; oxidizer stream, $O_2 = 877$ cc/min and $Ar = 2760$ cc/min). Filled symbols are OH rotational temperatures and open symbols are thermocouple measurements.

por pressure at room temperature. (Gaseous precursors such as SiH_4 also are convenient to use.) A small part of the fuel or the oxidizer stream is bubbled through a gas washing bottle that contains the precursor. This forms a saturated vapor of the precursor that is then sent to the flame together with the rest of the feed stream. There the precursor reacts with the oxygen present as oxygen molecules (O_2), oxygen radicals ($O \cdot$), or hydroxyl radicals ($OH \cdot$). As Fig. 3 schematically illustrates, after chemical reaction, the oxide thus formed nucleates and grows by both surface condensation and aggregation to form chainlike structures. Fig. 4(A) shows a typical chainlike structure, obtained on adding a $VOCl_3$–PCl_3 mixture to the fuel stream of flame 2. As these chainlike structures flow upward in the burner they encounter higher temperatures. They then collapse into spherical particles like those shown in Fig. 4(B). As these particles continue to flow upward they encounter a particle stagnation plane. The particle stagnation plane is the plane in the flame where the particles have no mean axial velocity because thermophoretic forces prevent their further upward motion. There the particles are forced to flow out of the flame with the outflowing gases [28,35].

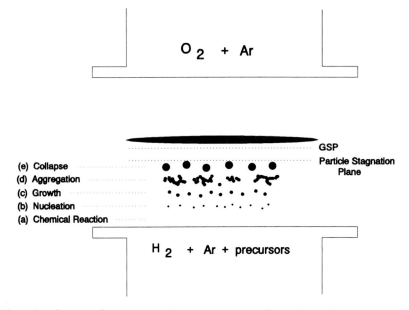

Figure 3. Schematic of oxide powder formation in counterflow diffusion flames (side view).

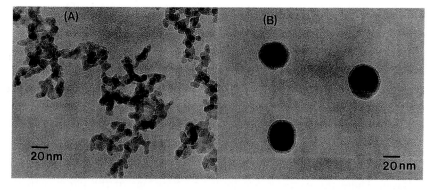

Figure 4. (A) aggregates collected on TEM grids on adding a $VOCl_3–PCl_3$ mixture to the fuel stream of flame 2. (B) Spherical particles resulting from collapsing aggregates.

20.3.2 *In Situ* Measurements

This burner geometry enables one to study oxide formation and growth in the flames using optical (therefore nonintrusive) techniques (i.e., light-scattering, and absorption or emission spectroscopy). For example, the temperature profiles shown in Fig. 2 were measured using two techniques with overlapping range [7]. Above 1500 K, the rotational fine structure in the UV absorption spectra of OH was measured using a 75-W xenon arc lamp (the light beam traverses the burner along the X direction). The distribution of the ground-state population of hydroxyl radicals was determined from the measured spectra and their rotational temperature, from this distribution. Below 2000 K, silica-coated Pt–Pt 10% Rh thermocouples were used and the temperatures thus measured were corrected for radiation losses. In Fig. 2, temperatures measured using the optical method are shown as solid symbols and those measured by thermocouples are shown as open symbols. (Note that they agree well in the regions where both can be used.)

20.3.3 Particle Collection and Analysis

Particles also were collected on two stainless steel strips located at the front and back sides of the flame (at $Z = 0$ mm). The crystalline forms of the powders collected on the strips were determined using a Philips APD 3720 X-ray diffractometer. CuK_α radiation was used and the diffractometer was run over a 2θ angular range of $10°–50°$. Fourier transform infrared (FT–IR) spectra were recorded at 2 cm^{-1} resolution, using 32 scan averaging, on a Mattson (Polaris) Fourier transform infrared spectrometer using the KBr disk technique [9,10,28]. Surface areas were determined by a single point BET measurement of nitrogen desorption using a Micromeritics Flowsorb II 2300 apparatus. Subsequent thermal decomposition of the powders was accomplished by heating them at 390°C in an inert atmosphere (30% N_2/He), followed by heating at up to 750°C in flowing helium in a tube furnace.

20.4 Review of Results Obtained to Date

When precursors are added to a gas stream, oxide particles nucleate and grow as they flow vertically toward the particle stagnation plane, where they flow out of the burner and are collected on stainless steel strips. The crystalline structures of the particles thus collected on the stainless steel strips strongly depend on the operating conditions of the burner and thus on their growth history in the flame. We now will show how, by selecting appropriate flame temperature, precursors, and precursor concentration ratio, one can produce vanadium oxide particles whose morphologies and crystalline structures match those of the known active catalysts: V_2O_5–TiO_2, V_2O_5–Al_2O_3, β-$VOPO_4$, γ-$VOPO_4$, δ-$VOPO_4$, and $(VO)_2P_2O_7$.

20.4.1 Formation of Vanadium Oxide "Monolayer" on TiO_2 and Al_2O_3

V_2O_5–TiO_2 and V_2O_5–Al_2O_3 mixed oxides are well-known catalysts for the selective oxidation of hydrocarbons, particularly the oxidation of *o*-xylene to phthalic anhydride [1], and for the selective reduction of nitrogen oxide with NH_3 [36]. There is general agreement that high activity and high selectivity are achieved when the vanadia is present in the form of highly dispersed, amorphous species. This occurs when the amount of vanadium present corresponds to that necessary to form a bidimensional layer (called a monolayer) of vanadium oxide on the surface of the support [37–39]. With larger amounts of vanadium, the presence of crystalline V_2O_5 can be detected (crystalline V_2O_5 is thought to be catalytically inactive). The dependence of the activity and selectivity of such catalysts on the composition and phase of the support remains a controversial matter. Titanium and aluminum oxides are the most studied components for use as vanadium oxide supports.

These vanadium oxide monolayer catalysts are usually obtained by wet impregnation of the support with ammonium metavanadate solution [40,41], by mechanical mixing and grinding of the support with V_2O_5 powders [42,43], or by vapor-phase reaction of $VOCl_3$ with hydroxylated supports. However, these traditional preparation methods require a very high number of steps (i.e., pretreatment of the support, chemical reaction, filtration, purification and drying of the final product).

V_2O_5–TiO_2 and V_2O_5–Al_2O_3 powders were obtained in our burner using flame 2 [9]. $Al(CH_3)_3$, $TiCl_4$, and $VOCl_3$ were used as precursors in the fuel stream and the concentration of $VOCl_3$ was adjusted so as to collect powders having a monolayer-like structure of vanadium oxide on TiO_2 and on Al_2O_3.

The X-ray diffraction (XRD) patterns of the V_2O_5–TiO_2 powders collected on the stainless steel strips indicate that vanadium oxide is crystalline V_2O_5 when $VOCl_3$: $TiCl_4$ ratios of 1 : 3 or larger are used, whereas it is amorphous for smaller ratios. The FTIR spectra of these powders are shown in Fig. 5 along with their respective $VOCl_3$: $TiCl_4$ ratios and the FTIR spectra of pure V_2O_5 collected in the flame. At V : Ti ratios equal to

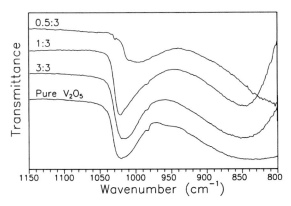

Figure 5. FTIR spectra of V$_2$O$_5$–TiO$_2$ powders using 0.5:3, 1:3, and 3:3 VOCl$_3$:TiCl$_4$ precursor mixtures, and of pure V$_2$O$_5$.

or lower than 0.5 : 3, the FTIR spectrum of the powders exhibit a band at 1029 cm^{-1}. Went *et al.* [41] reported that the Raman spectra of vanadium oxide catalysts supported on TiO$_2$ and Al$_2$O$_3$ exhibit a broad band at 1030 and 1026 cm^{-1}, respectively. They assigned this band to monomeric vanadyl species. Busca *et al.* [44] also showed, based on IR studies, that vanadium oxide catalysts exhibit a narrow feature at 1035 and 1045 cm^{-1} when supported on TiO$_2$ and Al$_2$O$_3$, respectively, at low vanadium coverage. They too attribute this band to monomeric vanadyl species. In the case of V$_2$O$_5$–TiO$_2$, in addition to the 1029 cm^{-1} band, three other bands at 1010, 995, and 830 cm^{-1} are found when a 0.5 : 3 ratio is used. This observation is in agreement with the results of Centi *et al.* [42]. They reported the FTIR spectra of V$_2$O$_5$–TiO$_2$ catalysts obtained by mixing a small quantity of V$_2$O$_5$ with pure anatase or pure rutile, followed by grinding and calcination of the powders. Their infrared spectra of catalytically active V$_2$O$_5$–TiO$_2$ shows a band at 995 cm^{-1} along with two other bands at 1020 and 1010 cm^{-1}. They assigned the 995 cm^{-1} band to the active catalytic phase because they observed a correlation between the intensity of this band and the selectivity to phthalic anhydride in *o*-xylene oxidation [43]. According to Centi *et al.*, the active phase in V$_2$O$_5$–TiO$_2$ catalysts is XRD amorphous and characterized by an IR band at 995 cm^{-1}. They postulated that this band is due to stretching of V^V=O in a distorted octahedral environment.

In the case of V$_2$O$_5$–Al$_2$O$_3$ powders, the X-ray diffraction patterns of powders collected on the strips indicated that vanadium oxide is amorphous for VOCl$_3$: Al(CH$_3$)$_3$ ratios of 1 : 3 or lower. The FTIR spectra of these powders are shown in Fig. 6, along with their respective

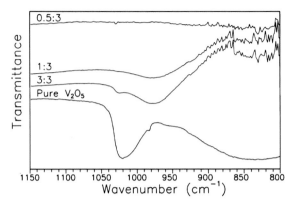

Figure 6. FTIR spectra of V_2O_5–Al_2O_3 powders using 0.5:3, 1:3, and 3:3 $VOCl_3$–$Al(CH_3)_3$ precursor mixtures, and of pure V_2O_5.

$VOCL_3 : Al(CH_3)_3$ ratios and the FTIR spectra of pure V_2O_5 collected in the flame. The presence of only a weak band at 1029 cm^{-1} (at a 0.5 : 3 V : Al ratio) indicates that vanadium oxide forms a monomeric vanadyl species at low $VOCl_3$ loading. However, when the 1 : 3 mixture is used, the spectrum features a broad band at 980 cm^{-1}. Roozeboom *et al.* [39] reported a similar band, based on Raman studies, in the spectrum of V_2O_5 catalysts supported on γ-Al_2O_3. They assigned this band to an octahedral polyvanadate species.

In the vanadium-based-mixed oxides produced in the burner, vanadium oxide seems to form monomeric vanadyl species at V : M (M = Ti, Al) ratios lower than 0.5 : 3, a polyvanadate species at V : M ratios lower than 1 : 3, and crystalline V_2O_5 at ratios of 3 : 3 and higher. This agrees with the results of Eckert and Wachs [45] who studied V_2O_5-based catalysts obtained by a wetness impregnation method. They observed that at low V-loading, vanadia forms a four-coordinated species; at high loading, it becomes six-coordinated; and at even higher loadings, it forms crystalline V_2O_5.

In summary, the FTIR spectra of our powders show peaks and band systems identical to those of the amorphous vanadium oxide "monolayer" obtained by traditional methods. Moreover, the presence of these peaks and band systems depends on the $VOCl_3$ concentration, and agrees with their dependence on vanadium loading as found by others. This strongly suggests the presence of the desired vanadium oxide species and confirms that one can use the burner to produce structured materials in a single-step operation.

20.4.2 Formation of VPO Powders

Vanadium–phosphorus oxides (VPO) are the most widely used catalysts for the selective oxidation of butene and n-butane to maleic anhydride. Maleic anhydride and its derivatives are produced with a worldwide capacity of about 1 billion pounds per year. This important chemical is used principally in the manufacture of unsaturated polyester resins, agricultural chemicals, food additives, lubricating oil additives, and pharmaceuticals.

Maleic anhydride was traditionally produced by the selective oxidation of benzene over V_2O_5–MoO_3, but environmental and economic reasons have favored the use of C_4 feedstocks, such as butane. Butane is a lower cost and environmentally more desirable material because its oxidation produces cleaner product streams, forming mainly maleic anhydride and carbon oxides [3]. The oxidation of n-butane to maleic anhydride is a very complex reaction, involving the abstraction of eight hydrogen atoms and the insertion of three oxygen atoms [46,47]. As noted by Bordes [48], the surface of the catalyst must undergo structural changes without structural collapse or irreversible changes. Catalysts that can meet this requirement are V_2O_5 and several phases of VPO. Among the first to describe the use V_2O_5–P_2O_5 catalysts for the formation of maleic anhydride from C_4 hydrocarbons were Ai *et al.* [49,50], who found increased selectivity when V_2O_5 with added P_2O_5 was used as the catalyst. Since this pioneering work, extensive research has been undertaken to better understand this particular selective oxidation reaction [4,48,51].

It is generally accepted that the activity and selectivity of these catalysts are related to the presence of a specific crystalline phase of VPO, that is, vanadyl pyrophosphate $(VO)_2P_2O_7$ [52], a mixture of β-$VOPO_4$ and $(VO)_2P_2O_7$, or a mixture of γ-$VOPO_4$ and $(VO)_2P_2O_7$ [4,53–55]. The performance of such catalysts depends not only on the crystalline phase present, but also on the P:V ratio and on the method of preparation. Moreover, mild oxidation reactions have been shown to be structure-sensitive or crystal phase-sensitive (i.e., the selectivity of a given product is related to the crystallographic plane exposed at the surface of the catalyst) [56,57]. Thus, research to date on VPO catalysts has focused not only on optimizing the catalytic performances, but also on understanding structure-sensitive reactions.

The solid-state chemistry of the vanadium–phosphorus oxide system is characterized by a large number of crystalline phases, whose structures have not all been solved. $VOPO_4$, for example, is believed to exist in five anhydrous phases (i.e., α_I, α_{II}, β, γ, and δ) and at least two hydrate phases (i.e., $VOPO_4 \cdot H_2O$ and $VOPO_4 \cdot 2H_2O$). Most of the existing crystalline phases interconvert on reduction and oxidation. They are structurally composed of distorted vanadium octahedra (VO_6), forming chains of vanadium–oxygen bonds, with the four equatorial oxygen generally shared with phosphorus tetrahedra (PO_4).

Two main routes have been followed in the literature for what will be referred to as traditional methods of preparation of VPO catalysts: in aqueous medium, by mixing of V_2O_5 and o-H_3PO_4 [58,59], by addition of o-H_3PO_4 to V_2O_5 reduced with HCl [60,61], or addition of o-H_3PO_4 to NH_4VO_3 reduced with oxalic acid [61]; in organic medium, by reduction of V_2O_5 with organic alcohols followed by the addition of o-H_3PO_4 [62], or by reduction of $VOPO_4 \cdot 2H_2O$ with 2-butanol [63–65]. To obtain catalytically active VPO by these traditional methods typically requires a mixing step, followed by a stirring step for several hours, then drying, and finally calcination or activation of the given phase.

In the preceding section, we showed that one can use the burner to produce a specific vanadium structure on TiO_2 and Al_2O_3, by selecting the appropriate precursor concentration ratio. One also can use the burner to select a specific crystalline phase in a mixed oxide system as complex as the vanadium–phosphorus oxide system, by varying the temperature [10]. Two vanadium precursors, namely, VCl_4 and $VOCl_3$, and one phosphorus precursor, PCl_3, were used as the source materials in Flame 1 and Flame 2.

20.4.3 Formation of β-$VOPO_4$, γ-$VOPO_4$, and δ-$VOPO_4$

PCl_3 and $VOCl_3$ were added to the fuel stream of Flame 1 and Flame 2, in a 1 : 1 ratio to produce VPO powders [10]. When Flame 1 (the high-temperature flame) was used, a mixture of $VOPO_4 \cdot 2H_2O$ and of a $VOH_xPO_4 \cdot yH_2O$ phase was obtained on the stainless steel strips. This mixture was heated at 390°C in an inert atmosphere and a pure δ-$VOPO_4$ phase was obtained. This δ-$VOPO_4$ phase was then heated at 750°C in flowing helium, and γ-$VOPO_4$ phase was obtained. When Flame 2 (the low-temperature flame) was used, the X-ray diffraction pattern of the powder collected on the stainless steel strips did not match any known VPO phases. We have assigned the FTIR spectrum of this powder to a VO-$H_xPO_4 \cdot yH_2O$ phase [10]. (The same phase was also obtained when the δ-$VOPO_4$ phase, obtained using Flame 1, was left in open atmosphere for several days.) This powder formed an amorphous α-$VOPO_4$ phase when heated at 390°C in an inert atmosphere, and converted to β-$VOPO_4$ on subsequent reheating at 750°C in flowing helium. The $VOH_xPO_4 \cdot yH_2O$ phase obtained in Flame 2 was characterized by X-ray diffraction pattern at d-spacings 7.05 (vs), 6.84 (w), 4.21 (w), 3.54 (m), 3.04 (vs), 2.60 (w), 1.95 (w), and 1.56 (m) [10]. This diffraction pattern does not match that of VPO powders formed by traditional methods. However, it does match that of a VPO powder also produced at high temperature by Moser [66], using a high-temperature aerosol reactor.

The results obtained in Flame 1 and 2 are summarized in Fig. 7. The two $VOH_xPO_4 \cdot yH_2O$ phases obtained also were shown to form two polytypes of $(VO)_2P_2O_7$ on decomposition in an oxygen atmosphere. These results show that one can obtain specific crystalline phase of VPO powders by varying the temperature. Moreover, the surface area of the powders is very high, often much higher than that obtained by traditional methods.

20.4.4 Formation of $(VO)_2P_2O_7$

The vanadium to phosphorus oxide phase that is active in the selective oxidation of *n*-butane to maleic anhydride is in dispute in the literature. However, there is general agreement that the presence of vanadyl pyrophosphate, $(VO_2)P_2O_7$, is necessary for the reaction to proceed. The best catalysts are believed to be related to the presence of $(VO)_2P_2O_7$ when they preferentially exhibit (100) faces [3], or to the presence of γ-$VOPO_4$/$(VO_2)P_2O_7$ or of β-$VOPO_4$/$(VO_2)P_2O_7$ mixtures [4]. $(VO_2)P_2O_7$ is traditionally obtained by the topotactic transformation of the hemihydrate of vanadyl acid phosphate ($VOHPO_4 \cdot 0.5H_2O$) in the temperature range 500–750°C, in an inert atmosphere [4].

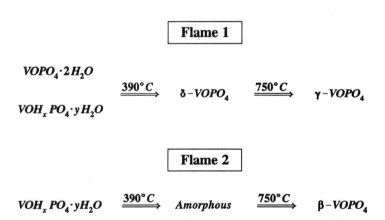

Figure 7. Crystalline phase of the powders collected in Flames 1 and 2 before and after post-thermal treatment at 390 and 750°C.

We currently are investigating the possibility of producing either $(VO)_2P_2O_7$ or $VOHPO_4 \cdot 0.5H_2O$ directly in the burner [67]. Because all the anhydrous VPO powders obtained in the burner used $VOCl_3$ as precursor (i.e., the vanadium in the V^{5+} oxidation state) and because $(VO)_2P_2O_7$ and $VOHPO_4 \cdot 0.5H_2O$ have the vanadium in the V^{4+} oxidation state, we are investigating the possibility of forming these phases using VCL_4.

PCl_3 and VCl_4 were added in a 1:1 ratio to the fuel stream of Flame 2 (the low-temperature flame). The X-ray diffraction pattern of the powder collected on the stainless steel strips is shown in Fig. 8, curve a. Five broad lines can be observed at d-spacings 7.1, 4.1, 3.6, 3.12, and 3.0 Å. The lines are very broad because the particles that compose this powder are very small in size. We attribute the 4.1-, 3.6-, and 3.12-Å lines to the presence of δ-$VOPO_4$ (▲), and the 7.1 and 3.0 Å to the presence of the same $VOH_xPO_4 \cdot yH_2O$ phase (●) obtained previously. FTIR measurements on this powder confirm this assignment. Furthermore, the diffraction lines attributed to the $VOH_xPO_4 \cdot yH_2O$ phase do not match the diffraction pattern of $VOHPO_4 \cdot 0.5H_2O$.

The powders collected on the strips were heated at 390°C in an inert atmosphere (Fig. 8, curve b). This heat treatment caused a sharpening of the diffraction lines corresponding to δ-$VOPO_4$ (▲) (i.e., 4.03, 3.67, and 3.12 Å, and the disappearance of the lines attributed to the $VOH_xPO_4 \cdot yH_2O$ phase. The presence of $(VO)_2P_2O_7$ is not detected. This powder was then heated at 700°C in a tube furnace in flowing helium. The diffraction pattern (Fig. 8, curve c) shows the presence of three VPO phases: β-

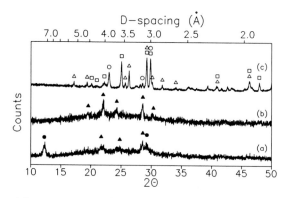

Figure 8. X-ray diffraction pattern of powders produced using PCl_3 and VCl_4 in the fuel stream: (a) before heat treatment, δ-$VOPO_4$ (▲), $VOH_xPO_4 \cdot yH_2O$ (●); (b) after heating at 390°C; and (c) after heating at 700°C, β-$VOPO_4$ (△), $α_{II}$-$VOPO_4$ (□), $(VO)_2P_2O_7$ (○).

VOPO$_4$ (($\triangle$) d-spacings 3.05, 3.39, 5.17, 4.58, and 2.97 Å), α_{II}-VOPO$_4$ (($\square$) d-spacings 3.05, 2.99, 3.55, 4.41, and 2.21 Å), and (VO)$_2$P$_2$O$_7$ (($\bigcirc$) d-spacings 3.86, 3.13, and 2.99 Å). (Note the low intensity of the 00l) lines in the diffraction pattern of α_{II}-VOPO$_4$ i.e., 4.41 and 2.21 Å, indicating some disorder along the c axis.)

The diffraction pattern of the powder collected in Flame 2 using PCl$_3$ and VOCl$_3$ also showed the presence of the VOH$_x$PO$_4 \cdot y$H$_2$O phase. However, in this case, when the powder was heated in the tube furnace at 750°C, (VO)$_2$P$_2$O$_7$ was not detected. It is the use of V^{4+} in the precursor (as VCl$_4$) that leads to its presence in VPO powders.

These initial results show that one can use the counterflow diffusion flame burner to produce VPO catalysts in a single-step operation. They also show that one can select various crystalline phases of VPO by selecting the appropriate flame temperature or oxide precursor. Moreover, because the powders produced are nanosized, they can be expected to exhibit novel or enhanced properties; and their availability may lead to a better understanding of mixed oxide catalysis.

References

1. Hucknall, D. J. (1974). *Selective Oxidation of Hydrocarbons*, Academic Press, New York.
2. Bond, G. C., and Tahir, S. F. (1991). *Appl. Catal*, 71, 1.
3. Centi, G., Trifiro, F., Ebner, J. R., and Franchetti, V. M. (1988). *Chem Rev.*, 88, 55.
4. Bordes, E. (1987). *Catal. Today*, 1, 499.
5. Chung, S. L., and Katz, J. L. (1985). *Combust. Flame*, 61, 271.
6. Katz, J. L., and Hung, C. H. (1992). *Combust. Sci. Technol.*, 82, 169.
7. Hung, C. H., and Katz, J. L. (1992). *J. Mater. Res.*, 7, 1861.
8. Hung, C. H., Miquel, P. F., and Katz, J. L. (1992). *J. Mater. Res.*, 7, 1870.
9. Miquel, P. F., Hung, C. H., and Katz, J. L. (1993). *J. Mater. Res.*, 8, 2404.
10. Miquel, P. F., and Katz, J. L. (1994). *J. Mater. Res.*, 9, 746.
11. Gleiter, H. (1990). *Phase Transitions*, 24–26, 15.
12. Brus, L. E., Siegel, R. W. *et al.* (1993). *J. Mater. Res.*, 6, 704.
13. Beck, D. D., and Siegel, R. W. (1992). *J. Mater. Res.*, 7, 2840.
14. Stucky, G. D., and McDougall, J. E. (1990). *Science*, 247, 669.
15. Iijima, S. (1991). *Nature (London)*, 354, 56.
16. Sei, T., Takeda, H., Tsuchiya, T., and Kineri, T. (1993). *Ann. Chim. (Paris)*, 18, 329.
17. Gurav, A., Kodas, T., Pluym, T., and Xiong, Y. (1993). *Aerosol Sci. Technol.*, 19, 411.
18. Gleiter, H. (1989). *Prog. Mater. Sci.*, 33, 223.
19. Pratsinis, S. E., and Kodas, T. T. (1992). *Aerosol Measurement* (K. Willeke and P. Baron, eds.), p. 721, Van Nostrand-Reinhold, New York.
20. Gleiter, H. (1981). *Deformation of Polycrystals: Mechanisms and Microstructures* (N. Hansen, A. Horeswell, T. Leffers, and H. Liholt, eds.), p. 15, Ris National Laboratory, Roskilde.
21. Zachariah, M. R., and Huzarewicz, S. (1991). *J. Mater. Res.*, 6, 264.

22. Lindquist, D. A., Kodas, T. T., Smith, D. M., Xiu, X., Hietala, S. L., and Paine, R. T. (1991). *J. Am. Ceram. Soc.*, 74, 3126.
23. Gurav, A. S., Duan, Z., Wang, L., Hampden-Smith, M. J., and Kodas, T. T. (1993). *Chem. Mater.*, 5, 214.
24. Nagashima, K., Himeda, T., and Kato, A. (1991). *J. Mater. Sci.*, 26, 2477.
25. Caillat, R., Cuer, J. P., Elston, J., Juillet, F., Pointud, R., Prettre, M., and Teichner, S. J. (1959). *Bull. Soc. Chim. Fr.*, 159.
26. Cuer, J. P., Elston, J., and Teichner, S. J. (1961). *Bull. Soc. Chim. Fr.*, 81.
27. Cuer, J. P., Elston, J., and Teichner, S. J. (1961). *Bull. Soc. Chim. Fr.*, 89.
28. Miquel, P. F. (1995), Ph.D. thesis, The Johns Hopkins University, Baltimore (available from University Microfilms).
29. Calcote, H. F., Felder, W., Keil, D. G., and Olson, D. B. (1990). *Twenty-Third Symposium (International) on Combustion*, p. 1739, The Combustion Institute, Pittsburgh, PA.
30. Revankar, V., Zhao, G. Y., and Hlavacek, V. (1991). *Ind. Eng. Chem. Res.*, 30, 2344.
31. Akhtar, M. K., Pratsinis, S. E., and Mastrangelo, S. V. R. (1994). *J. Mater Res.*, 9, 1241.
32. Stamatakis, P., Natalie, C. A., Palmer, B. R., and Yuill, Y. A. (1991). *Aerosol Sci. Technol.*, 14, 316.
33. Ulrich, G. D. (1984). *Chem. Eng. News*, 62, 22.
34. Sanders, H. J. (1984). *Chem. Eng. News*, 62, 26.
35. Gomez, A., and Rosner, D. E. (1993). *Combust. Sci. Technol.*, 89, 335.
36. Janssen, F. J. J. G., van den Kerkhof, F. M. G., Boss, H., and Ross, J. R. H. (1987). *J. Phys. Chem.*, 91, 6633.
37. Bond, G. C., Flamerz, S., and Shukri, R. (1989). *Faraday Discuss. Chem. Soc.*, 87, 65.
38. Bond, G. C., and Brückman, K. (1981). *Faraday Discuss. Chem. Soc.*, 72, 235.
39. Roozeboom, F., Mittelmeijer-Hazeleger, M. C., Moulijn, J. A., Medema, J., de Beer, V. H. J., and Gellings, P. J. (1980). *J. Phys. Chem.*, 84, 2783.
40. Inomata, M., Mori, K., Miyamoto, A., Ui, T., and Murakami, Y. (1983). *J. Phys. Chem.*, 87, 761.
41. Went, G. T., Oyama S. T., and Bell, A. T. (1990). *J. Phys. Chem.*, 94, 4240.
42. Centi, G., Pinelli, D., and Trifirò, F. (1990). *J. Mol. Catal.*, 130, 220.
43. Centi, G., Giamello, E., Pinelli, D., and Trifirò, F. (1990). *J. Mol. Catal.*, 130, 220. (1991). *J. Catal.*, 130, 220.
44. Busca, G., Centi, G., Marchetti, L., and Trifirò, F. (1986). *Langmuir*, 2, 568.
45. Eckert, H., and Wachs, I. E. (1989). *J. Phys. Chem.*, 93, 6796.
46. Schiott, B., Jørgensen, K. A., and Hoffmann, R. (1991). *J. Phys. Chem.*, 95, 2297.
47. Ziółkowski, J., Bordes, E., and Courtine, P. (1990). *J. Catal.*, 122, 126.
48. Brodes, E. (1993). *Catal. Today*, 16, 27.
49. Ai, M., Boutry, P., and Montarnal, R. (1970). *Bull. Soc. Chim. Fr.*, 8, 2775.
50. Ai, M., Boutry, P., Montarnal, R., and Thomas, G. (1970). *Bull. Soc. Chim. Fr.*, 8, 2783.
51. Hodnett, B. K. (1985). *Catal. Rev. Sci. Eng.*, 27, 373.
52. Busca, G., Cavani, F., Centi, G., and Trifirò, F. (1986). *J. Catal.*, 99, 400.
53. Bordes, E., and Courtine, P. (1985). *J. Chem Soc. Chem. Commun.*, 294.
54. Bordes, E., and Courtine, P. (1979). *J. Catal.*, 57, 236.
55. Harrouch Batis, N., Batis, H., Ghorbel, A., Vedrine, J. C., and Volta, J. C. (1991). *J. Catal.*, 128, 248.
56. Courtine, P. (1985). *Solid State Chemistry in Catalysis*, ACS Symposium Series 279, p. 37, American Chemical Society, Washington, DC.
57. Volta, J. C., and Portefaix, J. L. (1985). *Appl. Catal.*, 18, 1.
58. Bordes, E., Courtine, P., and Pannetier, G. (1973). *Ann. Chim. (Paris)*, 8, 105.
59. Hodnett, B. K., and Delmon, B. (1984). *Appl. Catal.*, 9, 203.

60. Centi, G., Maneti, I., Riva, A., and Trifirò, F. (1984). *Appl. Catal.*, **9**, 177.
61. Poli, G., Resta, I., Ruggeri, O., and Trifirò, F. (1981). *Appl. Catal.*, **1**, 395.
62. Cavani, F., Centi, G., and Trifirò, F. (1984). *Appl. Catal.*, **9**, 191.
63. Johnson, J. W., Johnston, D. C., and Jacobson, A. J. (1987). *Preparation of Catalysts IV* (B. Delmon, P. Grange, P. A. Jacobs, and G. Poncelet, Eds.), Elsevier, Amsterdam.
64. Bordes, E., Johnson, J. W., Raminosona, A., and Courtine, P. (1985). *Mater. Sci. Monogr.*, **28B**, 887.
65. Johnson, J. W., Johnston, D. J., Jacobson, A. J., and Brody, J. F. (1984). *J. Am. Chem. Soc.*, **106**, 8123.
66. Moser, W. R. (1993). *Catalytic Selective Oxidation* (S. T. Oyama, and J. W. Hightower, eds.), p. 244, ACS Symposium Series, Washington, DC.
67. Miquel, P. F., Bordes, E., and Katz, J. L. (1996). *J. Solid State Chem.*, **124**, 95.

CHAPTER 21

The Preparation of Advanced Catalytic Materials by Aerosol Processes

William R. Moser, John D. Lennhoff, Jack E. Cnossen,
Karen Fraska, Justin W. Schoonover, and Jeffrey R. Rozak

Department of Chemical Engineering
Worcester Polytechnic Institute
Worcester Massachusetts 01609

KEYWORDS: advanced catalyst synthesis, aerosol processes, Fischer–Tropsch process, high-temperature aerosol decomposition

21.1   Introduction

Following the discovery of aerosol techniques by Ebner in 1939 [1–3] for the preparation of advanced materials, little use was made of the technology until the late 1970s. Since that time aerosol processing has resulted in the synthesis not only of advanced catalytic materials [4–29], but also of high-performance ceramics, electronic materials, and superconducting metal oxides. Aerosol processing for ceramics and thin films was reviewed [30,31]. Several reactor configurations by Roy and co-workers [32] in 1977, Kato *et al.* [33] in 1979, Gorska *et al.* [34] in 1980, Moser and Lennhoff [10] in 1989, and Gurav *et al.*, [31] in 1993 have been developed and optimized for the synthesis of specific materials. The basic aerosol process is alternatively called evaporative decomposition of solutions (EDS), spray pyrolysis (SP), high-temperature aerosol decomposition (HTAD), mist decomposition (MD), aerosol thermolysis (AT), or aerosol high-temperature decomposition (AHTD). In all of these processes, the desired mixture of metal ions is dissolved in either an aqueous or organic solvent as their nitrates, chlorides, acetates, hydroxides, and several other salts. The metal salt solution, an ideal precursor for an advanced material synthesis, is converted to an aerosol and fed into a high-temperature furnace where the solvent is rapidly evaporated and the intimately mixed metal salts are rapidly decomposed to their finished metal oxides.

21.2   Historical Background

21.2.1 Advanced Catalyst Synthesis by Aerosol Techniques in Other Laboratories

Although many aerosol synthesis studies have described the preparation of materials that are known catalysts, few studies have actually examined their catalytic performance. For example, the early studies of Ebner [1–3] described the preparation of important catalytic support materials such as Al_2O_3 and MgO, as well as simple metal oxide catalysts like copper, iron, zinc, and manganese oxides. Bush [4] showed for the first time that the aerosol reactors had the capabilities to produce metal oxide catalysts, reduced metal oxides, or fine metallic materials. In his study, an ammonium para-molybdate solution was converted to MoO_3 when a downflow aerosol reactor was operated in air between 400 and 700°C. When a flow of hydrogen was passed into the reactor at 400–750°C, the same solution was converted to MoO_2. Finely divided molybdenum metal particles were produced when the reactor was operated above 900°C in a hydrogen environment. Operation under similar conditions resulted in the analogous series of tungsten compounds.

Berndt and Ksinsik [7] fabricated three-way auto exhaust catalysts by aerosol methods and compared the catalytic results to an incipient wetness prepared catalyst of the same composition. The aerosol catalyst was prepared by passing an aqueous solution of H_2PtCl_6, $Rh(NO_3)_3$, and $La(NO_3)_3$ into an aerosol reactor at 950°C. The fine powder that resulted was suspended in water and wash-coated onto a cordierite honeycomb. This device was examined for the catalytic conversion of a synthetic exhaust gas and compared to a parallel classically prepared catalyst. The aerosol catalyst resulted in a 90% conversion of the components in the synthetic fuel at 280–300°C, while the conventional catalyst required 360–380°C for the same conversion level.

A variety of studies have examined aerosol materials as photocatalysts. Wold and co-workers [20–24] examined several spray pyrolysis photocatalysts such as TiO_2 and determined the effect of further modifications with metals and metal oxides on their catalytic properties. Thin films of TiO_2 were fabricated by spray pyrolysis techniques [22] and examined for organic photodecomposition. Promoted titania photocatalysts [21] were prepared by passing nebulized ethanolic solutions of molybdenum acetylacetonate and dipropoxy(titanium)-bis-acetylacetonate into a high-temperature pyrolysis oven. Both the molybdenum and tungsten [27] modified titania samples were more active for the photodegredation of 1,4-dichlorobenzene than for titania itself, and all were more active than the identical catalysts prepared by impregnation of commercial nanophasic titania.

Thin, crack-free electrocatalysts were prepared [12,13,29,38] in a form that was more active than traditionally prepared catalysts by aerosol deposition of $NiCo_2O_4$ onto a Ni support [29]. Other oxygen evolving electrocatalysts in the spinel series were synthesized [38] by spray pyrolysis resulting in the substitutional series $Cu_{(1+x)}Mn_{2-x}O_4$. These catalysts were more active than the catalysts prepared from the thermal decomposition of the corresponding metal salts.

Hu and co-workers [14,15,28] synthesized methanol synthesis catalysts of the composition, $Cu:Zn:Al$ (8.1:0.9:1.0) by the AHTD process and showed that the materials were more active than classical, coprecipitated catalysts. The aerosol catalysts were shown to consist of separate phases of CuO and ZnO, along with a solid solution of CuO–ZnO that was claimed to change the redox properties of the material as compared to the classically prepared coprecipitation catalyst. The surface areas were in the range of $5–12$ m^2/g.

Although no catalytic results were reported, several studies have shown that nanostructured materials of potential catalytic importance could be synthesized by aerosol techniques. Fine metallic particles of Ag [35] and Pd [36] were prepared by decomposing aerosols of the corresponding metal salts. Process temperatures above 900°C led to Pd° while operation in the 400–800°C regime resulted in PdO. Similar studies [37] showed that nanocrystalline V_2O_5 resulted from aerosol processing temperature high enough to permit salt decomposition, but lower than that which caused grain growth.

21.2.2 Advanced Catalyst Synthesis by the High-Temperature Aerosol Decomposition (HTAD) Process

The aerosol studies conducted in our laboratory started in 1981 with the construction of a reactor similar to that of Ebner's [1], and it was evaluated for the synthesis of a wide variety of complex metal oxide and metal-supported catalysts. It was soon realized that each material synthesized required a precise set of reaction parameters to achieve a desired property. Many materials could not be synthesized having the desired phase purities, morphologies, and particle sizes. Consequently, the HTAD process, which evolved after several revisions, incorporates a variety of facilities that are necessary for the general aerosol synthesis of advanced catalytic materials. The present upflow reactor configuration [16,19,25] is described in the following section. One of our earlier configurations [10] used a downflow reactor section and was employed for the synthesis of some of the catalysts described in this chapter.

To illustrate the wide range of catalytic materials that were synthesized using the HTAD process since 1981, the following materials were synthesized and most were extensively evaluated for their catalytic performance in a variety of chemical processes: perovskites for Fischer–Tropsch synthesis of fuels; cubic spinels for lower alcohol synthesis; bismuth molybdates and iron-modified bismuth molybdates for selective oxidation of propylene to acrolein; perovskites for methane oxidative coupling to ethylene and ethane; cobalt molybdates for hydrodesulfurization; copper modified zinc oxide/alumina catalysts for methanol synthesis; palladium on a wide variety of supports for hydrocarbon total oxidation; VPO catalysts for maleic anhydride; Pt, Pt–Ir, Ir, and Pt–Sn on alumina prepared in one step for methylcyclohexane dehydrogenation to toluene; silver on alumina catalysts for ethylene oxidation; Co–Cu-modified zirconium oxides for lower alcohol synthesis from syngas; alkali-modified zinc oxide for methane oxidative coupling and a multicomponent catalyst supported on silica for acrylonitrile synthesis. As will be seen from the detailed description of the catalytic properties of aerosol-prepared catalysts that follows, the catalytic performance of these materials was in most cases superior to those observed using conventionally prepared catalysts. This effort represents the most extensive evaluation of aerosol-prepared catalysts reported to date.

21.2.3 Reactor Configuration

The optimum process configuration that has been incorporated into the Worcester Polytechnic Institute (WPI) version of the HTAD aerosol process reactor, shown in Fig. 1, uses the following facilities: (1) up-flow reactor configuration to permit the control of the aerosol contact time within the heated zone; (2) facilities to perform the synthesis under a slight vacuum to ensure the even flow of materials through the reactor; (3) facilities to permit a pulsed injection of the liquid feed to enable better temperature control within the reactor section; (4) three-zone furnace to control the heat-up rate of the aerosol particles; (5) effluent heat exchangers and

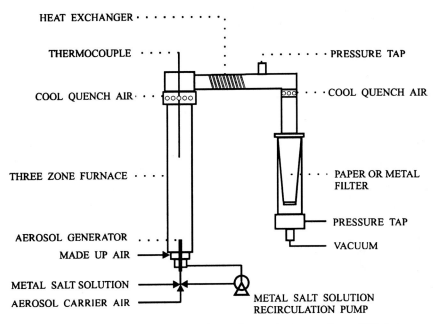

Figure 1. High-temperature aerosol reactor for catalyst synthesis according to WPI design.

cool air injection ports to thermally quench products; (6) preheated makeup air to control the heat-up rate of the aerosol that affects particle morphology; (7) nozzles of various orifice sizes and an ultrasonic aerosol generator to control the size of the aerosol particle; and (8) installation in a high-velocity hood to permit the synthesis of hazardous materials. Metal salt concentrations generally appropriate for the synthesis of hollow spheres was in the range of 0.1–0.6 M of total metal ions, whereas fragmented particles are formed using metal nitrate solutions having concentrations of 1.5 M. The range of experimental conditions under which the HTAD reactor was run was described in prior studies [10,16,19,25].

21.2.4 Scope of Advanced Catalyst Synthesis by Aerosol Processing

For advanced catalyst synthesis, the HTAD process has resulted in several properties that are desirable in modern catalyst synthesis: (1) high-phase purity, (2) high-surface areas, (3) variable surface areas, (4) low degree of microporosity, (5) synthesis of catalytically active components on inert supports, (6) synthesis as either dense or hollow spheres of high porosity, (7) synthesis of microstructures having strong metal to support interactions, (8) synthesis of nanostructured catalyst particles agglomerated into larger structures, (9) preparation of noble metal particles in an exceptionally high state of dispersion over any type of metal oxide support material, and (10) synthesis of multimetallic homogeneous solid solutions in high-phase purities.

As a starting point in advanced catalyst synthesis, it is advantageous to fabricate catalyst powders that contain only the desired phase. One often modifies this host catalyst with one or more promoter metal ions; thus, it is important to use a synthesis process that has the capability of affording homogeneous solid solutions over a wide substitutional range, again in high-phase purity. Depending on the specific application of the catalyst in a fluidized-bed, packed-bed, or slurry reactor, different particle morphologies are required. Generally, the active catalyst needs to be prepared in a high-surface-area form; and some applications require supporting the active phase on an inert support. The following studies on the HTAD process serve to illustrate the general capabilities of aerosol processes to afford most of the desired properties generally encountered in advanced catalyst synthesis.

21.3   Results and Discussion

The configuration of the aerosol process described previously and shown in Fig. 1 was used for the synthesis of a wide range of catalytic materials. The synthesis, characterization, and catalytic evaluation of their properties is later in the text. This configuration was also used in the formation of high-surface-area basic metal oxides, stabilized zirconias, superconductors, superconductors containing fine grains of silver, high-temperature electrically conducting ceramics, piezoelectrics, oxygen and electrical solid-state conductors, spherical cerium nickel oxides and other metal oxides for ultrasonic heart imaging, and a variety of metal oxide support materials.

Because the catalysis studies using aerosol materials prepared by the HTAD process configuration are quite extensive, this chapter will provide only a survey of our results. The purpose of this communication is to summarize the wide range of materials synthesized for catalytic applications and to describe their structural features that are important in catalysis.

21.3.1 $La_{(1-x)}M_xFeO_3$ Catalysts for the Fischer–Tropsch Process

These studies resulted in the synthesis of a wide range of perovskites, and their structures and properties were compared with the same materials from classical methods of synthesis. More than 50 compositions were synthesized in the perovskite family in the following series: $La_{(1-x)}M_xFeO_3$

where $x = 0.0$ to 1.0 and M was both Ca^{2+} and Sr^{2+}; a series of perovskites with nine different lanthanide ions of $La_{(1-x)}Ln_xFeO_3$, $La_{(1-x)}Sr_xCoO_3$ where $x = 0.0$ to 1.0; and a wide range of perovskites in the $LaCoO_3$ and $LaFeO_3$ series having two additional ions modifying both the A- and B-sites simultaneously. The syntheses were conducted at $1075°C$ using a heated air makeup gas. All of the materials were characterized and evaluated for their activity, selectivity, and stability in the Fischer–Tropsch reaction at $250°C$ and 20 atm pressure of $2:1\ H_2:CO$.

To illustrate the capabilities of aerosol techniques for the synthesis of catalytic materials of exceptionally high-phase purities and as fine particles, a wide range of perovskite catalysts was synthesized. Although only one set of the characterization data is illustrated here, all of the perovskites synthesized could be formed as single-phase powders of microcrystallites after a short HTAD process parameter variation study. The striking aspect about the synthesis results is the XRD study illustrated in Fig. 2 for the $La_{(x)}Ca_{(1-x)}FeO_3$ series. The figure shows that the perovskite phase was formed in all cases in high-phase purities over a wide range of Ca^{2+} fractional substitutions. High-phase purity is essential in the synthesis of any advanced catalyst where one is attempting to examine the effect of systematically changing a modifying ion in a host crystal structure on chemical performance. The surface areas of these perovskites could be varied from 10 to $50\ m^2/g$, and an analysis of their pore–size distributions showed that they had little to no micropore structure. The relatively high-surface areas resulted from the fine crystallite structure and thin-walled hollow spherical

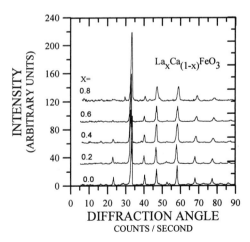

Figure 2. X-ray diffraction patterns for Ca^{2+} substitution into $La_{(x)}Ca_{1-x}FeO_3$ where x equals the fractional substitution of Ca^{2+} for La^{3+}.

morphologies. Classical synthesis of the same materials by high-tempera-ture fusion techniques resulted in surface areas in the range of 0.1–1 m^2/g. In addition, no perovskite synthesized by the conventional method could be obtained free from separate phases of the component metal oxides.

All of the perovskites were catalytically active for Fischer–Tropsch synthesis. Both activities and hydrocarbon molecular weight distribution (polymerization probability, α) varied in a systematic way with composi-tion for the strontium substitution series where x was varied over the wide range of x = 0.0 to 0.9 in $La_{(1-x)}Sr_xFeO_3$. The calcium series showed a simi-lar variation in selectivity with substitution, and the olefin to paraffin ra-tios for all products could be controlled by substitution. The surface area normalized rates of CO conversion in both the Sr and Ca series smoothly increased with greater Sr and Ca substitution by over a factor of 100. The cobalt-containing structures collapsed to carbidic materials under the cat-alytic conditions, but the iron-based perovskites were very stable and pro-vided XRD reflection after 40 h onstream that was superimposable on the XRD of the fresh catalysts.

The CO conversion data for the $La_{(1-x)}Ca_xFeO_3$ series with Ca frac-tional substitution from 0.0 to 0.8 in the La site is shown in Fig. 3. The

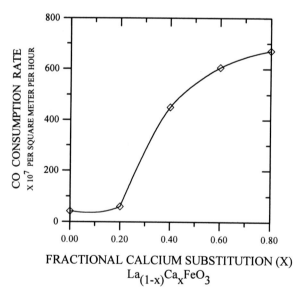

Figure 3. CO conversion reaction rates normalized to unit surface areas for the substitution of calcium, x-atom fraction, into $La_{(1-x)}Ca_xFeO_3$ at a reaction temperature of 250°C, and a pressure of 20 atm using 2:1 Co:H_2.

rate data were normalized to unit surface areas to take into account the fact that the surface areas of the catalysts increased as the degree of calcium substitution increased. The XRD data on this series of compounds indicated that the defect oxygen concentration becomes important at a fractional substitution of Ca between 0.3 and 0.4. The fact that the olefin to paraffin ratios for the calcium substitution series demonstrated a sharp transition at a Ca fractional substitution of 0.4 is also consistent with the introduction of a higher concentration of oxygen defect sites at this degree of substitution.

21.3.2 Iron Oxide Catalysts for the Fischer–Tropsch Process

The aerosol synthesis of iron oxide afforded hematite in variable surface areas of 20–88 m^2/g depending on the selected process conditions. Infrared analysis by diffuse reflectance of the aerosol-prepared materials showed that it was exceptionally pure, whereas that of hematite prepared by a classical precipitation technique demonstrated numerous peaks for undecomposed salts even after calcination at 240°C. An aerosol hematite preparation of 52 m^2/g and a precipitated hematite of 48 m^2/g were evaluated for their synthesis gas conversion properties. These studies showed that the precipitated material had about twice the steady-state activity after 50 h and gave twice the proportions of olefins to paraffins. The polymerization probability was practically the same at 0.57 (aerosol) and 0.63 (precipitated).

21.3.3 Higher Alcohol Synthesis Catalysts

21.3.3.1 Copper-Modified Zinc Chromites–$Zn_{(1-x)}Cu_xCr_2O_4$

A series of copper-modified zinc chromites were synthesized in the substitutional series, $Zn_{(1-x)}Cu_xCr_2O_4$, where x was varied between 0.0 and 1.0. Figure 4 shows the XRD of the entire series of compounds with the interesting observation that all materials, where x was varied between 0.0 and 0.9, demonstrated the unusual cubic spinel structure. The diffraction patterns further show that the zinc chromite structure dominated the compositions even at substitutions of copper to nearly 90%. An examination of all of the diffraction patterns in this series shows that all materials were formed in exceptionally high-phase purities. Figure 5 shows the systematic alteration in d-spacing for the crystal host as a function of substitution. The linear change as well as the observation of no reflections for extraneous phases in the XRD and a homogeneous EDX mapping analysis suggests that the copper modification occurred as a homogeneous solid solution. The surface areas of the solid-state materials were between 22 and 36 m^2/g.

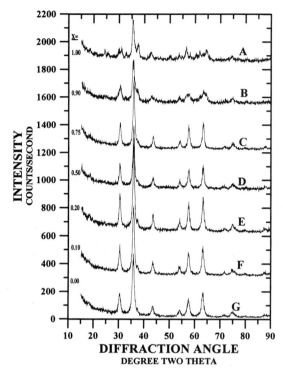

Figure 4. X-ray diffraction patterns for Cu^{2+} substitution into the $Zn_{(1-x)}Cu_xCr_2O_4$, where x equals the fractional substitution of Cu^{2+} for Zn^{2+}. Curve (x =): A (1.0); B (0.90); C (0.75); D (0.50); E (0.20); F (0.10), and G (0.00).

Arsene *et al.* [39,40] have previously reported the synthesis of the same substitutional series, utilizing a coprecipitation method, followed by calcination in air at 850°C for 6–8 weeks. These workers reported the formation of the cubic spinel at low copper concentrations and the tetragonal spinel at high concentrations, in sharp contrast to the results from the HTAD synthesis, noted previously. The reason for this difference may lie in the fact that the aerosol-prepared materials were formed as nanostructured grains that were then thermally quenched from 1000°C to 100–125°C within a few seconds, resulting in the thermodynamically stable high-temperature phase, i.e., the cubic spinel.

Another series of catalysts were prepared in the $Zn_{0.5}Cu_{0.05}Cs_{y/3}Cr_{(2-y/3)}O_4$ series. Energy dispersive X-ray (EDX) analysis showed that the cesium was well dispersed, but the surface areas were lowered to 7–13 m^2/g.

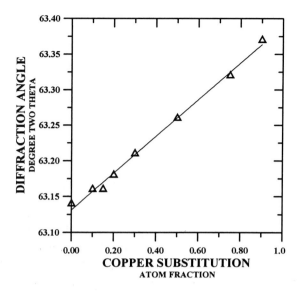

Figure 5. Change in the diffraction angle for the 63.13° 2θ reflection of $ZnCr_2O_4$ as a function of the fractional substitution, x, of Cu^{2+} for Zn^{2+} into $Zn_{(1-x)}Cu_xCr_2O_4$.

The synthesis of this series of catalysts with the desired structure and as homogeneous solid solutions required a systematic study of HTAD reactor parameters over a range of residence times and reactor temperatures. The conditions that resulted in the desired cubic spinels used synthesis conditions of 1000°C, a preheated air temperature of 600°C, residence times of 3–5 s, and a pulsed aqueous metal salt solution injection system operating at 2 s on and 4 s off. The Cu–Zn–O system was synthesized using the downflow configuration equipped with an aqueous scrubbing system to collect the catalysts. The cesium-modified series of catalysts used an upflow configuration with a dry particle collection system.

An evaluation of the catalytic properties of these materials under $H_2:CO$ of 2:1 at a constant pressure of 34 atm between 225 and 350°C resulted in the formation of linear alcohols having a Schulz–Flory product distribution. The polymerization probability and activity were systematically altered as a function of the degree of copper substitution in $Zn_{(1-x)}Cu_xCr_2O_4$. The cesium-modified series resulted mainly in branched alcohols.

21.3.3.2 $K_{0.05}Co_{0.15}Cu_xZr_{0.80-x)}O_n$, $K_{0.05}Co_xCu_{0.020}Zr_{(0.75-x)}O_n$, and $K_xCo_{0.20}Cu_{0.20}Zr_{(0.6-x)}O_n$

Several series of higher alcohol synthesis catalysts were synthesized in which K, Co, and Cu were systematically introduced into a zirconia host to examine their limits of homogeneous ion substitution. Pure zirconia prepared by the aerosol technique at 1000°C was tetragonal. XRD analysis of the series $K_{0.05}Co_{0.15}Cu_xZr_{(0.8-x)}O_n$ from $x = 0.10$ to 0.50 demonstrated through $x = 0.20$ no reflections for CuO or Cu_2O; however, the material having higher copper concentrations of $x = 0.30$ and 0.40 showed reflections for separate phase CuO. In addition, when x was 0.10 or 0.20, the sole reflections observed in the XRD were those due to tetragonal zirconia. Increasing the concentration of cobalt above 0.15 in the series, $K_{0.05}C_xCu_{0.20}Zr_{(0.75-x)}O_n$, led to the appearance of separate phase cobalt oxide. The surface areas of these materials ranged from 3 to 9 m²/g and were not affected very much by the degree of potassium substitution. Most particles were produced as spheres, but depending on composition, the surface of the sphere varied from very smooth to very rough. For example, the composition, $K_{0.05}Co_{0.15}Cu_{0.30}Zr_{0.50}O_n$, resulted in relatively smooth hollow spheres shown in Fig. 6. However, the material having a higher copper concentration, $K_{0.05}Co_{0.15}Cu_{0.50}Zr_{0.30}O_n$, resulted in the highly textured structure shown in Fig. 7. In general, the HTAD aerosol synthesis parameters may be changed to form products varying from hollow spherical structures of the type shown in Figs 6 and 7 highly fragmented particles. The main modification resulting in fragmentation is to run the synthesis using high solution concentration of the metal salts in the league of 1.5 total molarity of metal ions. The fragmentation results in higher surface areas, and it is caused by the high exotherm and rapid pressure increase within the aerosol particle caused when the high concentration of metal nitrates rapidly decomposes. The rapid heat release essentially causes the aerosol sphere to blow apart into small thin-walled particles.

21.3.4 Bismuth Molybdate Catalysts for Propylene Oxidation to Acrolein

A series of 14 catalysts were prepared in the bismuth molybdate series including the end members MoO_3 and Bi_2O_3. The aerosol syntheses were conducted at 900°C using both upflow and downflow [5] configurations of the HTAD reactor. Pure MoO_3, when prepared in the aerosol reactor, resulted in approximately equal amounts of α- and β-MoO_3. Calcination in air at 450°C led to the expected thermodynamically stable α-form. The aerosol bismuth oxide synthesis resulted in the nonequilibrium, β-Bi_2O_3 phase without phase impurities. Calcination of this material in air at 450°C resulted in complete conversion to the stable α-Bi_2O_3 phase. Compositions having the following Bi:Mo atom ratios (1:10, 1:3, 4:10, 2:3, 5:6, 1:1, 3:2, 2:1, 3:1, 4:1, 6:1, and 14:1) were synthesized for the purpose of determining whether phase compositions, dictated by the

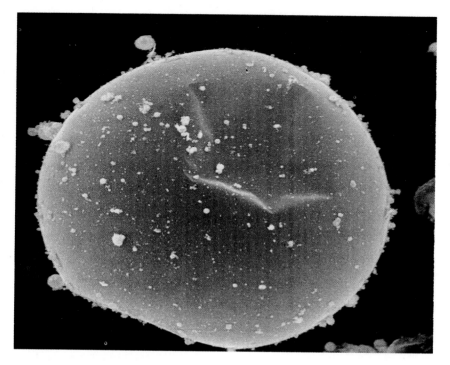

Figure 6. SEM of a particle of $K_{0.05}Co_{0.15}Cu_{0.30}Zr_{0.50}O_n$ at a magnification of 1000.

Bi–Mo–O phase diagram, dominated or whether nonequilibrium homogeneous solid solutions would be produced. Phase analysis for all materials synthesized at 900°C, after calcination to temperatures appropriate for propylene oxidation between 350 and 450°C, resulted in phase compositions that were generally in accord with the predictions of the phase diagram for materials synthesized at 900°C. Aerosol feed solutions containing Bi:Mo ratios of 2:3, 2:2, and 2:1 resulted in single phases of $Bi_2Mo_3O_{12}$, $Bi_2Mo_2O_9$, and Bi_2MoO_6, respectively. Intermediate ratios of Bi:Mo of 5:6 resulted in separate phases of $Bi_2Mo_3O_{12}$ and $Bi_2Mo_2O_9$. A higher intermediate feed ratio of Bi:Mo of 3:2 resulted in major phases of $Bi_2Mo_2O_9$ and α-Bi_2MoO_6. The freshly synthesized bismuth molybdates were microcrystallites as shown in the XRD for $Bi_2Mo_3O_{12}$ in Fig. 8. The top diffraction pattern shows all of the correct reflections for $Bi_2Mo_3O_{12}$ riding on top of a broad X-ray amorphous envelope. When this material was calcined at 450°C for 15 h, the lines sharpened and showed only the reflections contained in the calculated diffraction pattern for $Bi_2Mo_3O_{12}$. The surface area of the fresh catalyst was 0.82 and the calcined material was 0.43.

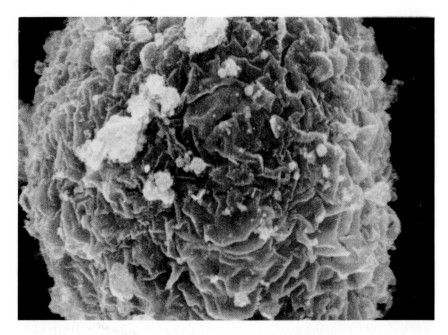

Figure 7. SEM of a particle of $K_{0.05}Co_{0.15}Cu_{0.50}Zr_{0.30}O_n$ at a magnification of 5000.

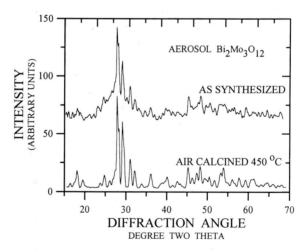

Figure 8. X-ray diffraction patterns for freshly prepared $Bi_2Mo_3O_{12}$ by aerosol synthesis at 900°C (top curve), and aerosol-synthesized $Bi_2Mo_3O_{12}$ after calcination in air at 450°C (lower curve).

The phase purities of materials obtained by the aerosol synthesis were compared to the same compositions prepared by a classical coprecipitation and a high-temperature fusion synthesis. The aerosol-prepared materials were in all cases of equal or moderately superior phase purities.

21.3.5 Iron-Modified Bismuth Molybdate Catalysts in the Series $Bi_{(2-2x)}Fe_{2x}Mo_3O_{12}$

Iron modifications were made to α-bismuth molybdate $(Bi_2Mo_3O_{12})$ involving 14 compositions varying the atom fraction of iron from $x = 0.0$ to 1.0 in the series, $Bi_{(2-2x)}Fe_{2x}Mo_3O_{12}$. In the low-iron-substitution regime of 0.0 to 0.05 atom fraction iron, the XRD analysis on the air-calcined materials at 450°C showed reflections solely for $Bi_2Mo_3O_{12}$. At higher iron concentrations, a separate phase of $Fe_2Mo_3O_{12}$ progressively became more apparent. An analysis of the phase composition of the catalysts is shown in Fig. 9. It is evident that if the iron substituted into the $Bi_2Mo_3O_{12}$ host, it occurred only in the low-iron-concentration ranges. When the series of catalysts were examined for the oxidation of propylene to acrolein, the activity and selectivity of the catalysts of $x = 0.03$ to 0.05 reached a sharp maximum. As the value of x increased to regions where the concentration of

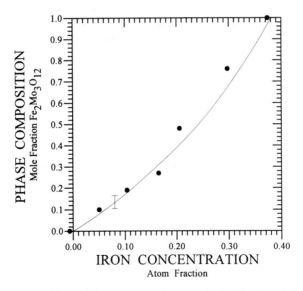

Figure 9. Phase composition for the aerosol catalysts synthesized in the substitutional series, $Bi_{(2-2x)}Fe_{2x}Mo_3O_{12}$. The $Fe_2Mo_3O_{12}$ phase composition resulting from XRD analysis and the iron concentration were obtained from atomic adsorption analysis. The iron concentration is the iron atom fraction of the total metals in the catalysts.

separate phase $Fe_2Mo_3O_{12}$ became dominant, the activity and selectivity steadily declined.

The interesting aspect of the catalytic data for both the Bi–Mo–O and Fe–Bi–Mo–O series is that the Arrhenius treatment of the catalytic reactivity data resulted in activation energies for all of the catalysts that were the same as that measured on pure MoO_3. This observation suggests that neither Bi nor Fe has an electronic effect on altering the activity and selectivity of the MoO_3 catalyst, and the sole effect of this type of ion modification is to modify the structure of the catalyst to provide a larger number of active sites that are also the selective sites for acrolein formation. Indeed, the variation in overall reaction rates was due to differences in the preexponential factors resulting from the Arrhenius treatment because all of the activation energies were the same.

21.3.6 Methanol Synthesis Catalysts

Methanol synthesis catalysts were synthesized in the series $Zn_{(0.90-x)}Cu_xAl_{0.1}O_y$. A member of this series was compared to catalysts made by a classical synthesis [41], both having the composition $Cu_{0.25}Zn_{0.65}Al_{0.10}O_y$. The XRD analysis of the aerosol and classical material, after the latter had been calcined in air at 240°C, is shown in Fig. 10. The diffraction pattern of the aerosol preparation, shown as the lower pattern in Fig. 10, reveals reflections mainly for zinc oxide. A small amount of separate phase CuO appears at 28.7° 2θ. A comparison of the lower curve

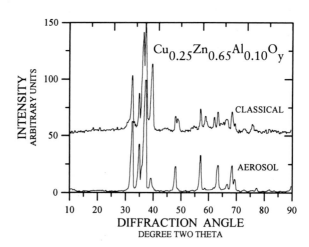

Figure 10. Coprecipitation synthesis of a methanol synthesis catalyst of composition $Cu_{0.25}Zn_{0.65}Al_{0.10}O_y$ (top curve) and the same composition by aerosol synthesis (lower curve).

to that of the upper diffraction pattern in the figure shows that the classical material resulted in the formation of much more separate phase CuO. These data indicate that the aerosol preparation results in a much better dispersion of the copper component, perhaps as a homogeneous solid solution. For methanol synthesis activity, this is an important structural feature.

21.3.7 Hydrodesulfurization Catalysts

Unsupported cobalt molybdate hydrodesulfurization catalysts were synthesized by the HTAD aerosol process [42] to understand the phase relationship of the synthesized materials as a function of $Co:Mo$ atom ratios and to determine the degree of nanometer-sized grains produced by the aerosol synthesis. The XRD analysis of three composition of $Co:Mo$ feed ratios of 1.03, 1.16, and 2.67 prepared at 900°C showed that all three samples were qualitatively similar. All three patterns exhibited reflections for monoclinic $CoMoO_4$. However, the selected area diffraction (SAD) studies showed that the three materials contained significant amounts of nanometer grains of other compounds not observable by XRD due to their small grain sizes. SAD analysis of these materials [43] resulted in well-formed electron diffraction rings that were indexed for hexagonal $Co_2Mo_3O_8$ at a $Co:Mo$ ratio of 1.03 and monoclinic $CoMoO_4$ for the 1.16 ratio. Indexing of other regions of the 1.16 sample by SAD showed material consisting of a separate CoO phase. CoO dominated the SAD analysis of the material of $Co:Mo$ ratio of 2.67. A TEM analysis for the 1.03 sample is shown in Fig. 11 and indicates that the grain size of the particles was between 5 and 60 nm. The selected area electron diffraction on several areas of the sample of $Co:Mo$ ratio of 1.03 demonstrated smooth ring patterns that are typical of materials consisting mainly of nanostructured grains. As noted previously, the diffraction rings were indexed as hexagonal $Co_2Mo_3O_8$. It is interesting to note that, although the aerosol reactor was run under identical conditions in all of these experiments, the grain sizes of the particles using a $Co:Mo$ feed composition of 1.16 resulted in much finer grain materials as compared to the experiment using a $Co:Mo$ ratio of 1.03. The TEM of the materials using the $Co:Mo$ feed ratio of 1.16 is shown in Fig. 12, and indicates that the grain sizes were in the range of 2–10 nm. The SEM of the material of $Co:Mo$ ratio of 1.03 is illustrated in Fig. 13, and it shows that the nanometer-sized grains are agglomerated into a much larger structure of irregular particles in the size range of 1–15 μm. These particles were formed as irregular agglomerates, whereas the preparation of the catalyst of $Co:Mo$ ratio of 1.16 resulted mainly in a hollow spherical morphology in the size range of 1–6 μm. Al-

Figure 11. TEM analysis of catalyst using a feed ratio of Co:Mo = 1.03 showing grain sizes of 5–60 nm.

though the HTAD process permits the deposition of active catalysts on an inert support during synthesis, these supported materials were not synthesized.

The importance of these results for hydrotreating catalysts is that substantial amounts of the Co–Mo–O material were formed in nanometer-size grains. This structural feature suggests that these catalysts would have enhanced hydrotreating activities because the edge-to-basal-plane site population increases [44] as the size of a crystallite decreases due to geometric considerations. Studies by Bouwens and co-workers [45] indicate that the most active site in a cobalt molybdate hydrodesulfurization catalyst is the decoration of cobalt on the edge site of the molybdenum sulfide part of the crystallite.

21.3.8 Methane Oxidative Coupling Catalysts

21.3.8.1 Perovskite Catalysts

A series of methane oxidative coupling catalysts were synthesized at 1300°C in the perovskite substitutional series of $SrCe_{(1-x)}Yb_xO_{(3-x/2)}$ where the atom fraction of Yb, x, was varied between 0.0 and 1.0. These

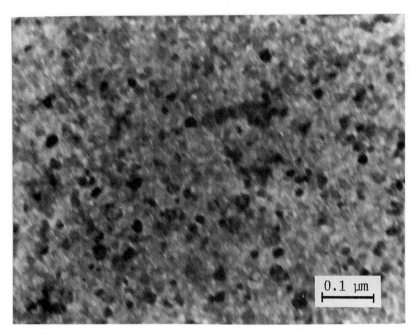

Figure 12. TEM analysis of catalyst using a feed ratio of Co:Mo = 1.16 showing grain sizes of 2–10 nm.

aerosol syntheses were conducted to determine the limits of homogeneous substitution of Yb into the perovskite lattice, and to determine if the individual component metal oxides appeared as separate phases in the products. Because Machida and Enyo [46] reported exceptionally high ethane and ethylene yields using this type of catalyst prepared by high-temperature fusion, it was expected that even higher yields would result if the HTAD process resulted in a higher phase purity and homogeneous solid substitution of Yb into the host. The XRD analysis of four of the aerosol products in this series is shown in Fig. 14. The alteration in d-spacing for the 3.04-Å reflection for the perovskite and the 2.02-Å reflection for the fluorite structure as a function of ytterbium fractional substitution is shown in Fig. 15. These data show that the perovskite structure is formed with Yb fractional substitutions of x = 0.0 to 0.5, and the fluorite structure results within the range of 0.5 to 1.0. Values near 0.5 resulted in products containing both structures. That the Yb substituted into the perovskite and fluorite as homogeneous solid solutions over the entire compositional range is shown by the smooth decrease in lattice spacings in Fig. 15. Extensive catalytic studies using the aerosol materials under a variety of conditions resulted in a maximum yield of 20%, which was lower than that

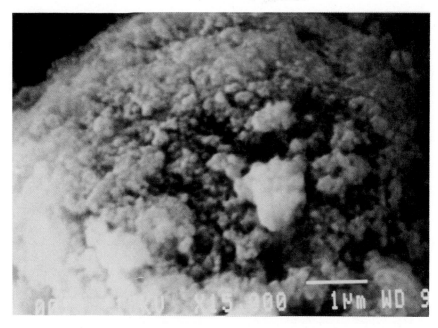

Figure 13. SEM of a catalyst resulting from a feed composition of Co:Mo = 1.03 at a magnification of 15,000 illustrating agglomerate sizes of 1–15 μm.

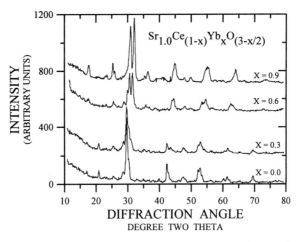

Figure 14. X-ray diffraction of four aerosol-prepared methane-coupling catalysts synthesized at 1300°C in the series $Sr_{1.0}Ce_{(1-x)}Yb_xO_{(3-x/2)}$. X denotes the fractional substitution of Yb^{3+} for Ce^{3+}. The values of x were 0.0, 0.3, 0.6, and 0.9 starting with 0.0 at the bottom of the figure.

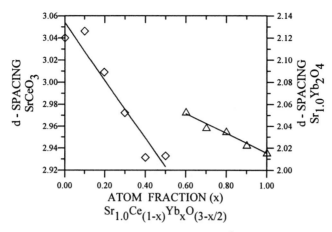

Figure 15. Left curve: change in lattice spacing for the 3.05-Å reflection in $SrCeO_3$ as a function of the fractional substitution, x, of Yb^{3+} for Ce^{3+} in $Sr_{1.0}Ce_{(1-x)}Yb_xO_{(3-x/2)}$. Right curve: change in lattice spacing for the 2.01-Å reflection in $Sr_{1.0}Yb_2O_4$ as a function of the fractional substitution, x, of Yb^{3+} for Ce^{3+} in $Sr_{1.0}Ce_{(1-x)}Yb_xO_{(3-x/2)}$.

reported [46] previously. However, preparation of these materials using the previously reported [46] fusion technique, and evaluating them for oxidative coupling of methane could not confirm the previously reported [46] catalytic performance results.

21.3.8.2 Alkali Metal Modified Zinc Oxide Catalysts

Three series of alkali and alkaline earth modified materials were synthesized at an aerosol reactor temperature of 1300°C. In the first series, the fractional lithium substitution into $Li_xZn_{(1-x)})O_{(1-x/2)}$, varied from 0.0 to 0.4. although no separate phase lithium oxide appeared in the XRD analysis until a 0.40 extent of fractional substitution, the d-spacing of the host hexagonal zinc oxide varied little with substitution. This indicates that the lithium oxide did not substitute into the lattice, but the aerosol method simply resulted in well-dispersed particles of separate nanophase Li_2O that were too small to be detected by XRD. The second substitutional series used Li, Na, K, Rb, and Cs at 0.1 fractional substitution into $M_{0.1}Zn_{0.9}O_{0.95}$. These aerosol syntheses resulted in well-dispersed alkali metals for Li and Na; and progressively more intense separate phase reflections for the individual (M_2O) metal oxides progressing down the alkali metal series for K, Rb, and Cs. In the case of the Rb and Cs materials, their separate phase oxides could be detected by XRD mixed in with the ZnO host. When these materials were synthesized using a water scrubber to collect the products instead of the usual dry filter, their SEM analysis indicated that the materials were formed as hollow spheres containing a high concentration of evenly spaced holes in the walls of the spheres. The surface area of the Cs-modified composition was 15 m^2/g when collected using the water scrubber, whereas that of the same composition collected dry was 5 m^2/g. Conditions in most of the study were adjusted so that the two series of materials were formed as hollow spherical particles of 0.5–6 μm, and their surface areas ranged from 4 to 8 m^2/g. The third series of syntheses examined the solubility of lithium into Li$_x$Mg$_{(0.5-x/2)}$Zn$_{0.5-x/2}$O$_{(1-x/2)}$ when x was varied from 0.0 to 0.4. The XRD of all of these materials showed separate phases for ZnO (hexagonal) and MgO (cubic periclase). A small amount of separate phase lithium oxide appeared only when x was increased to the limit of 0.4.

21.3.9 Vanadium–Phosphorous Oxidation Catalysts

The synthesis of VPO maleic anhydride oxidation catalysts was extensively examined as a part of a NASA program studying the effect of microgravity on aerosol synthesis of catalysts in space [19]. The catalysts were synthesized in the laboratory over a temperature range of 350–800°C using P:V ratios of 1.0:1.5. The synthetic studies used a feed solution composed of aqueous ammonium vanadate and the required amount of 85% phosphoric acid to obtain the desired stoichiometric ratio of P:V. These studies showed that the capability of the HTAD process reactor to control the residence time of the aerosol in the heated zone was essential to produce these catalysts. The materials obtained directly from the aerosol reactor were nanostructured catalysts that experienced grain growth when allowed to stand at ambient conditions in a sealed container for 14 days. The fresh material was similar to the product of a VPO flame synthesis which Miguel and Katz [47] concluded was VOH$_x$PO$_4$ yH$_2$O. The VPO catalyst obtained directly from the HTAD reactor using a furnace temperature of 700°C showed strong XRD reflections at 12.5° and 29.3° 2θ; and coupled with the Raman spectra, Michalakos and co-workers [18] assigned the dominant materials as VOPO$_4$·2H$_2$O. The optimum material for catalysis was formed at 700°C using a residence time of 8 s and had a P:V ratio of 1.2. Several experiments examined the effect of adding varying amounts of oxalic acid to the aerosol feed solution on the vanadium oxidation state. X-ray photoelectron spectroscopy (XPS) analyses on the products showed that as the amount of oxalic acid modifier increased, the valency of vanadium could be systematically decreased. Catalytic evaluations [19] of the aerosol VPO materials for the conversion of butane to maleic anhydride showed that the materials were active, selective, and stable. The XRD analysis of the catalyst after use in the catalytic experiments showed strong reflections at 21.3°, 28.9°, and 29.8° 2θ. This material was characterized as α_1-VOPO$_4$. Only when the used catalyst was heated at 900°C in helium did reflections for vanadyl pyrophosphate, (VO)$_2$P$_2$O$_7$, appear along with vanadyl metaphosphate, VO(PO$_3$)$_2$. The pyrophosphate is the equilibrium catalyst in conventional butane to maleic anhydride processes where the catalyst is prepared by precipitation using organic solvents.

The observation that the vanadium valency could be systematically varied by the addition of different amounts of oxalic acid to the feed solution appears to be an important capability of aerosol methods of catalyst preparation. If the technique is general for other metal oxide systems, controlling solid-state-oxidation states of component transition metals in the bulk of a solid-state material would likely have considerable utility in the synthesis of advanced catalysts where a precise oxidation state is desired for enhancing chemical performance. If a desired, lower oxidation state can be locked into a very stable host metal oxide and stabilized in this way, this could be an important aspect of aerosol preparations.

21.3.10 Hydrocarbon Dehydrogenation Catalysts

A series of Pt, Ir, and bimetallic Pt–Ir catalysts were synthesized on alumina by passing homogeneous aqueous solutions of their metal salts along with dissolved aluminum nitrate hydrate into the HTAD reactor at 1000°C. The catalysts were prepared in six compositions between 0.0 and 5.0% w/w of each metal on alumina. In all cases the noble metal was well dispersed as evidenced by the fact that neither Pt nor Ir could be observed by XRD in any of the compositions. The catalysts could be synthesized as perfectly formed hollow spheres or as highly fragmented particles. The surface areas for the Pt–Ir/alumina series were between 8 and 13 m^2/g. No reflections were observed by XRD for the alumina component; however, when a pure alumina, aerosol-prepared sample was calcined in air at 900°C, XRD analysis resulted in only reflections for α-alumina.

Catalysis studies on these materials showed that they were very active and selective for the dehydrogenation of methylcyclohexane to toluene [25]. Neither the Pt–Ir on alumina catalysts nor aerosol Pt–Sn on alumina catalysts showed any activity for reforming *n*-hexane to benzene, indicating their low acidity. This conclusion is also supported by the observation that the methylcyclohexane dehydrogenation studies resulted in virtually no cracked products. The preparation of reforming catalysts by the aerosol technique is regarded as potentially important because the aerosol method of preparation has the capability of homogeneously introducing acidic components into the metal-supported-catalyst composition in a single-step synthesis. In addition, the ability to control the degree of acidity by adjusting the concentration of acidic components introduced in the feed solution should enable the synthesis of reforming catalysts having sufficient acidity to isomerize hydrocarbon components without causing an unacceptable degree of cracking.

21.3.11 Silver on Alumina Partial Oxidation Catalysts

A wide range of compositions of silver on alumina catalysts [25] between 0 and 65% w/w Ag were prepared in the HTAD reactor at 1000°C where silver nitrate and aluminum nitrates in water were fed to the rector. Although no metallic silver or any other silver compounds could be detected by XRD analysis on the compounds directly obtained from the aerosol reactor, all of the compositions resulted in strong reflections for both metallic silver and α-alumina when calcined to 900°C in air. Pure alumina prepared by the HTAD process at 900°C was analyzed by electron diffraction with the results that no diffraction rings were observed. The 50% w/w silver/alumina and higher concentrations resulted in compositions that were electrical conductors after pressing into pellets and calcining in air at 600°C or higher. Although this concept has not yet been evaluated, it seems likely that such high-temperature electrically conducting ceramics would be useful in making electrical contacts in the fabrication of solid-state-electrolytic cells for oxygen conduction and fuel cells. The X-ray photoelectron analysis of these catalysts obtained directly from the aerosol reactor resulted in binding energies for silver that were consistent with metallic silver. Energy dispersive X-ray analysis of the uncalcined 15% Ag/alumina catalyst showed that the silver was evenly dispersed throughout the material. The Ag^O appears to be well dispersed and partially embedded within the alumina matrix although synthesized at 1000°C. That the matrix adds a significant degree of stabilization is suggested by the fact that each material was calcined in air at 600, 800, and 900°C with the appearance of silver reflections in the XRD occurring only at the 900°C level, slightly below the melting point of silver.

Catalytic evaluation of these materials for the partial oxidation of ethylene to ethylene oxide, without any further ion modifications, resulted in chemical performance data inferior to that of a commercial catalyst.

21.3.12 Palladium on Reactive Supports for Waste Hydrocarbon Total Oxidation

A single-step-aerosol synthesis was used for the fabrication of 1% w/w palladium supported catalysts where the supports were alumina, chromia, zirconia, lanthana, magnesium oxide, titania, and molybdena. The synthesis of these catalysts was conducted using the upflow reactor with a continuous flow of a single solution of the support metal salts and palladium nitrate through an ultrasonic aerosol generator into the reactor that operated at 1000°C. Another series of syntheses resulted in palladium on alumina of 0.5–20% w/w on alumina. Characterization of these catalysts by EDX mapping showed that the palladium was well dispersed in all cases. XRD analysis on samples calcined at 600°C in air for 15 h showed no reflections for either palladium metal or palladium oxide or any reflections caused by the alumina support. Electron diffraction studies on a 600°C air-calcined sample of 5% palladium on alumina showed smooth and discrete rings that were indexed solely for palladium oxide.

The electron diffraction pattern showed no diffraction rings for any alumina compound and no rings for any other material. The observation of smooth SAD rings, indexed for PdO, and the absence of PdO or Pd reflections in the XRD is evidence for the nanometer grain sizes of the well-dispersed palladium particles.

The evaluation of the catalytic performance for waste hydrocarbon removal was studied using 4% methane in air over a temperature range of 250–600°C. The sole products were CO_2 and water due to total oxidation, and the most effective catalyst was 1% w/w of Pd on alumina. The activities of the other catalysts were evenly spread out in which the temperature required for 50% conversion of the methane ranged from 400 to 525°C. When compared to a conventionally prepared 1% w/w Pd impregnated onto α-alumina, the 1% Pd on alumina aerosol prepared catalyst exhibited much greater activity and stability under high-performance conditions. Arrhenius treatment of the data on all of the catalysts showed a wide range of activation energies, suggesting that the synthesis technique resulted in differing degrees of palladium to support interactions.

21.4 Potential of Aerosol Processing for Commercial Catalyst Fabrication

Although aerosol processes for the fabrication of advanced materials were described in the literature starting in the mid 1930s, little industrial use has been made of this technique for the production of industrial advanced catalytic materials. However, studies (described earlier in the chapter) from a wide number of academic and industrial laboratories have provided new evidence relating to the materials properties and catalytic performance of aerosol-fabricated catalysts. The observation that many systems have resulted in nanometer grain sizes offers access to a wide range of catalysts with enhanced activities and in some cases higher selectivities. Nanostructured grains of catalysts offer higher activity due to the higher concentrations of sites of low coordination and high-edge-to-basal-plane sites [44]. In hydrocarbon partial oxidation catalysis [48,49] and in hydrotreating [45], higher concentrations of edge-to-basal-plane sites resulted in higher selectivity due to the more selective edge sites. Furthermore, a nanostructured catalyst grain in the range of $2-5$ nm has $>90\%$ of its metal atoms located on the surface [50] of the grain, making this type of catalyst more easily modified by an activity of selectivity-moderating dopant. The observation that aerosol processing generally results in high-phase-purity powders in reasonably high-surface areas represents a good starting point in catalyst research where a host metal oxide is further modified by other ions to increase chemical performance.

Aerosol-prepared-complex catalysts were shown in the preceding discussion to result in superior catalytic properties when compared to conventionally synthesized commercial catalysts. It is expected that a more extensive use of this technique in industrial laboratories will result in a wider application of the method at the discovery stage of new catalytic process research followed by its application in large-scale industrial catalyst synthesis.

21.5   Conclusions

The aerosol technique for the synthesis of catalysts generally provides high-surface areas and high-phase purities. The method is especially useful for systematic catalysis studies where one attempts to understand the effect of a homogeneous substituted modifying metal ion within a metal oxide host on the materials catalytic chemical performance. The method was shown to have a wide range of materials synthesis capabilities, and in many cases the catalytic performance studies resulted in more active and selective catalysts as compared to conventional methods of synthesis. Aerosol fabrication of advanced catalysts has the advantage of the synthesis of metastable materials, in some cases, that are not accessible by conventional methods of synthesis; the preparation of complex metal oxides such as spinels, scheelites, and perovskites in high-surface areas and as essentially pure phases; the synthesis of dense, hollow spheres or highly fragmented particles; the preparation of exceptionally well-dispersed supported metal catalysts; and the synthesis of nanostructured grains of catalysts in a 1-μm and larger agglomerate size. The combined effect of these materials properties qualifies the process as a prime candidate for commercial catalyst synthesis.

Acknowledgments

We acknowledge the National Science Foundation (Grant number, CBT 85-16935), American Cyanamid, Union Carbide Corporation, National Aeronautics and Space Administration, DuPont, Monsanto Chemical Company, Dow Chemical Company, and BP America for support for this research.

References

1. Ebner, K. (1939). U. S. Patent 2 155 119.
2. Ebner, K. (1951). West German Patent, 753 306.
3. Ebner, K. (1953). West German Patent 877 196.
4. Bush, W. E. (1970). U. S. Patent 3 510 291.
5. Moser, W. R., and Lennhoff, J. D. (1984). *Proceedings Materials Research Society,* Boston, MA, November.
6. Brown, D. E., Mahmood, M. N., Turner, A. K., Hall, S. M., and Fogarty, P. O. (1982). *Int. J. Hydrogen Energy,* 7, 405.
7. Berndt, M., and Ksinsik, D. (1986). U. S. Patent 4 624 941.
8. Ho, S. I., and Rajeshwar, K. (1987). *J. Electrochem. Soc.,* **134**, 768.
9. Ho, S. I., Whelan, D. P., Rajeshwar, K., Weiss, A., Murley, M., and Reid R. (1988). *J. Electrochem. Soc.,* **135**, 1452.
10. Moser, W. R., and Lennhoff, J. D. (1989). *Chem. Eng. Commun.,* **83**, 241.
11. Nomura, K. Ujihira, Y., Sharma, S. S., Fueda, A., and Murakami, T. (1989). *J. Mater. Sci.,* **24**, 937.
12. Singh, R. N., Hamdani, M., Koenig, J. F., Poillerat, G., Gautier, J. L., and Chartier, P. (1990). *J. Appl. Electrochem.,* **20**, 442.
13. Singh, R. N., Koenig, J. F., Poillerat, G., and Chartier, P. (1991). *J. Electroanal. Chem. Interfacial Electrochem.,* **314**, 241.
14. Hu, Y., Cai, J., Wan, H., Cai, Q. (1991). *Ranliao Huaxue Xuebao,* **19**, 181.
15. Hu, Y., Cai, J., Wan, H., and Cai, Q. (1992). *Chin. Sci. Bull.,* **37**, 262.
16. Moser, W. R., and Cnossen, J. E. (1992). *Am. Chem. Soc. Pet. Div. Chem. Prepr.,* **37**, 1105.
17. Lyons, S. W., Wang, L. M., and Kodas, T. T. (1992). *Nanostruct. Mater.,* **1**, 283.
18. Michalakos, P. M., Bellis, H. E., Brusky, P., Kung, H. H., Li, H. Q., Moser, W. R., Partenheimer, W., and Satek, L. C. (1995). *Indust. Eng. Chem. Res.,* **34**, 1994.
19. Moser, W. R. (1993). *Am. Chem. Soc. Symp. Ser.,* **523**, 244.
20. Lee, W., Shen, H. S., Dwight, K., and Wold, A. (1993). *J. Solid State Chem.,* **106**, 288.
21. Lee, W., Do, Y. R., Dwight, K., and Wold, A. (1993). *Mater. Res. Bull,* **28**, 1127.
22. Cui, H., Shen, H. S., Gao, Y. M., Dwight, K., and Wold, A. (1993). *Mater. Res. Bull.,* **28**, 195.
23. Papp, J., Shen, H. S., Kershaw, R., Dwight, K., and Wold, A. (1993). *Chem. Mater.,* **5**, 284.
24. Cui, H., Shen, H. S., Gao, Y. M., Dwight, K., and Wold, A. (1993). *Mater. Res. Bull.,* **28**, 195.
25. Moser, W. R., Knapton, J. A., Koslowski, C. C., Rozak, J. R., and Vezis, R. H. (1994). *Catal. Today,* **21**, 157.
26. Papp, J., Soled, S., Dwight, K., and Wold, A. (1994). *Chem. Mater.,* **6**, 496.

27. Do, Y. R., Lee, W., Dwight, K., and Wold, A. (1994). *J. Solid State Chem.*, **108**, 198.
28. Hu, Y.-H., Wan, H., and Tsai, K.-R. (1994). *J. Nat. Gas Chem.*, **3**, 280.
29. Tiwari, S. K., Samuel, S., Singh, R. N., Poillerat, G., Koenig, J. F., and Chartier, P. (1995). *Int. J. Hydrogen Energy*, **20**, 9.
30. Kodas, T. T. (1989). *Angew. Chem.*, **101**, 814.
31. Gurav, A., Kodas, T., Pluym, T., and Xiong, Y. (1993). *Aerosol Sci. Technol.*, **19**, 411.
32. Roy, D. M., Neurgaonkar, R. R., O'Holleran, T. P., and Roy, R. (1977). *Am. Ceram. Soc. Bull.*, **56**, 1023.
33. Kato, A., Ishimatsu, H., and Suyama, Y. (1979). *Funtai Oyobi Funmatsuyakin*, **26**, 131.
34. Gorska, M., Beaulieu, R., and Loferski, J. J. (1980). *Thin Solid Films*, **67**, 341.
35. Kodas, T. T., Ward, T. L., and Glicksman, H. D. (1993). European Patent Application, EP 591 882.
36. Kodas, T. T., Lyons, S. W., and Glicksman, H. D. (1993). European Patent Application, EP 591 881.
37. Lyons, S. W., Wang, L. M., and Kodas, T. T. (1992). *Nanostruct. Mater.*, **1**, 283.
38. Restovic, A., Poillerat, G., Chartier, P., and Gautier, J. L. (1994). *Electrochim. Acta*, **39**, 1579.
39. Arsene, J., Lenglet, M., and Jorgensen, C. K. (1984). *Mater. Res. Bull.*, **19**, 1281.
40. Lenglet, M., Lopitaux, J., and Arsene, J. (1983). *J. Solid State Chem.*, **50**, 294–303.
41. Herman, R. G., Klier, K., Simmons, G. W., Finn, B. P., and Bulko, J. B. (1979). *J. Catal.*, **56**, 407.
42. Rossetti, G. A., Jr. (1987). M. Sci. thesis, Worcester Polytechnic Institute, September.
43. Rossetti, G. A., Jr., Berger, J. L., and Sisson, R. D. (1989). *J. Am. Ceram. Soc*, **72**, 1811.
44. Van Hardwell, R., and Hartog, F. (1972). *Adv. Catal.*, **22**, 75.
45. Bouwens, S. N. M. N., Veen, J. A. R., Konnigsberger, D. C., De Beers, V. H. J., and Prins, R. (1991). *J. Phys. Chem.*, **95**, 123.
46. Machida, K., and Enyo, M. (1987). *J. Chem. Soc. Chem. Comm.*, **21**, 1639.
47. Miguel, P. F., and Katz, J. L. (1994). *J. Mater. Res.*, **9**, 746.
48. Volta, J. C., and Portefaix, J. L. (1985). *Rev. Appl. Catal.*, **18**, 1.
49. Desikan, A. N., and Oyama, S. T. (1992). *ACS Symposium Series* (D. J. Dwyer, and F. M. Huffman, eds.) **482**, 260.
50. Siegel, R. (1991). *Annu. Rev. Mater. Sci.*, **21**, 559.

CHAPTER 22

Use of an Aerosol Technique to Prepare Iron Sulfide Based Catalysts for Direct Coal Liquefaction

Dady B. Dadyburjor, Alfred H. Stiller, Charter D. Stinespring,
Ajay Chadha, Dacheng Tian, Stephen B. Martin, Jr.,
and Sushant Agarwal

Department of Chemical Engineering,
West Virginia University,
Morgantown, West Virginia 26506-6102

KEYWORDS: aerosol, catalysts, coal liquefaction, preparation

22.1   Introduction

Coal liquefaction refers to the formation of lower molecular weight products from coal. In so-called first-stage direct coal liquefaction (DCL), there may be relatively low yields of products in the oil range, suitable for liquid fuels; upgrading to high-quality liquid fuels is typically conducted in subsequent steps.

Catalysts used for first-stage DCL are generally not recovered. Hence, process economics requires that such catalysts be cheap and environmentally benign. Iron catalysts fit both criteria. In the presence of sulfur species under DCL conditions, virtually all iron compounds form a nonstoichiometric iron sulfide, pyrrhotite (FeS_x, x around 1). This species is taken to be the active catalytic species for DCL. The activity of iron catalysts for DCL can be improved by decreasing their effective particle size, for improved contact between coal and catalyst.

Our laboratory has focused on ferric sulfide (Fe_2S_3) based catalysts, with small particle sizes ensured by preparation in an aerosol reactor; and also on the characterization of these catalysts and their performance in DCL. (The preparation technique is of course not restricted to these catalysts, and their use is not restricted to DCL.) The Fe_2S_3 is prepared by reaction of a sulfide with a ferric salt. Ferric sulfide is typically unstable above 10°C [1], and disproportionates via

$$Fe_2S_3 \rightarrow \alpha\ FeS_2 + \beta\ FeS_x + \gamma\ S. \tag{1}$$

The relative amounts of pyrite (FeS_2, PY) and pyrrhotite (FeS_x, PH) (i.e., the values of α and β) and the composition of the nonstoichiometric PH (i.e., the value of x) depend on the time and temperature of disproportionation. The presence of the two forms of iron in close proximity is expected to be related to the efficacy of this material as a catalyst in DCL. When the disproportionation reaction is conducted in a hydrothermal bomb, optimal results for DCL (at 300–350°C and 1000 psi cold H_2) are obtained for the catalyst prepared at 200°C for 1 h, and such a material has a pyrrhotite to pyrite ratio (PH:PY) of approximately unity [2]. A value of PH:PY around unity is desirable for catalytic materials prepared by the aerosol technique as well.

In this chapter, we describe in some detail the preparation of PH–PY materials in an aerosol reactor. In previous work, reported in references [3,4], the materials were allowed to be exposed to air. However, work with catalysts prepared by an alternative technique indicates that surface oxidation of the particles may cause some degradation of catalytic performance; hence current preparation techniques go to some lengths to minimize contact of the particles with air, at least prior to characterization or DCL. The preparation of multi-metal sulfides, notably Fe–Cu–S, by this technique is also described. This section is followed by some mention of characterization techniques used on these materials, either in our laboratory or elsewhere. Finally, some results are reported on the use of these materials as catalysts in DCL. However, first we present a brief survey of current techniques for producing small iron particles, for use as catalysts or as engineered materials.

22.2   Background

Ultrafine particles are typically produced by one of two techniques: gas-phase reaction and condensation, or reactions in liquids. The former technique is well established and widely used, and will not be described here. The latter technique includes precipitation from homogeneous solutions, or chemical reactions in aerosol droplets.

In the precipitation technique, strict control of the process results in the generation of a uniform burst of nuclei in the homogeneous solution, and uniform growth of these to form solid particles. The technique is sensitive to parameters such as pH, temperature, concentration, stirring characteristics, aging time, and nature of the anion used. Due to the complex and interacting nature of these parameters, a frequent disadvantage of the precipitation process is the lack of reproducibility of size, shape, or composition of the particles produced. However, it is possible to produce a wide range of products with this technique.

Pradhan *et al.* [5] have produced nanometer-sized particles by precipitation techniques. The average crystallite size of the so-called ultrafine sulfated iron oxide is about 20 nm, and that for sulfated tin oxide is about 15 nm. These catalysts have been used successfully in DCL.

General reviews on the use of aerosol techniques can be found in, for example, References [6–8]. This technique is most effective in producing powders that are required to be synthesized at high temperature and pressure—superconducting oxides, polymer colloids, etc. To the best of our knowledge, this technique has not been used previously to generate iron sulfides.

In the aerosol technique, droplets of one reactant are generated as an aerosol in a bulk phase of a coreactant. A chemical reaction takes place in each droplet (i.e., each droplet acts as an individual, small reactor). The major disadvantage of this technique is the difficulty in obtaining sufficiently small droplets. Further, not all materials are amenable to this type of two-reactant synthesis. However, there are several advantages to this technique when it can be used. Uniform, spherical particles of known composition can be generated; reaction conditions can be relatively easily controlled; the size and composition can be pre-determined (and reproducibly obtained) for a given set of experimental conditions; and product purity can generally be assured.

Pratsinis [9] has reviewed the types of reactors used for this process. These include spray–pyrolysis reactors, flame reactors, laser reactors, plasma reactors, condensers, and furnace reactors. Of these, the spray–pyrolysis type is the one used in the present work. Examples of some of the other reactors are given as follows.

Andres *et al.* [10] investigated synthetic iron catalyst precursors of Fe_2O_3. These were prepared by flame decomposition of ferric chloride in the gas phase. The precursors were then sulfided *in situ* by carbon disulfide during the DCL reaction. Andres *et al.* reported that catalytic activity increases as the particle size is reduced to $0.05 \mu m$, relative to particle sizes of up to $5 \mu m$, and noted that the particle size is a more significant parameter than the crystallite size. This is because diffusional limitations may cause the crystallites to be poorly active. Hence, if the particle size could be reduced to the crystallite size, it might be possible to obtain higher values of the catalytic activity.

Eklund *et al.* [11] manufactured small-sized iron carbonyl crystallites using a laser reactor. Particle sizes in the range of 3–13 nm were produced. The improved catalytic activity of these particles is attributed to a combination of two factors. First, the increased specific surface area of the particles leads to an increased number of active sites per unit volume. Second, the properties of particles containing on the order of a few atoms or molecules may be radically different from the bulk properties expected for larger particles.

Gadalla and Hsuan [12] have used atomization techniques to produce hollow spherical particles of $NiFe_2O_4$, for use as ceramic precursors or superconductors. Solutions of nickel nitrate and ferric nitrate were atomized using a six-jet atomizer. The resulting aerosol was introduced into a streamlined tubular reactor. Here the solvent was evaporated, and decomposition and solid-state transformation reactions occurred. It was observed that lower concentrations of the precursor solutions led to smaller and more uniform products from the reactor. The hollow nature of the particles was confirmed by obtaining the density of the original powder and comparing it with that of the crushed powder.

Gadalla and Hsuan [12] proposed a mechanism whereby the preliminary evaporation of water from the aerosol droplet results in a reduction in size of the aerosol as well as a concentration gradient inside the droplet. The solute precipitates at the surface of the droplet, forming a spherical crust. An impermeable skin is formed, possibly by decomposition of the material of the crust and sealing of the pores. Finally, the crystallites can sinter and the hollow spheres may agglomerate.

Zhang *et al.* [13] used a similar technique to obtain a 1–2–3 superconductor, $YBa_2Cu_3O_{7-x}$, in microfine powder form. The precursors were organic acids. These were atomized separately, producing droplets having a diameter of 10–20µm. The droplets were introduced into a furnace, and the products were collected in a textile bag. Scanning electron microscopy (SEM) of the products shows that primary particles have sizes of 10–20µm. Shell fragments from these particles are also present, as are smaller particles formed by explosion of the primary particles. Zhang *et al.* [13] suggest that solid particles are formed by using lower concentrations of precursor, while hollow particles are formed when more concentrated solutions are used. For the same concentration of precursor, particles are more likely to be hollow if the precursor is of low viscosity, whereas solid particles are more likely to be formed if the precursor viscosity is high.

22.3 Preparation

The catalysts are made in the batch mode. Theoretically, 5–50 g can be made in a batch, which can take 2–5 h of run time. Typically, we recover 60–80% of the theoretical yield, with the rest being lost to filtration, etc.

First, ferric hydroxide is prepared by mixing ferric chloride and ammonium hydroxide. The precipitated ferric hydroxide is washed and rinsed with distilled–deionized water to remove the soluble chloride salt and any excess ammonium hydroxide. The precipitate is then stirred and heated with acetic acid. The resulting ferric acetate is kept at a nominal concentration of 0.1 N and maintained in solution at pH 4 by appropriate use of acetic acid or ammonium hydroxide. When Fe–Cu–S materials are required, the ferric acetate solution is mixed with cupric acetate solution. The cupric acetate is prepared analogously to ferric acetate, except that cupric chloride, sodium hydroxide, and acetic acid are used. (If ammonium hydroxide is used, amines are formed and these are difficult to separate.) The acetate solutions and hydrogen sulfide gas are the starting materials for the preparation of the catalyst.

The overall system for preparation of the PH–PY materials based on ferric sulfide by the aerosol technique is described schematically in Fig. 1. The reactor system of Fig. 1 can be divided into four basic components: a pump, a nozzle, the aerosol reactor, and a scrubbing system.

The pump is driven by high-pressure nitrogen and is capable of pumping liquid in 1-ml increments at flow rates from 40 to 12,000 ml/h (0.01–3.3 ml/s) at 8800 psi pressure.

The nozzle is a hollow-cone atomizer made from 303 stainless steel and backed up by a pressure-check valve. A fuel injector used in earlier systems was found to have problems with pitting, orifice plugging, and "leakage" of larger-than-design droplets between injections.

The preparation reactor (Fig. 2) consists of a cylinder with a sealable top flange (for the nozzle) and a side tap (for removing the catalyst materials produced after the reaction is completed). An ultrasonic horn is also sealed into the reactor near the bottom. The horn has been used to reduce particle size further, after the reaction is completed, but has been found to be unnecessary. Two external band heaters serve to raise the reactor temperature. The reactor has been designed to operate at up to 250°C and up to 300 psig; current operating conditions are well below these values.

The scrubbing system is an essential safety feature because H_2S is involved in the preparation process. The system consists of two sealed containers of aqueous NaOH through which outlet gas from the reactor is

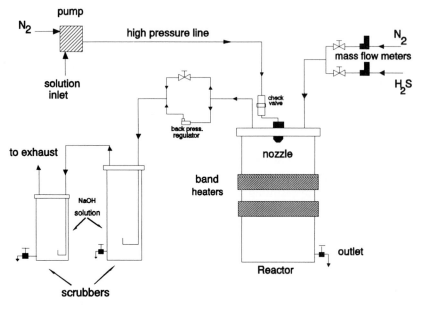

Figure 1. Schematic for preparation of catalysts by aerosol techniques.

bubbled in series. The second scrubber is monitored to ensure that no H_2S is in the vapor phase exiting the first scrubber. The exit stream from the second scrubber is vented into a hood.

Before the preparation process starts, the reactor contains N_2 at slightly above atmospheric pressure. The reactor is first flushed; next it is filled with N_2 from the gas cylinder to a given pressure, typically 75 psia; and then H_2S is added until the total reactor pressure typically reaches 108 psia. (These values correspond to a total pressure of 200 psia when the reactor temperature is 200°C.) The band heaters are then turned on and the temperature is controlled at the final value, typically 200°C. (The typical values here correspond to conditions for the optimum catalyst as prepared by hydrothermal disproportionation.) The reactor pressure is fine-tuned by opening and closing the manual valve on the reactor outlet. The concentration of the ferric acetate solution is adjusted; typical values used are in the range of 10^{-3}–10^{-1} M. The appropriate amount of cupric acetate solution is admixed. Generally 5 liter of total solution is prepared for a single run.

The solution is connected to the liquid inflow line of the pump. Nitrogen pressure to the pump is set at 95 psia and the liquid outlet pressure is around 5000 psia. The pump is set for 5–45 injections per minute, of 1 ml volume each.

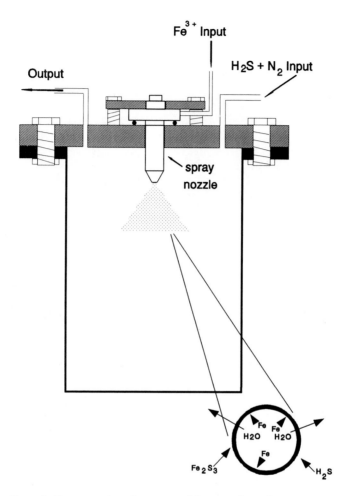

Figure 2. Representation of reactor used for aerosol–catalyst preparation.

The liquid solution enters the reactor as droplets (aerosols) of approximately 50-μm size (estimated from fuel oil calculations). The reaction products found at the bottom of the reactor (after the process is complete) have been found by scanning electron microscopy to consist of shells of varying thickness, with average diameters of submicron size; see Fig. 3. (The decrease from 50 μm to submicron size is because the inlet solution is relatively dilute, as described earlier). To explain the shells, we postulate that two processes occur in parallel during the process: reaction of ferric ions with sulfide to form ferric sulfide that subsequently disproportionates

Figure 3. SEM of typical aerosol catalyst.

to PH, PY, and S; and vaporization of water. The former process occurs primarily at the vapor–liquid interface, while the latter occurs from the center of the aerosol droplet. The relative rates of the two processes determine the thickness and external diameter of the resulting solid; see Fig. 2.

During the batch process, droplets spend relatively little time in the gas phase, on the order of 1 s. However, after deposition on the chamber walls and the bottom, the droplets could spend up to a few hours in the reactor before the batch process is complete. Some secondary reactions (e.g., interchange between PY and PH) occur during this time, and fine-tune the PH:PY ratio to the desired value.

While the process is occurring, injection of the high-pressure droplets leads to an increase in reactor pressure. The pressure is manually controlled by the valve at the vapor outlet from the reactor. (The relief valve, in parallel with the manual valve in Fig. 1, is an emergency safety device. Automatic control of the reactor pressure, say with a back-pressure regulator, has not been successful; the manual device appears to be necessary for sufficient control over the final product, at least to date). To ensure that hydrogen sulfide is not gradually depleted by the vapor bleed, additional hydrogen sulfide

(10% in nitrogen) is added to the reactor approximately every 20 min. Accordingly, hydrogen sulfide is always in excess in the reactor. The H_2S-H_2O mixture in the vapor phase is condensed and cleaned in the scrubber.

After the requisite amount of feed liquid has been processed, the heating elements are turned off and nitrogen is passed through the reactor to flush out as much H_2S as possible. To remove the liquid containing the catalyst particles, the vapor outlet is closed, the bottom tap is opened, and nitrogen is used to push the reactor contents into a collection vessel previously filled with N_2. A vapor outlet from the collection vessel is connected to the scrubbing unit. When all the reactor contents have been removed to the collection vessel, it is isolated and transferred to a glove box. There the liquid contents are poured into centrifuge tubes and sealed. Centrifugation for approximately 5 min at approximately 4500 rpm is required to deposit the particles at the bottom of the tubes. The particles are removed for characterization or DCL reaction. The reactor vessel is cleaned and prepared for the next run. The latter procedure consists of pressure-testing at 250-psig nitrogen, then depressurizing to 35-psia nitrogen for standby.

22.4   Characterization

A series of catalysts were earlier prepared without special attention to avoiding exposure to air. Conditions for these preparations are given in Table 1. Also shown in Table 1 are results of He pycnometry, yielding density; average diameter, as noted by laser light scattering; the bulk ratio S:Fe, by energy dispersive X-ray analysis (EDX); and the surface ratio S:Fe by Auger electron spectroscopy (AES). Although a more detailed discussion of these numbers can be found elsewhere [3], it should be noted that the diameters shown in Table 1 are relatively large; transmission electron microscopy (TEM) results obtained in the laboratory of Professor G. Huffman (University of Kentucky) show sizes around 40 Å (see Fig. 4). The difference is probably due to clustering of particles in the solvent used during the laser-based measurement. From Fig. 4 and other micrographs (not shown), some large cubic structures of (possibly) PY are clearly visible, while much smaller particles of (possibly) PH are clustered around them.

X-ray diffraction (XRD) studies of samples 7 and 8 are seen in Figs. 5 and 6, respectively. These results were obtained in Professor M. Seehra's laboratory (WVU). Sample 6 (not shown) was seen to contain PY and S; PH was not seen, probably because of the smaller crystallinity of these particles (see preceding). The S lines are sharp and well resolved; the PY lines are less so, with preferential orientations of (200) and (311). Sample 7 is similar to 6, except that some PH lines appear, although at low resolution.

TABLE 1

Characterization of Air-Exposed Aerosol-Generated Fe-S Catalysts. Based on (3).

Sample number	Precursor concentration [M]	Preparation pressure [psi]	Preparation temperature [°C]	Density (He Pyc) [g/cc]	Diameter (LSS) [nm]	S/Fe ratios Bulk (EDX)	S/Fe ratios Sfce (AES)
A7	0.01	200	200	—	—	2.5	3.6
A8	0.01	100	200	4.60	507	2.5	3.3
A9	0.10	100	200	4.27	1200	0.9	2.0
A12	0.01	100	250	4.66	—	1.9	2.5

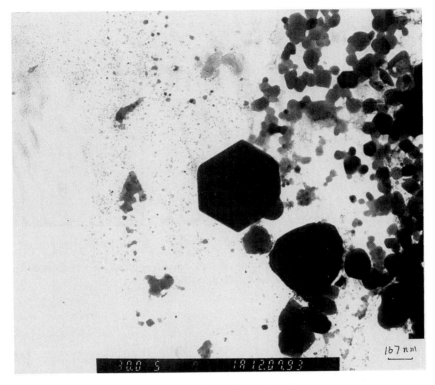

Figure 4. TEM of typical aerosol catalyst.

Sample 8 indicates PY and PH lines, with no lines for S seen. PH lines correspond to monoclinic pyrrhotite (Fe_7S_8, $x = 1.143$). Lines for marcasite (orthorhombic FeS_2) are also seen.

Figure 7, from the laboratory of Professor J. Renton (WVU), indicates an interesting XRD pattern when preparation conditions are changed. At higher precursor concentration (0.1 M) and a temperature of 165°C, peaks observed include not only PY and S but also PH in the form of greigite, Fe_3S_4 ($x = 1.333$). However, greigite is stable typically only below 100°C [14]. Perhaps the thermodynamically unstable form is trapped in the particle due to the rapid quenching in the aerosol reactor.

Figure 8 shows more recent results with a material that differs in two respects from those discussed earlier: air (oxygen) was carefully excluded, and the catalyst contains Cu as well as Fe cations. In all three cases, the conditions in the aerosol reactor were precursor concentration of 0.1 M, temperature of 200°C, and pressure of 200 psig. The diffractograph corresponding to no Cu present ($f_{Cu} = Cu/(Cu + Fe) = 0.0$) can be compared with sample 7,

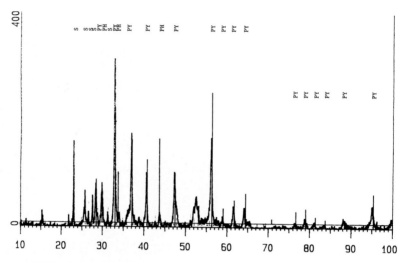

Figure 5. XRD of aerosol sample A7. See Table 1 for details of preparation. Peaks of elemental sulfur (S), pyrite (PY), and pyrrhotite (PH) are as indicated.

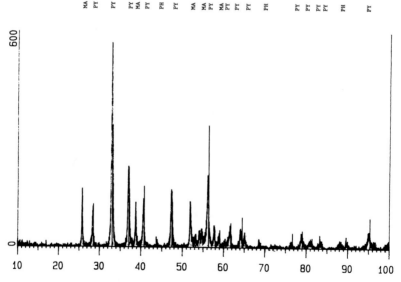

Figure 6. XRD of aerosol sample A8. See Table 1 for details of preparation. Peaks of pyrite (PY), marcasite (MA), and pyrrhotite (PH) are as indicated

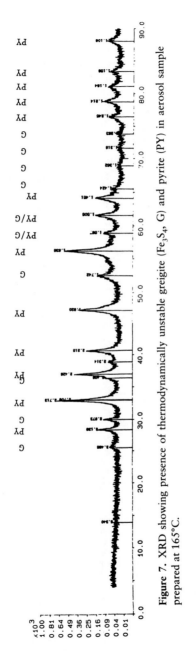

Figure 7. XRD showing presence of thermodynamically unstable greigite (Fe₃S₄, G) and pyrite (PY) in aerosol sample prepared at 165°C.

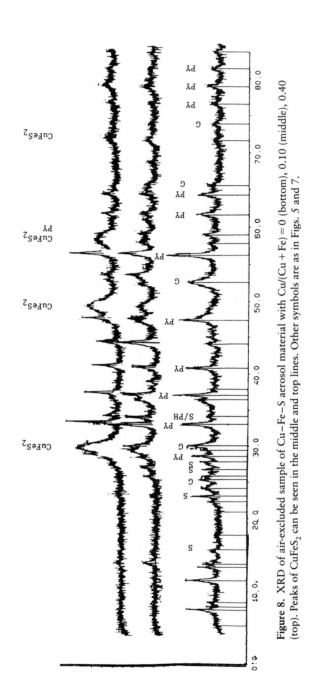

Figure 8. XRD of air-excluded sample of Cu–Fe–S aerosol material with Cu/(Cu + Fe) = 0 (bottom), 0.10 (middle), 0.40 (top). Peaks of CuFeS₂ can be seen in the middle and top lines. Other symbols are as in Figs. 5 and 7.

Fig. 5. More greigite is seen in the curve of Fig. 8, relative to monoclinic pyrrhotite, again possibly due to the higher concentration of precursor.

The other two diffractographs of Fig. 8 correspond to increasing amounts of Cu present, $F_{Cu} = 0.10$ and 0.40. Many of the peaks corresponding to $F_{Cu} = 0$ are repeated. In addition, peaks corresponding to $CuFeS_2$ are found, and the peak area increases with F_{Cu}. Hence the Cu is clearly incorporated into some of the pyrrhotitic structure of the aerosol material.

22.5   Catalytic Performance

Table 2 summarizes earlier data for DCL using samples 8, 9, and 12 discussed earlier. Thermal results are also noted. The experiments were conducted at 350°C and 1000 psi (cold) H_2 for 1 h. Blind Canyon coal, characterized as DECS-6, was used together with tetralin as a solvent and with 0.1 ml of CS_2 as a sulfiding agent. The catalyst loading was 5% of the coal. The reactor was of the tubing-bomb type agitated vertically at 500 rpm. All three catalysts show considerable improvement in conversion (to tetrahydrofuran (THF)-soluble materials), although the yield to oils (THF- and hexane solubles) is not significantly affected. This is not a particular disadvantage for iron-based catalysts for first-stage DCL.

Figure 9 shows results using the Fe–Cu–S catalyst with $f_{Cu} = 0.10$, as described previously. As mentioned earlier, air (oxygen) was carefully excluded in the preparation of these catalysts. These results were also obtained with DECS-6 coal using 0.1 ml of CS_2 at 1000 psi (cold) H_2 with 500-rpm agitation in a tubing-bomb reactor. However, phenanthrene is used as the solvent, the reaction time is 30 min, and a temperature range of 350–440°C is used. Accordingly, the thermal results are repeated, as are

TABLE 2

Results of DCL with Air-Exposed
Aerosol-Generated Fe-S Catalysts. DCL Conditions:
350°C, 1000 psi (Cold) H_2, 1h; DECS-6 coal, CS_2,
Tetralin. NC: No Catalyst, Thermal Run.
Based on (3).

Sample number	Conversion [%]	Oil + gas yield [%]
NC	54.9	13.6
A8	64.9	9.3
A9	63.0	10.8
A12	64.4	10.9

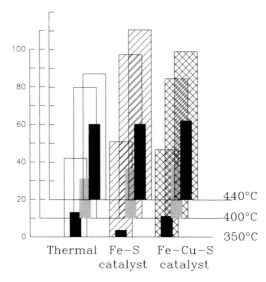

Figure 9. Results of DCL with unexposed aerosol-generated Fe–Cu–S, catalysts at various liquefaction temperatures. Other DCL conditions are 1000 psi (cold) H$_2$, 0.5 h, DECS-6 coal, CS$_2$, phenanthrene. Here blank blocks represent the conversion resulting from thermal runs, single-hatched blocks represent the conversions from catalytic runs with f$_{Cu}$ = 0 (Fe-alone catalysts), and double-hatched blocks represent the conversion from catalysts with f$_{Cu}$ = 0.10. Solid blocks represent the corresponding yields of oil + gas.

results for the Fe-alone catalyst (f$_{Cu}$ = 0). In all cases, conversions and oil yields increase with temperature. The Cu-containing catalysts exhibit conversions intermediate between thermal values and those for the Fe-alone catalyst. However the oil + gas yields are definitely superior to those of the Fe-alone catalyst and the thermal values. Clearly the addition of 10 at % Cu increases the yield of oil at the expense of conversion. It is conceivable that an optimum atomic fraction of Cu could be found that would increase the oil yield with little or no loss of total conversion.

22.6   Conclusions

The aerosol technique can be used to prepare DCL catalysts. These are mixtures of pyrite and pyrrhotite in intimate contact, present in the form of small, hollow particles and the corresponding fragments. The relative amount of pyrite and pyrrhotite depends on the preparation conditions; the type of nonstoichiometric pyrrhotite also depends on these parameters. A second cation can also be introduced into this process.

The catalysts prepared to date have been characterized by a variety of techniques. XRD yields gross results of the species present. AES and EDX yield atomic ratios on the surface and the bulk, respectively. He pycnometry yields information on the density of the particles, and therefore on the extent of the shell-like behavior of the particles. Average sizes are obtained by laser light-scattering and by TEM. The two measurements do not agree, due to clumping effects expected in the presence of the solvent used for the light-scattering measurements.

The use of the iron(-alone) sulfide material as a catalyst for first-stage DCL results in improvements in activity, but little or no improvement in the yield of oil–range products. When a Fe–Cu–S catalyst containing 10 at % Cu is tested, the yield is improved, especially at the highest temperature.

Acknowledgments

This work was conducted under U.S. Department of Energy contract DE-FC22-93PC93053 under the cooperative agreement to the Consortium for Fossil Fuel Liquefaction Science. We gratefully acknowledge AA analysis by Jianli Yang and technical assistance by John Trent and Dr. Albert Brennsteiner.

References

1. Stiller, A. H., McCormick, B. J., Russel, P., and Montano, P. A. (1978). *J. Am. Chem. Soc.,* **100**, 2553.
2. Stansberry, P. G., Wann, J.-P., Stewart, W. R., Yang, J., Zondlo, J. W., Stiller, A. H., and Dadyburjor, D. B., (1993). *Fuel*, **72**, 793.
3. Dadyburjor, D. B., Stiller, A. H., Stinespring, C. D., Zondlo, J. W., Wann, J.-P., Sharma, R. K., Tian, D., Agarwal, S., and Chadha, A. (1994). *Am. Chem. Soc. Div. Fuel Chem. Preprt.,* **39**(4), 1088.
4. Stiller, A. H., Agarwal, S., Zondlo, J. W., and Dadyburjor, D. B. (1993). *Proceedings 10th International Pittsburgh Coal Conference* (S.-H. Chiang, ed.) 258.
5. Pradhan, V. R., Tierney, J. W., Wender, I., and Huffman, G. P. (1991). *Energy Fuels*, **5**, 497.
6. Matijevic, E. (1991). *Chemtech*, **21**(3), 176.
7. Gaurav, A., Kodas, T., Pluym, T., and Xiong, Y. (1993). *Aerosol Sci. Technol*, **19**, 411.
8. Kodas, T. (1989). *Angew. Chem. Int. Ed. Engl.,* **101**, 814.
9. Pratsinis, S. E. (1987). *AIChE Symp. Ser.,* **85**(270), 57.
10. Andres, M., Charcosset, H., Chiche, P., Davignon, L., Djega-Mariadassou, G., Joly, J. P., and Pregermain, S. (1983). *Fuel* **62**, 69.
11. Eklund, P. C., Stencel, J. M., Bi, X.-X., Keogh, R. A., and Derbyshire, F. J. (1991). *Am Chem. Soc. Div. Fuel Chem. Prepr.,* **36**(2), 551.
12. Gadalla, A. M., and Hsuan, F.-Y. (1990). *J. Mater. Res.,* **5**(12).
13. Zhang, S. C., Messing, G. L., Huebner, W. (1991). *J. Aerosol Sci.,* **22**(5), 585.
14. Power, L. F., and Fine, H. A. (1976). *Miner. Sci. Eng.,* **8**, 106.

Index

581